Preconcentration Techniques for Natural and Treated Waters

T0199504

Preconcentration Techniques for Natural and Treated Waters

High Sensitivity Determination of Organic and Organometallic Compounds, Cations and Anions

T.R. Crompton

CRC Press
Taylor & Francis Group
Boca Raton London New York

CRC Press is an imprint of the
Taylor & Francis Group, an **informa** business
A SPON PRESS BOOK

CRC Press
Taylor & Francis Group
6000 Broken Sound Parkway NW, Suite 300
Boca Raton, FL 33487-2742

First issued in paperback 2019

© 2003 T.R. Crompton
CRC Press is an imprint of Taylor & Francis Group, an Informa business

No claim to original U.S. Government works

ISBN-13: 978-0-415-26811-0 (hbk)
ISBN-13: 978-0-367-44692-5 (pbk)

Publisher's note:
This book was prepared from camera-ready copy supplied by the author.

British Library Cataloguing in Publication Data
A catalogue record for this book is available from the British Library

Library of Congress Cataloging in Publication Data
A catalog record for this book has been requested.

Visit the Taylor & Francis Web site at
http://www.taylorandfrancis.com

and the CRC Press Web site at
http://www.crcpress.com

Contents

Preface

Despite the great strides forward in analytical instrumentation that have been made in the last decade, the analyst working in the fields of potable water analysis and environmental analysis of non saline and saline waters finds that, frequently, the equipment has insufficient sensitivity to be able to detect the low concentrations of organic and inorganic substances present in his samples with the consequence that he has to report less than the detection limit of the method. Consequently, trends upwards or downwards in the levels of background concentrations of these substances in the environment cannot be followed. This is a very unsatisfactory situation which is being made worse by the extremely low detection limits being set in new directives on levels of pollution, issued by the European Community and other international bodies. To overcome the problem, there has been a move in recent years to apply preconcentration to the sample prior to analysis so that, effectively, the detection limit of the method is considerably reduced to the point that actual results can be reported and trends followed.

The principle of preconcentration is quite simple. Suppose that we need to determine $5ng\ L^{-1}$ of a substance in a sample and that the best technique has a detection limit of $1\mu g\ L^{-1}$ ($1000ng\ L^{-1}$). To reduce the detection limit to $5ng\ L^{-1}$, we might, for example, pass 1L of the sample down a small column of a substance that absorbs the substance with 100% efficiency. We would then pass down the column 5ml of a solvent or reagent which completely dissolves the substance from the column thereby achieving a preconcentration of $1000/5 = 200$. Thus, if the detection limit of the analytical method without preconcentration were $1000ng\ L^{-1}$, then with preconcentration it would be reduced to approximately $5ng\ L^{-1}$.

The use of a column is but one of many possible methods of achieving preconcentration. Each chapter of the book discusses a different method of preconcentration and its application to the preconcentration of cations, anions, organic substances and organometallic compounds. The book is based on a survey of the recent world literature dealing with preconcentration and its application to saline and non saline waters.

A combination of preconcentration with the newest, most sensitive, and, by definition, most expensive analytical techniques now becoming available is achieving previously undreamt of detection limits at the very time when the requirements for such sensitive analysis is increasing at a rapid pace. Thus the combination of preconcentration with graphite furnace, Zeeman or inductively coupled plasma atomic absorption spectrometry and, particularly, the combination of the latter technique with mass spectrometry is enabling exceedingly low concentrations of metals in the ng L^{-1} or lower range to be determined. Preconcentration prior to gas or high performance liquid chromatography is achieving similar results in the analysis of organic compounds.

Another aspect of preconcentration is, however, worthy of mention, particularly in the case of smaller laboratories which cannot afford to purchase the full range of modern analytical instrumentation. Using older, less sensitive instrumentation preconcentration will still achieve very useful reductions in detection limits which will be adequate in many but not all instances. Thus, if conventional atomic absorption spectrometry achieves detection limits of 1 and 5mg L^{-1} for cadmium and lead in water, then a 200-fold preconcentration will reduce these limits to approximately 5 and 25µg L^{-1} and a 1000-fold preconcentration will achieve 1 and 5µg L^{-1}.

Throughout the book emphasis is laid on providing practical experimental detail so that the reader can, in many cases, apply the methods without reference to source literature and will be in a position to adopt procedures to his or her particular requirements. This is a field where much remains to be discovered and it is hoped that this book will assist chemists to further develop procedures.

While the book has been written with the interest of water chemists in mind, many of the preconcentration procedures discussed could with little or no modification be applied to improving detection limits in laboratories in a wide range of other industries including electronics, semiconductors, metallurgy, pharmaceuticals, organic chemicals, petroleum and polymers.

The book commences in Chapter 1 with a review of the principles of various preconcentration techniques that have been used for organic and organometallic compounds, cations and anions.

For organic and organometallic compounds preconcentration methods discussed include chelation solvent extraction (Chapter 2), direct solvent extraction (Chapter 13), adsorption on ion exchange resins (Chapters 4 and 5), coprecipitation techniques (Chapter 11), adsorption of various solids including polymeric adsorbents (Chapter 7), macroreticular resins (Chapter 3), polyurethane foam (Chapter 8), octadecyl resins (Chapter 9), chelating solids (Chapter 6) and active carbons (Chapter 10). Also discussed are the applications of dynamic headspace analysis (Chapter 16) and static headspace analysis (Chapter 15), supercritical fluid chromatography (Chapter 14) and various other techniques (Chapter 17).

For cations, techniques include chelation–solvent extraction (Chapter 20), adsorption on ion exchange resins (Chapters 24 and 25), coprecipitation techniques (Chapters 27 and 28), adsorption on metals and metal oxides (Chapter 26), carbons (Chapter 23), chelating resins (Chapter 21), non-chelating resins (Chapter 22) and various miscellaneous methods (Chapter 29).

For anions, in addition to chelation–solvent extraction methods (Chapter 34), ion exchange methods are considered (Chapter 36), also coprecipitation methods (Chapter 38) and adsorption on metals, metal oxides (Chapter 37), chelating solids (Chapter 35) and various miscellaneous methods (Chapter 39).

Modern on-line methods for the preconcentration of organic compounds (Chapter 12) and cations (Chapter 31) are also considered.

In Chapters 18 and 19, 32 and 40 respectively a rationale of available methodology for organic/organometallic compounds, cations and anions is presented. It is hoped that these tables will enable the reader to quickly locate in the book particular items in which he or she is interested.

The following groups of people will find much to interest them: management and scientists in all aspects of the water and other industries, river management, fishery industries, sewage effluent treatment and disposal and land drainage and water supply, as well as management and scientists in all branches of industry which produce aqueous effluents. It will also be of interest to agricultural chemists; agriculturalists concerned with the ways in which chemicals used in crop and soil treatment permeate the ecosystem; to the biologists and scientists involved in fish, insect and plant life; and to the medical profession, toxicologists, public health workers and public analysts. Other groups of workers to whom the book will be of interest include oceanographers, fisheries experts, environmentalists and, not least, members of the public who are concerned with the protection of the environment. The book will also be of interest to practising analysts, and, not least, to the scientists and environmentalists of the future who are currently passing through the university system and on whom, more than ever previously, will rest the responsibility of ensuring that we are left with a worthwhile environment to protect.

T R Crompton
May 2002

Chapter 1

Introduction

Many factors can affect the efficiency of a preconcentration procedure, and the more important of these are discussed below by means of illustrative examples concerning different preconcentration techniques. One very important factor to be borne in mind when applying preconcentration procedures is the contribution made to the final analytical result by concentrations of the element to be determined present in the reagents used in the analysis. Obviously, this has an important bearing on the detection limit that can be achieved by preconcentration.

Limits of detection are discussed in detail in the Appendix.

1.1 Purity of reagents and detection limits

Reagent purity is a very important consideration in all preconcentration procedures. It is considered in the example discussed below, but similar considerations apply to all preconcentration methods in which reagents are used.

Consider the preconcentration and determination of traces of a substance in water by extraction with an organic solvent, prior to determination in the solvent extract by a suitable analytical finish (see Chapter 13). In such procedures various reagents such as mineral acids and buffers might be added to a large volume of the aqueous sample. The substances to be determined are extracted into a much smaller volume of an organic solvent, thereby achieving the necessary preconcentration. As well as the final extract containing a contribution of the substance to be determined originating in the sample there is also a contribution from the reagents used. As a rule of thumb, in these circumstances when the weight of this substance contributed by the reagents equals the weight contributed by the aqueous sample then the detection limit of the method has been reached. When low detection limits are required then contributions of impurities from reagents must be kept as low as possible by using very pure reagents and, in some cases, by pre-purifying some or all of the reagents immediately prior to using them in the analysis. Some idea of the

Table 1.1 Specification analysis of concentrated nitric acid grades available from Rhone–Poulenc

	Normatom ultra pure (69%) mg kg⁻¹	Normapure (69%) Analysis of Cd, Hg, Pb mg kg⁻¹	Normapure (69%) ISO–ACS reagent mg kg⁻¹	Normapure (68%) RP mg kg⁻¹	Rectapure (68%) mg kg⁻¹
Ag	0.001				
Al	0.005				
As	0.001	0.01	0.01	0.01	
Au	0.010				
B	0.020				
Ba	0.005				
Be	0.001				
Bi	0.001				
Ca	0.010		1		
Cd	0.001	0.005	0.01		
Co	0.001				
Cr	0.001				
Cu	0.001		0.05		
Fe	0.010	0.2	0.2	1	10
Ga	0.005				
Hg	0.001	0.001			
In	0.002				
K	0.020				
Li	0.005				
Mg	0.010				
Mn	0.001		0.4	0.4	
Mo	0.001				
Na	0.005				
Ni	0.001		0.05		
Pb	0.005	0.005	0.05	0.05	
Sn	0.005				
Sr	0.005				
Ti	0.001				
Tl	0.005				
V	0.001				
Zn	0.005		0.1		
Ce	0.05	0.005	0.5	0.5	10
PO₄	0.010		2	2	
SO₄	0.1	1	1	2	

Source: Own files

variation in levels of impurities in commercial supplies of reagents can be obtained by considering the example of concentrated nitric acid.

In Table 1.1 are given the specification analyses (not actual levels) of a range of nitric acids obtainable from Rhone–Poulenc. These range from the

Normatom ultra pure grade where analyses are quoted for 26 elements in the 0.001–0.005mg kg^{-1} (1410–7050ng L^{-1}) range and eight elements in the 0.01–0.1mg kg^{-1} (14100–141000ng L^{-1}) range. The intermediate Normapure ranges of concentrated nitric acid specify the analyses in the case of the ISO–ACS reagent quality 12 elements only occurring in the range 0.01–2mg kg^{-1} (14100–2820000ng L^{-1}). The Rectapure ordinary grade specifies the analysis of only two impurities, both present at the 10mg kg^{-1} level. Clearly, then the Normatom grade of acid is the only one worth considering for use in preconcentration techniques. Similar considerations will apply to any other reagents used in the analysis.

Consider, as an example, the preconcentration and determination of a metal by a solvent extraction procedure of the type discussed above. To the sample volume (up to 1L) is added N ml of concentrated nitric acid of specific gravity 1.41 containing Mmg kg^{-1} of copper as impurity. The weight (W) of the metal in Vml of concentrated acid is therefore:

$$W(ng) = M \times N \times 1.41 \times 10^3 \qquad (1.1)$$

In order for detection of the element to be achieved the weight of the element in Vml of sample should be $2W$ (ng) $= 2 \times M \times N \times 1.41 \times 10^3$ng per Vml of sample, ie $(1000 \times 2 \times M \times N \times 1.41 \times 1000)/V$ng element per litre which is taken as the detection limit (LD) for the element concerned, ie

$$LD = \frac{2820000MN}{V} \qquad (1.2)$$

where M is impurity concentration (mg kg^{-1}) in concentrated nitric acid; N = volume (ml) of concentrated nitric acid used in determination; V = volume (ml) of aqueous sample taken for preconcentration.

If 0.5ml of concentrated nitric acid is used then

$$LD = \frac{1410000M}{V} \qquad (1.3)$$

If the concentration of the element in the concentrate of nitric acid is 0.001mg kg^{-1} (M) and 1L sample (V) is taken for analysis

$$LD = \frac{1410000 \times 0.001}{1000} = 1.41\text{ng L}^{-1} \qquad (1.4)$$

In Table 1.2 are tabulated the expected limits using three grades of nitric acid of differing purities when 5ml, 200ml, 1000ml and 5000ml aqueous sample are preconcentrated to 5ml of final extract, ie preconcentration factors, respectively of ×1 (no preconcentration), ×40, ×200 and ×1000. Taking cadmium as an example it is seen that detection limits obtainable with ultra pure Normatom nitric acid range from 282ng L^{-1} (no preconcentration) to 0.28ng L^{-1} (×1000 preconcentration). For the Normapure acids detection limits for cadmium are much poorer ranging

Table 1.2 Detection limits expected using various grades of concentrated nitric acid and preconcentration factors

Initial volume of aqueous sample (Vme)	Volume of sample extract	Preconcentration factor	Grade of nitric acid					
			Normanton ultrapure (65%)		Normapure (69%) (for analysis) of Cd, Hg, Pb		Normapure (69%) ISO–ACS reagent	
			Element	LD* (ng L⁻¹)	Element	LD* (ng L⁻¹)	Element	LD* (ng L⁻¹)
5	5	1	Cd	282	Cd	1410	Cd	2820
			Hg	282	Hg	282	Hg	–
			Pb	1410	Pb	1410	Pb	14100
			As	282	As	2820	As	2820
			Cu	282	Cu	–	Cu	14100
			Fe	2820	Fe	56400	Fe	56400
			Mn	282	Mn	–	Mn	112800
			Ni	282	Ni	–	Ni	14100
200	5	40	Cd	7.05	Cd	35.2	Cd	70.5
			Hg	7.05	Hg	7.05	Hg	–
			Pb	35.2	Pb	35.2	Pb	352.5
			As	7.05	As	70.5	As	70.5
			Cu	7.05	Cu	–	Cu	352.5
			Fe	70.5	Fe	1410	Fe	1410
			Mn	7.05	Mn	–	Mn	2820
			Ni	7.05	Ni	–	Ni	352.5
1000	5	200	Cd	1.41	Cd	7.05	Cd	14.10
			Hg	1.41	Hg	1.41	Hg	–
			Pb	7.05	Pb	7.05	Pb	70.5
			As	1.41	As	14.10	As	14.10

Table 1.2 continued

Initial volume of aqueous sample (Vme)	Volume of sample extract	Preconcentration factor	Grade of nitric acid					
			Normanton ultrapure (65%)		Normapure (69%) (for analysis) of Cd, Hg, Pb)		Normapure (69%) ISO-ACS reagent	
			Element	LD* (ng L⁻¹)	Element	LD* (ng L⁻¹)	Element	LD* (ng L⁻¹)
			Cu	1.41	Cu	–	Cu	70.50
			Fe	14.1	Fe	282	Fe	282
			Mn	1.41	Mn	–	Mn	564
		1000	Ni	1.41	Ni	–	Ni	70.5
			Cd	0.28	Cd	1.41	Cd	2.82
			Hg	0.28	Hg	0.28	Hg	–
			Pb	1.41	Pb	1.41	Pb	14.10
5000	5		As	0.28	As	2.82	As	2.82
			Cu	0.28	Cu	–	Cu	14.10
			Fe	2.82	Fe	56.4	Fe	56.4
			Mn	0.28	Mn	–	Mn	112.8
			Ni	0.28	Ni	–	Ni	14.10

*From equation LD (ng L⁻¹) = $\dfrac{2 \times M \times N \times 1.41 \times 10^6}{V}$ = $\dfrac{2 \times M \times 0.5 \times 1.41 \times 10^6}{V}$ = $\dfrac{1410000M}{V}$

where M = concentration (mg kg⁻¹) of impurity in concentrated nitric acid (see Table 1.1); N = volume (ml) of concentrated nitric used in determination (0.5ml); V = volume (ml) of aqueous sample taken for preconcentration

Source: Own files

Table 1.3 Grades of acetone available from Rhone–Poulenc

Grade	Non-volatile residues at 100°C (mg kg^{-1})	Acidity as CH$_3$COOH (mg kg^{-1})
Acetone for pesticide analysis	3	–
Acetone for high performance liquid chromatography	5	18
Acetone for ultraviolet spectroscopy	5	–
R P Normapure A R	10	20
Acetone Rectapure	50	–
Acetone, laboratory grade	–	–

Source: Own files

from 1410ng L^{-1} (no preconcentration) to 1.41ng L^{-1} (×10000 preconcentration). Clearly then, if a detection of down to 2ng L^{-1} of cadmium is required then Normaton nitric acid must be used and a preconcentration factor of 200 adopted, eg 1L of sample preconcentrated to 5ml of extract.

Similar considerations apply in the preconcentration and determination of organic substances. Thus Rhone–Poulenc quote six grades of acetone ranging in non-volatile residue contents from 3 to 50mg kg^{-1} (Table 1.3). Clearly, if, for example, pesticides are being preconcentrated and acetone is the solvent then the pesticide analysis grade is preferable to the others, simply because with the purer acetone there is less chance of interference with the determination of the substance of interest.

1.2 Pre-purification of reagents

If reagents of adequate purity to meet detection limit requirements in preconcentration methods are to be available then laboratory purification of reagents may be necessary. Indeed many workers prefer to clean up their own reagents to stipulated degrees of purity so that they can achieve better control on the environmental contamination factors controlling the results obtained in analyses. In this approach some or all of the reagents are purified immediately prior to their use in the analyses or as part of the analytical procedure. A good example of this is the cleaning up of ammonium pyrrolidinedithiocarbamate–diethylammonium dithiocarbamate mixed chelating reagent before its use in the preconcentration of nickel, copper, cadmium and lead, also organometallic compounds.

In this procedure 1g of each chelating agent is dissolved in 50ml water in a separatory funnel. Freon (1, 1,2–trichlorotrifluoroethane) (100ml) is added. After agitation the lower freon layer containing metallic impurities is separated and rejected, and a similar extraction with 100ml Freon performed. The resulting aqueous solution of chelating agents is now virtually free from metallic contamination and is suitable for use in the preconcentration of samples.

1.3 Effect of pH

In a typical chelation–solvent extraction procedure a small volume of an aqueous solution of ammonium pyrrolidinediethyldithiocarbamate is added to a large volume of aqueous sample, and the metal chelates extracted with a relatively small volume of methyl ethyl ketone [1,2]. The pH of the aqueous sample has a profound effect on the efficiency with which various metals are extracted as illustrated in Table 1.4 and Figs. 1.1 and 1.2. Large decreases in recovery were observed for all metals at pH values below 3.0 and small decreases at pH values greater than 8.0.

Table 1.4 Influence of pH on extraction of metal pyrrolidine diethyldithriocarbamates from water with methyl ethyl ketone

Element	Element recovery
Co, Cu, Fe, Ni, Mn, Zn	100% at pH 4.0–4.6
Cd, Pb	100% at pH 4.0–4.6 but complex only stable 2–3
Ag	100% at pH 1–2
Cr	100% at pH 1.8–3.0
Mo	100% at pH 1–1.5

Source: Own files

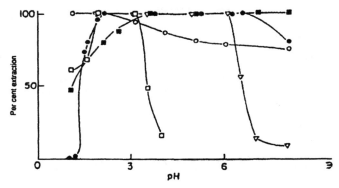

Fig. 1.1 Effect of pH on the extraction of some trace metals using the APDC–MIBK procedure (aqueous/organic = 5); O Ag, 4µg L $^{-1}$; □ Cr, 20µg L $^{-1}$; ∇ Fe, 20µg L $^{-1}$; • Mn 6µg L $^{-1}$; ■ Pb, 8µg L $^{-1}$
Source: Reproduced with permission from Reggeks, G. and Van Grieken, R. [3] Springer Verlag

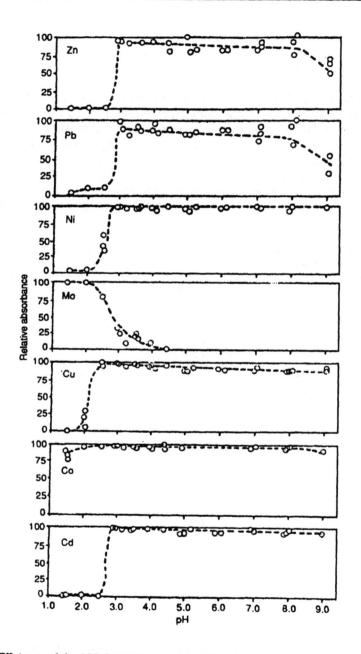

Fig. 1.2 Efficiency of the APDC/MIBK method as a function of sample pH
Source: Reproduced with permission from Subramanian, K.S. and Meranger, J.C. [1]
Gordon AC Breach

2,2–Diaminodiethylamine cellulose powder has particularly good chelation properties for metals achieving a chelation capacity of 1.5mol equivalent g $^{-1}$ resin [3]. Fig. 1.3 illustrates the effect of pH on the metal collection efficiency of this resin. At pH 7, 90–100% recovery is obtained for chromium, zinc, cadmium, iron, cobalt, europium, and mercury while only a 30% recovery of manganese is obtained. In all cases recoveries decrease sharply at pH values below 6.

1.4 Effect of chelating agent–metal ratio

Another factor which affects recovery in chelation–solvent extraction procedures is the ratio of the molar concentration of the chelating agent to the molar concentration of metal present in the aqueous sample. The results in Fig. 1.4 illustrate that in the methylethyl ketone extraction of metal pyrrolidinedithiocarbamates from water samples 100% recovery of cadmium, lead, silver, nickel and copper is obtained when the chelating agent:metal ratio (log (APDC/metal)) is 3.0 while a ratio of 5.0 is required to full recovery of iron, chromium and cobalt [4]. It is necessary, therefore, when developing such procedures to carry out a preliminary systematic examination of the effect of molar excess of chelating agent used on metal recovery in the preconcentration step.

1.5 Matrix interference effects

Water samples might contain either naturally occurring chelating agents (eg humic acids) or man made chelating agents (eg nitriloacetic acid or linear alkylbenzene sulphonates) that form strong complexes with metals present in the samples. Such complexes might be stronger than the complex formed between metals and the chelating agent added to the sample in chelation–solvent extraction methods of preconcentration which would under these conditions give low metal recoveries. Such effects can be overcome by using a standard addition method for calibration or by giving the water sample a preliminary treatment with agents such as ozone or by exposure to ultraviolet light to decompose such complexes before proceeding with the preconcentration.

Another type of matrix effect is covered by the presence in the sample of relatively high concentrations of inorganic solids such as occurs, for example, in the case of sea or estuary water samples. Such an effect is illustrated in Fig. 1.5 which shows the effect of the concentration of sodium chloride on the partition coefficient of metals between the water sample, adjusted to pH 11, and cellulose. In the absence of sodium chloride K_d is in the range 500–2000 for iron, nickel, cadmium, copper, cobalt and zinc. The presence of 1.5–2.0g L $^{-1}$ of sodium chloride enhances K_d by a factor of 30 to 15000–60000. This indicates that the deliberate

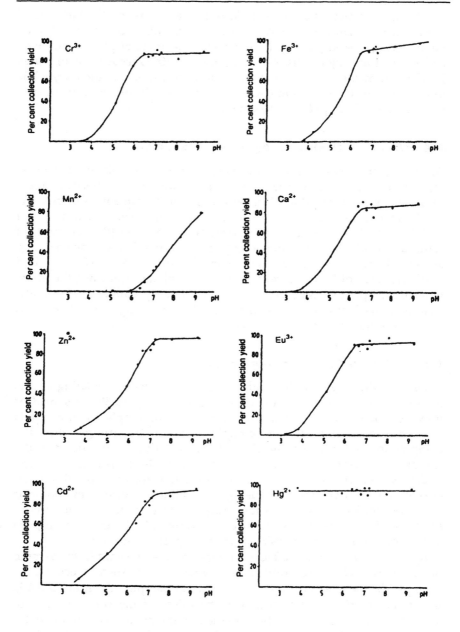

Fig. 1.3 Influence of the pH on the collection efficiency of cellulose DEN powder for Cr^{3+}, Mn^{2+}, Fe^{3+}, Ca^{2+}, Zn^{2+}, Cd^{2+}, Eu^{3+} and Hg^{2+}

Source: Reproduced with permission from Reggeks, G. and Van Grieken [3] Springer Verlag

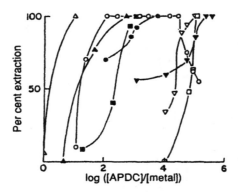

Fig. 1.4 Effect of APDC on the extraction of some trace metals using the APDC–MIBK procedure: ▲ Ag, 4µg L^{-1}; ● Cd, 0.2µg L^{-1}; ∇ Co, 20µg L^{-1}; ▼ Cr, 20µg L^{-1}; Δ Cu, 20µg L^{-1}; □ Fe, 20ng mL^{-1}; ■ Ni, 50µg L^{-1}; O Pb, 8µg L^{-1}

Source: Reproduced with permission from Burba, P. and Willman, P.G. [4] Elsevier Science, UK

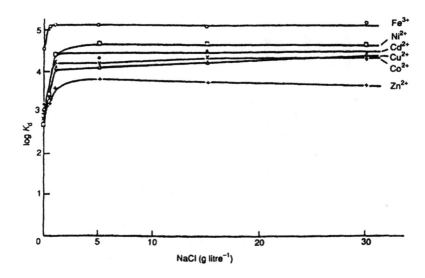

Fig. 1.5 Influence of electrolyte solution concentration (0–30g L^{-1} NaCl) on heavy metal sorption (Cd^{2+}, Co^{2+}, Cu^{2+}, Fe^{3+}, Ni^{2+}, Zn^{2+} each 20µg L^{-1}) on cellulose

Source: Own files

Table 1.5 Adsorption efficiency of organics on macroreticular resins

	% recovery carbon disulphide desorption		% recovery thermal desorption	
	200ml water sample containing 2–10μg L^{-1} of organic	200ml water sample containing 100μg L^{-1} of organic	20ml water sample containing 10μg L^{-1} of organic	200ml water sample containing 200μg L^{-1} of organic
Aromatic hydrocarbons	72 ± 7	76 ± 11	87 ± 8	90 ± 11
Polyaromatic hydrocarbons	–	–	88 ± 3	85 ± 2
Aliphatic chloro hydrocarbons	90 ± 2	56 ± 1	–	–
Aromatic halo hydrocarbons	94 ± 12	–	88 ± 3	84 ± 7
Alcohols	91 ± 8	72 ± 18	98 ± 6	89 ± 7
Ketons/aldehydes	99 ± 3	98 ± 6	97 ± 2	84 ± 7
Esters	78 ± 8	50 ± 18	97 ± 2	87 ± 10
Phenols	–	–	84 ± 20	77 ± 10

Source: Reproduced with permission from Ryan, J.P. and Fritz, J.S. [6] Preston Publications Inc

addition of salt to an aqueous sample might have beneficial effects in preconcentrations which involve adsorption of metals on to an adsorbent, followed by desorption with a small volume of a reagent.

1.6 Adsorption efficiency of solid adsorbents

The efficiency of adsorption of organic substances from water on to columns of microreticular resins is influenced by several factors such as the flow rate of the water through the column and sample pH. The subsequent desorption of the preconcentrated organic, by either extraction with a relatively small volume of organic solvent such as acetone or carbon disulphide or thermally, can be affected by desorption time and temperature. In a typical thermal preconcentration technique [5] based on these principles the water sample is passed down a column of XAD–2 or XAD–4 resin at a controlled flow rate. At the end of this stage the XAD–2 column is connected to a small column of Texax GC and the organics swept on to the latter column with a purge of helium at 230°C. The Texax GC column is then heated to 45°C and swept with helium to remove residual water. Finally, the Texax GC column is heated to 220°C and purged with nitrogen on to a gas chromatograph or gas chromatograph–mass spectrometer for quantification and identification.

In a typical solvent extraction preconcentration technique [6] the water is passed down a column of XAD–2 or XAD–4 resin and then a relatively small volume of carbon disulphide or acetone is passed down the column to elute organics prior to gas chromatographic analysis. Some typical recovery data for various types of organic compounds from macroreticular resins are given in Table 1.5. As might be expected adsorption efficiency of the organic from the water sample on to the resin and the desorption efficiency for removal of the organic from the resin either by solvent extraction or thermally differ appreciably from one type of organic compound to another. When devising such methods for the preconcentration of particular organic compounds carefully controlled and systematic recovery studies should be conducted to check on recovery and on reproducibility of recovery in order to establish experimental conditions for maximum recovery. If the recoveries are acceptably reproducible but low, then the preconcentration step should be included in the method calibration procedure adopted.

A further method of metal preconcentration involves passage of the aqueous sample through a bed of a metal oxide, usually manganese dioxide, titanium dioxide, zirconium dioxide, alumina, hydrated iron oxide or silica or C_{18}–bonded silica. The adsorbed metal is then desorbed with a volume of an aqueous reagent in order to achieve preconcentration. Generally speaking, such preconcentrations are carried out on relatively thin beds rather than columns of the metal oxide. In Fig. 1.6

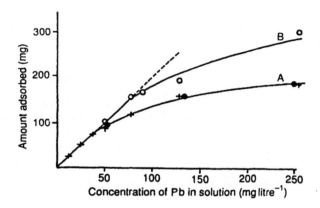

Fig. 1.6 The amount of lead adsorbed from solution versus amount in solution. Curve A: potable water (+) and sea water (•), 11cm diameter filter. Curve B: potable water, 4.7cm filter system. The dashed line represents quantitative (100%) adsorption from solution Source: Reproduced with permission from Matthews, K.M. [7] Marcel Dekker Inc., New York

(curve A) is shown the relationship between the weight of lead adsorbed on a 0.25cm thick disc of manganese dioxide from 2L of saline and non-saline aqueous solutions containing between 10 and 500mg lead, ie (5–250mg L^{-1} lead) [7]. The flow rate used in this experiment was 500ml min^{-1}, ie 4min to pass through a 2L sample. It is seen that 100% adsorption of lead on manganese dioxide occurs when the sample contains up to 38mg L^{-1} lead (75mg lead), and then drops dramatically to 50% recovery when the sample contains 150mg L^{-1} lead (300mg lead). Increasing the thickness of the manganese dioxide layer eightfold from a 0.25cm to a 2cm thick disc and using the same sample flow rate extended the range over which quantitative recovery of lead from non-saline acid saline water samples was obtained, from 38mg L^{-1} (75mg lead) to 80mg L^{-1} (160mg lead). Increasing the pH from 2.0 to 7.4 did not affect recovery.

In such concentration methods the factors which most affect recovery are the flow rate of the aqueous sample and the thickness and possibly area of the collecting bed. Such parameters should be carefully studied when devising procedures.

1.7 Ion exchange resin theory

When an ion exchange reaction is carried out by a 'batch' method – that is, by putting a quantity of resin into a certain volume of solution – the reaction begins at once, but a certain time elapses before the equilibrium

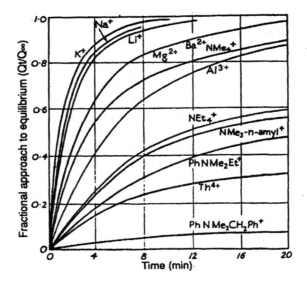

Fig. 1.7 Rate of exchange of phenolsulphonic acid resin (in ammonium form) with various cations
Source: Own files

state is reached. It is generally a simple matter to determine the rate of exchange, for example, by sampling and analysing the solution at intervals or by making use of some physical property, such as electrical conductivity, which changes as the reaction proceeds. Such experiments show that rates of exchange can vary very much from one system to another, the times of half-exchange ranging from fractions of a second to days or even months in certain extreme cases. Fig. 1.7 shows the progress of exchange with a number of different cations exchangers with equal samples of the ammonium form of a phenolsulphonic acid resin under the same conditions.

Clearly, a knowledge of the rate of exchange is a prerequisite for the most effective use of resins. The factors which influence the rate of exchange and show how they account quantitatively for the form of kinetic curves such as those shown in Fig. 1.7 are listed below. By a series of controlled experiments in which one factor is varied at a time, it can readily be shown that, other things being equal, a high rate of exchange is generally favoured by the following choice of conditions:

(1) a resin of small particle size;
(2) efficient mixing of resin with the solution;
(3) high concentration of solution;

(4) a high temperature;
(5) ions of small size;
(6) a resin of low cross-linking.

The exchange reactions are generally much slower than reactions between electrolytes in solution, but there is no evidence to indicate the intervention of a slow 'chemical' mechanism such as is involved in most organic reactions, where covalent bonds have to be broken. The slowness of exchange reactions can be satisfactorily accounted for by the time required to transfer ions from the interior of the resin grains to the external solution, and vice versa. The rate of exchange is therefore seen to be determined by the rate at which the entering and leaving ions can change places. The process is said to be 'transport-controlled'; it is analogous, for instance, to the rate of solution of a salt in water. All the factors (1) to (6) listed above are such as to facilitate the transport of ions to or from or through the resin.

In preconcentration techniques the batch procedure, ie putting a quantity of resin into a certain volume of solution, is not usually used. Most practical applications of ion exchange resins in preconcentration involve column processes wherein a large volume of sample is passed down a small column of resin. For example if cation B is to be preconcentrated from water by means of a sodium form resin

$$NaR + B^+ (aq) \rightarrow BR + Na^+ (aq)$$

it would be inefficient to shake a quantity of resin with a given volume of water, as an equilibrium would be set up, and some of the cation B would be left in solution. Instead the water is passed through a column of the resin, and as the water becomes depleted of cation B by contact with the first layers, it passes to fresh resin containing little or no cation B, and so the equilibrium is constantly displaced. With a long enough column the effluent is entirely free from cation B.

In order to arrange conditions for efficient operation of columns some regard must be paid to the 'local' kinetics of the exchange reactions. The quantitative theory [8] of column processes is somewhat specialised, but a qualitative understanding of the principal factors is readily obtained.

The usual methods of operating columns can be divided into (A) displacement, in which the ion on the column is sharply displaced by another, more strongly sorbed, and (B) elution, in which the ion is more gradually moved down the column by treatment with a more weakly sorbed element. (A) leads to sharp bands travelling at a rate determined by the flow of incoming solution. (B) leads to bands with somewhat diffuse fore and aft boundaries, travelling at a rate dependent on the relative affinity of the resin for the two ions. (A) is useful for dealing with large quantities ('heavy loading' of the column), while (B) is the preferred

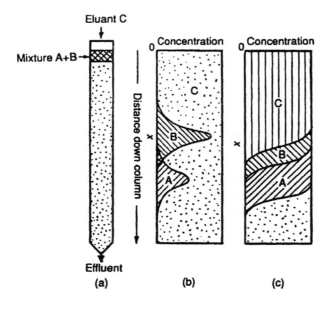

Fig. 1.8 Form of bands in elution (b) and displacement (c) chromatography: (a) represents the column, *x* distance down column, and figures (b) and (c) show the concentrations of ions A, B and C along the column after a certain time
Source: Own files

process for obtaining optimum separation of substances, albeit in small quantities. Fig. 1.8 shows diagrammatically the types of band obtained with a ternary mixture of A, B and C in the two cases.

Whatever the purpose of the process to be carried out on the column – absorption, displacement or chromatographic analysis or separation – the highest efficiency would be reached if the liquid passing down the column came to full equilibrium with each layer of resin grains. This condition, however, would require an infinitesimal rate of flow. At all practical rates of flow non-equilibrium conditions prevail with reduced efficiency. The essence of column operation is to choose an appropriate compromise between efficiency and speed, compatible with the required result.

The factors militating against attainment of local equilibrium are:

(a) non-uniformity of the solution in the interstices, ie film–diffusion,
(b) non-uniformity within the grains of resin, ie particle–diffusion resistance; and
(c) irregular flow of the solution down the column, spoiling sharpness of the moving bands of solute.

The obvious steps to be taken in the direction of improving efficiency are:

(1) the use of resins of small particle size;
(2) low flow-rate;
(3) an elevated temperature (to increase diffusion coefficients); and
(4) great care in the packing of the column

but these are offset by the extra pressure, time or experimental difficulty incurred.

Assuming that factor (c) could be ignored (which, however, is practically never the case) it is possible to calculate the performance of a column under conditions where film- or particle-diffusion are rate determining [8]. A more practicable approach, however, which covers any form of non-equilibrium, is to use a semi-empirical treatment similar to the 'plate theory' employed in the theory of distillation columns [9,10]. According to the plate theory, the column can be considered as consisting of a number (N) of sections ('theoretical plates') in each of which the average concentration (c) of solution in the pores can be considered as effectively in equilibrium with the average amount of solute sorbed by the resin. The effectiveness of the column can be judged by the number of theoretical plates it appears to contain, and once N is known the elution curve can be predicted.

Consider by way of example, the most important case of the application of the plate theory – the separation of two similar species, A and B, by elution chromatography with an eluant, C. On preconcentration it may be required, for example, to retain species A on the column (ie the preconcentrated species) and to allow mobile species B to be completely eluted from the column (as this substance might interfere in the determination of A in the final analytical step on the eluted pre-concentrate). A small quantity of the mixture of A + B is first sorbed on the top of the column (previously in form C), and then 'developed' by elution with the solution of C. As the mixed band moves down the column, A and B gradually separate or in the ideal case species A hardly moves and species B moves relatively quickly down the column, because of their different sorption affinities. If the 'loading' of the column is light, the sorption of A and B can be considered linear with concentration, and each is characterised by an equilibrium distribution coefficients, K_d, defined by

$$K_d = \frac{\text{conc. of the ion in resin (mol per g)}}{\text{conc. of the ion in solution (mol per ml)}}$$

The rate of movement of the band down the column is inversely proportional to K_d, and, hence, the rate of separation of A and B depends on the ratio of their distribution coefficients, $(K_d)_A/(K_d)_B$, which can be determined by simple batch equilibrium experiments, using the appropriate solution of C as the medium in excess. For a high rate of movement K_d is low and for a low rate of movement K_d is high, so to separate substance B with a high rate of movement from substance A with little or no movement, $(K_d)_A$ should be high and $(K_d)_B$ low, ie $(K_d)_A/(K_d)_B$ should be high.

The effective number of theoretical plates, N, in the column (which depends on the flow-rate) can be determined by a study of the elution curve for a suitable species (say A). An elution curve is a graph of the concentration (c) of the species in the effluent flowing from the column, as a function of the volume (v) of elutriant passed through. According to Glueckauf's theory [8,10] the shape of the elution curve is given approximately by

$$c = c_{max}\exp\left[-\frac{N}{2}\frac{(v_{max}-v)^2}{vv_{max}}\right] \tag{1.5}$$

where c_{max} and v_{max} are the co-ordinates of the centre of the band; the position of the maximum is given by

$$c_{max} = \frac{m}{v_{max}}\sqrt{\frac{N}{2\pi}} \tag{1.6}$$

where m is the total quantity (mol equiv.) of the given solute in the band and is equal to the area under the elution curve; and the band width (at the concentration level $c_{max}/e = 0.368c_{max}$) is given by

$$\text{band width} = \frac{64v_{max}}{N^2} \tag{1.7}$$

N can be determined conveniently from either eqn (1.5) or (1.7). If two similar solutes, A and B, are being separated, N will be common to both, but v_{max} will be approximately inversely proportional to K_d, ie $(v_{max}K_d)_A = (v_{max}K_d)_B$.

If the distribution coefficients are rather similar, there will be a significant overlap of the two bands, and the purity of the fractions can be calculated. For equal purity of the separated bands, the 'cut' should be made at the volume given by

$$v = \sqrt{_Av_{max\,B}\,v_{max}} + \frac{2_Av_{max\,B}\,v_{max}\,(m_A^2-m_B^2)}{N(_Bv_{max}-_Av_{max})\,(m_A^2+m_B^2)} \tag{1.8}$$

where m_A and m_B are the total quantities of A and B present. The proportion of impurity of A in B (or vice versa) is given by

$$\frac{\Delta m_A}{m_B} = \frac{2m_Am_B}{m_A^2+m_B^2}\left[\frac{1}{2} - S\left(\frac{\sqrt{N}\,(\sqrt{_Bv_{max}} - \sqrt{_Av_{max}})}{\sqrt[4]{_Av_{max\,B}v_{max}}}\right)\right] \tag{1.9}$$

where S is the area under the tail of the overlapping curve, and is given by the appropriate integral of the normal errors curve (obtained from tables), ie

$$S = \frac{1}{2\pi} \int_0^x \exp - \left(\frac{x^2}{2}\right)^2 dx, \quad \text{with } x = \frac{v_{max} - v}{\sqrt{v_{max}v}} \sqrt{N} \quad (1.10)$$

Since N is proportional to the length of the column, the improvement of separation obtainable by increasing the length can be calculated. Increase of cross-section of the column increases its handling capacity.

Glueckauf [10] has provided a chart showing the purity obtainable with different separation factors, $(K_d)_A/(K_d)_B$, and different numbers of plates; for example with a separation factor of 1.2 and a column of 1000 plates, a purity of 99.9% is obtainable (starting with equal quantities of A and B).

This brief explanation of the plate theory should suffice to show how a few well-conceived exploratory measurements make it possible to plan a critical preconcentration step. These factors which have a profound effect on the binding capacity of ion exchange resins include the pH of the aqueous sample and the presence of natural chelating agents such as humic acid as discussed previously.

1.8 Coprecipitation techniques

In this technique to the organic substance or metal to be preconcentrated is added a substance such as an iron or lanthanum salt which upon the addition of a alkaline precipitant coprecipitates the substance to be determined on to the iron or lanthanum hydroxide precipitates. Subsequent dissolution of the total precipitate in a small volume of an acidic reagent gives a preconcentration of the substance that is required to be determined.

This technique has been used to preconcentrate phenols and humic and fulvic acids (coprecipitation with ferric hydroxide), aminoacids (coprecipitation with zirconium hydroxide); polyaromatic hydrocarbons and organolead compounds (coprecipitation with barium sulphate) and polyaromatic hydrocarbons and polychlorobiphenyls (coprecipitation with magnesium hydroxide) and also a wide range of metals.

An example of this technique is the preconcentration of metals by the addition of iron followed by pH adjustment to an alkaline pH by addition of sodium hydroxide. The mixture is left for an equilibrium period during which the iron precipitates as hydroxide with some or all of the oxides of the metals to be preconcentrated occluded on to it. Following filtration of the combined metal oxides, they are dissolved in a relatively small volume of an acidic reagent and this preconcentrated solution analysed. One of the most important factors affecting recovery to be controlled

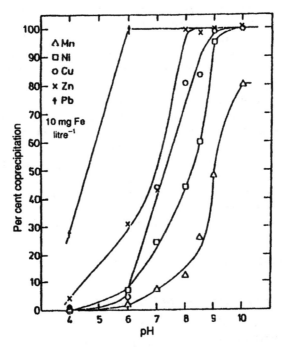

Fig. 1.9 pH dependence on the coprecipitation of Mn, Ni, Cu, Zn and Pb in seawater
Source: Reproduced with permission from Chakravarty, R. and Van Grieken, R. [11]
Gordon AC Breach, Amsterdam

during such preconcentrations is the pH to which the solution is adjusted
with sodium hydroxide.

In Fig. 1.9 is shown metal recovery–pH curves obtained in the
preconcentration of manganese, nickel, copper, zinc and lead [11] in
200ml aqueous sample to which had been added 2mg of ferric iron. Lead
and zinc are fully coprecipitated at a pH above 7–8, copper and nickel
require a pH above 9, whilst manganese is only 60% coprecipitated at a
pH in excess of 10. Therefore, with the exception of manganese the
optimum pH is 9. Using this procedure in conjunction with an analytical
finish involving X-ray fluorescence spectrometry enabled preconcentration factors of up to 1.5×10^4 to be achieved.

1.9 Preconcentration of organics by direct solvent extraction

The following two factors affect the efficiency of such preconcentration
procedures:

(a) *Mutual solubility effects.* When a water sample containing a solute to be preconcentrated is shaken with a relatively small volume of organic solvent then at the end of the extraction the organic phase contains dissolved water and the water phase contains dissolved solvent, low recovery of solute in the organic extract will be obtained depending on the solubility of solute containing water in the solvent. Low recoveries can be overcome by carrying out two or three successive solvent extractions with fresh portions of solvent, but at the expense of decreasing the preconcentration factor achieved hence the detection limit of the overall procedure.

(b) *Partition effects.* When an aqueous phase containing a dissolved solute is shaken with an organic solvent, then, depending on the partition coefficient of the solute under the particular experimental conditions the solute will redistribute itself between the two phases. It is necessary to choose solvents which have a high preference for the solute.

The theory of these two effects is discussed below.

1.9.1 Theory of mutual solubility effects

If we start with $V_w$ml of water sample containing Aμg of solute to be preconcentrated (ie $A \times 1000/V_w\mu$g L^{-1} solute in water) and $V_s$ml of organic solvent, then, assuming complete transfer of solute from water to solvent and taking into account the solubilities of solvent in water and water in solvent, the volumes of solvent and water at the end of the extraction are:

$$\left(V_w + \frac{V_w S_2}{100} - \frac{V_s S_1}{100} \right) = V_w \left(1 + \frac{S_2}{100} \right) - \frac{V_s S_1}{100} \text{ml water} \quad (1.11)$$

and

$$\left(V_s + \frac{V_s S_1}{100} - \frac{V_w S_2}{100} \right) = V_s \left(1 + \frac{S_1}{100} \right) - \frac{V_w S_2}{100} \text{ml solvent} \quad (1.12)$$

where S_1 is the per cent v/v solubility of water in solvent and S_2 is the percent v/v solubility of solvent in water.

The term $V_w S_2/100$ represents the volume of solvent still containing solute dissolved in the aqueous phase, as $V_s$ml of solvent contains Aμg solute, $V_w S_2/100$ml of organic solvent contains A $V_w S_2/100V_s\mu$g solute. Therefore at the end of the first extraction we have $AV_w S_2/100V_s\mu$g solute dissolved in $[V_w(1 + S_2/100) - V_w S_2/100]$ml water, and $A - AV_w S_2/100V_s =$ A1 $- V_w S_2/100V_s$, $[V_s(1 + S_1/100) - V_w S_2/100]$ml solvent.

It now remains to extract the $AV_w S_2/100V_s\mu$g residual solute from the water phase by carrying out a second extraction with a further $V_s$ml of fresh organic solvent. The water obtained at the end of the first extraction is now already saturated with solvent and, hence, does not dissolve further solvent in the second extraction. The $V_s$ml organic solvent used in the second extraction is not saturated with water and dissolves $S_1V_1/100$ml water. The volumes of solvent and water at the end of the second extraction are:

$$\left(V_w + \frac{S_2V_w}{100} - \frac{S_1V_s}{100}\right) - \frac{S_1V_s}{100} = V_w\left(1 + \frac{S_2}{100}\right) - 2\frac{S_1V_s}{100}\text{ml water} \quad (1.13)$$

and

$$V_s + \frac{S_1V_s}{100} = V_s\left(1 + \frac{S_1}{100}\right) \text{ ml solvent} \quad (1.14)$$

At the end of the second extraction the $V_s(1 + S_1/100)$ml of solvent contains the residual $A\,V_wS_2/100\,V_s$µg solute, ie at the end of the second extraction the combined organic extracts contain the theoretical initial solute content (Aµg) of the original water sample:

$$A\left(1 - \frac{V_wS_2}{100\,V_s}\right) + \frac{A\,V_wS_2}{100\,V_s} = A \quad (1.14)$$

This data is summarised below:

	Initially	End of first extraction	End of second extraction	Combined organic extracts
Volume of water	V_w	$V_w\left(1 + \frac{S_2}{100}\right) - \frac{V_sS_1}{100}$	$V_w\left(1 + \frac{S_2}{100}\right) - \frac{2S_1V_s}{100}$	
Volume of solvent	V_s	$V_s\left(1 + \frac{S_1}{100}\right) - \frac{V_wS_2}{100}$	$V_s\left(1 + \frac{S_1}{100}\right)$	$2V_s\left(1 + \frac{S_1}{100}\right) - \frac{V_wS_2}{100}$
Wt of solute in water	A	$\frac{AV_wS_2}{100V_s}$	Nil	
Wt of solute in solvent	Nil	$A\left(1 - \frac{V_wS_2}{100V_s}\right)$	$\frac{AV_wS_2}{100V_s}$	A

The preconcentration factor, ie $\dfrac{\text{volume of water sample}}{\text{volume of total organic extracts}} = \dfrac{V_w}{V_s(1 + S_1/100) - V_wS_2/100}$

(single extraction) or $\dfrac{V_w}{2V_s(1 + S_1/100) - V_wS_2/100}$ (two extractions)

Some theoretical results are illustrated in Table 1.6. If 1L water is extracted with solvent then, depending on the solubility of solvent in the aqueous phase, some 50–95% of solute in the aqueous phase is extracted in a single solvent extract and 100% in two solvent extractions. Concentration factors achieved are between 100 and 200 using a single extraction

Table 1.6 Mutual solubility effects in extraction of solutes from water with organic solvents

A. 1L of water extracted with 10ml solvent

Volume (ml) Water V_w	Organic solvent V_s	Original weight (µg) of solution in aqueous phase A	Solubility (%) of solution in aqueous phase S_2	Weight of solute µg in organic phase	% recovery of solute	Preconcentration factor One extraction	Two extractions
First extraction				$A\left(1 - \dfrac{V_w S_2}{100 V_s}\right)$	$A\left(1 - \dfrac{V_w S_2}{100 V_s}\right)\dfrac{100}{A}$	$\dfrac{V_w}{V_s\left(1 + \dfrac{S_1}{100}\right) - V_w \dfrac{S_2}{100}}$	$\dfrac{V_w}{2V_s\left(\dfrac{1 + S_1}{100}\right) - V_w \dfrac{S_2}{100}}$
1000	10	20				assuming $S_1 = 0.1\%$	
			0.05	19.0	95	105	51
			0.1	18.0	90	111	52
			0.5	10.0	50	200	66
Second extraction				$\dfrac{AV_w S_2}{100 V_s}$	$\dfrac{100 A V_w S_2}{100 V_s A}$		
	10	—	0.05	1.0	5	—	—
			0.1	2.0	10		
			0.5	10.0	50		—
Combined first and second extractions				$A\left(1 - \dfrac{V_w S_2}{100 V_s}\right) + \dfrac{A V_w S_2}{100 V_s} = A$		—	$100 V_s$
			0.05	20	100		
			0.1	20	100		
			0.5	20	100		

Table 1.6 continued

Volume (ml) Water V_w	Organic solvent V_s	Original weight (µg) of solution in aqueous phase A	Solubility (%) of solution in aqueous phase S_2	Weight of solute µg in organic phase	% recovery of solute	Preconcentration factor One extraction	Two extractions
B. 1 L of water extracted with 1ml solvent							
First extraction							
1000	1	20	0.05	10	50	1996	–
			0.1	Nil*	–	–	–
Second extraction							
	1	–	0.05	10	50	–	–
			0.1	Nil*	–	–	–
Combined first and second extracts							
			0.05	20	100	–	666
			0.1	Nil*	–	–	–
C. 10L of water extracted with 5ml solvent							
First extraction							
10000	10	20	0.02	16.0	80	1250	–
			0.05	10.0	50	2000	–
			0.01	Nil*	Nil	–	–
Second extraction							
	10		0.02	4.0	20	–	–
			0.05	10.0	50	–	–
			0.1	Nil*	Nil*	–	–
Combined first and second extracts							
			0.02	20.0	100	–	555
			0.05	20.0	100	–	667
			0.1	Nil	–	–	–

*Organic solvent completely dissolved in organic phase

Source: Own files

Table 1.7 Interaction of preconcentration factor and detection limit

Volume of water	Volume of solvent (ml)	Preconcentration factor	If 1 µg L^{-1} can be detected in the aqueous phase then the following concentration can be determined by preconcentration
1000	10	105–200	0.0095–0.005
	2 × 10	51–66	0.0196–0.0151
1000	1	1996	0.00050
	2 × 1	666	0.00151
10000	10	1250–2000	0.0080–0.0005
	2 × 10	555–667	0.00180–0.00149

Source: Own files

approximately halving to 51–66 when using a double extraction. Appreciably higher preconcentration factors of 1996 (single extraction) or 666 (double extraction) are obtained when 1L of water is extracted with 1ml of solute, 50% of the solute is obtained in a single extract and 100% in a double extraction. When the volume of water taken is increased to 10L and two 10ml solvent extracts are prepared, depending on the solubility of the solvent in the aqueous phase, 50–80% of solute is recovered in a single solvent extract and 100% in two solvent extracts. Preconcentration factors are between 1250 and 2000 (single solvent extraction) and 555 and 667 (double solvent extraction).

The highest preconcentration factors of all are obtained in those circumstances where the volume of solvent used is small relative to the volume of water and where the solubility of the solvent in the aqueous phase is lowest. Multiple solvent extractions, due to the larger solvent volume always reduce preconcentration factors, with consequent adverse effects on detection limits (Table 1.7). Consequently, in devising such procedures compromises have often to be reached.

1.9.2 Theory of partition coefficient effects

When an organic solvent is shaken with an aqueous phase containing a solute the solute distributes itself between the aqueous and organic phases. If a volume of water containing A_1µg solute is shaken with a volume of organic solvent, then at the end of the extraction W_sµg of solute is present in the organic phase and W_wµg remains in the aqueous phase. This partitioning of the solute at a fixed ratio of water to solvent volumes can be represented as a partition coefficient K as follows:

$$K = \frac{\text{weight of solute in solvent } (A_1 - W_w \mu g)}{\text{weight of solute in aqueous phase } (W_w \mu g)} \tag{1.15}$$

The higher W_s $(A_1 - W_w)$ is relative to W_w, ie the higher K, the better the extraction from the aqueous to the organic phase and the better the preconcentration achieved. At the end of the extraction we are left with aqueous phase still containing $W_w \mu g$ unextracted solute. We have at this stage

$$W_{w1} + W_{s1} = A_1 \quad \text{and} \quad K = \frac{W_{s1}}{W_{w1}} \tag{1.16}$$

(the suffixes 1 indicating the extraction number).
    Putting in terms of W_{w1}

as $W_{s1} = KW_{w1}$

$$W_{w1} + KW_{w1} = A_1, \text{ie } W_{w1} (1 + K) = A_1 \tag{1.17}$$

$$\therefore W_{w1} = \frac{A_1}{1 + K}$$

Putting in terms of W_{s1}

as $$W_{w1} = \frac{W_{s1}}{K}$$

$$\frac{W_{s1}}{K} + W_{s1} = A_1, \text{ie } W_{s1} \left(\frac{1}{K} + 1 \right) = A_1 \tag{1.18}$$

$$\therefore W_{s1} = \frac{A_1}{(1 + K)/K} = \frac{A_1 K}{1 + K}$$

Similarly, when we come to perform a second extraction of the residual water phase with a further equal volume of fresh organic solvent the $A_1/(1 + K)$ $(= W_{w1})$ µg solute in the aqueous phase redistributes itself between the organic and aqueous phases. If $A_2 \mu g$ represents the total weight of solute available for redistribution (where the suffix 2 indicates the second extraction)

as $$K = \frac{W_{s2}}{W_{w2}} \qquad W_{s2} = KW_{w2}$$

and $$A_2 = KW_{w2} + W_{w2} = \frac{A_1}{1 + K}$$

it follows that

Table 1.8 Influence of partition coefficient on solute recovery

K	Weight of solute in initial water sample, A_1 (μg)	Weight (μg) of solute in organic extract				Recovery (%)			
		1st WS_1 $\dfrac{A_1K}{1+K}$	2nd WS_2 $\dfrac{A_1K}{(1+K)^2}$	3rd WS_3 $\dfrac{A_1K}{(1+K)^3}$	Cumulative WS_{cum} $\dfrac{A_1K[(1+K)^2+(1+K)+1]}{(1+K)^3}$	1st $\dfrac{(A_1K)\,100}{(1+K)A_1}$	2nd $\dfrac{(A_1K)\,100}{(1+K)^2A_1}$	3rd $\dfrac{(A_1K)100}{(1+K)^3A_1}$	Cumulative
2	20	13.33	4.444	1.481	19.25	66.7	22.22	7.41	96.2
20	20	19.05	0.907	0.043	20.00	95.2	4.54	0.22	100.0
200	20	19.90	0.099	0.00099	20.00	99.5	0.50	0.005	100.00
2000	20	19.99	0.010	0.0000	20.00	100.0	0.050	0.000	100.0

Source: Own files

$$W_{w2} (1 + K) = \frac{A_1}{1 + K} \qquad (1.19)$$

$$\therefore W_{w2} = \frac{A_1}{(1 + K)^2}$$

also $\quad W_{w2} = \dfrac{W_{s2}}{K}$

$$\therefore A_2 = W_{s2} + \frac{W_{2s}}{K} = \frac{A_1}{1 + K} \qquad (1.20)$$

$$\therefore W_{s2} \left(1 + \frac{1}{K} \right) = \frac{A_1}{1 + K}$$

ie $\quad W_{s2} = \dfrac{A_1}{1 + K/(1 + 1/K)} = \dfrac{A_1 K}{(1 + K)^2} \qquad (1.21)$

Summarising:

	μg solute in aqueous phase	μg solute in organic phase
First extraction:	$W_{w1} = \dfrac{A_1}{1 + K}$	$W_{s1} = \dfrac{A_1 K}{1 + K}$
Second extraction:	$W_{w2} = \dfrac{A_1}{(1 + K)^2}$	$W_{s2} = \dfrac{A_1 K}{(1 + K)^2}$
Similarly, for a third extraction:	$W_{w3} = \dfrac{A_1}{(1 + K)^3}$	$W_{s3} = \dfrac{A_1 K}{(1 + K)^3}$

Some results in Table 1.8 illustrate that a K value of 2 gives only 66.7% recovery of solute in the organic phase in a single extraction, improving to 90% in two extractions and 96% in three extractions. When the K value is 20 then 95% solute recovery is obtained in a single solvent extraction and above 99% in two extractions. For K values above say 100 a single solvent extraction is sufficient to achieve 99% plus solute recovery in the organic phase.

Summarising, in solvent extraction preconcentrations, highest solute recoveries and preconcentration factors result under the following circumstances:

Table 1.9 Recoveries of organics in hexane extraction of water*

Organic	% recovery	
	One extraction	Three extractions
Aldrin	47.9	89.4
Heptachlorexpoxide	58.5	91.3
α-Cis–Chlordane	59.2	92.0
Dieldrin	62.2	92.6
n-Decane	69.3	98.6
n-Dodecane	60.3	97.3
n-Tetradecane	56.0	96.7
n-Hexadecane	52.9	96.4
Di-n-butyl phthalate	65.6	86.6

*1L water extracted with 0.2ml hexane, 0.05ml hexane recovered

Source: Reproduced with permission from Murray, D.A.J. [12] Kluwer/Plenum Publishing

1. Volume of solvent used small relative to volume of water sample.
2. Choosing a solvent with a low solubility in the aqueous phase.
3. Choosing a solvent with a high K value (where K = wt of solute in solvent/wt of solute in water).

The overall effect of the number of solvent extractions on solute recovery in the organic phase is illustrated in Table 1.9 [12], which shows recoveries obtained from a range of organics by extracting 1L of a water sample with 0.2ml hexane. Due to solubility effects 0.05ml of hexane was recovered, ie 0.15ml of hexane had dissolved in 1L of water, solubility of hexane in water 0.015%. Both solubility and partition effects are reflected in these results. Both effects are felt in a single extraction where recoveries are between 47.9 and 69.3%, but only the partition effect has any influence when three solvent extractions were performed with recoveries in the range 86.6 to 98.6%.

Having obtained the organic extract some workers in an attempt to further improve preconcentration factors evaporate the extract to a smaller volume prior to final chemical analysis. In Table 1.10 are shown some results obtained [12] by concentrating 10ml hexane extract to 1ml by two methods. Clearly such procedures are to be avoided as they introduce further serious losses of organics in addition to those discussed above.

1.10 Preconcentration of organics by static headspace analysis

In this technique (discussed in Chapter 15) the aqueous sample is enclosed in a container under a head space of a relatively small volume of inert gas.

Table 1.10 Losses of organics in evaporating hexane solutions tenfold

Organic	% losses in evaporating 10ml hexane to 1ml		Combined % partition effect loss and % losses on evaporating 10ml hexane to 1ml[a]	
	Synder evaporator at 90°C	Rotary evaporation at 50°C	Synder evaporator at 90°C	Rotary evaporation at 50°C
Aldrin	77	84	69	75
Heptachlorexpoxide	78	90	71	74
α-Cischlordane	78	87	72	80
Dieldrin	79	85	73	79
n-Decane	57	25	56	25
n-Dodecane	59	39	57	38
n-Tetradecane	61	49	59	47
Di-n-butyl phthalate	84	100	71	87

[a]Using recoveries in Table 1.9

Source: Reproduced with permission from Murray, D.A.J. [12] Kluwer/Plenum Publishing

Volatile organics redistribute themselves between the liquid and the gas phases during an equilibration period and then a portion of the headspace gas is withdrawn for analysis by a suitable technique. Three factors affect the enrichment obtained in the gas phase as expressed by the partition coefficient K. These are the water temperature, the presence of added inorganic salts in the water which have an appreciable effect on K, and the pH of the water sample which has a small but important effect on K. Table 1.11 shows the effects of these factors on the preconcentration of a solution of methyl ethyl ketone in water [13]. It is apparent from these results that by raising sample temperature to 50°C and adjusting the sodium sulphate content to 3.35M for pH 7.1 a $260/3.9 = 67$ improvement in preconcentration factor has been very simply achieved.

1.11 Preconcentration of organics by purge and trap analysis

In this procedure an inert gas is purged through the water sample, sometimes using a closed loop system, and then the gas is passed through a suitable adsorbent usually activated carbon which retains organics removed from the water sample. The volatiles are then removed from the carbon by a relatively small volume of solvent, or thermally, and analysed

Table 1.11 Effect of sample temperature, added salt content, and pH on partition coefficient between aqueous methyl ethyl ketone and nitrogen

pH	Temperature (°C)	$k \times 10^{-3}$ for salt concentration (mol Na_2SO_4)		
		0.6	1.41	3.35
4.5	30	4.19	21.3	118
	50	21.3	39.8	234
7.1	30	3.9	20.0	109
	50	20.0	37.6	260
9.1	30	4.56	18.7	105
	50	18.7	35.0	229

Source: Reproduced with permission from Friant, S.L. and Suffet, T. [13] American Chemical Society

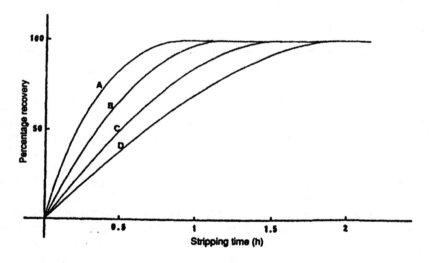

Fig. 1.10 Percentage recovery of stripped compounds with increasing stripping time: A, n-hexane; B, n-octane; C, n-decane; D, n-dodecane
Source: Reproduced with permission from Colenutt, B.A. and Thorburn, S. [15] Gordon AC Breach, Amsterdam

by a suitable technique. Several factors affect the efficiency with which carbon retains organics including flow rate of stripping gas, stripping time (in closed loop systems), particle size of carbon, and choice of desorption solvent. The effect of stripping time on recovery is illustrated

Table 1.12 Effect of particle size of carbon on recovery of organics

Organic	% recovery for particle size (mm dia):				
	0.35	0.25	0.18	0.15	0.12
n-Octane	92	95	98	99	99
Toluene	91	95	99	100	100
Ethanol	86	90	93	95	96
Phenol	90	93	95	96	97
Methyl ethyl ketone	86	90	92	94	94
Mean	91	93	95	97	97

Source: Reproduced with permission from Colenutt, B.A. and Thorburn, S. [15] Gordon AC Breach

Table 1.13 Effect of desorption solvent on efficiency of removing adsorbed organics from carbon

Adsorbate	% recovery with adsorption solvent:					
	n-Hexane	Toluene	CS$_2$	Ether	Carbon tetrachloride	Methanol
n-Octane	99	97	99	97	98	90
Toluene	97	–	100	97	96	89
Ethanol	90	90	95	92	88	95
Phenol	94	91	96	92	90	96
Methyl ethyl ketone	91	90	94	93	91	92

Source: Reproduced with permission from Colenutt, B.A. and Thorburn, S. [15] Gordon AC Breach

in Fig. 1.10 [14,15] which shows that stripping times of 1h suffice for relatively volatile hydrocarbons such as n-hexane but that 2h are required for less volatile n-dodecane. Clearly, such studies are always required before the technique can be applied to any particular organic. Table 1.12 illustrates the effect of carbon particle size on the recovery of various organics showing that smaller particle size carbons are more effective.

Table 1.13 shows the effect of the desorption solvent chosen on the efficiency of recovery of various adsorbed organics from carbon. Clearly hydrocarbons (n-hexane and toluene), carbon disulphide, ether and carbon tetrachloride are more efficient in desorbing hydrocarbons (n-octane and toluene) than they are in desorbing polar compounds such as ethanol, phenol and methyl ethyl ketone. The situation is reversed with methanol desorbent which desorbs polar compounds more efficiently than hydrocarbons.

References

1 Subramanian, K.S. and Meranger, J.C. *International Journal of Environmental Analytical Chemistry*, **7**, 25 (1979).
2 Tessier, A., Campbell, P.G.C. and Bisson, M. *International Journal of Environmental Analytical Chemistry*, **7**, 41 (1979).
3 Reggeks, G. and Van Grieken, R. *Fresenius Zeitschrift für Analytische Chemie*, **317**, 520 (1984).
4 Burba, P. and Willman, P.G. *Talanta*, **30**, 381 (1983).
5 Tateda, A. and Fritz, J.S. *Journal of Chromatography*, **152**, 329 (1978).
6 Ryan, J.P. and Fritz, J.S. *Journal of Chromatographic Science*, **16**, 448 (1978).
7 Matthews, K.M. *Analytical Letters*, **16**, 633 (1972).
8 Glueckauf, A. *Ion Exchange and its Analytical Applications*. Society of Chemical Industry, London (1955).
9 Mayer, A. and Tomkins, C. *Journal of the American Chemical Society*, **69**, 2866 (1947).
10 Glueckauf, A. *Transactions of Faraday Society (London)*, **51**, 34 (1955).
11 Chakravarty, R. and Van Grieken, R. *International Journal of Environmental Analytical Chemistry*, **11**, 67 (1982).
12 Murray, D.A.J. *Journal of Oceanography*, **177**, 135 (1979).
13 Friant, S.L. and Suffet, T. *Analytical Chemistry*, **51**, 2167 (1979).
14 Colenutt, B.A. and Thorburn, S. *International Journal of Environmental Analytical Chemistry*, **7**, 23 (1980).
15 Colenutt, B.A. and Thorburn, S. *International Journal of Environmental Studies*, **15**, 25 (1980).

Chapter 2

Organics: Chelation–solvent extraction techniques

This technique has only found limited applications in the preconcentration of organic and organometallic compounds which are reviewed below.

Organic compounds

2.1 Non saline waters

2.1.1 Amines

Batley et al. [1] extracted the aliphatic amines into chloroform as ion association complexes with chromate, then determined the chromium in the complex spectrophotometrically with diphenylcarbazide. The chromium might also be determined by atomic absorption spectrometry. With the colorimetric method, the limit of detection of a commercial tertiary amine mixture was 15µg L $^{-1}$. The sensitivity was extended to 0.2µg L $^{-1}$ by extracting into organic solvent the complex formed by the amine and Eosin Yellow. The concentration of the complex was measured fluorometrically.

Organometallic compounds

2.2 Non saline waters

2.2.1 Organolead compounds

Chau et al. [2,3] preconcentrated low levels of organolead compounds (Me_2Pb^{2+}, Me_3Pb^+, Et_2Pb^{2+}, Et_3Pb^+) in lake water by extraction with a benzene solution of diethyldithiocarbamate, followed by butylation with Grignard reagent to produce the tetrabutyl–lead derivative. Gas chromatography of the extract using an atomic absorption detector enabled down to 0.01µg L $^{-1}$ of these substances to be determined in lake water samples.

2.2.2 *Organomercury compounds*

Dithiocarbamate resin has been used to preconcentrate organomercury compounds in non saline waters at the 40–300µg L^{-1} level in 1L water samples [4].

2.3 Seawater

2.3.1 *Organotin compounds*

Meinema *et al.* [5] studied the effect of combinations of various solvents with 0.05% tropolone on the preconcentrative recoveries of mono-, di-, and tri-butyltin species either individually or simultaneously present in seawater. The results obtained by gas chromatography–mass spectrometry after methylation show that Bu$_3$Sn and Bu$_2$Sn recoveries appear to be almost quantitative both for neutral and hydrobromic acid–acidified aqueous solutions. BuSn recovery appears to be influenced by the presence of hydrobromic acid in that, in general, recovery rates are higher from solutions acidified with hydrobromic acid than from non-acidified solutions. Bu$_3$Sn and Bu$_2$Sn recoveries remain fairly constant with ageing of an aqueous solution of these species over a period of several weeks. BuSn recoveries, however, do decrease with time to a notable extent (20–40%) most likely as a result of adsorption/deposition of BuSn species to the glass wall of the vessel. Addition of hydrobromic acid obviously affects the desorption of these species as recovery of BuSn species increase to almost the same values as obtained from hydrobromic acid–acidified freshly prepared aqueous solutions of BuSn species.

2.3.2 *Organomercury compounds*

Schintu *et al.* [6] preconcentrated manganese and methylmercury by extraction with dithizone solution. Organic mercury was then recovered from the solvent phase by extraction with aqueous sodium thiosulphate prior to analysis by gold trap cold vapour atomic absorption spectrometry.

Graphite furnace atomic absorption spectrophotometry has been used for the determination down to 5ng L^{-1} of inorganic and organic mercury in seawater [7]. The method used a preliminary preconcentration of mercury using the ammonium pyrrolidinedithiocarbamate–chloroform system. Recovery of mercury of 85–86% was reproducibly obtained in the first chloroform extract and consequently it was possible to calibrate the method on this basis. A standard deviation of 2.6% was obtained on a seawater sample containing 529ng L^{-1} mercury. The relative standard deviation of ten repeated determinations of 500ml distilled water containing 10ng mercury(II) chloride was 17.4%.

References

1 Batley, G.E., Florence, T.M. and Kennedy, J.R. *Talanta*, **20**, 987 (1973).
2 Chau, Y.K., Wong, P.T.S. and Kramer, O. *Analytica Chimica Acta*, **146**, 211 (1983).
3 Chau, Y.K., Wong, P.T.S. and Goulden, P.D. *Analytica Chimica Acta*, **85**, 421 (1976).
4 Emteborg, H., Baxter, D.C., Sharp, M. and French, W. *Analyst (London)*, **120**, 69 (1995).
5 Meinema, H.A., Burger, N. and Wiersina, I. *Environmental Science and Technology*, **12**, 288 (1978).
6 Schintu, M., Kauri, T., Contu, A. and Kudo, A. *Ecotoxicology and Environmental Safety*, **14**, 208 (1987).
7 Filippi, M. *Analyst (London)*, **109**, 515 (1984).

Chapter 3

Organics: Macroreticular non-polar resins

Organic compounds

3.1 XAD–1 resins

The principle of adsorption techniques for preconcentrating samples is quite simple. A large volume of sample is contacted with or passed down a column of a solid material which removes organic material from the water into the solid. Many mechanisms can be responsible for this; adsorption phenomenon is one. In the next stage, the organics and in a limited number of cases organometallic compounds are removed from the solid into a volume of a solvent or a chemical reagent which is relatively low compared to the original volume of sample taken, and subsequently examined by gas chromatography or high performance liquid chromatography either of which might be coupled to a mass spectrometer. Alternatively, the adsorbent might be heated to release the organics which are then swept into a cryogenic trap for subsequent analysis by the aforementioned chromatographic techniques or swept directly into a gas chromatograph. In favourable circumstances preconcentration factors of up to 10^4 or even higher are commonly achieved by these techniques. Naturally, the types of resins from which preconcentrated substances can be subsequently released either by solvent extraction or thermal desorption must themselves have no polarity. Only the non-polar resins are discussed in this chapter. Polar (ie ion exchange) resins are discussed in Chapters 4 and 5). Non-polar resins, as opposed to polar resins, operate by forming various kinds of loose physical bonds (rather than chemical bonds) with the adsorbed substance, or, in the case of larger adsorbed molecules, may operate on the principle of acting as microfilters, ie the pore sizes in the resin are of molecular dimensions.

The first non-polar macroreticular adsorbent to be manufactured was Amberlite XAD–1. Since then, improved non-polar resins (XAD–2 and XAD–4) have become available and the applications of these are discussed below. Tenax GC is a further very popular non-polar macroreticular resin. It is based on 2,6–diphenyl-p-phenylene oxide.

3.1.1 Sewage effluents

3.1.1.1 Chlorophyll, coprostanol

Wun *et al.* [1] have described a method for the simultaneous extraction of the water quality marker algal chlorophyll *a* and the faecal sterol coprostanol from water. This method utilises a column of Dunberlite XAD–1 for the simultaneous extraction of both markers. Chlorophyll content was determined by the trichromatic method [2] using a double beam spectrophotometer. Gas–liquid chromatographic analyses of faecal sterols were performed with a Perkin–Elmer 900 gas chromatograph equipped with flame ionisation detectors. Conditions for the gas–liquid chromatograph analyses have been described by Wun [3].

Wun *et al.* [1] using ^{14}C-labelled cholesterol *c*, showed that 100% of this substance is adsorbed from a 500–1000ml water sample by XAD–1 resin. When the column was subsequently eluted with 50ml of basic methanol and 30ml of benzene, up to 97% of the added sterol was recovered in the benzene eluate.

To ascertain the suitability of the column method for the simultaneous extraction of coprostanol and chlorophyll *a* in actual field testing situations, various unialgal cultures were mixed with sewage samples. The effectiveness of the neutral resin column extraction for these markers was evaluated with that of the conventional procedures. Representative results are presented Tables 3.1 and 3.2. Results presented in Table 3.1 indicate that the efficiency of the resin column for the simultaneous extraction of coprostanol and phytoplankton chlorophyll *a* was comparable or better than the conventional extraction procedures for the respective compounds. Dilution of the samples and the presence of extraneous materials did not affect the recovery efficiency significantly. Data in Table 3.2 reveal that the column technique was effective in isolating chlorophyll *a* from various algae. The superiority of the column method was more pronounced when the small green alga (*ocystis* sp.) and the blue–green alga (*oscillatoria* sp.) were used as test organisms. The coprostanol extraction efficiency was again shown to be comparable to that of the hexane liquid–liquid partitioning process.

3.2 XAD–2 and XAD–4 resins

3.2.1 Non saline waters

3.2.1.1 General discussion

Daignault *et al.* [4] reviewed the literature on the use of XAD resins to concentrate organic compounds in water including the occurrence of artifacts released into water from these resins; methods of cleaning the resin and/or solvent before use; factors affecting adsorption on the resin

Table 3.1 Efficiency of the XAD–1 (60–120) resin column for the simultaneous isolation of phytoplankton chlorophyll *a* and coprostanol

Sample[a]	Chlorophyll a[b] (µg/sample)	Coprostanol[c] (µl/sample)
50mL sewage + 50mL distilled water	17.01	11.0
50mL sewage + 450mL distilled water	16.99	10.0
50mL sewage + 950mL distilled water	17.01	10.5
100mL sewage + 900mL distilled water	16.78	26.0
200mL sewage + 800mL distilled water	16.36	50.0
500mL sewage + 500mL distilled water	16.05	125.0

Notes
[a] 2.0mL aliquot of unialgal culture of *Chlorella vulgaris* (total cell count: 9.0×10^7) was added to each of the test samples containing 50–500mL of sewage. Photosynthetic pigments and coprostanol were extracted by the XAD–1 (60–120) resin column procedure
[b] Chlorophyll *a* content of the 2.0mL *Chlorella vulgaris*, as determined by the conventional aqueous acetone extraction method, was 11.75µg. When analysed by the column method the sewage was found to contain approximately 0.64µg chlorophyll *a* dL $^{-1}$. This amount was subtracted from the total chlorophyll *a* recovered.
[c] The coprostanol concentration of the sewage as determined by hexane liquid–liquid partitioning was 219ppb.

Source: Reproduced with permission from Wun *et al.* [1] Kluwer/Plenum Publishing

(pH value, ionic strength, flow rate, resin capacity and bead size, and presence of humic substances); eluants used and compound recoveries; and resin regeneration. Wigilius *et al.* [5] discussed a systematic approach to adsorption on XAD–2 resin for the preconcentration and analysis of organics in water below the pg L $^{-1}$ level. The effects of solvent changes on the adsorption of trace organics from water by XAD–2 resin were investigated. Methanol slurried resin was packed into a column and washed with methanol and diethyl ether. Tap water was passed through the column to condition the resin before passing 10–20L of an aqueous sample containing 32 model compounds. Tap water or river water containing 200ng L $^{-1}$ of each compound was used. The effects of changing the solvent sequence to methanol/diethyl ether/water and to methanol/ acetone/water were investigated. Diethyl ether was used to desorb the organics. For a large number of non-polar compounds, recovery was limited by evaporation losses. For polar compounds (alcohols and phenols), adsorption efficiency was critical. Desorption losses were made negligible, except for naphthalene and anthracene. Humic acids gave no decrease in recovery of low molecular weight compounds and no increased breakthrough. The limit of detection was below 20ng L $^{-1}$ for a 10L sample volume. Acceptable blanks were achieved.

Table 3.2 A comparison of column and conventional extraction procedures for the recovery of chlorophyll a and coprostanol

	Extraction method			
	XAD-1 column		90% Aqueous acetone	Hexane partitioning
Sample	Chloro-phyll a[a]	Copros-tanol[b]	Chlorophyll a[a]	Coprostanol[b]
Sewage[c] + 4.0mL Chlorella vulgaris	33.3	350.0	32.0	330.0
Sewage + 4.0mL Chlamydomonas moewuii	13.5	326.3	9.2	250.5
Sewage + 2.0mL Chlorococcum hypnosporum	16.6	330.00	15.1	270.0
Sewage + 7.0mL Oocystis marssonii	51.4	590.0	171	560.0
Sewage + 8.0mL Oscillutoria tenius	26.9	912.00	16.0	912.0

Notes
[a]Chlorophyll a = μg per sample
[b]Coprostanol = ppb
[c]Sewage samples were collected from the sewage treatment plant (a primary treatment plant) on different days. Samples were each divided into two equal aliquots of 500mL each. Coprostanol content of one aliquot was extracted by hexane partitioning. The other, after adding a known amount of plankton (chlorophyll a content predetermined by 90% aqueous acetone extraction of an equal amount of the algal sample) was extracted by the XAD-1 (60-120) resin column procedure.

Source: Reproduced with permission from Wun et al. [1] Kluwer/Plenum Publishing

Dietrich et al. [6] have made an intercomparison study of the applications of adsorption on XAD resins and methylene dichloride liquid–liquid extraction from the preconcentration of numerous organic compounds from river water. Capillary gas chromatography–mass spectrometry was used to detect and identify compounds. Of the 48 individual chemicals identified over 13 months the most frequently observed were Atrazine, and methyl atraton, dimethyl dioxan, 1,2,4–trichlorobenzene, tributylphosphate, triethylphosphate, trimethylindolinone, and the three isomers of tris(chloropropyl) phosphate. Both extraction methods gave similar results in the ng L^{-1} range but the XAD resin method showed more artefacts. Stephan and Smith [7] showed that adsorption on macroreticular resins is a suitable method for extracting trace organic pollutants from water prior to determination by gas chromatography. The flow rate and pH value of the sample should be controlled to obtain the best results. Extraction efficiencies obtained for various compounds on XAD-2 and XAD-7 resins (ie non-polar and mildly polar) are in the range 27–93% (Table 3.3).

Table 3.3 Effect of polarity of resin on extraction efficiency of polar and non-polar compounds

Compound	Extraction efficiency (Averages of 5 extractions) ±2%)	
	XAD–2 (%)	XAD–7 (%)
Cumene	67	67
Ethyl benzene	60	59
Naphthalene	90	93
n-Hexane	82	83
Phenol	27	45
Octanoic acid	58	81
o-Cresol	59	67
Chlorophenol	70	85

Sample flow: 20ml min $^{-1}$; pH: 5.7; concentration 10ppm

Source: Reproduced with permission from Stephan, S.F. and Smith, J.F. [7] Elsevier Science UK

Cheng and Fritz [8] pointed out that the XAD–2 resin sorption method was a reliable and thoroughly tested method for determining trace organic pollutants in potable water. However, the solvent evaporation step in this procedure results in a partial or complete loss of volatile compounds. To avoid these drawbacks a procedure was developed by them in which organic compounds that are sorbed on a resin are thermally desorbed directly on to a gas chromatographic column for analysis. They desorbed the organics on to a Tenax tube from a tube containing XAD–2. This eliminates virtually all of the water entrained in the XAD–2 tube. Then the organics are thermally desorbed from the Tenax tube directly on to the gas chromatographic column. The method is as follows:

Resins: XAD–2 resin (Rohm and Hass) ground and sieved in the dry state Tenax GC, 60–80 mesh, used without further purification.

Apparatus and equipment: Hewlett Packard 5756 B gas chromatograph equipped with a linear temperature programmer and FID detector or equivalent. Stainless steel tubing for all columns and injection port liners. The injection port was enlarged by drilling to accommodate the glass sorption tubes. The column for separations was either 180cm × 3mm od packed with 5% OV–17 on Chromosorb W AW DMCS (80–100 mesh) or 120cm × 3mm packed with Tenax GC (60–80 mesh).

Sorption tubes were 2mm id, 10cm in length. They were filled with 7cm of either XAD–2 or Tenax GC, held in place with a plug of glass wool at either end. The XAD 2 tubes were conditioned by thermal desorption at

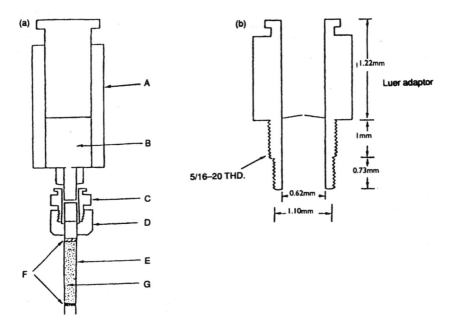

Fig. 3. 1 (a) Minisampler for small water volumes: A, 20ml glass hypodermic syringe; B, water sample; C, coupler for attaching minicolumn (E) to the syringe; D 1mm Swagelok nut; E, 2mm id Pyrex tube; F, glass wool plugs; G, 80mg of 60–80 mesh resin (b) Kel–F–Coupler

Source: Reproduced with permission from Cheng, R.C. and Fritz, J.S. [8] Elsevier Science

240°C and repetitive blanks were run until a low, tolerable, background was achieved. A blank run involved wetting the resin with triply distilled water and following steps two and three of the analytical procedure, ie adsorption on Tenax and thermal desorption. The Tenax GC tube required only a simple conditioning by heating for 1h at 275°C to achieve a tolerable blank level.

A minisampler for water in constructed from a 20ml glass syringe connected to an XAD–2 sorption tube (Fig. 3.1(a)). The connection is made with the special Kel–F coupler shown in Fig. 3.1(b).

Procedure:
1. *Sampling:* Pass a 15ml (or larger) sample through the XAD–2 sorption tube by using a minisampler with hand pressure to force the water through (about 1–2min). Then force as much residual water as possible from the tube, using 20ml of air. Cap the sorption tube at both ends unless the analysis is to continue without delay.

Table 3.4 Overall recovery efficiency of resin extraction and thermal desorption method of analysis on XAD–2 for organics in water at the 3–100µg L $^{-1}$ level

Compounds	Recovery efficiency (%)	Compounds	Recovery efficiency (%)
Alkanes		Polynuclear aromatics	
Octane	88	Naphthalene	98
Heptane	81	2–Methylnaphthalene	97
Tridecane	90	Chlorobenzenes	
Alkylbenzenes		Chlorobenzene	90
Benzene	90	o-Dichlorobenzene	82
Toluene	97	Ketones	
o-Xylene	90	Acetone	55
Cumene	82	2–Octanone	83
Ethers		Undecanone	86
Hexyl	80	Acetophenone	92
Benzyl	70	Alcohols	
Esters		1–Butanol	<40
Benzyl acetate	95	1–Pentanol	<40
Methyl decanoate	88	1–Octanol	85
Methyl hexanoate	86	1–Decanol	84
Haloforms*		Phenols and acids	
Chloroform	87	No quantitative results	
Bromodichloromethane	95		

*Chlorinated methanes were tested at a concentration of 200µg L $^{-1}$ in water

Source: Reproduced with permission from Cheng, R.C. and Fritz, J.S. [8] Elsevier Science UK

2. *Thermal desorption*: Connect the XAD–2 tube to a Tenax sorption tube. Place the XAD–2 tube in a heated zone, maintained at 230°C with helium flowing at 50ml min $^{-1}$ into the XAD–2 tube and out of the Tenax tube (the Tenax tube is held at approximately 45°). Continue desorption from the XAD–2 on to the Tenax tube for 15min or until the Tenax is visibly dry.

3. *Separation*: Disconnect the two tubes and place the Tenax tube in the modified gas chromatography injection port, held at 220°C and apply a helium carrier gas flow of 20ml min $^{-1}$ (the oven section of the chromatograph is at approximately room temperature). Allow 10min for transfer of organic solutes from the Tenax tube to the column, then separate by raising the temperature at 20°C min $^{-1}$ up to 200°C finally holding the temperature for 10min.

4. *Measurement*: Inject 2–5µL of standard organic solution in methanol through a zone heated to 220°C on to a Tenax tube at room

temperature (or slightly above). Desorb and chromatograph the organic compounds as in step 3. Compare the heights of the sample and standard peaks to calculate the amounts of sample constituents.

Recoveries obtained by this procedure are tabulated in Table 3.4. The recoveries of carboxylic acids were not reproducible. This is believed to be due to the difficulty of direct determination of acids by gas chromatography. A selective method for concentration of the acids with subsequent gas chromatography of their derivatives would be more desirable. The recoveries of phenols varied from 4–90%. These inconsistent results are probably due to the high affinity of Tenax GC for phenols. All attempts to improve the yield for phenols failed and the method is not satisfactory for these compounds.

Ryan and Fritz [9] also used thermal desorption of the organics on the resin directly into the gas chromatograph. In this method a small water sample (20–250ml) is passed through a small tube containing XAD–4 resin; this effectively retains the organic impurities present in the water. This tube is connected to the permanent apparatus and the sorbed organics are thermally transferred to a small Tenax pre-column while the water vapour is vented. The pre-column is closed off, preheated to 275–280°C and then a valve is opened to plug inject the vaporised sample into a gas chromatograph.

Adsorption tubes: The tube is made from standard Pyrex glass tubing 8cm × 2mm id and filled with a bed of XAD–4 resin, 120–140 mesh, about 2.5cm in length. The resin is held in place approximately at the centre of the tube by plugs of silanised glass wool. Connections were made with 1mm Swagelok nuts and PTFE reducing ferrules. A pair of each were dedicated to the absorption tube.

Purification of XAD–4 resin: First, pass 5ml of distilled water through the XAD–4 absorption tube, and then place the tube in a desorption chamber held at 200°C. During this time, pass helium through the tube at about 20ml min $^{-1}$. After about 4min, raise the temperature to 240°C for 15 to 20min. Repeat the entire procedure three more times. This purification method resulted in a tolerably low blank when the tube was tested in blank thermal desorption runs.

Thermal desorption apparatus: The apparatus is shown in Fig. 3.2 and consists of the following components:

A. An aluminium block, movable by sliding on its mount. (a) Thermocouple connected to external pyrometer; (b) one 200 watt cartridge heater, controlled by variable transformer; (c) XAD–4 minicolumn.
B. Heated zone insulated with Temp–mat Glass Insulation (courtesy of Pittsburgh Corning, Pittsburgh, Pennsylvania). (d) Two zero–volume high temperature valves, one 4 port and one 6 port as shown (Valco Valve Co., Houston, Texas); (e) high temperature heating cord, 200W

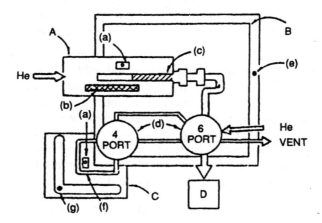

Fig. 3.2 Thermal desorption apparatus
Source: Reproduced with permission from Ryan, J.P. and Fritz, J.S. [9] Preston Publications Inc., Niles, US

(Class Col. Apparatus, Co., Terre Haute, Indiana) controlled by variable transformer.

C. A removable insulated (as B) sheet metal enclosure containing the Tenax desorption heating zone. (f) Tenax precolumn (stainless steel 0.5mm × 18cm, Tenax GC 80–100 mesh); (g) high temperature heating cord, 200W (as 3) controlled by temperature controller (Ames Lab, Instrument Group).

D. Gas chromatograph (Tracer 560).

Gas chromatography: Carry out separations on glass columns 180cm × 2mm id 10% Carbowax 20M on Chromosorb W 80–100, 5% FFAP on Chromosorb W 80–100, and 5% OV–1 on Chromosorb W 80–100 stationary phase were used. Run temperature programs at 10°C min^{-1} with initial temperature ranging from 50–75°C. Final temperature was usually 190°C. The gas chromatograph was equipped with dual FID detectors. The attenuation was 4 × 10. This was ideal for samples ranging from 100–300ng of individual organic impurities.

Thermal desorption procedure: Connect the minicolumn to the desorption apparatus by means of the two (Swagelok nut) PTFE ferrule connections on the XAD–4 tube. Then thermally desorb the organics from the XAD–4. Typical conditions are: temperature 180–200°C; time, 10min; gas flow, helium, 5ml min^{-1}. The vapour passes through a heated (200–220°C) zone to a Tenax precolumn (temp. 45°C). Water passes through the Tenax to vent while the organics are retained.

Close the Tenax precolumn off by means of a 4-port valve equipped with zero volume fittings, and heat to 275–280°C. Then open the valve to inject the vapour phase into the gas chromatography by diverting the carrier gas to back-flush through the hot precolumn. The total volume of the precolumn, including Tenax volume, is 1.0ml so the void volume is somewhat less than 1.0ml.

Open the Tenax desorption chamber while the sample is being chromatographed. This cools the precolumn in preparation for the next sample. During the chromatographic run the desorption block reaches 230–240°C effectively regenerating the minicolumn.

On completion of the run, remove the XAD–4 tube and place it on a stainless steel sheet which serves as a heat sink. This rapidly cools the XAD–4 tube for the next analysis and minimises oxygenation of the resin.

The thermal desorption procedure was tested by analysing water samples, each containing a known concentration of a model organic compound. The model compounds were selected to include different organic functional groups and compounds of varying volatility. The percentage recovery for each model compound was calculated by comparing the peak height for the compound in the sample with the peak height of the same compound in a methanol or acetone standard injected directly on to an XAD–4 tube. The recoveries reported in Table 3.5 are generally quite good at both 10 and 1µg L^{-1} levels in water although there is some apparent loss of the halocarbons tested at the lower concentration level.

The recovery (as defined in Table 3.5) of high boiling compounds was found to be dependent on desorption time and temperature. The reason for this is not entirely clear. Methyl nonyl ketone, for example, requires a higher temperature than normal for efficient desorption. Perhaps the methanol or acetone in the standard is caught by the XAD–4 and helps sweep out the methyl nonyl ketone (the sample has almost no methanol or acetone, so this effect is absent); or the methyl nonyl ketone may be dispersed in a broader band on the XAD–4 tube when taken up from a water sample and a higher temperature is needed to desorb it in a back-flush mode. The higher temperature (225°C) required for thermal desorption of methyl nonyl ketone greatly enhances the resin blank. However, even under these strenuous conditions a 1.0µg L^{-1} solution of this ketone can be observed with a signal to noise ratio of better than 10. Volatile ketones give low recoveries at the higher temperature (225°C) when compared to a standard desorbed at 180°C. Of course the standard and sample should be desorbed under similar conditions but this comparison does indicate some loss of more volatile compounds if the desorption time and temperature are too great. The partial loss of highly volatile compounds to vent is a consequence of keeping the Tenax column at a slightly elevated temperature. However, the Tenax must be heated slightly in order to ensure efficient passage of water vapour.

Table 3.5 Recovery of organic compounds by thermal desorption from XAD–4 resin

| Compound | Recovery (%)[a] | | Desorption | |
	10μg L^{-1} 20ml	1μg L^{-1} 200ml	Time (min)	Temp (°C)
Toluene	88	90	15	210
Ethylbenzene	79	79	15	210
Indene	96	102	10	180–200[b]
Naphthalene	90	81	15	210
1–Methylnaphthalene	95	97	10	180–200
Hexane	88	65	8	175
Chloroform	93	56	4	175
Dibromomethane	88	57	4	175
Cyclohexanol	98	90	5	200
n-Heptyl alcohol	100	98	6	200
Benzyl alcohol	83	54	13	220
Methyl isobutyl ketone	100	98	8	200
Amyl isopropyl ketone	102	99	8	200
Methyl nonyl ketone	96	92	13	220
p-Methyl acetophenone	99	104	13	220
Ethyl heptanoate	96	68	10	200
Octyl acetate	61	32	10	200
Bromobenzene	106	–	10	200
o-Dichlorobenzene	102	–	10	200

[a]The recovery of compounds from water by XAD–4 is based on the comparison of spiked water sample determinations with standards injected directly onto an XAD–4 minicolumn using the 'wet' column procedure. Standards were 1×10^{-17}g^{-1} μL in methanol or acetone. Spiked water samples contained 2μL of the standard solution
[b]The two numbers represent the initial and final temperatures of the desorption block. This crude temperature programming procedure seemed to minimise resin bleed while maintaining good desorption efficiency

Source: Reproduced with permission from Ryan, J.P. and Fritz, J.S. [9] Preston Publications Inc

The efficiency of the thermal desorption procedure was determined for several of the model compounds. This was done by comparing the peaks of compounds in standards taken through the entire procedure with peaks of the same compounds in standard injected directly into the gas chromatograph. The data in Table 3.6 show that except for methyl nonyl ketone the recovery efficiencies of all compounds tested are quite good. Thus for quantitative analysis of actual water samples, the results could be compared with standards injected directly into the gas chromatograph (as in Table 3.6) or with water standards passed through the XAD–4 tube.

Tateda and Fritz [10] used solvent extraction with carbon disulphide or acetone, rather than a thermal method to desorb adsorbed organics from

Table 3.6 Thermal desorption efficiency

Compound	Mini column condition	Desorption time (min)	Desorption temperature (°C)	Boiling point (°C)	Fractional[b] recovery
Cyclohexanol	Dry	10	180–200	161	0.94
Cyclohexanol	Wet[a]	10	180–200	161	1.02
Bromobenzene	Wet	8	200–215	156	0.97
Bromobenzene	Wet	10	180–200	156	0.97
Undecanone	Wet	10	210–225	232	0.66
Undecanone	Wet	10	180–200	232	0.50
Chloroform	Wet	10	180–185	62	0.93
Indene	Wet	10	180–200	183	0.99
1–Methylnaphthalene	Wet	13	200–235	245	0.96
1–Methylnaphthalene	Wet	10	180–200	245	0.86

[a]Wet columns are wetted after the sample has been injected into them
[b]Fractional recovery is the ratio of peak height of injection onto the XAD–4 minicolumn to the peak height of direct injection into the gas chromatograph. Samples were 2×10^{-7}g

Source: Reproduced with permission from Ryan, J.P. and Fritz, J.S. [9] Preston Publications Inc.

XAD resins. These workers used a minicolumn 12–1.8mm × 25mm containing XAD–4 resin or Spherocarb to adsorb organic contaminants from a 50–100ml potable water sample. The sorbed organics are eluted by 50–100μL of an organic solvent and the organic solutes separated by gas chromatography. The procedure is simple, it requires no evaporation step and gives excellent recoveries of model organic compounds added to water. The average recovery was 89% at 2–10μg L^{-1} levels and 83% at the 100μg L^{-1} level. The average standard deviation is 6.3%. Errors other than sorption and desorption would be included, like evaporation losses from the sample, decomposition, sorption on glassware, calibration of eluate volume and errors in the gas chromatographic determination.

Tateda and Fritz [10] compared the adsorptive properties of XAD–4 resin with those of a spherical carbon molecular sieve of large surface area (Spherocarb). It was found that 100μg carbon disulphide will elute most organic compounds tested with Spherocarb except for phenols and some strongly sorbed compounds like naphthalene. Spherocarb has one major advantage over XAD–4 for analytical use; stronger retention of low molecular weight polar organic compounds. The minicolumn method while directly applicable for waste water analysis where organics contents are relatively high, does not have the sensitivity needed for the analysis of organics in most potable water samples. Large scale desorption methods such as that described by Junk *et al.* [11] have better sensitivity but suffer from a small fraction of the sample being used for the gas chromatographic analysis (typically a 2μL aliquot of 1000μL of diethyl ether extract). Tateda and Fritz [10] combined their procedure with that of Junk *et al.* [11] by adding 1ml of the ether concentrate from the larger scale procedure to 50ml of pure water and then proceeding according to the minicolumn procedure. A large fraction of the sample is thus taken for analysis (2μL of the 100μL carbon disulphide eluate) so that the original ether concentrate is further concentrated by a factor of 10. By the standard procedure only small peaks are obtained, Spherocarb shows much larger peaks at the same attenuation (10 × 8). With XAD–4 most peaks are close to the expected 10-fold increase in peak height.

The sensitivity of the minicolumn method for most organic compounds is about 2μg L^{-1} which is quite adequate for analysis of waste water and badly contaminated potable water. The major components in a well water sample were indene, methyl indene, methylnaphthalene, acenaphthalene and acenaphthene. The other peaks were not identified. Using naphthalene as a standard the total concentration of the five major peaks in the well water was estimated to be 260μg L^{-1}. The total concentration of all peaks was roughly 325μg L^{-1}.

Workers at the Water Research Centre, UK [12], passed up to 5L of potable water samples through a column of XAD–2 resin and removed the adsorbed organics with 40μg of diethyl ether. The ether extract is then concentrated 2500 times and examined by gas chromatography–mass spectrometry. Several hundred organic compounds were examined in this survey, many of which were identified in potable water samples.

The sample (5L) was spiked in a pre-cleaned ground glass stopper bottle with a solution (5μL) of the internal standards in acetone (100mg L^{-1}) and passed (40ml min^{-1}) through a column (13cm × 1.2cm id) containing the XAD–2 resin (bed height 6cm). Deuterated internal standards were used to provide, if necessary, quantitative information on some of the compounds identified. The standards (chlorobenzene–d_5, *p*-xylene–d_{10}, phenol–d_5, naphthalene–d_8, hexadecane–d_{34} and phenanthrene–d_{10}) were added to each sample immediately before extraction. The resin required extensive purification before use. This involved an initial washing with water, heat desorption (200°C for 16h) under a flow (100ml min^{-1}) of purified nitrogen, and a Soxhlet extraction (6h with methanol). After the sample had passed through the resin, purified diethyl ether (15ml) was added to the column head and allowed to flow through the column until solvent first appeared at the bottom of the column. The flow of ether was stopped and the ether allowed to remain in contact with the resin for 10min. A further aliquot of ether (10ml) was added to the column, and all of the ether allowed to flow through into a collection flask. This extract was dried by storing overnight in a freezer (–18°C) and decanting the ethereal solution off any ice formed. The extract was concentrated (to

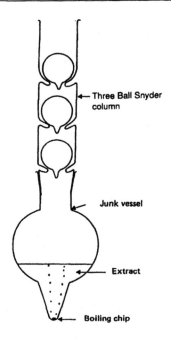

Fig. 3.3 Apparatus for concentration of XAD–2 resin diethyl ether extract
Source: Reproduced with permission from Fielding, M. *et al.* [12] Water Research Centre, UK

250μL) by evaporation in a Junk vessel fitted with a three-ball Snyder column (Fig. 3.3), and to the final volume (100μg) by evaporation using a stream of dry, purified nitrogen. The extract was then suitable for gas chromatography–mass spectrometry.

Stephan *et al.* [13] have described an apparatus for on-site extraction of organic compounds from non saline water samples on to XAD–2, XAD–7 and Tenax GC resins. The sampler consist of two peristaltic pumps driven by a battery powered motor. The maximum lift of the peristaltic pumps is approximately 10. Hence the sampler may be used to sample ground water, provided the water table is no more than 10m below the surface. Operation of the sampler is controlled by an electronic timer. A voltage stabilisation circuit is used to control the pumping speed. The rate of sampling may be varied continuously between 0 and 60ml min $^{-1}$ by variation of the pump speed. The electronic timer is designed to allow continuous or automatic intermittent sampling. In the continuous mode, at a sample flow of 25ml min $^{-1}$ and 1m lift, the maximum sampling time is approximately 14h. In the intermittent mode, with timer set to one hour off–one hour on, with a sample flow of 25ml min $^{-1}$ and 1m lift, the maximum sampling time is approximately 30h. Organics were later desorbed in the laboratory from the resin filters in the field apparatus using diethyl ether and methanol and examined by gas chromatography. Very similar results were obtained by the field method and conventional laboratory procedures.

Applications of XAD resins

3.2.1.2 Polyaromatic hydrocarbons

Benoit *et al.* [14] have investigated the use of macroreticular resins, particularly Amberlite XAD–2 resin, in the preconcentration of Ottawa potable water and river water samples prior to the determination of 50 different polyaromatic hydrocarbons by gas chromatography–mass spectrometry.

Water samples were prepared as follows: sampling cartridges, containing 15g Amberlite XAD–2 macroreticular resin that had been previously cleaned by the method of McNeil *et al.* [15], were rinsed with 250ml acetone and washed with at least 1L of purified water. The cartridges were attached to a potable water supply and the flow of water was controlled at *c.* 70ml min $^{-1}$. When 300ml of water had been passed through the cartridge, the cartridge was disconnected from the tap and as much water as possible was removed from the cartridge by careful draining followed by the application of vacuum from a water aspirator. The XAD–2 resin was eluted with 300ml of 15:85 v/v acetone: hexane solution at a flow rate of *c.* 5ml min $^{-1}$ (all solvents were of 'distilled in glass' quality and were redistilled in an all glass system). The organic layer was dried by passage through a drying column containing anhydrous sodium sulphate over a glass wool plug. Both the sodium sulphate and the glass wool plug were cleaned by successive washings with methylene chloride, acetone, and hexane prior to use. The dried solution was concentrated to a volume of *c.* 3ml using a rotary evaporator, then quantitatively transferred with acetone to a graduated vial and was further concentrated, using a gentle stream of dry nitrogen gas, to a final volume of 1ml.

To analyse the solvent extracts a 10µL aliquot of the concentrated extract was injected into a Finnigan 4000 gas chromatography–mass spectrometry instrument coupled to a 5110 data system. A 3% OV–17 provided the best separation of the detectable polyaromatic hydrocarbons. A 1.8m × 2mm id glass column packed with 3% OV–17 of 80–100 mesh Chromosorb 750, was operated at an initial temperature of 100°C for 1min and was programmed to a final temperature of 225°C at a rate of 3°C min $^{-1}$ and held at that temperature for the remainder of the analysis. The flow of helium carrier gas was set at 20ml min $^{-1}$ and the injection port temperature set at 200°C. The glass jet separator and the ion source temperatures were set at 260°C and 250°C respectively. Data acquisition was under the control of the Finnigan 6110 data system. The mass range, 35–400 atomic mass units, was scanned at a rate of 2.1s per scan and the mass spectra (c. 1000) stored on magnetic disc for subsequent analysis.

To test the effectiveness of their method of analysis for polyaromatic hydrocarbons in potable water Benoit et al. [14] prepared and analysed a control blank and carried out a recovery study of 32 selected polyaromatic hydrocarbons from XAD–2 resin. None of the compounds contained in the standard solution was detected in the concentrated extract from the control blank. This indicates that the XAD–2 resin is effective for the removal of these compounds from water samples and that none of the reference compounds originates from the precleaned XAD–2 resin. However, when the amounts of polyaromatic hydrocarbons loaded and recovered are compared (Table 3.7) an average recovery of 84% is observed with recoveries ranging from 57–100% of the loaded material. The weighted average recovery was 88% of loaded material. The fate of the unrecovered material was not established although, based on the results of the control blank, it is not likely that these materials were carried away by the effluent water.

Ottawa potable water samples were analysed in order to obtain some indication of whether the results are representative of the general background level of anthropogenic contamination. Aliquots of the reference standard solution (50 polyaromatic hydrocarbons and five oxygenated aromatic hydrocarbons) and the concentrated extracts from XAD–2 resin were analysed consecutively by gas chromatography–mass spectrometry under identical operating conditions. As is evident from Fig. 3.4(a) the XAD–2 resin extract concentrate, as reconstructed from the total ion current, contained a multitude of poorly defined peaks. Complete mass spectra, free of extraneous ions, could rarely be obtained from such data despite the background subtraction routine possible with the data system and, hence, individual components of the concentrate were achieved from mass chromatograms (Fig. 3.4(b)) which were reconstructed from selected ion currents rather than the total ion current. Mass chromatograms for selected ions that were characteristic of the compound of interest were obtained by searching the accumulated data for the ion of interest and recording the abundance of this ion as a function of retention times. As an example, the mass chromatograms of three ions, m/e 128, m/e 142 and m/e 154, are superimposed in Fig. 3.4(b). The location of the peaks corresponding to the compounds of interest are indicated by asterisks in each chromatogram. For m/e 128 the asterisked peak corresponds to the molecular ion of naphthalene, for m/e 142 to the molecular ions of 2–methylnaphthalene and 1–methylnaphthalene,

Table 3.7 The recovery of selected polycyclic hydrocarbons from Amberlite XAD–2 macroreticular resin

Compound	Amount loaded (ng)	Fraction recovered (ng)
Naphthalene	625	0.57
2–Methylnaphthalene	1200	0.88
1–Methylnaphthalene	625	0.71
2–Ethylnaphthalene	>2300	>0.86
2,6–Dimethylnaphthalene		
Biphenyl	775	0.66
1,3–Dimethylnaphthalene	975	0.81
2,3–Dimethylnaphthalene	>2250	>0.82
1,4–Dimethylnaphthalene		
4–Phenyltoluene	600	0.85
Diphenylmethane	>2075	>0.76
3–Phenyltoluene		
Acenaphthene	625	0.60
Bibenzyl	975	0.65
1,1–Diphenylethylene	1625	0.98
cis-Stilbene	575	1.0
2,3,5–Trimethylnaphthalene	700	0.75
3,3'–Dimethylbiphenyl	1625	1.0
Fluorene	750	0.89
4,4'–Dimethylbiphenyl	700	0.90
trans-Stilbene	>1075	>0.84
9,10–Dihydrophenanthrene		
Phenanthrene	>1700	>1.0
Anthracene		
Triphenylmethane	800	0.87
Fluoranthene	650	1.0
Pyrene	725	1.0
1,2–Benzfluorene	>1550	>0.99
2,3–Benzfluorene		
Triphenylene		
Benz(a)anthracene	>1650	>1.0
Chrysene		
Average	1110	0.84

Source: Reproduced with permission from Benoit, F.M. *et al.* [14] American Chemical Society

respectively, in order of increasing retention, and for *m/e* 154 to the molecular ions of biphenyl, 2–vinylnaphthalene, and acenaphthalene, respectively, in order of increasing retention times. The retention times for each standard were established by analysis of the reference standard solution and the data from the XAD–2 resin extracts were then searched

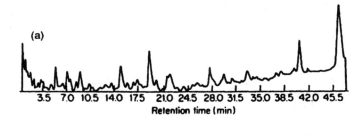

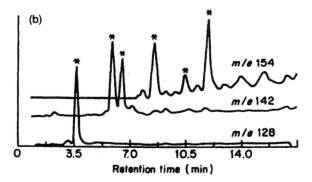

Fig. 3.4 Reconstructed gas chromatogram (total ion current) from Ottawa potable water, Amberlite XAD–2 resin extract
Source: Reproduced with permission from Benoit, F.M. et al. [14] American Chemical Society, Washington

for the ion of interest within the appropriate time region. In all instances, the molecular ion and the next most abundant ion were selected as the characteristic ions which are listed in Tables 3.8 and 3.9.

A compound was considered identified if the two characteristics of the compound of interest were found to elute from the column within the retention time window (±0.1min) of the reference standard and to be in the relative abundance ratio (±20%) observed in the mass spectrum of the pure compound. For most, but not all, compounds screened unique identification was possible. In some instances however, coeluting isomers yielding similar mass spectra could not be resolved sufficiently to allow unequivocal identification. Such coeluting isomers are grouped together in Table 3.8 and are indicated by > beside the concentration values which is the sum of the contributions from all coeluting isomers. Furthermore, it is emphasised that, because of the large number of compounds contained in the field sample extract, it was not possible to eliminate entirely from all the ion peaks of interest, contributions from possible interfering

Table 3.8 Polycyclic aromatic hydrocarbons detected in Ottawa potable water samples

Compound	Ions monitored		Rel. ret. time[a]	Concentration (ng L^{-1}) 1	2
Naphthalene	128	102	1.00	6.8	4.8
2–Methylnaphthalene	142	141	1.59	2.4	4.6
1–Methylnaphthalene	142	141	1.75	1.0	2.0
Azulene	128	102	1.90	nd	nd
2–Ethylnaphthalene	156	141	2.26	>0.70	2.1
2,6–Dimethylnaphthalene	156	141	2.32		
Biphenyl	154	153	2.30	0.70	1.1
1,3–Dimethylnaphthalene	156	141	2.51	1.9	1.1
2–Vinylnaphthalene	154	153	2.68	nd	nd
2,3–Dimethylnaphthalene	156	141	2.69	>0.68	14
1,4–Dimethylnaphthalene	156	141	2.69		
3–Phenyltoluene	168	167	2.74	0.20	1.5
Diphenylmethane	168	91	2.88	1.4	2.8
4–Phenyltoluene	168	167	2.94	0.20	3.7
Acenaphthylene	152	151	3.00	0.05	nd
Acenaphthene	154	153	3.25	0.20	1.8
Bibenzyl	182	91	3.41	1.9	1.5
1,1–Diphenylethylene	180	179	3.48	>7.4	nd
cis-Stilbene	180	179	3.59		
2,2–Diphenylpropane	196	181	3.62	nd	nd
2,3,5–Trimethylnaphthalene	170	155	3.71	0.65	5.2
3,3′–Dimethylbiphenyl	182	167	4.03	0.31	5.2
Fluorene	166	165	4.15	0.15	2.2
4,4′–Dimethylbiphenyl	182	167	4.18	0.57	7.0
4–Vinylbiphenyl	180	178	4.56	nd	nd
Diphenylacetylene	178	89	4.90	0.05	nd
9,10–Dihydroanthracene	180	179	5.03	0.66	nd
trans-Stilbene	180	179	5.20	>0.47	9.2
9,10–Dihydrophenanthrene	180	179	5.29		
10,11–Dihydro–5H– dibenzo(a,d)cycloheptane	194	179	6.03	0.40	nd
Phenanthrene	178	89	6.08	>0.52	2.2
Anthracene	178	89	6.14		
1–Phenylnaphthalene	204	203	6.77	nd	nd
1–Methylphenanthrene	192	191	7.06	nd	11
2–Methylanthracene	192	191	7.25	0.51	0.70
9–Methylanthracene	192	191	7.59	nd	0.70
9–Vinylanthracene	204	203	7.82	nd	nd
Triphenylmethane	244	167	8.04	nd	nd
Fluoranthene	202	101	8.41	0.55	1.9
Pyrene	202	101	8.90	0.53	1.7
9,10–Dimethylanthracene	206	191	8.99	0.19	nd
Triphenylethylene	256	178	9.25	>0.08	nd
p-Terphenyl	230	115	9.44	nd	nd
1,2–Benzfluorene	216	108	9.64	nd	nd

Table 3.8 continued

Compound	Ions monitored		Rel. ret. time*	Concentration (ng L⁻¹)	
				1	2
2,3–Benzfluorene	216	108	9.73	nd	nd
Benzylbiphenyl	244	167	9.75	nd	nd
1,1'–Binaphthyl	254	126	10.95	nd	nd
Triphenylene	228	114	11.5		
Benz(a)anthracene	228	114	11.6	>8.1	3.3
Chrysene	228	114	11.8		

*Retention times are relative to the retention time of naphthalene (3.81 min)

Source: Reproduced with permission from Benoit, F.M. et al. [14] American Chemical Society

Table 3.9 Oxygenated polycyclic aromatic hydrocarbons detected in Ottawa potable water samples

Compound	Ions monitored		Rel. ret. time*	Concentration (ng L⁻¹)	
				1	2
Xanthene	182	181	4.83	0.20	0.10
9–Fluorenone	180	152	5.93	0.90	1.5
Perinaphthenone	180	152	7.70	0.28	0.15
Anthrone	194	165	7.90	1.4	nd
Anthraquinone	208	180	8.11	2.4	1.8
Naphthalene	128	102	1.00		

*Retention times are relative to the retention time of naphthalene (3.81 min)

Source: Reproduced with permission from Benoit, F.M. et al. [14] American Chemical Society

species. This was particularly true for methyl substituted polyaromatic hydrocarbons for which numerous positional isomers may elute within a narrow time window. In many cases only a small number of the possible positional isomers were available commercially and could be included in the reference standard. Hence, unequivocal identification of positional was often not possible.

Quantitative estimates of the detectable polyaromatic hydrocarbons and oxygenated polyaromatic hydrocarbons in Ottawa potable water were obtained by comparison of the areas of the two characteristic peaks

(Tables 3.8 and 3.9), in the mass chromatograms of the reference standard and the field sample, respectively. The average of the concentrations for the two ions is presented in Table 3.8 (polyaromatic hydrocarbons) and Table 3.9 (oxygenated polyaromatic hydrocarbons) for the two water samples analysed. No corrections were made for incomplete recovery. Of the 50 polyaromatic hydrocarbons in the standard used by Benoit *et al.* [14] 38 are detected in at least one of the two water samples tested. In sample 1 (February 1978) 16 polyaromatic hydrocarbons ranging in concentration from 0.05 to 8.1ng L^{-1} and in sample 2 (January 1978) 30 polyaromatic hydrocarbons (Table 3.8) ranging in concentration from 0.05 to 14ng L^{-1} were detected. Twenty-eight polyaromatic hydrocarbons and four oxygenated polyaromatic hydrocarbons were detected in both samples analysed. The lower concentration of 0.05ng L^{-1} represents the lower limit of detection of this method of analysis. There was an appreciable variation in the concentrations of most of the polyaromatic hydrocarbons detected in the two water samples; however, all compounds detected were found to be in the low ng L^{-1} range. This suggests that the observed concentrations of polyaromatic hydrocarbons and oxygenated polyaromatic hydrocarbons are representative of the background level of contamination. The mean concentration of poly-aromatic hydrocarbons in samples 1 and 2 was 1.4ng L^{-1} and 50.4ng L^{-1}.

Of the five oxygenated polyaromatic hydrocarbons in the standard, all five, ranging in concentration from 0.20 to 2.4ng L^{-1} are detected in sample 1 and four, ranging in concentration from 0.1 to 1.8ng L^{-1} are detected in sample 2 (Table 3.9). The mean concentrations and total weights of detected oxygenated polyaromatic hydrocarbons in the two samples were 1.0 and 5.2ng L^{-1} for sample 1 and 0.91 and 3.7ng L^{-1} for sample 2, respectively. It is noteworthy that for three of the oxygenated compounds (anthrone, anthraquinone, and 9–fluorenone) detected, the parent compound is also detected in the water sample. Thus, the oxygenated species could possibly originate from the oxidation of the parent compound in the aqueous media.

Junk [16] compared adsorption on XAD–2 resin and solvent extraction procedures for the preconcentration of polyaromatic hydrocarbons from groundwater and surface water samples. Compounds adsorbed on XAD–2 resin were desorbed with small volumes of ethyl acetate or benzene prior to gas chromatography.

3.2.1.3 Coprostanol

Wun *et al.* [3] used XAD–2 resin for the analysis of coprostanol in river and lake water and secondary sewage treatment plants. They showed that extraction of coprostanol from water by adsorption on a column of Amberlite XAD–2 resin is as efficient as conventional liquid–liquid extraction. Maximal recovery depends on the pH value of the sample, flow rate, resin mesh size and concentration of the coprostanol. The final determination was carried out by gas chromatography.

In further work, Wun *et al.* [17] improved the efficiency of the extraction of coprostanol using an XAD–2 resin column by decreasing the extraction time using a 'closed' column technique and by determining the effects of sample pH on adsorption processes.

Coprostanol was strongly adsorbed to polystyrene XAD–2 adsorbents, at pH2, with 100% retention, and the adsorbed sterol was easily eluted with acetone adjusted to pH8.5–9.0 with ammonium hydroxide. It was also shown that with a closed column method, large volumes of water can be extracted in a relatively short time and with higher sensitivity than that of the liquid–liquid partitioning procedure.

3.2.1.4 Non ionic surfactants

Non ionic detergents. Nickless and Jones [18] and Musty and Nickless [19,20] evaluated Amberlite XAD–4 resin as an extractant for down to 1mg L^{-1} of polyethylated or secondary alcohol ethoxylates (R(OCH$_2$CH$_2$)$_2$OH) surfactant and their degradation products from water samples. This resin was found to be an effective adsorbent for extraction of polyethoxylated compounds from water except for polyethylene glycols of molecular weight less than 300. Flow rates of 100ml min^{-1} were possible using 5g of resin, and interfering compounds can be removed by a rigorous purification procedure. Adsorption efficiencies of 80–100% at 10µg L^{-1} were possible for non-ionic detergents using distilled water solutions. The main purpose of the work of Nickless and Jones [18] was the investigation of secondary alcohol ethoxylate as it proceeds through the water system. Associated with this is polyethylene glycol, a likely biodegradation product. Alkylphenol ethoxylate was also considered but only as a possible interferent that should be differentiated in order to allow fuller characterisation of secondary alcohol ethoxylate residues. Nickless and Jones [18] used in their study fairly pure secondary alcohol ethoxylate standards (R(OCH$_2$CH$_2$)$_n$OH) where n is 3, 5, 7, 9 and 12 also alkylphenol ethoxylate standards RC$_6$H$_4$(OCH$_2$CH$_2$)$_n$OH where n is 2, 3, 9 and 22. The barium chloride phosphomolybdic acid spectrophotometric method [21,22] and also thin-layer chromatography [18] were used to determine polyoxyethylated compounds in column effluents. Nickless and Jones [18] tested and efficiencies of three column materials for adsorption of secondary alcohol ethoxylate from solutions and found that Amberlite XAD–4 resin combined a high adsorption on the column with a high subsequent desorption of the ethoxylate from the column with ethanol. Subsequently acetone was found to be a superior desorption solvent.

The XAD–4 results were very encouraging but the measured desorption was well over 100% and was an indication of interfering compounds still present in the resin from its manufacture. These interfering compounds were polyanionic in nature and Nickless and Jones [18] devised a procedure involving successive washing with acetone–hexane (1:1), acetone, and ethanol of XAD–4 resin to remove impurities before its use in the analyses of samples. Results obtained by this procedure indicated a fairly rapid drop in adsorption efficiency of the resin for polyethylene glycols between 9EO and 3EO. It is not clear if the drop in efficiency is roughly linear with a shortening in chain length, but it appears that the resin might be limited to the study of polyethylene glycols in water for chain length greater than 7EO.

The ability of XAD–4 resin to adsorb with high efficiency polyethoxy-lated compounds at very low concentrations is very important, if it is to be used for the examination of all types of water systems. When the concentration is very low then this necessitates the processing of large volumes of water in order to obtain sufficient material for characterisation using methods such as infrared, ultraviolet and NMR spectroscopy, as well as liquid chromatography. The performance of the resin in this respect was evaluated by using solutions of a mixture of three model compounds, secondary alcohol ethoxylate 9EO, alkylphenol ethoxylate 9EO, and polyethylene glycol 9EO. Recoveries varied from 82% at the 0.01mg L $^{-1}$ level to 98% at the 1mg L $^{-1}$ level.

Nickless and Jones [18] continued their study of Amberlite XAD–4 resins by examining polyethoxylated materials before and after passage through a sewage works. Samples from the inlet and outlet of the sewage works and from the adjacent river were subjected to a three stage isolation procedure and the final extracts were separated into a non ionic detergent and a polyethylene glycol. The non ionic detergent concentration was 100 times lower in the river than in the sewage effluent. Thin layer chromatography and ultraviolet, infrared and NMR spectroscopy were used to identify, in the non ionic detergent component, alkylphenol ethoxylates (the most persistent), secondary alcohol ethoxylates and primary alcohol ethoxylate.

The three stage isolation procedure was carried out as follows.

Stage 1. The water sample was first passed down a column of XAD–4 resin to extract surface active materials on to the resin. The organics adsorbed from the water sample by the XAD–4 resin were eluted off with four solvent systems to give two fractions:

(1) 2ml of methanol–water (1:1) followed by 20ml of methanol;
(2) 50ml acetone, followed by 20ml of acetone–*n*–hexane (1:1).

Each fraction was collected and evaporated to dryness in a stream of filtered air on a hot water bath. Fraction 2 contained most of the polyethoxylated material but significant amounts were also present in fraction 1. The residue from fraction 1 was treated with 10ml of acetone, decanted into a small tube, centrifuged, and the clear acetone layer poured into the beaker containing fraction 2. The combined extracts were again evaporated to dryness.

Stage 2 – liquid–solid chromatography. The polyethoxylated material contained in fraction 2 from the XAD–4 elution will still be a relatively minor part of the residue. The indications were that most of the unwanted organic compounds extracted from the river water would be medium to non-polar in character. There would therefore be overlap with the medium polar oligomers of secondary alcohol ethoxylate and alkylphenol ethoxylate (ie those with only 1, 2 or 3 ethylene oxide units per molecule). Silica gel was found to be the best chromatographic adsorbent and the correct choice of eluting solvent strength is important for efficient separation, a (7:3) ethyl acetate–benzene extract of fraction 2 was poured down a column of silica gel. The ethyl acetate–benzene insoluble residue contained highly polar polyethoxylated material and was dissolved and loaded on to the column when more polar solvents were used later in the eluting procedure. Most of the unwanted organic compounds came through in fractions 1 and 2 and were discarded. Fractions 3 and 4 (less than 1% of total) were combined and kept for thin layer chromatographic examination. Fractions 5 and 6 were combined and usually contained greater than 98% of the total polyethoxylated material present in the original extract, together with some highly polar compounds imparting a faint yellow colour to the solution.

Stage 3. The combined fractions 5 and 6 obtained from liquid solid chromatography needed further separation for two reasons:

(1) the fraction still contained significant amounts of other organic compounds which were found to be mainly acidic in character, and
(2) meaningful results could only be obtained from spectroscopic examination if the polyethoxylated material is divided into the non ionic detergent components and the polyethylene glycol component.

Jones and Nickless [23] solved both of these problems in one operation using liquid chromatography based on a procedure by Nadeau and Waszeciak [24]. With this procedure using water as a stationary phase (on Celite) and chloroform–benzene as the mobile phase, very good separations of polyethylene glycol from secondary alcohol ethoxylate and alkylphenol ethoxylate can be obtained. By simply changing the stationary phase to dilute sodium hydroxide solution, unwanted acidic compounds can be removed with the polyethylene glycol fraction.

Having separated the mixture into a secondary alcohol ethoxylate plus alkylphenol ethoxylate and polyethylene glycol components, Jones and Nickless examined these fractions using thin layer chromatography [23] and ultraviolet infrared, and NMR spectroscopy [25].

In the thin layer chromatographic separations ethyl acetate–acetic acid–water, 4:3:3, was used for quantitative information, where a compact spot is obtained for non ionic detergents (secondary alcohol ethoxylate and alkylphenol ethoxylate combined) and an elongated spot of lower R_f value for polyethylene glycol. Ethyl acetate–acetic acid–water, 70:15:15, was used to obtain information on the molecular weight distribution of secondary alcohol ethoxylate and alkylphenol ethoxylate. Alkylphenol ethoxylate gives a 'string' of well resolved spots and secondary alcohol ethoxylate a long unresolved streak. These patterns will be superimposed if both are present in the residue.

Ultraviolet spectroscopy was used to evaluate alkylphenol ethoxylates as only this type of compound gives a peak at 277nm with a characteristic shoulder at 285nm. Chloroform was the solvent used and after measurement the sample can be reconcentrated in a stream of air back to 0.5ml. The infrared spectra of secondary alcohol ethoxylates and alkylphenol ethoxylates isolated from samples are very similar with the broad strong peak at 1100–1120cm $^{-1}$ characteristic of a polyethoxylate grouping. The only clearly recognisable difference between them is the very sharp aromatic peak present in the alkylphenol ethoxylate spectrum at 1500–1505cm $^{-1}$. NMR spectroscopy was able to distinguish between secondary alcohol ethoxide and primary alcohol ethoxylate.

3.2.1.5 Organophosphorus insecticides

Paschal et al. [26] have discussed the preconcentration and determination of Parathion–ethyl and Parathion–methyl in run-off water using high performance liquid chromatography. The organic compounds are concentrated on an XAD–2 resin before analysis by reverse phase, high performance liquid chromatography. Detection limits were found to be approximately 2–3mg L $^{-1}$. These workers examined the possible interference in the method from other agricultural chemicals and organic compounds commonly occurring in water. This method is based on the use of Rohm and Haas XAD–2 macroreticular resins. Organics in water can be sorbed on a small column of resin, and the sorbed organics then eluted by diethyl ether. After evaporation of the eluate, the concentrated organics can be determined by chromatography. In addition to the obvious benefit of 100- to 100-fold concentration, this method offers the possibility for on-site sampling.

Apparatus: Gas chromatograph, six port injector, a Whatman prepacked micro-particle reverse phase column (Partisil ODS) and a variable wavelength detector.

Reagents and materials: Macroreticular resin XAD–2, purified by Soxhlet extraction as described by Junk et al. [11] and stored under pure methanol.

Pesticide grade acetonitrile as received.

Diethyl ether, glass distilled before use.

Distilled water, glass distilled before use.

Preparation of standards: Make up microlitre amounts of 100mg L^{-1} stock solutions of organophosphorus insecticides in methanol and dilute with distilled water to 100ml. Pass the diluted standards through a 10cm column of purified XAD–2 resin, at a rate of 4–6ml min^{-1}. After the last of the dilute aqueous standards had passed through the column, remove most of the water clinging to the resin by gentle vacuum aspiration. Pass 30ml of glass distilled diethyl ether through the column at 2–3ml min^{-1} then remove the last of the ether by passing dried purified nitrogen through the column. Dry the ether by shaking with 2g of anhydrous sodium sulphate and evaporate to dryness using a rotary evaporator at temperatures not exceeding 35°C. Dissolve the residue in 100ml of pesticide grade acetonitrile, and chromatograph the resulting solution on a Partisil ODS reverse phase column at 2.40ml min^{-1} with 50% acetonitrile/water mobile phase. An average recovery of 99% was obtained through the procedure.

Procedure: Prepare extracts of 2L samples as in the above procedure and evaluate chromatograms to establish calibration curves for the parathions.

To evaluate the efficiency of extraction of XAD–2 resin for trace organics in run-off water, Paschal *et al.* [26] ran a sample of water through the procedure and obtained a large number of peaks in the 2–3min region. A number of relatively polar compounds elute early in the chromatograph, with relatively few peaks in the 3–10min region of the chromatogram. On changing from 50% acetonitrile to 100% acetonitrile to regenerate the column, several more peaks were eluted, apparently consisting of less polar materials strongly adsorbed under the conditions of the procedure. No interference was obtained from these strongly adsorbed compounds, although it was found to be useful to regenerate the column with 100% acetonitrile after every five to six runs to ensure reproducibility.

The retention times for Parathion–methyl and Parathion–ethyl, obtained from volumetric dilutions of methanolic standards with acetonitrile, were 3.45 and 4min, indicating no interference from naturally occurring organics in the run-off water. Spiked samples of run-off water were prepared containing Parathion–ethyl and Parathion–methyl. A typical chromatogram for such a spiked sample is shown in Fig. 3.5. The parathions are well separated, with no observed interference from organics already present in the water. Calibration curves were prepared from a set of standards containing 20–120µg L^{-1} Parathion in methanol. Atrazine was added as an internal standard to the concentrated extract. Ratio of speak heights or areas of parathions to those of Atrazine were plotted versus concentration. Good linearity was obtained over the range of concentration examined for both parathions. In order to evaluate the

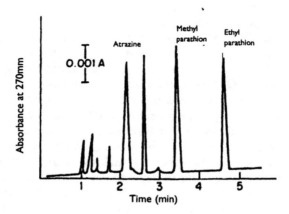

Fig. 3.5 Chromatogram of spiked run off water extract
Source: Reproduced with permission from Paschal, D.C. *et al.* [26] American Chemical
Society, Washington, US

Table 3.10 Reproducibility of method

Taken (μg L⁻¹)	Found (μg L⁻¹)ᵃ	sd sd (%)	Relative
Parathion–methyl			
15.0	14.8	0.45	3.0
37.5	37.1	1.07	2.8
75.0	75.9	0.73	1.0
112.5	112.7	2.56	2.3
Parathion–ethyl			
10.0	9.9	0.37	3.7
25.0	24.6	1.40	5.6
50.0	49.3	0.97	1.9
75.0	75.0	2.40	3.2

ᵃAverage of six determinations

Source: Reproduced with permission from Paschal, D.C. *et al.* [26] American Chemical
Society

accuracy and reproducibility of the method, a series of solutions was
prepared in run-off water with concentrations of parathions in the range
of the calibration curve. The results of this study are given in Table 3.10.
The lower limit of detection was calculated from those data to be 3.1 and
2.9ng for Parathion–methyl and Parathion–ethyl, respectively.

Table 3.11 Interference study

Compound	Relative retention (parathion–methyl = 1.00)	Wavelength measured (nm)
Aroclor 1260	3.94–5.88 multiple peaks	225
Atrazine	0.75	265
Azinphos–ethyl	1.14	285
Alachlor	0.89	235
Carbaryl (Sevin)	0.69	280
Carbofuran	0.61	270
Chloramben	0.26	240
Chlorpyrifos	2.01	290
p,p'–DDT	2.78	235
DEHP	1.59	235
Dialifor	1.61	290
Diazinon	1.30	245
Dyfonate (fonofos)	1.36	240
Fenitrothion	1.18	265
Methoxychlor	1.72	225
p-Nitrophenol	0.72	310
Phosmet	0.93	230
Phorate	1.30	220
Propachlor	0.67	260
2,3,5–T	0.28	250
Trifluralin	0.58	270

Source: Reproduced with permission from Paschal, D.C. et al. [26] American Chemical Society

Table 3.11 shows the effect of potential interference by other agricultural chemicals and organics commonly occurring in non saline water. Wavelengths chosen for measurement were at or near the absorbance maxima for the compounds as determined by ultraviolet scans from 350 to 200nm. If a potentially interfering compound showed a retention time near one of the parathions, then chromatography was performed with detection at 270nm. Of the compounds investigated only Fonofos (ethyl–S–phenylethyl phosphonothiolothionate) interferes at 270nm. However, if the wavelength of detection is changed to 280nm, the interference is overcome.

Various workers [27–33] have reported on the use of XAD–2 for the extraction of Fenitrothion from water. Mallet et al. [30] used an automated gas chromatographic system which consisted of a gas chromatograph mounted with an automatic sampler interfaced to an integrator. A Melpar flame photometric detector (phosphorus mode) was connected with the flame gas inlets in the reverse configuration to prevent solvent flame–out. The detector was maintained at 185°C and flame gases were optimised with flow rates (ml min $^{-1}$) as follows: hydrogen, 80; oxygen, 10; air, 20. A 1.8m × 4.0mm id U-shaped glass column packed with 4% (w/w) OV–101 and 6% (w/w) OV–210 on Chromosorb W AW DMCS, 80–100 mesh, was used. Nitrogen was used as carrier gas at a flow rate of 70ml min $^{-1}$. A column temperature of 195°C sufficiently resolved Fenitrooxon from its parent compounds. The injection port temperature was set at 225°C. The water sample was passed through an XAD–2 column, which was subsequently eluted with ethyl acetate. The ethyl acetate extract was examined by gas chromatography.

In an alternative method involving thin layer chromatography, and fluorometric analysis, a 1L water sample was extracted with two 50ml portions of chloroform, which were collectively dried through an anhydrous sodium sulphate (50g) column. The chloroform was replaced with ethyl acetate on a flash evaporator, carefully reduced to 4–5ml and made up to the mark with ethyl acetate in a 10ml volumetric flask. Aminofenitrothion was recovered by this method from environmental water using an XAD–2 column. Recoveries were in the range 87–118% at an average flow rate of 153ml min $^{-1}$. The relative error, c. 10%, is normal at a concentration of 50μg L $^{-1}$ when using *in situ* fluorometry. Under similar conditions, Fenitrooxon can also be recovered with good yields. The procedure was adapted to the simultaneous analysis of the parent compound and its two derivatives by gas liquid chromatography. With XAD–2 resin, conditions such as flow rate and column length are crucial to obtain good recoveries. If a 10 × 1.9cm column is used the maximum flow rate is limited to c. 50ml min $^{-1}$, which is easily sustained by gravity flow.

Various good recoveries are obtained when using XAD–2 for extracting spiked lake water in the case of Fenitrooxon. Relative standard deviations of 5.1–6.4% are very good. The overall average recovery (99.7%) of the three compounds by the conventional serial solvent extraction procedure is somewhat better than by the XAD method (95.3%). Reproducibilities are all better with the exception of Fenitrooxon. Volpe and Mallet [33] developed a method for preconcentrating and determining down to 0.5ng of Fenitrothion and five fenitrothion derivatives in water by adsorption on XAD–4 and XAD–7 resins, followed by solvent elution and gas liquid chromatography of the extract.

Puijker *et al.* [34] preconcentrated and separated organic compounds containing phosphorus and sulphur from water on to XAD resin, then reduced the compounds with hydrogen at 1100°C. The resulting phosphine and hydrogen sulphide were separated on chromatographic column, and detected at 526 and 384nm respectively, in a flame photometer. The detection limits for phosphorus and sulphur were 0.1ng and 1ng respectively. Frobe *et al.* [35] preconcentrated organophosphorus insecticides on Amberlite XAD–4 prior to determination of phosphorus by oxygen flask combustion.

Tolosa *et al.* [36] studied the extraction efficiencies of XAD–2, XAD–7, C_{18} discs and poly(styrene–divinylbenzene) discs for 11 organochlorine and 24 organophosphorus compounds. Macroreticular resins provided low recoveries and C_{18} and poly(styrene–divinylbenzene) produced good recoveries of analytes with log Kw72.

3.2.1.6 Organochlorine pesticides PCBs, organophosphorus pesticides, triazine herbicides, chlorophenoxyacetic acid herbicides and phthalate esters

XAD–2 resin has been used to preconcentrate a range of organochlorine pesticides including Dieldrin [37,38] prior to desorption with hot solvents (eg acrylonitrile [38]) and gas chromatography. Levesque and Mallet [39] used preconcentration on XAD–4 and XAD–7 resins followed by solvent extraction of adsorbate and gas chromatography using a nitrogen phosphorus detector to determine Aminocarb herbicide and some of its degradation products in water. Similarly, Aldicarb and its oxidation products, aldicarb sulphoxide and Aldicarb sulphone, have been determined using XAD–2 resin and high performance liquid chromatographic finish [40]. Harris *et al.* [41] preconcentrated Kepone at low concentrations using XAD–2 resin. Sundaram *et al.* [42] preconcentrated carbamate herbicides on XAD–2 resin and desorbed them with a small volume of ethyl acetate or acetone prior to gas chromatography using an NP specific detector. Chloromethoxynil, Bifenox, and Butachlor have been preconcentrated by similar techniques which are capable of achieving detection limits of $0.2\mu g\ L^{-1}$.

Rees and Au [43] used small columns of XAD–2 resin for the recovery of ambient trace levels of pesticides (organochlorine also organophosphorus pesticides, Triazine and chlorophenoxy acid herbicides, phthalate esters, and polychlorinated biphenyls) from water samples at concentrations ranging from 0.001 to $50\mu g\ L^{-1}$.

Reagents: Solvents distilled in glass, glass wool and sodium sulphate, solvent washed prior to use.

Diethyl ether, redistilled daily over metallic sodium to remove alcohol preservative.

XAD–2 resin, 20–60 mesh, purify by sequential extraction with methanol, acetonitrile, and diethyl ether in a Soxhlet extractor for 8h per solvent. Store purified resin under methanol. Florisil, PR grade, 60–80 mesh, store at 130°C.

Standard solutions made up in hexane or benzene. Appropriate dilutions for fortifications made in acetone.

Apparatus: Resin column, Pyrex 1.0 × 20cm with Teflon stopcock and 1L integrator reservoir.

Concentration flask, 250ml flask with 10ml graduated conical extension. Florisil column, Pyrex, 0.6 × 20cm with Teflon stopcock and 100ml reservoir.

Gas chromatographic conditions: Organochlorine pesticides: Pyrex column, 190cm × 2mm, packed with 11% OV 17/QFI on Gas Chrom Q operated at 215°C, Ni 63 detector operated at 300°C.

Chlorophenoxy Acids: The same conditions as above except for the column temperature (170°C) while for PCBs the column packing was 3% Dexsil 300 on Chromosorb W.

Phthalates: Flame–ionisation detector at 300°C, column temperature programmed from 90°C to 250°C at 10°C min $^{-1}$. Packing Carbowax 20M on Chromosorb W.

Organophosphates: Flame photometric detector in the reversed flow mode (air: 40ml min $^{-1}$; oxygen, 20ml min $^{-1}$; hydrogen: 2.00ml min $^{-1}$) operated at 200°C. Column (3% Dexsil 410) programmed from 100 to 220°C at 7°C min $^{-1}$.

Triazines: Column packing 6% Carbowax 20M on Chromosorb W, conditioned at 260°C. Oven temperature, programmed from 190 to 250°C at 4°C min $^{-1}$. AFID detector, operated at 260°C.

Procedure: Set up the resin column and plug the tower end with glass wool. Add XAD–2 resin as a methanol slurry until a 6cm bed is formed. Insert a second glass wool plug to cap off the bed. Drain the methanol until the level reaches the top of the bed; then rinse the bed with 3 × 30ml portions of pre-purified redistilled water in order to wash off the methanol and aid in reduction of air bubble formation. For higher concentrations of pesticides, add 1L of pre-purified distilled water to the reservoir, fortified with the appropriate acetone solution (1ml) swirl to mix for 1min, then drain through the column at approximately 35–40ml min $^{-1}$. Further rinse the reservoir with 30ml pre-purified distilled water, which is also drained through the column. For lower concentrations, where 20L of water are used, carry out fortifying in one gallon pre-rinsed bottles as alternating supplies. Transfer fortified water to the reservoir by syphoning through 0.2mm Teflon tubing.

After all the water has passed through the resin, allow the bed to drain for 5min, then run absolute diethyl ether through the resin and collect in a 250ml separatory funnel, until no more water is coeluted. Then close the column tap and allow the dry ether (approx. 20ml) to stand in the resin bed for approximately 10min. Then run off the ether and add to the contents of the separatory funnel. Repeat the equilibration with two further 20ml portions of ether, adding each eluate to the separatory funnel.

After separation, run off the water in the funnel and discard. Dry the remaining ether by passing through sodium sulphate and collect in a

250ml concentrating flask. Vacuum rotary evaporate the extract (30°C) to 0.5ml, blow gently to dryness, make up in hexane, and submit for gas chromatographic analysis under the conditions mentioned above. In the case of organochlorine pesticides and PCBs clean up extracts using a 16cm hexane slurry packed Florisil column prewashed with 40ml hexane. Transfer the sample to the column then elute with 15ml 25:75 (v/v) dichloromethane: hexane, containing PCBs, Lindane, Heptachlor, Aldrin, DDT group, and 15ml dichloromethane containing heptachloroepoxide, Dieldrin, Endrin. For chlorophenoxy acid recoveries, acidify fortified solutions to pH2, before resin extraction and methylate the concentrated eluates with diazomethane/diethyl ether prior to Florisil clean up and gas chromatography.

The adsorption efficiency of the resin was tested by running large samples (47L) of natural river water through two successive cartridges. Individual desorption of each cartridge showed that there was no trace of carry through and the organochlorines and PCBs were completely adsorbed on the first cartridge, at levels of 3ng L $^{-1}$ for PCBs and 9ng L $^{-1}$ of pp–DDD. The concentrations of 0.1ng L $^{-1}$ of PCB and 0.1ng L $^{-1}$ of organochlorinated pesticides can be detected in non saline waters.

Musty and Nickless [19,20] used Amberlite XAD–4 for the extraction and recovery of chlorinated insecticides and PCBs from water. In this method a glass column (20cm × 1cm) was packed with 2g of XAD–4 (60–85 mesh) and IL of tap water (containing 1ng L $^{-1}$ of insecticides) was passed through the column at 8mL min $^{-1}$. The column was dried by drawing a stream of air through, then the insecticides were eluted with 100ml of ethyl ether–hexane (1:9). The eluate was concentrated to 5ml and was subjected to gas chromatography on a glass column (1.7m × 4mm) packed with 1.5% OV–17 and 1.95% QF–1 on Gas–Chrom Q (100–200 mesh). The column was operated at 200°C, with argon (10ml min $^{-1}$) as carrier gas and a ^{63}Ni electron-capture detector (pulse mode). Recoveries of BHC isomers were 106–114%; of Aldrin, 61%; of DDT isomers, 102–144%; and of PCBs 76%.

3.2.1.7 Carbamate insecticides

Sundaram *et al.* [45] have described a rapid and sensitive analytical technique to quantify carbamate insecticides at nanogram levels using sorption on an Amberlite XAD–2 resin column and desorption followed by nitrogen–phosphorus–gas–liquid chromatographic analysis. The carbamates were extracted from non saline water by percolation through a column of Amberlite XAD–2 followed by elution with ethyl acetate. The carbamate residues were then analysed directly. Recoveries between 86 and 108% were obtained for the following insecticides: Aminocarb (4-dimethyl-amino-*M*-tolyl)*N*-methylcarbamate), Carbaryl (1-naphthyl-*N*-methyl-carbamate), Mexacarbate (4-dimethylcarbamate) and Propoxur (o-isopropoxyphenol-*N*-methylcarbamate). Only 41–58% recovery was obtained for Methomyl(*S*-methyl-*N*-(methylcarbamoyl)–oxy)thiocetimidate.

3.2.1.8 Sulphur containing insecticides

Methods [46,47] to determine Fenitrothion in water using gas–liquid chromatography with a flame photometric detector have been reported. An *in situ* fluorometric method [32] to detect simultaneously Fenitrothion, Fenitrooxon, Aminofenitrothion and nitrocresol on a thin layer chromatogram has been developed. Fenitrothion, Fenitrooxon and Aminofenitrothion have been analysed simultaneously by gas chromatography using an SE–30 plus QF–1 column and a flame photometric detector in the phosphorus mode [48].

Various other workers [27–32] have used macroreticular resin XAD–2 to recover Fenitrothion from river water samples and compared this method with solvent extraction techniques.

3.2.1.9 Polychlorinated biphenyls

XAD–2 and XAD–4 macroreticular resins have been used to preconcentrate polychlorobiphenyls prior to solvent desorption and analysis in amounts down to 0.4ng L $^{-1}$ using either electron capture gas chromatography [11,20,27,38,49–54] or higher performance liquid chromatography [55]. Musty and Nickless [19,56] used XAD–4 resin to preconcentrate polychlorobiphenyls from 1L samples of non saline water prior to desorption with diethyl ether:hexane (1:1) and gas chromatography.

3.2.1.10 Polychlorodibenzo–*p*–dioxins and furans

Le Bel *et al.* [57] developed a method for the extraction and preconcentration of polychlorodibenzo-*p*-dioxins and polychlorodibenzofurans using XAD–2 resin.

3.2.1.11 Hydrocarbons

Gomez-Belinchon *et al.* [58] compared liquid–liquid extraction and adsorption on polyurethane foam and XAD–2 resin for the isolation of hydrocarbons, PCBs and fatty acids from non saline waters.

Green and Le Pape [59] observed that XAD–2 resin had a preservative effect on hydrocarbons stored on the resin for long periods of time since no biological degradation of radiolabelled standards on the resin was observed for up to 100d. The macroreticular structure of the resin is thought to be responsible for the mechanism by which the extracted hydrocarbons were preserved.

3.2.1.12 Chlorinated aliphatic hydrocarbons

Martinsen *et al.* [60] preconcentrated aliphatic organochlorine compounds on activated carbon and XAD–4 resin prior to determination by neutron activation analysis, thin layer and gel permeation chromatography.

Chloromethanes have been preconcentrated on XAD–4 macroreticular resin [61].

3.2.1.13 Haloforms

Two approaches to this analysis are possible. In one, the water [62] sample is purged with purified nitrogen or helium and the purged trihaloforms are trapped on a column of macroreticular resin such as Amberlite XAD–2 or XAD–4. The trapped trihaloforms are then desorbed from the column with a small volume of a polar solvent prior to gas chromatography. In the second approach the water sample is contacted directly with the resin and then the trihaloforms desorbed as before. Alternatively the resin can be injected directly into the gas chromatograph injection port [63]. Both methods provide a very useful built-in concentration factor which improves method sensitivity.

3.2.1.14 Chlorophenols and chloroamines

Malalyandi *et al.* [64] developed a method for the extraction and preconcentration of chlorophenols, 4–chloroaniline, 3,3,dichlorobenzidine and α hexachlorocyclohexane in non saline waters using XAD–2 and XAD–4 resins.

3.2.1.15 Carboxylic acids

Richard and Fritz [65] employed macroreticular XAD–4 resin aminated with trimethylamine for the concentration, isolation and determination of acidic material from aqueous solutions. Acidic material is separated from other organic material by passing the water sample through a resin column in hydroxide form; other organic compounds are removed with methanol and diethyl ether. The acids are eluted with diethyl ether saturated with hydrogen chloride gas. After concentration the eluate is treated with diazomethane and the esters formed are separated by gas chromatography.

3.2.1.16 Carbanilides

Audu [66] used Amberlite XAD–2 to preconcentrate traces of carbanilides in non saline waters. High recoveries were generally observed but depended on the symmetry around the carbonyl group.

3.2.1.17 Cytokinins

Synak *et al.* [67] developed a method for the extraction and preconcentration of cytokinins in non saline waters using XAD–4 resin.

3.2.2 Seawater

3.2.2.1 General discussion

Macroreticular resins have also been used for the collection of trace organics in seawater. An excellent early review of the properties of the various XAD resins, along with comparisons with EXP–500 and activated carbon, can be found in work discussed by Gustafson and Paleos [68]. Riley and Taylor [69] have studied the uptake of about 30 organics from seawater on to XAD–1 resin at pH2–9. At the 2–5µg L $^{-1}$ level none of the carbohydrates, amino acids, proteins or phenols investigated were adsorbed in any detectable amounts. Various carboxylic acids, surfactants, insecticides, dyestuffs, and especially humic acids are adsorbed. The humic acids retained on the XAD–1 resin were fractionated by elution with water at pH7, 1M aqueous ammonia, and 0.2M potassium hydroxide.

Osterroht [70] studied the retention of non-polar organics from seawater on to macroreticular resins. Each of the XAD resins has slightly different properties and should collect a slightly different organic fraction from seawater. The major differences between the resins is in the degree of their polarity. The macroreticular resins should be useful in the analysis of particular classes of compounds; the stumbling block will be the determination of efficiencies. Earlier experience, admittedly with early versions of the resins, was that the collection of organic materials from water was far from complete. Once the properties of the resins are well understood, the analysis of at least some classes of compounds may quickly become routine.

3.2.2.2 Polychlorinated biphenyls and chlorinated insecticides

Musty and Nickless [19] used Amberlite XAD–4 for the extraction and recovery of chlorinated insecticides and PCBs from seawater. In this method a glass column (20 × 1cm) was packed with 2g XAD–4 (60–85 mesh), and 1L of water (containing 1 part per 10^9 of insecticides) was passed through the column at 8ml min $^{-1}$. The column was dried by drawing a stream of air through, then the insecticides were eluted with 100ml ethyl ether–hexane (1:9). The eluate was concentrated to 5ml and was subjected to gas chromatography on a glass column (1.7m × 4mm) packed with 1.5% OV–17 and 1.95% QG–1 on Gas–Chrom Q (100–120 mesh). The column was operated at 200°C, with argon (10ml min $^{-1}$) as carrier gas and a ^{63}Ni electron capture detector (pulse mode). Recoveries of BHC isomers were 106–114%; of Aldrin, 61%; of DDT isomers, 102–144%; and of polychlorinated biphenyls 76%.

Elder [49] determined PCBs in Mediterranean coastal waters by adsorption on to XAD–2 resin followed by electron capture gas chromatography. The overall average PCB concentration was 13ng L^{-1}.

Amberlite XAD–2 resin is a suitable adsorbent for polychlorinated biphenyl and chlorinated insecticides (DDT and metabolites, Dieldrin) in seawater. These compounds can be suitably eluted from the resin prior to gas chromatography [50]. Picer and Picer [71] evaluated the application of XAD–2; XAD–4 and Tenax macroreticular resins for concentrations of chlorinated insecticides and polychlorinated biphenyls in seawater prior to analysis by electron capture gas chromatography. The solvents used eluted not only the chlorinated hydrocarbons of interest but also other electron capture sensitive materials, so that eluates had to be purified. The eluates from the Tenax column were combined and the non-polar phase was separated from the polar phase in a glass separating funnel. Then the polar phase was extracted twice with n-pentane. The n-pentane extract was dried over anhydrous sodium sulphate, concentrated to 1ml and cleaned on an alumina column using a modification of the method described by Holden and Marsden [72]. The eluates were placed on a silica gel column for the separation of PCBs from DDT, its metabolites, and Dieldrin using a procedure described and Synder and Reinert [73] and Picer and Abel [74].

Picer and Picer [71] investigated the preconcentration from 10L samples of seawater of 0.1–1.0µg L^{-1} chlorinated pesticides (DDT, DDE, TDE and Dieldrin), and 1–2µg L^{-1} PCB (Aroclor 1254). Interestingly, for the elution of all chlorinated hydrocarbons 25ml polar solvent were required, and 50ml n-pentane was used as the re-extractant. Mirex was used as an internal standard, added to the eluate after the percolation of the polar solvent through the resin column. Hence this internal standard shows only the loss of chlorinated hydrocarbons during the re-extraction, alumina clean-up, and silica gel separation. The recovery of Mirex during these steps varied between 80% and 90%. Losses of the investigated chlorinated hydrocarbons during these steps were 10–30% for about 10ng pesticides. The experimental set-up is obviously capable of determining hydrocarbons in a 10L seawater sample at levels far below 1.0ng L^{-1} for pesticides and 10ng L^{-1} for PCBs.

Of the three macroreticular resins investigated XAD–2 was the best using methanol as elution solvent. Very unsatisfactory solvent blanks were obtained using Tenax resin. These workers conclude that application of macroreticular resins for the adsorption of chlorinated hydrocarbons from water samples and their determination after elution with different solvents has revealed several limitations. When water samples were spiked at levels close to the reported concentrations in seawater, the recovery of the investigated chlorinated hydrocarbons was low and unpredictable.

Organochlorine insecticides have been preconcentrated on XAD–2 resin prior to gas chromatographic analysis [74–77]. Gomez-Belinchon *et al.* [58] carried out an intercomparison study of three methods involving adsorption on XAD–resin, liquid–liquid extraction, and adsorption on polyurethane foam for the preconcentration of polychlorobiphenyls, hydrocarbons, and fatty acids from seawater. Sample sizes of 300–400L were used enabling very high preconcentration factors to be achieved. All three methods gave similar quantitative results, but there were qualitative differences which suggested that in this instance at least liquid–liquid extraction should be the method of choice.

3.2.2.3 Organophosphorus insecticides

The degradation products of the organophosphorus insecticides can be concentrated from large volumes of water by collection on Amberlite XAD–4 resin for subsequent analysis [29]. These are certainly not the only references to methods for the phosphorus-containing insecticides in non saline waters; however, most of the work has been done in fresh water and at concentrations very much higher than those to be expected in seawater.

3.2.2.4 Chlorinated aliphatic compounds

Dawson *et al.* [78] have described samples for large volume collection of seawater samples for chlorinated hydrocarbon analyses. The samplers use the macroreticular absorbent Amberlite XAD–2.

3.2.2.5 Azarenes

Shinohara *et al.* [79] have described a procedure based on gas chromatography for the determination of traces of two, three and five ring azarenes in seawater. The procedure is based on the concentration of the compounds on Amberlite XAD–2 resin, separation by solvent partition [11], and determination by gas chromatography–mass spectrometry with a selective ion monitor. Detection limits by the flame thermionic detector were 0.5–3.0ng and those by gas chromatography–mass spectrometry were in the range 0.02–0.5ng. The preferred solvent for elution from the resin was dichloromethane and the recoveries were mainly in the range 89–94%.

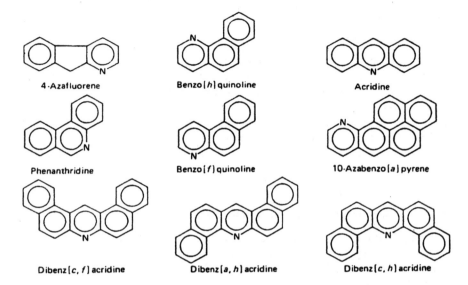

4-Azafluorene Benzo[h]quinoline Acridine

Phenanthridine Benzo[f]quinoline 10-Azabenzo[a]pyrene

Dibenz[c, f]acridine Dibenz[a, h]acridine Dibenz[c, h]acridine

3.2.2.6 Humic and fulvic acids

Several workers have used macroreticular resins, usually XAD–2 to collect high molecular weight humics from seawater [80–82].

3.2.2.7 Sterols

The sterols differ from the other compounds in that no class reaction has been proposed for the measurement of total sterols. Instead, various fractionation methods, usually derived from the biochemical literature, have been adapted to the concentrated materials collected from seawater. Certain of the more important sterols, particularly those used in the evaluation of water quality, have been determined by the use of a compound-specific reaction, after concentration from solution. Thus Wun et al. [3,83] measured coprostanol, a faecal sterol in seawater, after collection and separation, by extraction using liquid–liquid partitioning or extraction on a column of Amberlite XAD–2 resin.

3.2.2.8 Organosulphur compounds

Adsorption on XAD–2 and XAD–4 resins followed by solvent desorption and head space gas chromatography has been employed for the preconcentration and determination of volatile organosulphur compounds in estuary and seawaters [84].

3.2.3 Waste waters

3.2.3.1 Polyaromatic hydrocarbons

Owing to their very low solubility in water, PAHs occurring in industrial waste water are largely adsorbed on suspended solids. For example, Kadar et al. [85] showed that waste water from the aluminium industry has a PAH content of 10–150µg L^{-1} after filtering through a Micropore filter. Adsorption on a solid phase, such as Amberlite XAD, porous polyurethane resin and Tenax (a porous polymer based on 2, 6-diphenylphenylene oxide) was used to prepare an extract for gas chromatography.

3.2.4 Potable waters

3.2.4.1 Haloforms and aliphatic chlorocompounds

Renberg [86] has reported a resin adsorption method for the determination of trihalomethanes and choroethanes and dichloroethane in water. In this method halogenated hydrocarbons are determined by adsorption on to XAD–4 polystyrene resin and eluted with ethanol. The extract is analysed by gas chromatography and is sufficiently enriched in hydrocarbon to be suitable for other chemical analysis or biological tests. Volatile hydrocarbons yielded recoveries of 60–95%. By using two series-connected columns Renberg [86] was able to study the degree of adsorption and the chloroethane was found to be more strongly adsorbed than the haloalkanes. To study the adsorption character of substances described here, as well as other compounds, Renberg [86] connected two columns in a series and after the passage of the water sample the columns were eluted separately. The recovery of the substances of the second column, compared to those of the first one, will indicate the leakage of substances which is dependent on, for example, water flow rate and polarity of the substances. The ratio (a_2/a_1) of the amounts found in the second (a_2) and the first (a_1) column respectively, will predict the degree of adsorption, which can be regarded as a measurement of the lipophilicity of the substances studied. The values of the ratios found in the recovery experiments, shown in Table 3.12, indicate tetrachloroethane and 1,2–dichloroethane as the most and the least lipophilic of the substances tested.

3.2.5 Miscellaneous waters/miscellaneous organocompounds

Various other applications of non-polar types of XAD resins to the preconcentration of organics in various types of water are reviewed in Table 3.13.

Ryan and Fritz [9] used thermal desorption on a XAD–4 resin injected directly into the gas chromatograph to analyse mixtures of alkanes, alkyl benzenes, chloro and bromo compounds, alcohols, ketons and esters. In

Table 3.12 Results of recovery experiments from five spiked samples

Substance	Amount of spiked substance ($\mu g\ 250\mu L^{-1}$)	Recoveries[a]	
		Mean value (%) ± rel. sd	Ratio a_2/a_1 ± sd[b]
Chloroform	2.9	85 ± 4.1	0.27 ± 0.03
Bromodichloromethane	0.63	86 ± 9.6	0.22 ± 0.08
Chlorodibromomethane	1.3	91 ± 6.7	0.12 ± 0.03
1,2–Dichloroethane	120	95 ± 9.0	0.44 ± 0.09
Trichloroethane	2.5	74 ± 3.6	0.07 ± 0.03
Tetrachloroethane	1.4	60 ± 4.5	0.06 ± 0.02

Notes
[a]Sum of the amounts, found in the two series–connected columns ($a_1 + a_2$)
[b]Ratio of the amounts found in the second and first column, respectively
To prepare solutions purified water (250ml) was transferred to a 250ml separatory funnel and 250μL of an ethanol solution containing chloroform, bromodichloromethane, chlorodibromomethane, 1,2–dichloroethane, and tri- and tetrachloroethane was added. The funnel was shaken and the spiked water allowed to pass the columns. The flow was stopped when the level just reached the top of the upper bed. In order to rinse the glass walls from substances adsorbed, additional water (3 × 250ml) was passed through the funnel columns and the columns were eluted as described. The separatory funnel was shaken with ethanol (25ml) for the determination of substances adsorbed on the glass walls.

Source: Reproduced with permission from Renberg, L. [86] American Chemical Society

this method a small water sample (10–250ml) is passed through a small tube containing XAD–4 resin; this effectively retains the organic impurities present in the water. This tube is connected to the permanent apparatus and the sorbed organics are thermally transferred to a small Tenax precolumn while the water vapour is vented. The precolumn is closed off, pre-heated at 275–280°C and then a valve is opened to plug inject the vaporised sample into a gas chromatograph. The recoveries are generally quite good at both 10 and 1μg L^{-1} levels in water although there is some apparent loss of the halocarbons tested at the lower concentration level. The recovery of high boiling compounds was found to be dependent on desorption time and temperature. The reason for this is not entirely clear.

The efficiency of the thermal procedure was determined for several of the model compounds. This was done by comparing the peaks of compounds in standards taken through the entire procedure with peaks of the same compounds in standards injected directly into the gas chromatograph.

The recovery efficiencies of all compounds tested are quite good. Thus for quantitative analysis of actual water samples, the results could be compared with standards injected directly into the gas chromatograph or with water standards passed through the XAD–4 tube.

Table 3.13 Application of non-polar XAD resins (XAD–2, XAD–4) to preconcentration of organics from waters

Substance	Resin	Method of desorption	Analytical finish	Type of water sample	Detection limit*	Ref.
Total organic halogen	XAD	Solvent elution	Schoniger combustion-potentiometric titration of halogen	Non saline	1–2μmol L^{-1}	[87]
Chloroethanes	XAD–4	–	–	Non saline	–	[86]
Humic substances	XAD–2	–	–	Non saline	–	[82]
Haloforms	XAD–2, XAD–4	Solvent elution	GLC	Non saline	–	[88]
Haloforms	Acetylated XAD–2	Pyridine	GLC	Non saline	–	[89,90]
Haloforms	XAD–4	Ethanol	GLC	Non saline	–	[91]
Alkylphosphates, aklythiophosphates	XAD–4	Solvent	GLC	Non saline	–	[29]
Phenols	XAD–2	Pyridine	Silation GLC	Non saline	0.2mg L^{-1}	[92]
Phenols	XAD–2	Pyridine	–	Non saline	–	[11]
Phenols	XAD–4	Diethyl ether (aminated with trimethylamine)	Methylation–GLC	Non saline	–	[65]
Phenol, p-chlorophenol	XAD–4	–	–	Non saline	–	[93]
Nitrilo acetic acid	Dowex 1	Formic acid	Propylation–GLC	Non saline	0.01μg L^{-1}	[94]
Chlorobenzenes	XAD–2	Solvent	GLC	Non saline	1–10μg L^{-1}	[95]
Miscellaneous hydrocarbons, phenols, alkyl phthalates, polyaromatic hydrocarbons	XAD–2	Solvent elution	GLC	Non saline	–	[96]

Table 3.13 continued

Substance	Resin	Method of desorption	Analytical finish	Type of water sample	Detection limit*	Ref.
Miscellaneous carboxylic acids, alcohols, amines, sucrose, amino acids, quinaldic acid	XAD-2	–	–	Fresh, non saline	–	[11,75, 97,98]
Miscellaneous, alcohols, esters, aldehydes alkyl-benzenes, polynuclear hydrocarbons, chlorocompounds	XAD-4	–	–	Waste and potable	–	[99]
Hydrocarbons	XAD-2	–	–	Non saline	–	[13]
Miscellaneous alkanes, alkylbenzenes, haloforms, polynuclear hydrocarbons, ketones, alcohols, phenols, carboxylic acids	XAD-2	–	–	Potable	–	[8]
Miscellaneous alkanes, alkylbenzenes, chlorocompounds, alcohols, ketones, esters, bromocompounds	XAD-4	–	–	Potable waste	–	[9]
Miscellaneous alkylbenzenes, alkanes, phenols, carboxylic acids, chlorocompounds	XAD-2	–	–	Non saline	–	[7]

Table 3.13 continued

Substance	Resin	Method of desorption	Analytical finish	Type of water sample	Detection limit*	Ref.
Miscellaneous phenols, carboxylic acids, aldehydes, alcohols	XAD–2	–	–	Waste, non saline	–	[100]
Miscellaneous	Misc XAD	Solvent extraction	–	–	–	[101]
Miscellaneous	XAD–2	Solvent extraction	–	Potable	–	[102]
Miscellaneous	Misc. XAD	Solvent extraction	–	Trade effluents	–	[103]
Miscellaneous	Misc. XAD	–	–	Non saline	–	[104]
Miscellaneous	Misc. XAD	Thermal desorption	GLC	Potable	–	[8]
Miscellaneous	XAD–2	–	GLC, GLC–MS	Pulp effluent	–	[105]
Miscellaneous	XAD–2	–	–	River	–	[106]
Miscellaneous	XAD–4	Solvent extraction (CS$_2$)	GLC–MS	Surface waters	0.1 µg L^{-1}	[107]
Miscellaneous	XAD–2	–	–	Well water	mg L^{-1}	[75]
Miscellaneous	XAD–2	CS$_2$ extraction	GLC–MS	Surface water	–	[107]
	XAD–4					
Miscellaneous	XAD (Misc)	Solvent extraction	–	Non saline	–	[108]
Miscellaneous	XAD–1	–	–	Non saline	–	[69]
Miscellaneous	XAD–2	–	–	Non saline	–	[75]
Miscellaneous acidic, humic and neutral types	XAD–2	–	–	Non saline	–	[98]

Table 3.13 continued

Substance	Resin	Method of desorption	Analytical finish	Type of water sample	Detection limit*	Ref.
PCBs	XAD (Misc)	–	–	Non saline	–	[20,52,53]
Chlorinated insecticides, PCBs	XAD–2	–	–	Non saline	–	[38,50,51]
Organosulphur compounds	XAD–2, XAD–4	Solvent extraction	Headspace gas chromatography	Non saline	–	[84]
Mineral, animal and vegetable oils	XAD–2	–	–	Non saline	–	[109]
Miscellaneous	XAD (Misc)	–	–	Trade effluents	–	[103,110]
Alcohols, esters, aldehydes, alkylbenzenes, polyaromatic hydro-carbons and chlorocompounds	XAD–4	–	–	Non saline	–	[99]
Alkanes, alkylbenzenes, chlorocarbons, alcohols, ketones, esters, bromocompounds	XAD–4	–	–	Non saline	–	[9]
Mutagenic organic compounds	XAD (Misc)	Solvent extraction	–	Potable	–	[111]
Mutagenic organic compounds	XAD–2 XAD–4	–	–	Non saline	–	[112]

Source: Own files

Tateda and Fritz [10,99] studied the adsorption and acetone desorption of 30 different organic compounds on minicolumns of XAD–4 resin. The recovery of model compounds by using XAD–4 at concentrations of 2–10µg L^{-1} and 100µg L^{-1} was 89% and 83% respectively. The average standard deviation is 6.3%. Errors other than sorption and desorption would be included, like evaporation losses from the sample, decomposition, sorption on glassware, calibration of eluate volume and errors in the gas chromatography determination.

Tateda and Fritz [10,99] compared the adsorptive properties of XAD–4 resin with those of a spherical carbon molecular sieve of large surface area (Spherocarb). It was found that 100µL carbon disulphide will elute most organic compounds tested with Spherocarb except for phenols and some strongly sorbed compounds like naphthalene. At the 100µg L^{-1} concentration level average recoveries with Spherocarb are 77% from water and from 10% methanol. Under similar conditions average recovery with XAD–4 was 79% from water and 76% from 10% methanol. The average standard deviation was 5.1% for Spherocarb and 4.0% for XAD–4. Spherocarb has one major advantage over XAD–4 for analytical use, namely a stronger retention of low molecular weight polar organic compounds.

Cheng and Fritz [8] pointed out that the XAD–2 resin sorption method is a reliable and tested method for determining trace organic pollutants in potable water. However, the solvent evaporation step in this procedure results in a partial or complete loss of volatile compounds. To avoid these drawbacks a procedure was developed by them in which organic compounds that sorbed on a resin are thermally desorbed directly onto a gas chromatographic column for analysis. They desorbed the organics on to a Tenax tube from a tube containing XAD–2. This eliminates virtually all of the water entrained in the XAD–2 tube. Then the organics are thermally desorbed from the Tenax tube directly on to the gas chromatographic column. Recoveries were 88–90% (alkanes), 90–97% (alkylbenzenes), 70–80% (ethers), 86–95% (esters), 97–95% (haloforms), 97–98% (polyaromatic hydrocarbons), 82–90% (chlorobenzenes), 55–92% (ketones), 40–85% (alcohols) and 4–90% (phenols).

Workers at the Water Research Centre, UK [12] passed up to 5L of potable water samples through a column of XAD–2 resin and removed the adsorbed organics with 40µL of diethyl ether. The ether extract was then concentrated 2500 times and examined by gas chromatography–mass spectrometry. Several hundred organic compounds were examined in this survey, many of which were identified in potable water samples.

Hunt [113] has discussed the types of chemical contamination which might be associated with the use of polymeric XAD resins in the preconcentration of organic compounds in water. The types of contamination were classified as:

(1) residual monomers
(2) artifacts in the polymer synthetic pathway and
(3) chemical preservatives used to inhibit degradation.

Gibs [114] has shown that exhaustive solvent extraction of XAD resins did not completely eliminate resin artefacts in distilled water blanks.

Gibs and Suffett [115] have developed a base extraction procedure aimed at minimising deterioration in the performance of capillary chromatography columns used to analyse XAD extracts of water samples.

Vartainen et al. [111] compared XAD resin, liquid–liquid extraction and Blue Cotton adsorption for concentrating mutagenic organic compounds from water. The highest yields of these compounds were obtained by XAD adsorption and liquid–liquid extraction of the acidified samples.

Tian et al. [116] found that CAD–40 resin and XAD–2 resin exhibited similar recoveries of aromatic organic compounds of diverse functionality.

Le Bel et al. [117] have described a procedure utilising a large volume XAD–2 resin cartridge for the extraction of organic compounds in 1.5L of non saline water.

Baird et al. [118] have described a four column concentrator consisting of a series of columns packed with MP–1 anionic resin, MP–50 cationic resin, and XAD–2 and XAD–7 resins. The concentration recovered about 70% of hydrophobic organic compounds in a 500ml water sample. Hydrophillic organic compounds were not adequately recovered. The technique focused primarily on the removal of humic materials from water.

Junk and Richard [119] have compared columns packed with C_{18} bonded silica and XAD–2 resins and found use of the former preferable because they did not require as much solvent for efficient extraction.

Dalgnault et al. [4] reviewed the literature concerned with the use of XAD resins for isolating and preconcentrating organic compounds from non saline waters.

3.3 Tenax GC

The Tenax materials manufactured by Akzo Research Laboratories are a range of porous polymers based on various polymers. For example, Tenax GC is based on 2,6–diphenyl-*p*-phenylene oxide. After adsorption of the organics from a large volume of water on to the resin, desorption can be achieved either by solvent extraction or thermal elution. Organic porous polymers have recently become popular in concentrating air and water pollutants. Among these porous polymers Tenax GC has become widely accepted in air and water analysis. The sorption power of Tenax is nearly unaffected by water and repeated re-use of the sorbent is possible. Also, it has high thermostability (thermal elution) and its compatability with alcohols, amines, amides, acids and bases with good recovery characteristics make Tenax very suitable as sorbent medium in water analysis. A disadvantage, however, is small break through volumes for light hydrocarbons and low molecular weight polar compounds thus limiting the sample volumes. A solution to this problem is to use another more active sorbent in series with Tenax for collecting low molecular weight compounds breaking through on Tenax.

Organic compounds

3.3.1 Non saline waters

3.3.1.1 Semi-volatile organic compounds

Pankow *et al.* [120] proposed adsorption/desorption on a small bed of Tenax cartridges and direct desorption onto a fused–silica capillary for the preconcentration of a wide range of semi-volatile compounds in non saline waters.

3.3.1.2 Miscellaneous organic compounds

Other applications of Tenax to natural non saline waters include the preconcentration of chlorinated and organophosphorus insecticides [121,122], *p*-dichlorobenzene, hexachloro-1, 3-butadiene, 2-chloronaphthalene [123], organohalides [124], trace organic compounds in ground water [125], gasoline [109], phenols [120], alkyl phthalates [120], nitrobenzene [126], 2,4–dinitrophenol [126], traces of organic substances in rain water [127] and chlorinated insecticides and polychlorobiphenyls in non saline waters.

3.3.2 Seawater

3.3.2.1 Polychlorinated biphenyls and organochlorine and organophosphorus insecticides

Leoni [129] observed that in the extractive preconcentration of organochlorine insecticides and PCBs from surface and coastal waters in the presence of other pollutants such as oil, surface active substance etc., the results obtained with an absorption column of Tenax Celite are equivalent to those obtained with the continuous liquid–liquid extraction technique. For non saline waters that contain solids in suspension that absorb pesticides, it may be necessary to filter the water before extraction with Tenax and then to extract the suspended solids separately. Analyses of river and estuarine sea waters, filtered before extraction, showed the effectiveness of Tenax and the extracts obtained for the pesticide analysis prove to be much less contaminated by interfering substances than the corresponding extracts obtained by the liquid–liquid technique. Leoni *et al.* [128] showed that for the extraction of organic micropollutants such as pesticides and aromatic polycyclic hydrocarbons from waters, the recoveries of these substances from unpolluted waters (mineral and potable waters) when added at the level of 1μg L $^{-1}$ averaged 90%.

Water samples were passed through the peristaltic pump into the absorption column at a flow rate of about 3L h $^{-1}$. When the absorption was completed the pesticides were eluted with three 10ml volumes of diethyl ether, in such a way that the solvent also passed through the section of hose through which the water reached the column. Finally, the diethyl ether was dried over anhydrous sodium sulphate. The water container was washed with light petroleum to remove pesticide adsorbed on the glass and this solution, after concentration, was added to the column eluate. For the analysis of naturally polluted water the mixed diethyl ether and light petroleum extract was evaporated, the residue dissolved in light petroleum and the solution purified by partitioning with acetonitrile saturated with light petroleum [129,130]. The resulting solution was evaporated just to dryness, the residue dissolved in 1ml of *n*-hexane and insecticides and polychlorobiphenyls were separated into four fractions by deactivated silica gel microcolumn chromatography (silica gel type Grace 950, 60–200 mesh [131]). The various eluates from the silica gel were then analysed by gas chromatography [132]. In order to evaluate the effectiveness of extraction from non saline waters with the Tenax Celite column, the samples were also extracted simultaneously by the liquid–liquid technique.

In the adsorption with Tenax alone satisfactory results were obtained, while in the presence of mineral oil a considerable proportion of the organophosphorus pesticides (particularly Malathion and Parathion-methyl) was not adsorbed and was recovered in the filtered water. This drawback can be overcome by adding a layer of Celite 545 which, in order to prevent blocking of the column, is mixed with silanised glass wool plugs. A number of analyses of surface and estuarine sea waters were carried out by this modified Tenax column and simultaneously by the liquid–liquid extraction technique. To some of the samples taken, standard mixtures of pesticides were also added, each at the level of 1μg L $^{-1}$ (ie in concentration from 13 to 500 times higher than that usually found in the waters analysed). One recovery trial also specifically concerned polychlorobiphenyls. The results obtained in these tests show that the two extraction methods, when applied to surface waters that were not filtered before extraction, yielded very similar results for many insecticides, with the exception of compounds of the DDT series, for which discordant results were frequently obtained.

Leoni *et al.* [128,133] conclude that the extraction of insecticide from waters by adsorption on Tenax, yields results equivalent to those by the liquid–liquid procedure when applied to mineral, potable and surface waters that completely or almost completely lack solid matter in suspension. For waters that contain suspended solids that can adsorb some insecticides in considerable amounts, the results of the two methods are equivalent only if the water has previously been filtered. In these instances, therefore, the analysis will involve filtered water as well as the residue of filtration.

3.3.3 Potable water

3.3.3.1 Triazine herbicides

Lintelman *et al.* [134] described a semi-automatic high performance liquid chromatographic method to determine Desethyl–Atrazine, Simazine, Atrazine and Terbethylazine at 15–32µg L $^{-1}$ in potable water. A 100ml sample was concentrated on Tenax TA and introduced into the analytical column.

3.3.4 Waste waters

3.3.4.1 Polyaromatic hydrocarbons

Kadar *et al.* [135] studied the efficiencies of Tenax for the preconcentration of polyaromatic hydrocarbons from standard water solutions. The method was applied to waste water samples from an aluminium plant. In this method the water samples were passed through the Tenax column at the rate of about 5ml min $^{-1}$. Residual water was removed from the Tenax by passing nitrogen gas through the column. The Tenax material was then transferred to a Soxhlet apparatus. Polyaromatic hydrocarbons end other organic compounds were extracted by reflux for 4h with 35ml of acetone. The Tenax can then be dried and reused. Preliminary experiments indicate that the extraction time can be reduced to 10–15min by the use of ultrasonic extraction. Residual water is removed from the acetone extract by passing the solution through a small column of anhydrous sodium sulphate and washing with 5ml of acetone. The extract is evaporated on a water bath at 50°C in a stream of nitrogen to a volume of approximately 0.5ml.

The residual solution was transferred to a thin layer plate. The time required for the development is 30–35min. A single development of the plate is sufficient for separating polyaromatic hydrocarbons from other compounds such as paraffins, naphthenes, acids and phenols. The spots were located visually under an ultraviolet lamp and the areas containing aromatic compounds were marked. The appropriate portions of the thin layer were scraped into a glass flask. Polyaromatic hydrocarbons were extracted by vigorous shaking for 2h with 5ml of chloroform. Extraction of polyaromatic hydrocarbons from the thin layer material is a critical step in the procedure. Extraction into approximately 1ml of chloroform by treatment in an ultrasonic bath for about 10min was shown to be the best procedure. The internal standard was added to the chloroform in advance so that the samples can be analysed directly after centrifugation. After filtration or centrifugation, the solution may be used for spectrophotometric determination of total polyaromatic hydrocarbons and gas chromatographic determination of individual polyaromatic hydrocarbon components.

The conditions for the gas chromatography were as follows:

Column: glass capillary, 25m × 0.28mm coated with SE–30. Detection by flame ionisation detector.

Carrier gas: helium at a flow rate of 2ml min^{-1}.

Injector temperature: 300°C.

Detector temperature: 300°C.

Volume injected: 1μL.

Column temperature: initially 100°C held for 4min before programme starts:

Programming: 3°C min^{-1}.

Final temperature: 260°C.

Kadar *et al.* [135] found that the separation of polyaromatic hydrocarbons from water on the Tenax column was the most critical step in the analytical procedure (Fig. 3.6). Unsatisfactory results are often due to neglect of details at this step. The most important operating parameters are the height of the Tenax column and the flow rate of the water sample through the column. In this work, the height of the column was set at 10cm without any appreciable decrease in polyaromatic hydrocarbon recovery. The optimum flow rate was 8–10ml min^{-1} for a Tenax column of 10cm height and 13mm diameter. The polyaromatic hydrocarbon values obtained by gas chromatography are presented in Table 3.14.

As indicated in Table 3.14 excellent agreement is obtained at the 100μg L^{-1} ml^{-1} level between the method and a much more time consuming liquid–liquid extraction technique. At the 10μg L^{-1} level the recovery decreases to 70–90%. At or below the 10μg L^{-1} level, the amount of polyaromatic hydrocarbons recovered may be increased by using a larger volume of water for instance 5 or 10L.

Table 3.15 shows results obtained for waste water sample from an aluminium plant. Parallel determinations were carried out on samples of 500ml each. The standard deviation for the series is approximately 1μ.g L^{-1}.

Unused Tenax may contain small amounts of low molecular weight components which can interfere with polyaromatic hydrocarbon components of the group chrysene, benzofluoranthene, benzo(*e*)pyrene on a thin layer chromatographic plate. It is therefore recommended that unused Tenax be extracted with acetone overnight. The acetone extract of

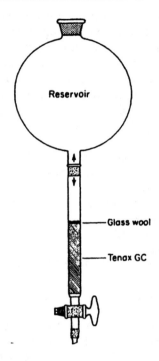

Fig. 3.6 Sorption column with Tenax bed
Source: Reproduced with permission from Kadar, R. et al. [135] Elsevier Science, UK

Table 3.14 PAH yield from water samples

Hydrocarbon	Amount ($\mu g\ L^{-1}$)	Liquid–liquid extraction[a] ($\mu g\ L^{-1}$)	Kadar method[135] undiluted sample ($\mu g\ L^{-1}$)	±sd[b]	Kadar method[135] 10 times diluted sample $\mu g\ L^{-1}$	±sd
Anthracene	11.0	10.3	10.2	0.5	0.8	0.1
Pyrene	32.0	30.8	30.2	0.7	2.6	0.1
Chrysene	43.0	41.3	40.3	0.6	3.2	0.3
3–Methylcholanthrene	12.0	10.9	11.0	0.3	0.9	0.1
Total	98.0	93.3	91.7		7.5	
Total yield		95	94		77	

[a]Glass capillary column gas chromatography method
[b]All standard deviations ($\mu g\ L^{-1}$) are based on 4 parallel results

Source: Reproduced with permission from Kadar, R. [135] Elsevier Science, UK

Table 3.15 Typical results for waste water from an aluminium plant

| Substance | Sample 1 | | Sample 2 | |
	Run 1 (μg L^{-1})	Run 2 (μg L^{-1})	Run 1 (μg L^{-1})	Run 2 (μg L^{-1})
Biphenyl	3.2	3.6	<1	<1
Fluorene/fluorenone	6.3	8.4	2.9	3.2
Phenanthrene	16.9	23.1	14.2	14.0
Anthracene	2.8	2.8	1.1	1.2
Fluoranthene	20.8	18.9	10.8	12.4
Pyrene	15.3	12.7	5.6	6.0
Benzo(a)fluorene	3.2	3.4	1.6	1.5
Benzo(b)fluorene	2.8	3.0	1.3	1.3
Benzdiphenylensulphide	3.4	3.5	2.0	1.7
Benzo(a)anthracene	2.5	2.8	5.6	5.5
Chrysene/triphenylene	5.8	6.0	15.6	16.0
Benzo(b,k)fluoranthene	6.8	6.8	38.1	38.0
Benzo(e)pyrene	2.6	2.7	16.2	16.4
Benzo(a)pyrene	1.3	1.5	7.0	7.4
Dibenzoanthracene	3.4	4.3	8.2	8.0
Anthanthrene	>1	>1	3.2	3.2
Coronene	>1	>1	1.9	2.0
Dibenzopyrene	>1	>1	4.0	4.3

Source: Reproduced with permission from Kadar, R. [135] Elsevier Science, UK

unused Tenax gives rise to fluorescent spots on the thin layer plate with a retention factor of about 0.4. Since several polyaromatic hydrocarbons components have about this retention factor, these artefacts will interfere in the thin layer procedure. Judging from their intensity, this interference will be small at the 100μg L^{-1} level but may become dominant at lower concentrations. These artefacts will also interfere in the gas chromatographic analysis.

When analysing industrial waste water, it is advisable to protect the Tenax column against irreversible contamination. This can be achieved by using a short Celite 545 prefilter. The thin layer plate must be pre-conditioned in chloroform before use. This preconditioning is particularly important when the polyaromatic hydrocarbon level is very low, 10μg L^{-1} or less. One dimensional thin layer chromatography will give a good separation of semi-polar polyaromatic hydrocarbon components from non-polar paraffins and naphthenes which will follow the solvent front, and from the more polar acids and phenols which will remain at the bottom of the plate. Fig. 3.7 shows a thin layer chromatogram of four polyaromatic hydrocarbon components. Under ultraviolet light the chromatographic spots may serve as a visual indication of the amount of

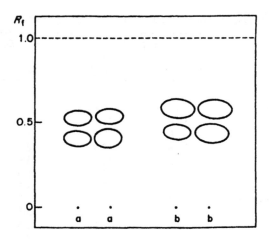

Fig. 3.7 Thin layer chromatogram of PAH on Kiesegel 60 F254 Merck; (a) sorbed on Tenax and extracted into acetone (initial concentration corresponds to 25µg of PAH per spot); (b) initial acetone solution 25µg of PAH per spot
Source: Reproduced with permission from Kadar, R. *et al.* [135] Elsevier Science, UK

polyaromatic hydrocarbon. In Fig. 3.7 it is noticeable that the total area of the chromatographic spots on the left hand side (a) of the plate is less than the area to the right (b). Both sets of chromatographic spots correspond to the same initial amount of polyaromatic hydrocarbons (50µg). The spots to the left, however, correspond to polyaromatic hydrocarbon components which have passed through the analytical procedure and thus have suffered a loss of about 10%. This loss is visible on the plate.

Organometallic compounds

3.4 Tenax GC

3.4.1 Seawater

3.4.1.1 Organotin compounds

Brinckmann and co-workers [136] used a gas chromatographic method with or without hydride derivatisation for determining volatile organotin compounds (eg tetramethyltin), in seawater. For non-volatile organotin compounds a direct liquid chromatographic method was used. This system employs a 'Tenax–GC' polymeric sorbent in an automatic purge and trap (P/T) sampler coupled to a conventional glass column gas chromatograph equipped with a flame photometric detector. Fig. 3.8 is a schematic of the P/T–GC–FPD assembly with typical operation

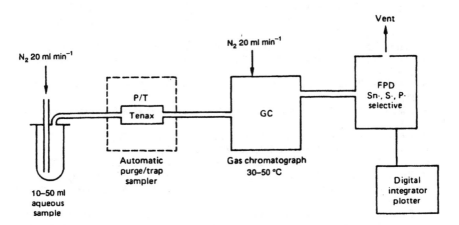

Fig. 3.8 The purge/trap GC–FPD system and operating conditions
Source: Reproduced with permission from Brinckmann, F.E. [136] Kluwer/Plenum
Academic Publications, New York

conditions. Flame conditions in the FPD were tuned to permit maximum response to SnH emission in a H–rich plasma, as detected through narrow band–pass interference filters (610 ± 5nm) [137]. Two modes of analysis were used:

(1) volatile stannanes were trapped directly from sparged 10–50ml water samples with no pretreatment; and

(2) volatilised tin species were trapped from the same or replicate water samples following rapid injection of aqueous excess sodium borohydride solution directly into the P/T sparging vessel immediately prior to beginning the P/T cycle [138].

Jackson *et al.* [138] devised trace speciation methods capable of ensuring detection of tin species along with appropriate preconcentration and derivatisation without loss, decomposition, or alteration of their basic molecular features. They describe the development of a system employing a Tenax GC filled purge and trap sampler, which collects and concentrates volatile organotins from water samples (and species volatilised by hydrodisation with sodium borohydride), coupled automatically to a gas chromatograph equipped with a commercial flame photometric detector modified for tin specific detection [137,139,140].

References

1 Wun, C.K., Rho, J., Walker, R.W. and Litsky, W. *Water, Air and Soil Pollution*, **11**, 173 (1979).

2 American Public Health Association, Standard Methods for the Examination of Water and Wastewater, p. 1007, American Public Health Association, Inc., New York (1976).
3 Wun, C.K., Walker, L.W. and Litsky, W. *Water Research*, **10**, 955 (1976).
4 Daignault, S.A., Noot, K.K., Williams, D.T. and Huck, P.M. *Water Research*, **22**, 803 (1988).
5 Wigilius, B., Boren, H., Carlberg, G.E. *et al. Journal of Chromatography*, **391**, 169 (1987).
6 Dietrich, A.D., Millington, D.S. and Seo, Y.H. *Journal of Chromatography*, **436**, 229 (1988).
7 Stephan, S.F. and Smith, J.F. *Water Research*, **11**, 339 (1977).
8 Cheng, R.C. and Fritz, J.S. *Talanta*, **25**, 659 (1978).
9 Ryan, J.P. and Fritz, J.S. *Journal of Chromatographic Science*, **16**, 448 (1978).
10 Tateda, A. and Fritz, J.S. *Journal of Chromatography*, **152**, 392 (1978).
11 Junk, G.A., Richard, J.J., Grieser, D. *et al. Journal of Chromatography*, **99**, 745 (1974).
12 Fielding, M., Gibson, T.M., James, H.A., McLoughlin, L. and Steel, C.P. In Water Research Centre Technical Report TR 159. Organic micropollutants in drinking water. February (1981).
13 Stephan, S.F., Smith, J.F., Flego, U. and Renkens, J. *Water Research*, **12**, 447 (1978).
14 Benoit, F.M., Label, G.L. and Williams, D.T. *International Journal of Analytical Chemistry*, **6**, 277 (1979).
15 McNeil, E.E., Olson, R., Miles, W.F. and Rajabalee, R.J.M. *Journal of Chromatography*, **132**, 277 (1977).
16 Junk, G.A. and Richard, J.J. *Analytical Chemistry*, **60**, 451 (1988).
17 Wun, C.K., Walker, L.W. and Litsky, W. *Health Laboratory Science*, **15**, 67 (1978).
18 Nickless, G. and Jones, P. *Journal of Chromatography*, **156**, 87 (1978).
19 Musty, P.R. and Nickless, G. *Journal of Chromatography*, **89**, 185 (1974).
20 Musty, P.R. and Nickless, G. *Journal of Chromatography*, **120**, 369 (1976).
21 Heatley, N.G. and Page, E.J. *Water Sanitation*, **3**, 46 (1952).
22 Rosen, M.J. and Goldsmith, H.A. In *Systematic Analysis of Surface Active Agents*, 2nd edn., Wiley, New York (1972).
23 Jones, P. and Nickless, G. *Journal of Chromatography*, **156**, 99 (1978).
24 Nadeau, H.G. and Waszeciak, P.H. In *Non-ionic Surfactants*, Marcel Dekker, New York, p. 906 (1967).
25 Patterson, S.J., Hunt, E.C. and Tucker, K.B.E. *Journal of Proceedings of Institute of Sewage Purification*, **190**, 00 (1966).
26 Paschal, D.C., Bicknell, R. and Dresbach, D. *Analytical Chemistry*, **49**, 1551 (1977).
27 Coburn, J.A., Valdamanis, J.A. and Chau, A.S.Y. *Journal of Association of Official Analytical Chemists*, **60**, 224 (1977).
28 Berkane, K., Caissie, G.E. and Mallet, V.N. In Proceedings of Symposium on Fenithrothion, NRC (Canada) Assoc. Comm. Sci. Crit. Environ. Quality, Report 16073 NRCC, Ottawa, 95 (1977).
29 Daughton, G.C., Crosby, D.G., Garnos, R.L. and Ssieh, J. *Journal of Agricultural and Food Chemistry*, **24**, 236 (1976).
30 Mallet, V.N., Brun, G.L., MacDonald, R.N. and Berkane, K. *Journal of Chromatography*, **160**, 81 (1978).
31 Berkane, K., Caissie, G.E. and Mallet, V.N. *Journal of Chromatography*, **139**, 386 (1977).
32 Zakrevsky, J.G. and Mallet, V.N. *Journal of Chromatography*, **132**, 315 (1977).
33 Volpe, G.G. and Mallet, V.N. *International Journal of Environmental Analytical Chemistry*, **8**, 291 (1980).

34 Puijker, L.M., Veenendaal, G., Jaonssen, H.M.J. and Griepin, B. *Fresenius Zeitschrift für Analytische Chemie*, **306**, 1 (1981).
35 Frobe, Z., Drevenkar, V., Stengl, B. and Stefanac, Z. *Analytica Chimica Acta*, **206**, 299 (1988).
36 Tolosa, A., Redman, J.W. and Mee, L.D. *Journal of Chromatography*, **725**, 93 (1996).
37 Akhing, B. and Jensen, S. *Analytical Chemistry*, **42**, 1483 (1970).
38 Richards, J.F. and Fritz, J.S. *Talanta*, **21**, 91 (1974).
39 Levesque, D. and Mallet, V.N. *International Journal of Environmental Analysis*, **16**, 139 (1983).
40 Narange, A.S. and Eadon, G. *International Journal of Environmental Analytical Chemistry*, **11**, 167 (1982).
41 Harris, R.L., Huggett, R.J. and Stone, H.D. *Analytical Chemistry*, **52**, 779 (1980).
42 Sundaram, K.M.S., Szeto, S.Y. and Hindle, R. *Journal of Chromatography*, **177**, 20 (1979).
43 Rees, G.A.V. and Au, L. *Bulletin of Environmental Toxicology*, **22**, 561 (1979).
44 Tolosa, I., Redman, J.W. and Mee, L.D. Private communication.
45 Sundaram, K.M.S., Szeto, S.Y. and Hindle, R. *Journal of Chromatography*, **177**, 29 (1979).
46 Ripley, B.D., Hall, J.A. and Chau, A.S.Y. *Environmental Letters*, **7**, 97 (1974).
47 Grift, N. and Lockhart, W.L. *Journal of Association of Official Analytical Chemists*, **57**, 1282 (1974).
48 NRC Associate Committee on Scientific Criteria for Environmental Quality, Fenithrothion: The Effects of its use on Environmental Quality and its Chemistry, NRCC No. 14104, p. 106 (1975).
49 Elder, D. *Marine Pollution Bulletin*, **7**, 63 (1976).
50 Harvey, G.R. In Report US Environment Protection Agency, EPA–R2–73–177, 32 pp. (1973).
51 Niederschulte, U. and Ballschmiter, K. *Fresenius Zeitschrift für Analytische Chemie*, **269**, 360 (1974).
52 Chriswell, C.D., Ericson, R.L., Junk, G.A. et al. *Journal of the American Waterworks Association*, **69**, 669 (1977).
53 Lawrence, J. and Tosine, H.M. *Environmental Science and Technology*, **10**, 381 (1976).
54 Le Bel, K. and Williams, D.T. Bulletin of Environmental Contamination and Toxicology, **24**, 397 (1980).
55 Norsdsij, A., Van Beveren, J. and Brandt, A. H_2O, **17**, 242 (1984).
56 Musty, P.R. and Nickless, G. *Journal of Chromatography*, **100**, 83 (1974).
57 Le Bel, G.L., Williams, D.T., Ryan, J.J. and Lau, B.P.Y. *Chlorinated Dioxins, Dibenzofurans, Perspectives*. Eds. C. Rappe, G. Choudhary, L.H. Keith, and M.L. Lewis Chelsea, pp. 329–341 (1986).
58 Gomez-Belinchon, J.I., Grimalt, J.O. and Albaiges, J. *Environmental Science and Technology*, **22**, 677 (1988).
59 Green, D.R. and Le Pape, D. *Analytical Chemistry*, **59**, 699 (1987).
60 Martinsen, K., Krigstad, A. and Carlberg, G.E. *Water Science and Technology*, **20**, 13 (1988).
61 Renberg, L. *Analytical Chemistry*, **60**, 1836 (1978).
62 Jones, L.K. *Chemical Engineering News*, **54**, 35 (1976).
63 Ligon, V.V. and Johnson, R.J. *Analytical Chemistry*, **48**, 481 (1976).
64 Malalyandi, M., Wightman, R.H. and La Ferriere, C. *Adv. Chem. Ser. (Org. Polut. Water)*, **214**, 163 (1987).
65 Richard, J.J. and Fritz, J.S. *Journal of Chromatographic Science*, **18**, 35 (1980).
66 Audu, A.A. *Acta Chim. Hungary*, **124**, 603 (1987).

67 Synak, R., Zarbska, I. and Kentzer, T. *Biol. Plant,* **28**, 412 (1988).
68 Gustafson, R.L. and Paleos, J. Interactions responsible for the selective adsorption of organics on organic surfaces in *Organic Compounds in Aquatic Environments,* eds. S.J. Faust and J.V. Hunter, Marcel Dekker, New York, pp. 213–237 (1971).
69 Riley, J.P. and Taylor, D. *Analytica Chimica Acta,* **46**, 307 (1969).
70 Osterroht, C. Kiel, *Meeresforsch,* **28**, 48 (1972).
71 Picer, N. and Picer, M. *Journal of Chromatography,* **193**, 357 (1980).
72 Holden, A.V. and Marsden, K. *Journal of Chromatography,* **44**, 481 (1969).
73 Synder, D.E. and Reinert, R.E. *Bulletin of Environmental Contamination and Toxicology,* **6**, 385 (1971).
74 Picer, M. and Abel, M. *Journal of Chromatography,* **150**, 1191 (1978).
75 Burnham, A.K., Calder, G.V., Fritz, J.S. *et al. Analytical Chemistry,* **44**, 139 (1972).
76 Ahling, B. and Jenson, S. *Analytical Chemistry,* **42**, 1483 (1970).
77 Harvey, G.R. In *Absorption of chlorinated hydrocarbons from seawater by a crosslinked polymer.* Woods Hole Oceanographic Institute, Woods Hole, Massachusetts, published manuscript (1972).
78 Dawson, R., Riley, J.P. and Tennant, R.H. *Marine Chemistry,* **4**, 83 (1976).
79 Shinohara, R., Kido, A., Okomoto, Y. and Takeshita, R. *Journal of Chromatography,* **256**, 81 (1983).
80 Stuermer, D.H. and Harvey, G.R. *Nature (London),* **250**, 480 (1974).
81 Stuermer, D.H. and Harvey, G.R. *Deep Sea Research,* **24**, 303 (1977).
82 Mantoura, R.F.C. and Riley, J.P. *Analytica Chimica Acta,* **76**, 97 (1975).
83 Wun, C.K., Walker, R.W. and Litsky, W. *Health Laboratory Science,* **15**, 67 (1978).
84 Przyazny, A. *Journal of Chromatography,* **346**, 61 (1988).
85 Kadar, R., Nagy, K. and Fremstad, D. *Talanta,* **27**, 277 (1980).
86 Renberg, L. *Analytical Chemistry,* **50**, 1836 (1978).
87 Sjostrom, L., Radestrom, R., Carlberg, G.E. and Kringstad, A. *Chemosphere,* **14**, 1107 (1985).
88 Anon, *Chemical Engineering News,* **54**, 35 (1976).
89 Kissinger, L.D. In Report No IS–T–845, US National Technical Information Service, Springfield, Virginia, VA., 161 pp. (1979).
90 Kissinger, L.D. and Fritz, L.S. *Journal of the American Water Works Association,* **68**, 435 (1976).
91 Renberg, L. *Analytical Chemistry,* **50**, 1836 (1978).
92 Prater, W.A., Simmons, H.C. and Mancy, K.M. *Analytical Letters,* **13**, 205 (1980).
93 Noll, K.E. and Gounaris, A. *Water Research,* **22**, 815 (1988).
94 Chau, Y.K. and Fox, M.E. *Journal of Chromatographic Science,* **9**, 271 (1971).
95 Oliver, B.G. and Bothen, K.D. *Analytical Chemistry,* **52**, 2066 (1980).
96 Cabrident, R. and Solika, A. *TSM L'Eau,* 285, July (1975).
97 Fritz, J.S. *Industrial. Engineering Produce Development Research,* **14**, 95 (1975).
98 Leenheer, J.A. and Huffman, E.W.D. *Journal of Research, US Geological Survey,* **4**, 737 (1976).
99 Tateda, A. and Fritz, J.S. *Journal of Chromatography,* **152**, 382 (1978).
100 Ishangir, L.M. and Samuelson, O. *Analytica Chimica Acta,* **100**, 53 (1978).
101 More, R.A. and Karasek, F.W. *International Journal of Environmental Analytical Chemistry,* **17**, 187 (1984).
102 James, H.A., Steel, C.P. and Wilson, I. *Journal of Chromatography,* **208**, 89 (1981).
103 Webb, R.G. National Technical Information Service, Springfield, Virginia, PB 245674, 22 pp. (1975).

104 Chudaba, J., Chlebkova, E. and Tucek, F. *Vodni Hospodarstvi*, **27B**, 236 (1977).
105 Fox, M.E. *Journal of the Fisheries Research Board of Canada*, **34**, 798 (1977).
106 Ishiwatari, R., Hamara, H. and Machihara, T. *Water Research*, **14**, 1257 (1980).
107 De Groat, R. *H₂O*, **12**, 333 (1979).
108 Noordsij, A. *H₂O*, **12**, 167 (1979).
109 Belkin, F. and Hable, M.A. *Bulletin of Environmental Contamination and Toxicology*, **40**, 244 (1988).
110 Fox, M.E.J. *Fisheries Res. Bd. Canada*, **34**, 798 (1977).
111 Vartainen, T., Liimtianene, A., Jaaskelainen, S. and Kaurenen, P. *Water Research*, **21**, 773 (1987).
112 Koal, H.J., Van Krelft, C.F. and Veriaan-De Vries, M. *Advanced Chemistry Series (Organic Pollution Water)*, **214**, 605 (1987).
113 Hunt, G. *Advances in Chemistry Series (Org. Pollution Water)*, **214**, 247 (1987).
114 Gibs, J., Brenner, L., Cognet, L. and Suffett, I.H. *Advances in Chemistry Series (Org. Pollution Water)*, **214**, 267–294 (1987).
115 Gibs, J. and Suffett, I.H. *Advanced Chemistry Series (Org. Pollut. Water)*, **214**, 327 (1987).
116 Tian, J., Zhao, F., Zheng, Q., Su, T. and Ryan, L. *Xui Kuanging Kexue*, **8**, 77 (1987).
117 Le Bel, G.L., Williams, D.T. and Benoit, F.M. *Advanced Chemistry Series (Organic Pollution Water)*, **214**, 309 (1987).
118 Baird, R.B., Jacks, C.H. and Nelsess, L.B. *Advanced Chemistry Series (Organic Pollution Water)*, **214**, 557 (1987).
119 Junk, G.A. and Richard, J.J. *Journal of Research of the National Bureau of Standards, US*, **93**, 274 (1988).
120 Pankow, J.F., Ligocid, M.P., Rosen, M.E., Isabelle, L.M. and Hart, K.M. *Analytical Chemistry*, **60**, 40 (1988).
121 Agostiano, A., Caselli, M. and Provenzano, M.A. *Water, Air and Soil Pollution*, **19**, 309 (1983).
122 Tator, V., Popl, M. *Fresenius Zeitschrift für Analytische Chemie*, **322**, 419 (1985).
123 Parkow, J.F. and Isabelle, L.M. *Journal of Chromatography*, **237**, 25 (1982).
124 Sekerka, L. and Lechner, J.F. *International Journal of Environmental Analytical Chemistry*, **11**, 43 (1982).
125 Jankow, J.F., Isabelle, L.M., Hewetsen, J.P. and Cherry, J.A. *Ground Water*, **23**, 775 (1985).
126 Patel, S.C. *Analytical Letters*, **21**, 1397 (1988).
127 Pankow, J.F., Isabelle, L.M. and Asher, W.E. *Environmental Science and Technology*, **18**, 310 (1984).
128 Leoni, V., Pucelti, G. and Grella, A.J. *Journal of Chromatography*, **106**, 119 (1975).
129 Leoni, V. *Journal of Chromatography*, **62**, 63 (1971).
130 Johnston, L.V. *Journal of Association of Official Analytical Chemists*, **48**, 668 (1965).
131 Claeys, R.R. and Inman, R.D. *Journal of Association of Official Analytical Chemists*, **57**, 399 (1974).
132 Leoni, V. and Pucetti, G. *Journal of Chromatography*, **43**, 388 (1969).
133 Leoni, V., Pucetti, G., Columbo, R.J. and Oviddo, O. *Journal of Chromatography*, **125**, 399 (1976).
134 Lintelman, J., Mengel, C. and Kettrup, A. *Fresenius Zeitschrift für Analytische Chemie*, **346**, 752 (1993).
135 Kadar, R., Nagy, K. and Fremstad, D. *Talanta*, **27**, 227 (1980).
136 Brinckmann, F.E. In *Trace Metals in Seawater*, Proceedings of a NATO Advanced Research Institute on Trace Metals in Seawater, 30/3–3/4/81, Sicily, Italy. Eds. C.S. Wong *et al.* Plenum Press, New York (1981).

137 Aue, W.A. and Flinn, C.S. *Journal of Chromatography*, **142**, 145 (1977).
138 Jackson, J.A., Blair, W.R., Brinckmann, F.E. and Iverson, W.P. *Environmental Science and Technology*, **16**, 110 (1982).
139 Huey, C., Brinckmann, F.E., Grim, S. and Iverson, W.P. In *Proceedings of the International Conference on the Transport of Persistent Chemicals in Equatic Ecosystems*, eds A.S.W. Freilas, D.J. Kushner and D.S.U. Quadri, National Research Council of Canada, Ottawa, Canada, pp. 11–73–11–78 (1974).
140 Nelson, J.D., Blair, W. and Brinckmann, F.E. *Applied Microbiology*, **26**, 321 (1973).

Chapter 4

Organics: Cation exchange resins

In addition to the non-polar macroreticular Rohm and Hass and Amberlite XAD resins (XAD–2 and XAD–4) discussed in Chapter 3 a wide range of polar resins exist which are very useful for the preconcentration of anionic and cationic species and have some applications in the preconcentration of ionic organic substances. Intermediate and highly polar types of resins are commonly referred to as ion exchange resins. These may be subdivided into cationic types (cation exchange resins) which are discussed in this chapter and anionic types (anion exchange resins) which are discussed in Chapter 5. Cationic exchange resins carry a negative charge and this reacts with positively charged cationic species. Anionic ion exchange resins carry a positive charge and this reacts with negative anionic species.

Strong acid cation exchange resins manufactured by the sulphonation of polystyrene or polydivinyl benzenes undergo the following reaction with cations:

$$\text{Resin } SO_3^-H^+ + M^+X^- \rightarrow \text{Resin } SO_3^- M^+ + H^+ + X^-$$

Weak acid cation exchange resins manufactured, eg by the polymerisation of methacrylic acid undergo the following reaction with cations:

$$\text{Resin } COO^- H^+ + M^+X^- \rightarrow \text{Resin } COO^- M^+ + H^+ + X^-$$

Some basic properties of the various types of cation exchange resins available and their suppliers are tabulated in Table 4.1.

Preconcentration is achieved by. passing a large volume of water sample, suitably adjusted in pH and reagent composition down a small column of the resin. The adsorbed ions are then desorbed with a small volume of a suitable reagent in which the metals or metal complexes or anionic species dissolve. This preconcentrated extract can then be analysed by any suitable means.

Table 4.1 Properties of cation exchange resins

Resin type	Functional group	Water content (approx)* (g g⁻¹ dry resin)	Exchange capacity (approx)* (mol equiv g⁻¹ dry resin)	Packing density (approx)* (g ml⁻¹)	Regeneration	Washing of salt forms	Trade volumes of some commercial examples
Strong acid types	$-SO_3H$	0.7	4 at all pH values	0.8	Excess strong acid	Stable	(c) Dowex 50 Dowex 50W–X8 Dowex 50W–X4 Dowex A1 (b) Amberlite IR–120 Amberlite GC–120 Amberlite XAD–12 (a) Zeocarb 225 Cationite KB–4P–2
Weak acid types	$-COOH$	1	9–10 at high pH	0.7	Readily regenerated	Cation slowly hydrolyses off	(b) Amberlite IRC–50 (a) Zeocarb 226 (c) Dowex XAD–7 Dowex XAD–8

*Depends on grade and does not necessarily include recently developed resins available from (a) Permutit Co., London W4; (b) Rohm and Haas Co., Philadelphia, USA; (c) Dow Chemical Co., Midland, Michigan, USA

Source: Own files

Organic compounds

4.1 Non saline and other waters

4.1.1 Amino acids

Free amino acids in surface water have been preconcentrated on Amberlite IR120 resin and converted into the n-butyl-N-trifluoroacetyl esters (for gas liquid chromatography) argenine, cystine and lutidene were not eluted from an OV–175 column, and coelution of glycine, serine, leucine and isoleucine was observed [1].

4.1.2 Miscellaneous organic compounds

Gardner and Lee [2] preconcentrated free and combined amino acids from lake water on an ion exchange column, prior to desorption with a small volume of acid, conversion to N-trifluoroacetyl methyl esters, and gas chromatography. In determinations of organic impurities in waste water, eg in industrial effluents, treatment with a cation exchange resin in free acid form is often used to remove metal cations and cationic organic solutes such as amino acids. Possible losses of compounds such as fatty acids and aromatic compounds have been discussed [3–8].

Jahangir and Samuelson [9,10] have discussed the sorption of cyclohexane derivatives with a hydroxyl or carbonyl group separated by one or two methylene groups from aqueous media by these resins. They found that these compounds adsorbed more strongly than the corresponding aromatic compounds both on sulphonated styrene–divinylbenzene cation exchange resins and on non ionic styrene–divinylbenzene resins. These observations and the lower temperature dependence observed for the cyclohexane derivatives indicate that hydrophobic interactions have a marked influence on the adsorption.

Kaczvinsky et al. [11] give details of the use of a cation exchange resin (polystyrene–divinylbenzene type) for preconcentration of trace organic compounds from water. Neutral and acid compounds can be eluted from the resin by washing with methanol and ethyl ether; basic compounds can be eluted with the same solvents after treatment of the resin with ammonia gas. Over 50 organic bases were recovered from water at 1mg L^{-1} to 50μg L^{-1} levels at rates of over 85% for most compounds studied. Nielen et al. [12] used a strongly acidic cation exchange resin for on-line preconcentration of polar anilines in water. The method could be automated and a detection limit for the nine anilines examined corresponding to 0.02–0.5μg L^{-1} was obtained in river water samples.

Capacity factors have been evaluated [13] for the adsorption of organic solutes on Amberlite XAD–8 resin. Some other applications of cation

Table 4.2 Applications of intermediate polarity (XAD–7, XAD–8) and high polarity (XAD–12) cation exchange resins to preconcentrations of organics from various types of water

Substance	Resin	Method of desorption	Analytical finish	Detection limit*	Type of water sample	Ref.
Miscellaneous	XAD–7	–	–	$mg\ L^{-1}$	Well water	[14]
Miscellaneous	XAD–8	CS_2 extraction	GLC–MS	$0.1\mu g\ L^{-1}$	Surface waters	[15]
Miscellaneous	XAD–7	Solvent extraction	GLC	–	Non saline	[16]
Miscellaneous	XAD–7	Diethyl ether/methanol extraction	GLC	–	Non saline	[17]
Fenitrothion	XAD–7	Solvent extraction	GLC	0.5ng	Non saline	[18]
Aminocarb	XAD–7	Solvent extraction	GLC with N–P detector	–	Non saline	[19]
Urea (as indophenol)	XAD–7	Aqueous extraction	–	–	Non saline	[20]
Alkylethoxylated sulphates	XAD–7	Methanolic hydrochloric acid	Derivatisation to alkylbromides–GLC	–	Waste and surface waters	[21]
Phenols	Ambersorb XE–340	Diethyl ether or dichloromethane	GLC	$0.5\mu g\ L^{-1}$	Non saline	[22]
Nitrosamines	Ambersorb XE–340	Solvent extraction	GLC	–	Non saline	[23]
Alkylbenzene sulphonates	XAD–8	Methanol	^{13}C NMR	–	Sewage effluents	[24]

Table 4.2 continued

Substance	Resin	Method of desorption	Analytical finish	Detection limit*	Type of water sample	Ref.
Phenols	XE34	Diethylether	–	–	Non saline	[22]
Naphthalene, anthracene	XAD–7	Solvent extraction	–	–	Non saline	[25]
3-phenyl 4-hydroxy 6-chloropyridazine	XAD–7	Solvent extraction	–	–	Ground water	[26]
3,4 dichloropropanide, 5–ethyl-N-hexa-methylene–imino carbamate (Yolan)	KU28 KU23	–	–	–	Non saline	[27]
Humic and fulvic acids	–	–	–	–	Non saline	[38]

Source: Own files

exchange resins to the preconcentration of organics from water samples are listed in Table 4.2.

4.2 Seawater

4.2.1 Amino acids

Although free amino acids are present only at very low concentrations in oceanic waters, their importance in most biological systems has led to an inordinate amount of effort toward their determination in seawater. A sensitive, simple and easily automated method of analysis, the colorimetric ninhydrin reaction, has been known in biochemical research for many years. In order for the method to be useful in seawater, the amino acids had to be concentrated. This concentration was usually achieved by some form of ion exchange [28].

Ligand exchange was used as a concentrating mechanism by Clark et al. [29] followed by TLC for the final separation. The formation of 2,4-dinitro–1–fluorobenzene derivatives, followed by solvent extraction of these derivatives and circular TLC was suggested by Palmork [30]. Ligand exchange has been a favoured method for the concentration of amino acids from solution because of its selectivity [31]. Gardner [32], isolated free amino acids at the 20nmol L $^{-1}$ level in from as little as 5ml of sample, by cation exchange, and measured concentrations on a sensitive amino acid analyser equipped with a fluorometric detector.

The classical work of Dawson and Pritchard [33] on the determination of α-amino acids in seawater uses a standard amino acid analyser modified to incorporate a fluorometric detection system. In this method the seawater samples are desalinated on cation exchange resins and concentrated prior to analysis. The output of the fluorometer is fed through a potential divider and low-pass filter to a compensation recorder. Dawson and Pritchard [33] point out that all procedures used for concentrating organic components from seawaters, however mild and uncontaminating, are open to criticism, simply because of the ignorance as to the nature of these components in seawater. It is, for instance, feasible that during the process of desalting on ion exchange resins under weakly acidic conditions metal chelates dissociate and thereby larger quantities of 'free' components are released and analysed.

4.3 Waste waters

4.3.1 Phenolic compounds

Lee et al. [34] separated phenolic compounds in industrial waste water on a column of Dowex 1–XS resin (Cl-form) using a 0.05m aqueous cupric chloride solution of ethanol, methanol or propanol as eluant.

Compounds with low dissociation constants (eg ε-aminophenol) were only weakly adsorbed on the column, whereas strongly ionised compounds (such as salicylic and 4–aminosalicylic acids) were strongly adsorbed. Some compounds (eg catechol) form complexes with cupric copper ions in the eluant so that they are eluted more quickly than the others.

4.3.2 Chlorolignosulphonic acids

Van Loon et al. [35] used XAD–8 resin to extract chlorolignosulphonic acids and liganosulphonic acids from pulp mill effluents.

4.3.3 Alkyl ethoxylated sulphates

Preconcentration in XAD–7 resin has been used in the case of alkyl ethoxylated sulphates followed by derivativisation to alkyl bromides and gas chromatography to determine these substances in surface and waste waters [21,24]. Alkylbenzene sulphonate surfactants have been preconcentrated on XAD–8 resin and desorbed with methyl alcohol prior to examination by ^{13}C nuclear magnetic resonance spectrometry.

Organometallic compounds

4.4 Non saline waters

4.4.1 Organotin compounds

Neubert and Andreas [36] preconcentrated tri- and di-butyltin species from water on a cation exchange column, then desorbed them with a small volume of diethyl ether–hydrogen chloride prior to their conversion to methyltin species and gas chromatographic analysis.

4.5 Seawater

4.5.1 Organoarsenic compounds

Dimethylarsinic acid is preconcentrated on a strong cation exchange resin [37] by optimising the elution parameters, dimethylarsinic acid can be separated from other arsenicals and sample components, such as group I and II metals, which can interfere in the final determination. Graphite furnace atomic absorption spectrometry is used as a sensitive and specific detector for arsenic. The described technique allows for dimethylarsinic acid to be determined in sea water (20ml) containing 10^5-fold excess of inorganic arsenic with a detection limit of 0.02µg L $^{-1}$.

References

1 Benonfelln, F. and Gold, A. *Trib. Cebedeu*, **39**, 23 (1988).
2 Gardner, W.S. and Lee, G.E. *Journal of Environmental Science and Technology*, **7**, 719 (1973).
3 Bhattacharyya, S.S. and Das, A.K. *Atomic Spectroscopy*, **9**, 68 (1988).
4 Kingston, H. and Pella, P.A. *Analytical Chemistry*, **53**, 223 (1981).
5 Samuelson, S.V. *Kem. Tidskr.*, **54**, 170 (1942).
6 Vaisman, A. and Yampolskaya, M.M. *Zavod. Lab.*, **156**, 621 (1950).
7 Seki, T. *Journal of Chemical Society of Japan Pure Chemistry Section*, **75**, 1297 (1954).
8 Nomura, N., Hirski, S., Yamada, M. and Shiho, D. *Journal of Chromatography*, **59**, 373 (1971).
9 Jahangir, L.M. and Samuelson, O. *Analytica Chimica Acta*, **85**, 103 (1976).
10 Jahangir, L.M. and Samuelson, O. *Analytica Chimica Acta*, **100**, 53 (1978).
11 Kaczvinsky, J.R., Saitoh, K. and Fritz, J.S. *Analytical Chemistry*, **55**, 1210 (1983).
12 Nielen, M.W.F., Frei, R.W. and Brinkman, U.A.T. *Journal of Chromatography*, **317**, 557 (1984).
13 Thurman, E.M., Malcolm, R.L. and Aiken, G.R. *Analytical Chemistry*, **50**, 775 (1978).
14 Burnham, A.K., Calder, G.V., Fritz, J.S. *et al. Analytical Chemistry*, **44**, 139 (1972).
15 De Groot, R. *H₂O*, **12**, 333 (1979).
16 Stephan, S.F. and Smith, J.F. *Water Research*, **11**, 339 (1977).
17 Stephan, S.F., Smith, J.F., Flego, U.L. and Renkers, J. *Water Research*, **12**, 447 (1978).
18 Volpe, G.G. and Mallet, N.N. *International Journal of Environmental Analytical Chemistry*, **8**, 291 (1980).
19 Levesque, D. and Mallet, V.N. *International Journal of Environmental Analysis*, **16**, 139 (1983).
20 Moreno, P., Sanchez, E., Pons, A. and Palon, A. *Analytical Chemistry*, **58**, 585 (1986).
21 Neubecker, T.A. *Environmental Science and Technology*, **19**, 1232 (1985).
22 Cavelier, C., Gilber, M., Vivien, L. and Lamblin, P. *Revue Francais des Sciences de l'eau*, **3**, 19 (1984).
23 Kimoto, W.I., Dooley, C.J., Carre, J. and Fiddler, W. *Water Research*, **15**, 109 (1981).
24 Thurman, E.M., Willoughby, T., Barber, L.B. and Thorn, K.A. *Analytical Chemistry*, **59**, 1798 (1987).
25 Wigglius, B., Boren, H., Carleberg, G.E. *et al. Journal of Chromatography*, **391**, 169 (1987).
26 Auer, W. and Mallesa, H. *Chromatographia*, **25**, 817 (1988).
27 Tsitovich, I.K., Kuz'menko, E.A., Demidenko, O.A. and Shlorova, I.I. *Gidrokhim. Mater*, **97**, 116 (1987).
28 Semenov, A.D., Ivleva, I.N. and Datsko, V.G. *Trudy Komiss, Anal. Khim. Akad, Nauk. SSSR*, **13**, 162 (1964).
29 Clark, M.E., Jackson, G.A. and North, W.J. *Limnology and Oceanography*, **17**, 749 (1972).
30 Palmork, K.H. *Acta Chimica Scandanavia*, **17**, 1456 (1963).
31 Siegal, A. and Degens, E.T. *Science*, **151**, 1098 (1966).
32 Gardner, W.S. *Marine Chemistry*, **6**, 15 (1975).
33 Dawson, R. and Pritchard, R.G. *Marine Chemistry*, **6**, 27 (1978).

34 Lee, K.S., Lee, D.W. and Chung, Y.S. *Analytical Chemistry*, **45**, 396 (1973).
35 Van Loon, W.N., Boon, J.J., De Jong, R.J. and De Groot, B. *Environmental Science and Technology*, **27**, 332 (1993).
36 Neubert, G. and Andreas, H. *Fresenius Zeitschrift für Analytische Chemie*, **280**, 31 (1976).
37 Persson, J.A. and Irgum, K. *Analytica Chimica Acta*, **138**, 111 (1982).
38 Boening, P.H., Beckmann, P. and Snoeyink, V.I. *Journal of the American Water Works Association*, **72**, 54 (1980)

Chapter 5

Organics: Anion exchange resins

Strong base anion exchange resins are manufactured by chloromethylation of sulphonated polystyrene followed by reaction with a tertiary amine:

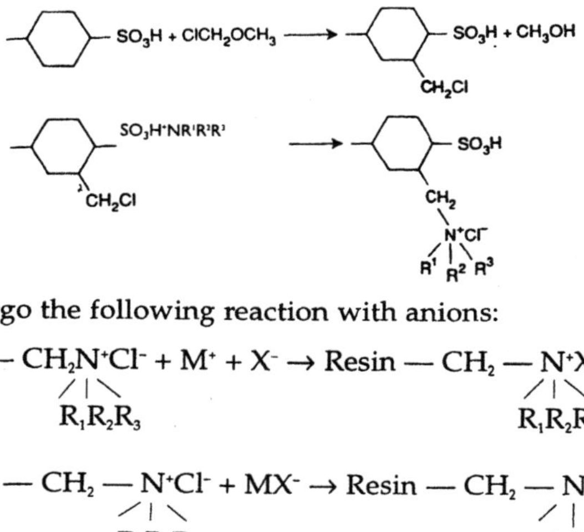

They undergo the following reaction with anions:

$$\text{Resin} - CH_2\underset{R_1R_2R_3}{N^+}Cl^- + M^+ + X^- \rightarrow \text{Resin} - CH_2 - \underset{R_1R_2R_3}{N^+}X^- + Cl^- + M^+$$

or

$$\text{Resin} - CH_2 - \underset{R_1R_2R_3}{N^+}Cl^- + MX^- \rightarrow \text{Resin} - CH_2 - \underset{R_1R_2R_3}{N^+}MX^- + Cl^-$$

where MX^- is a metal containing anion.

Weak base anion exchange resins are manufactured by chloromethylation of sulphonated polystyrene followed by reaction with a primary or secondary amine:

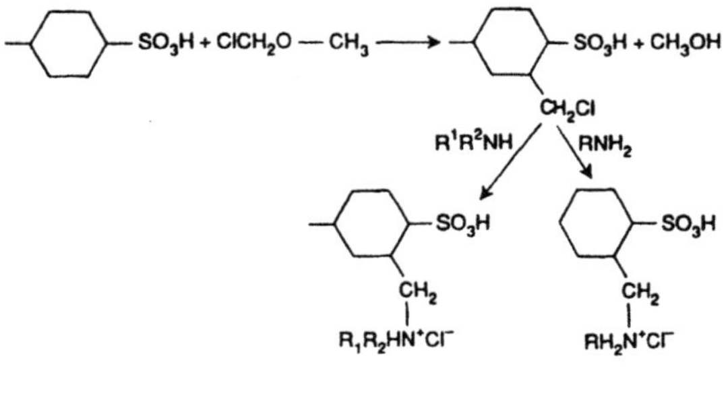

They undergo the following reaction with anions:

eg   Resin — N$^+$Cl$^-$ + X$^-$ + M$^+$ → Resin — N$^+$X$^-$ + Cl$^-$ + M$^+$
         ╱ | ╲   ╱ | ╲
         R^{1}R^{2}H   R^{1}R^{2}H

or

      Resin N$^+$Cl$^-$ + MX$^-$ → Resin — N+MX$^-$ + Cl$^-$
         ╱ | ╲   ╱ | ╲
         R^{1}R^{2}H   R^{1}R^{2}H

where MX$^-$ is a metal containing anion. Anionic exchange resins carry a positive charge and this reacts with negatively charged or anionic organic species.

Some basic properties of the various types of anion exchange resins and suppliers are tabulated in Table 5.1.

Organic compounds

5.1 Non saline waters

5.1.1 Carboxylic acids

Haddad and Jackson [1] have discussed the preconcentration of low molecular weight organic acids in non saline waters using an anion exchange column.

5.1.2 Nitriloacetic acid

Nitriloacetic acid has been preconcentrated from water at pH3 by passage through an anion exchange column and subsequent elution with sodium chloride [2] prior to estimation by polarography. Longbottom [3] preconcentrated nitriloacetic acid on a strong anion exchange resin, then desorbed it with sodium tetraborate at pH9, prior to estimation by ultraviolet spectroscopy.

5.1.3 Humic and fulvic acids

Weber and Wilson [4] used anion and cation exchange resins to preconcentrate fulvic and humic acids from water. Adsorption rates of various organic compounds on this gel have been studied by Tombo et al. [5] (G–15 gel), Hiraide et al. [6] (Sephadex A25) and Hine and Birsill [7]. Hiraide et al. [6] studied complexes of humic and fulvic acid with divalent copper, lead and cadmium in pond water. The negatively charged humic substances were selectively and quantitatively sorbed from 1L samples on a small column of the macroreticular, weak-base, anion exchanger, diethylaminoethyl–Sephadex A–25. The complexed metals were then quantitatively desorbed with a small volume of 4M nitric acid and

Table 5.1 Properties of anion exchange resins

Resin type	Functional group	Water content (approx)[a] (g g^{-1} dry resin)	Exchange capacity (approx)[a] (mol equiv g^{-1} dry resin)	Packing density (approx)[a] (g ml^{-1})	Regeneration	Washing of salt forms	Trade names of some commercial examples
Strong base types	Quaternary ammonium $-CH_2NR_3 + Cl^-$	1	4 at all pH values	0.7	Excess strong base	Stable	(c) Dowex 1 Dowex 2 Dowex AG1-X2 Dowex AG1-X8 Dowex AG1-X4 (b) Amberlyst PI-27 Amberlite IRA-400 Amberlite GC-400 Amberlite IRA-410 (a) Deacidite FF Lewatite M5080 Biorad AG1 Biorad 140-AG1-X2 Biorad X8
Weak base types	Secondary or primary amine $-CH_2-N^+HR_2Cl^-$ or $-CH_2-N^+H_2R\ Cl^-$	0.3	4 at low pH values	0.7	Readily regenerated with sodium carbonate	Anion slowly hydrolyses	(b) Amberlite IR-45 (a) Deacidite G

[a]Depends on grade and does not necessarily include recently developed resins available from (a) Permitit Co., London W4; (b) Rohm and Haas Co., Philadelphia, USA; (c) Dow Chemical Co., Midland, Michigan, USA

Source: Own files

determined by graphite furnace atomic absorption spectrometry. The method was simple, rapid and gave reproducible and reliable results. About 80% of the humic substances sorbed on the Sephadex column could be eluted with 0.5M sodium hydroxide and determined spectrophotometrically at 400nm. Boening et al. [8] studied the use of styrene divinylbenzene and acrylic ion exchange resins for the removal of humic substances from water.

Hezzlar [9] studied the effect of inorganic salts on the adsorption of organic substances on Sephadex gel from samples of peat–bog water, river water and secondary sewage effluent. Most of the organic compounds possessed pH-dependent reversible adsorption affinity to the gel. This effect was attributed to the presence of inorganic salts; the possible mechanism involved is discussed.

5.1.4 Aldicarb insecticide

Condo and Janauer [10] give details of a method of preconcentrating and determining traces of Aldicarb in water. The method involves decomposition of Aldicarb by passage through a strongly basic anion exchange resin to produce an oximate which was adsorbed on the resin. Subsequent addition of sulphuric acid resulted in in situ formation and elution of hydroxylamine, which was used for quantitative reduction of trivalent iron to the ferrous state; this was then determined spectrophotometrically.

5.1.5 Organophosphorus insecticides

Insecticides containing PH_3 after acid hydrolysis to free methyl-phosphoric acids have been preconcentrated on anion exchange resins prior to desorption, conversion to their dimethyl esters and estimation by gas chromatography [11].

5.1.6 α, α dichloropropionate, 2,3,6 trichlorobenzoate, 2–methoxy 3,6–dichlorobenzoate and 2,4D

Tsitovich et al. [12] used the anion exchange resin AV–17–8, a highly alkaline monofunctional polymer (NMe_3^+) in the chloride form to pre-concentrate α, α dichloropropionate, 2,3,6 trichlorobenzoate, 2-methoxy-3,6 dichlorobenzoate and 2,4D from non saline waters.

5.2 Seawater

5.2.1 Organophosphorus insecticides

Minear [13] and Minear and Walonski [14] used Sephadex gel to preconcentrate naturally occurring soluble organophosphorus compounds from seawater.

5.3 Surface waters

5.3.1 Nitriloacetic acid

Galassi *et al.* [15] have reported a procedure for the determination of nitriloacetic acid in surface waters involving concentration on an ion exchange column, recovery of the analyte from the exchange column, formic acid esterification and gas chromatography analysis of the butylesters. Concentrations down to 0.71µg L^{-1} were detected.

5.4 Potable water

5.4.1 Paraquat and Diaquat

In a method [16] for the preconcentration of Paraquat and Diaquat from potable water the sample is passed through an ion exchange column, followed by desorption, reduction with sodium dithionite and measurement of the reduced forms at 390nm for Paraquat and 379nm for Diaquat.

Organometallic compounds

5.5 Non saline waters

5.5.1 Organomercury compounds

Organomercury compounds have been preconcentrated by conversion to tetrachloromercury derivative $HgCl_4^{2-}$ and passage through a filter disc loaded with SB–2 anion exchange resin [17]. Mercury was then determined in the resin by neutron activation analysis. Ahmed *et al.* [18] preconcentrated methylmercury and inorganic mercury from rain water on an anion exchange column prior to analysis by cold vapour atomic absorption spectrometry. Methylmercury was determined by passing a 500ml sample (in 5% hydrochloric acid) through an anion exchange column on which ionic mercury was retained. Methylmercury passing the column was decomposed by ultraviolet and the methylmercury concentration determined by difference. Recovery of ionic mercury increased as acid concentration increased and with ultraviolet irradiation.

References

1 Haddad, P.R. and Jackson, P.E. *Journal of Chromatography*, **447**, 155 (1988).
2 Haberman, J.P. *Analytical Chemistry*, **43**, 63 (1971).
3 Longbottom, J.E. *Analytical Chemistry*, **44**, 418 (1972).

4 Weber, J.H. and Wilson, S.A. *Water Research*, **9**, 1079 (1975).
5 Tombo, N., Kameri, T., Nishimura, T. and Fukushi, K. *Japan Water Works Association*, **No 532**, 37 (1979).
6 Hiraide, E.M., Tillkeratue, K., Otsuka, K. and Mizuike, A. *Analytica Chimica Acta*, **172**, 215 (1985).
7 Hine, P.T. and Bursill, D.B. *Water Research*, **18**, 1461 (1984).
8 Boening, P.H., Beckmann, P. and Snoeyink, V.I. *Journal of the American Waterworks Association*, **72**, 54 (1980).
9 Hezzlar, J. *Water Research*, **21**, 1311 (1987).
10 Condo, D.P. and Janauer, G.E. *Analyst (London)*, **112**, 1027 (1987).
11 Verweij, J.A., Regenhardt, C.E.A. and Boter, H.L. *Chemosphere*, **8**, 115 (1970).
12 Tsitovich, I.K., Kuz'menko, E.A., Demidenko, O.A. and Shlorova, I.I. *Gidrokhim. Vlater*, **97**, 116 (1987).
13 Minear, R.A. *Environmental Science and Technology*, **6**, 431 (1972).
14 Minear, R.A. and Walonski, K.I. In *Investigation of the Chemical Identity of Soluble Organophosphorus Compounds found in Natural Waters* Research Report, UILUWRC–74–0086 (1974).
15 Galassi, A., Rosencrance, A.B. and Bruejgemann, E.E. Reprint USAMBROL–TR 8601 20pp. (1988).
16 HMSO. Methods for the examination of Waters and Associated Materials 1988. Determination of Diaquat and Paraquat in River and Drinking Water. Spectrophotometric methods. HMSO, London (1987).
17 Becknell, D.E., Marsh, R.H. and Allie, W. *Analytical Chemistry*, **43**, 1230 (1971).
18 Ahmed, R., May, K. and Stoeppler, M. *Fresenius Zeitschrift für Analytische Chemie*, **326**, 510 (1987).

Chapter 6

Organics: Organic and inorganic solid adsorbents and chelators

6.1 Sulphydryl cotton fibre

6.1.1 Non saline waters

6.1.1.1 Organomercury compounds

Lee [1] preconcentrated methyl- and ethyl–mercury in sub-nanogram amounts from non saline waters on to sulphydryl cotton fibre. Sulphydryl cotton fibre was prepared by soaking cotton for 4 to 5d in a mixture of thioglycolic acid, acetic anhydride, acetic acid, sulphuric acid, and water at 40 to 45°C. Adsorbed mercury compounds were eluted from the cotton fibre with hydrochloric acid/sodium chloride and extracted with benzene. Using gas chromatography with electron capture detection, the detection limits were 0.04ng L^{-1} using a 20L sample volume. Precision was approximately 20%. The method was applied to snow and freshwater samples of varying humic acid content. Methylmercury concentrations varied between 0.09 and 0.22ng L^{-1}, and recoveries of spiked methylmercury varied between 42 and 68% and were strongly correlated with the concentration of humic substances. Methylmercury concentration in snow was 0.28ng L^{-1} and the recovery was 79%.

In further work, Lee and Mowrer [2] preconcentrated methylmercury from non saline waters on to a sulphydryl cotton fibre adsorbent using the column technique or the batch–column two-stage technique. A small volume of 2M hydrochloric acid was used to elute methylmercury and to separate it from inorganic mercury; 0.4–0.6ml of benzene was used to extract methylmercury from the eluate. Analysis was performed by capillary gas chromatography with electron–capture detection. The detection limit for methylmercury was <0.05ng L^{-1} in a 4L water sample. Four surface waters were analysed to test the agreement of methylmercury concentration between the two preconcentration methods, and to test the interference of humic substances on the filtered and unfiltered surface water. The methylmercury concentrations found in different surface water samples ranged from 0.08 to 0.48ng L^{-1}.

6.2 Cellulose acetate

6.2.1 Non saline waters

6.2.1.1 Chlorinated insecticides and PCBs

Kurtz [3], Musty and Nickless [4] and Mantoura et al. [5] have used cellulose triacetate to preconcentrate polychlorinated biphenyls.

6.2.1.2 Miscellaneous organics

Jenkins et al. [6] observed that significant sorption losses occurred on cellulose acetate and on Gelman Acro LCI5 from aqueous and organic solvents (for trinitrotoluene, 2,4 DNT, RDX and HMX).

6.3 Cellulose

6.3.1 Non saline waters

6.3.1.1 Urea herbicides

Some of the urea herbicides have been preconcentrated on cellulose columns following hydrolysis to amines [7].

6.4 Chromosorb W

6.4.1 Non saline waters

6.4.1.1 Polychlorobiphenyls

Chromosorb W has been used to preconcentrate polychlorinated biphenyls prior to gas chromatographic analysis [8]. Ahling and Jensen [9] used a column of Carbomex 4000 monostearate/Chromosorb W to selectively preconcentrate chlorinated insecticides from samples also containing polychlorinated biphenyls. The insecticides were then desorbed with a small volume of organic solvent prior to analysis by gas chromatography.

6.5 Chromosorb 105

6.5.1 Non saline waters

6.5.1.1 Phenols

Pilipenko et al. [10] have reported that Chromosorb 105 is the most effective sorbent for the preconcentration of phenols in non saline waters.

6.6 Lipidex 5000

6.6.1 Non saline waters

6.6.1.1 Polychlorophenols

Pentachlorophenol has been preconcentrated from non saline water on a Lipidex 5000 bed [11]. Pentachlorophenol was eluted with acetone and simultaneously separated from lipophilic compounds. It was analysed by gas chromatography with electron capture detection after derivatisation to pentachlorophenol acetate. Less polar compounds could also be determined. For example, 1,1–bis(4–chlorophenyl)–2,2,–trichloroethane (p,p'-DDT) and 2,3,7,8–tetrachlorodibenzo-p-dioxin (TCDD) were collected in a separate fraction from the gel. Average recovery of 0.1–0.15ng pentachlorophenol per ml water was 96% and that of labelled p,p'-DDT at 5µg L^{-1} and TCDD at 0.001µg L^{-1} were 95 and 92%, respectively.

6.7 Support bonded silicones

6.7.1 Non saline waters

6.7.1.1 Organochlorine insecticides

Aue et al. [12] used support bonded silicones for the preconcentration of 10–20ng L^{-1} of Lindane, Heptachlor, Aldrin, Heptachlor epoxide and Dieldrin prior to gas chromatography. This absorbent was packed on to a glass tube (35 × 1cm) and 10L of sample (or of pure water treated with known compounds) was passed through the column at 50–55ml min^{-1}; the column was then dried by passage of nitrogen. The sorbed compounds were eluted with pentane (2 × 5ml) and a portion of the elute was injected directly into a borosilicate glass column (170cm × 3.5mm) packed with 1.5% of QF–1 plus 2% of OV–17 on Chromosorb W–HP (100–120 mesh). The column was operated at 185°C for chlorinated insecticides with nitrogen (60ml min^{-1}) as carrier gas and a ^{63}Ni electron capture detector.

6.8 Polytetrafluoroethylene (Chromosorb T)

6.8.1 Non saline waters

6.8.1.1 Linear alkyl benzene sulphonates

Miliotis et al. [13] used a porous PTFE membrane impregnated with a water immiscible organic solvent to extract linear alkylbenzenesulphonates at low ppb concentrations from non saline waters. These compounds were then made to form ion pairs with a tertiary amine and the ion pairs were transported through the membrane.

Josefson *et al.* [14] investigated the adsorption of trace organic compounds from water by Chromosorb T, an aggregate of polytetrafluoroethylene, in column chromatography. The differences in adsorption capacity between Chromosorb T and another polytetrafluoroethylene aggregate, Fluoropak 80, was attributed to differences in surface morphology. The adsorption of 33 organic solutes in aqueous solution was studied by frontal chromatography and the capacity was shown to be inversely related to solubility of solute for polycyclic aromatic hydrocarbons with the exception of the xanthines. The use of Chromosorb T columns to recover hydrophobic solutes from synthetic hard water containing such solutes was found to be effective.

6.9 Polyethylene and polypropylene

6.9.1 Non saline waters

6.9.1.1 Chlorinated insecticides

Thin polyethylene film has been employed for the preconcentration of traces of chlorinated insecticides from water. Weil [15,16] used 20–25µm thick film, and presents results for adsorption of γ-BHC, Heptachlor epoxide, Methoxychlor, Dieldrin, and DDT. Experiments [17] with ^{14}C–Lindane and ^{14}C–DDT showed that adsorption of these substances on polyethylene film was not likely to be of practical use in river water analyses because of the effects of foreign matter.

6.9.1.2 Miscellaneous organics

Low surface area polypropylene has been evaluated [18] as an adsorbent for preconcentrating aqueous organic compounds. An extraction process was developed and the adsorbent capacity of polypropylene was tested with 19 compounds (polycyclic aromatic hydrocarbons, ketones, esters, alcohols, alkanes). Results indicated an inverse relation between solubility and retention on the adsorbent. The surface micro environment of the polymer was investigated by using pyrene as a molecular fluorescence probe.

6.10 Polysorb–1

6.10.1 Mineral waters

6.10.1.1 Miscellaneous organics

Korenman *et al.* [19] preconcentrated organics from mineral waters by passage through a bed of Polysorb–1 presaturated with 1.5–2.0 parts of *n*-amyl acetate per weight of sorbent. The extracted organics were eluted with diazotised sulphanilic acid and determined photometrically. Infrared and ultraviolet spectroscopy, gas chromatography and luminescence analysis were all used to detect and identify extracted organics.

6.11 Poropak

6.11.1 Trade effluents

6.11.1.1 Nitro compounds

A method has been described [20] for the preconcentration of nitro-compounds in munitions works effluents by adsorption on Poropak resins. Following desorption with acetone the nitro-compounds (nitro-amines, nitrotoluenes and nitroaliphatics) were measured by high performance liquid chromatography at a gold/mercury electrode. The results of analysis of munitions in test samples show that Poropak resins were superior in performance to RDX resins. Analysis time was approximately 2h per sample for isolation and quantification. Detection limits of $1\mu g \ L^{-1}$ were approached.

6.12 Carbopak B

6.12.1 Non saline waters

6.12.1.1 Volatile organics

Mosesman et al. [21] found that a trap containing 200mg of Carbopak B backed with 50mg of carbon molecular sieves, traps and releases volatile organic compounds better than or equal to those recommended in the US Environmental Protection Agency method. These adsorbents were found to have limited water retention and low bleed levels during thermal desorption.

6.13 Separalyte

6.13.1 Non saline waters

6.13.1.1 2,3,7,8 tetrachlorodibenzo-p-dioxin

O'Keefe et al. [22] described a cartridge containing the reverse phase adsorbent Separalyte for the preconcentration of $pg \ L^{-1}$ levels of 2,3,7,8 tetrachlorodibenzo-p-dioxin in non saline waters.

6.14 CN bonded SPE

6.14.1 Non saline waters

6.14.1.1 Chlorinated insecticides

Russo et al. [23] have shown that CN bonded solid phase extraction cartridges efficiently extracted chlorinated pesticides at µg L^{-1} concentrations from non saline waters.

6.15 Dowex in [Co(SCN)$_4$]$^{2-}$ form

6.15.1 Non saline waters

6.15.1.1 Tween K100 and Tween 20

Gorenc et al. [24] preconcentrated Tween K100 and Tween 20 on Dowex resin in the [Co(SCN)$_4$]$^{2-}$ form and determined these substances spectrophotometrically in aqueous solution using potassium zinc thiocyanate (K$_2$Zn(SCN)$_4$). Recoveries of 90% were reported.

6.16 Glass beads

6.16.1 Non saline waters

6.16.1.1 Pentachloronitrobenzene and trichloroacetic acid

Baykut and Aroguz [25] used glass beads corroded with hydrochloric acid as adsorbents for pentachloronitrobenzene and trichloroacetic acid.

6.17 Graphite fluoride

6.17.1 Non saline waters

6.17.1.1 Miscellaneous organics

Yao and Zlatkis [26] have shown that graphite fluoride is a useful adsorbent for trace analysis of non saline waters due to its large breakthrough volume.

6.18 Cyclohexyl bonded phase resin

6.18.1 Groundwater

6.18.1.1 Creosote and pentachlorophenol

Rostad et al. [27] used a cyclohexyl bonded phase resin column to isolate creosote and pentachlorophenol in ground water samples taken at a waste site.

6.19 Apiezon L

6.19.1 Non saline waters

6.19.1.1 Dimethylbenzene

Apiezon L (APL) coated on to the surface of a nickel wire has been reported as an effective means of preconcentrating dimethylbenzene from non saline waters [28]. The wire was subjected to Curie-point pyrolysis–gas chromatographic analysis. A precolumn effectively trapped Apiezon L.

6.20 Silicic acid

6.20.1 Non saline waters

6.20.1.1 Phenols

Columns containing silica gel chemically linked to different groups have been compared for their suitability for preconcentration of selected phenols from dilute aqueous solution [29]. The chemically bonded phases comprised octyl, octadecyl, phenyl, diol and cyanide radicals of which the phenol-bonded silica gel gave the best recoveries following reversed-phase high performance liquid chromatographic analysis. Recoveries for *m*-cresol, *p*-chlorophenol, 2,5–dimethylphenol, 2,6–dichlorophenol and *o*-phenylphenol were all in excess of 90%, when the water sample was first acidified to pH2 and 30g sodium chloride per 100ml added before passage through the column. For phenol, however, a recovery of only 42% was achieved under these conditions.

6.20.2 Surface and coastal waters

6.20.2.1 Chlorinated insecticides

Sackmauereva *et al.* [30] used columns filled with silicic acid–Celite to separate organochlorine insecticides from PCBs. The PCBs were eluted with petroleum ether. To elute insecticides from the column they used a mixture of acetonitrile and hexane and methylene chloride.

6.21 Potassium aluminium silicate

6.21.1 Non saline waters

6.21.1.1 Phenyl urea herbicides

Aluminosilicates (Partisil ODS) have been used for the preconcentration of substituted phenylurea herbicides in non saline and waste waters [31].

6.21.1.2 Mineral, animal and vegetable oils

Belkin and Hable [32] used zeolites to preconcentrate mineral, animal and vegetable oils from non saline waters.

6.21.2 Potable waters

6.21.2.1 Polyaromatic hydrocarbons

Cannavacinolo *et al.* [33] in an attempt to improve analytical sensitivity, evaluated a sorbent material for preconcentrating PAHs from potable water. The material is a natural aluminium potassium silicate, thermally treated to increase its surface area and processed with silicone oil to make it hydrophobic, which also makes the sorption process a partition one. High pressure liquid chromatography was used to analyse samples passed through a 5cm bed of the material. Samples containing 5µg L^{-1} PAHs were eluted with *n*-pentane as follows.

Aliquots (200cm^3) of samples after percolation through the column were extracted with three equal portions (3 × 20ml) of *n*-pentane in a separating funnel. Each *n*-pentane eluate was dried by passing through a column of granular anhydrous sodium sulphate (2g). Then the sodium sulphate columns were washed with *n*-pentane (5ml). The combined eluates were then evaporated, after adding the internal standard (4–methylpyrene) with a rotary evaporator (1ml). Chromatography was carried out by the EPA method 14 on a column of octadecylsilane (Michrosorb RP–18 Merk) using absolute methanol as mobile phase. Detection was achieved using a photometric detector set at 254mm.

Fluoranthene is the compound least retained by the sorption column (62% recovery). The sorbent efficiency is extremely dependent on the pollution present in the water samples. For waters containing surfactants, the use of aluminium potassium silicate absorbent is not possible.

6.21.2.2 Aldehydes

Zeolite 25 H–5 has been used for the preconcentration of low molecular weight aldehydes as their 2,4–dinitrophenylhydrazones in potable water [35]. In the latter method the 2,4–dinitrophenylhydrazones were then separated by liquid chromatography and determined by gas chromatography using flame ionisation detection. Results are presented for the separation and detection of a range of aldehydes and ketones in aqueous solution at concentrations of 100mg L^{-1}. Recoveries approximated to 100% for model compounds tested, with the exception of formaldehyde.

6.22 Molecular sieves

6.22.1 Non saline waters

6.22.1.1 Mineral, animal and vegetable oils

Belkin and Hable [32] and Uichiyama [35] used molecular sieves to pre-concentrate mineral, animal and vegetable oils from non saline waters.

Adsorption on a molecular sieve 5A is a technique that has been examined for the preconcentration of oil in water samples. Uichiyama [35] gives details of a procedure for the separation and determination of mineral oil, animal oil and vegetable oil in water. After extraction with carbon tetrachloride, the extract is treated with molecular sieve 5A, on which animal and vegetable oils are adsorbed. The oil is then determined by infrared analysis.

6.23 Carbograph–4

6.23.1 Non saline waters

6.23.1.1 Polar organic compounds

Crescenzi et al. [36] used Carbograph materials to extract organic polar compounds from 4L samples of non saline waters thereby achieving a very useful preconcentration factor. These workers compared results obtained using Carbograph–4 with those obtained using other carbon-based materials.

6.23.2 Potable waters

6.23.2.1 Urea based herbicides

Di Corcia et al. [37] have discussed preconcentration methods for trace analysis of sulphonyl urea herbicides in potable water.

Trifensulphuron methyl, Metsulphuron methyl, Trisulphuron, Chloro-sulphuron, Rimusulphuron, Tribenzuron methyl and Bensulphuron methyl were extracted from water by off-line solid-phase extraction with a Carbograph–4 cartridge. Sulphonylurea herbicides were then isolated from both humic acids and neutral contaminants by differential elution. Analyte fractionation and quantification were performed by liquid chromatography with ultraviolet detection. Recoveries of sulphonylurea herbicides extracted from 4L of potable water (10ng L^{-1} spike level), 2L of ground water (50ng L^{-1} spike level) and 0.2L of river water (250ng L^{-1} spike level) were not lower than 94%. Depending on the particular sulphonylurea herbicides, method detection limits were 0.6–2ng L^{-1} in drinking water, 2.9ng L^{-1} in ground water and 13–40ng L^{-1} in river water. A preservation study of sulphonylurea herbicides stored on the

Carbograph–4 cartridge was conducted. Over 2 weeks of cartridge storage, no significant analyte loss was observed when the cartridge was kept frozen. Comparing this method with one using a C_{18} extraction cartridge the former appeared to be superior to the latter in terms of sensitivity and, chiefly, of selectivity. This method involves confirmatory analysis by liquid chromatography–electrospray–mass spectrometry instrumentation equipped with single–quadruple mass filter. Mass spectrometer data acquisition was performed by a time scheduled three-ion selected ion-monitoring programme. The necessary structure-significant fragment ions were obtained by controlled decomposition of sulphonylurea herbicides adduct ions after suitably adjusting the electrical field in the desolvation chamber. Under three-ion selected ion monitoring condition, limits of detection (S/N = 3) calculated from the ion current profiles of those fragment or parent ions giving the lowest S/N values ranged from between 0.5ng (Tribenzuron methyl) and 3ng (Metsulphuron methyl, Thifensulphuron methyl) injected into the liquid chromatographic column.

6.24 Thermotrap TA

6.24.1 Non saline waters

6.24.1.1 Polyaromatic hydrocarbons

Ghaoui [38] compared Thermotrap TA an organic adsorbent, with a variety $(C_8–C_{18})$ of reversed phase adsorbents normally used to preconcentrate polynuclear aromatic hydrocarbons from aqueous samples. An aliquot of aqueous solution containing 0.4mg L^{-1} each of fluorene, anthracene, phenanthrene, flouranthene, pyrene, chrysene, and benzo(a)pyrene in 2% aqueous tetrahydrofuran was passed, with suction, through each of the adsorbent cartridges. Thermotrap TA gave the highest recoveries of all phases tested for fluorene (98%), anthracene (92%), phenanthrene (less than 40%), fluoranthene (73%), and pyrene (70%) and better recoveries than $C_8–C_{18}$ for chrysene. However, lower recoveries than the phenyl sorbent were obtained for chrysene and benzo(a)pyrene.

6.25 Silica gel modified cellulose

6.25.1 Non saline waters

6.25.1.1 Phenols

Columns containing silica gel chemically linked to different groups have been compared for their suitability for preconcentration of selected phenols from dilute aqueous solution. The chemically bonded phases comprised octyl, octadecyl, phenyl, diol and cyanide radicals of which the phenyl bonded silica gel gave the best recoveries following reversed phase high performance liquid chromatography analysis.

6.26 C8 reversed phase adsorbent

6.26.1 Non saline waters

6.26.1.1 Chlorinated insecticides and polychlorobiphenyls

Noroozian et al. [39] used a simple on-line technique involving sorption on a liquid chromatograph microprecolumn packed with a reversed-phase sorbent (C8) followed by direct elution into a gas chromatograph with hexane, to preconcentrate and determine a series of chlorinated pesticides and polychlorinated biphenyls in aqueous samples. Determinations were made at the ng L $^{-1}$ level for sample volumes of 1.0ml. In experiments using chemical standards, recoveries of 95% and more were observed for the majority of these compounds.

6.27 Hydroxygraphite

6.27.1 Non saline waters

6.27.1.1 Nucleic acids

Natural levels of nucleic acids in lake waters have been preconcentrated on hydroxygraphite [40].

6.28 Silver foil

6.28.1 Non saline waters

6.28.1.1 Organomercury compounds

Kalb [41] decomposed organomercury compounds in non saline water with nitric acid. Liberated mercury was collected on a silver foil preconcentrator. Upon heating to 350°C the foil released a concentrated pulse of mercury which was then determined by cold vapour atomic absorption spectrometry.

6.29 Dithiocarbamate resin

6.29.1 Non saline waters

6.29.1.1 Organomercury compounds

Emteborg et al. [42] used dithiocarbamate resin to sample organomercury compounds at concentrations of 40–300µg L $^{-1}$ in 1L water samples.

6.30 Calcite

6.30.1 Seawater

6.30.1.1 Miscellaneous organics

The adsorption of organic matter on any surface presented to seawater has been well documented. Neihof and Loeb [43] have demonstrated this adsorbance by following the change in surface charge of newly immersed surfaces. There has even been an attempt to use this phenomenon as a means of measuring dissolved organic carbon. Chave [44] found an association between calcite and dissolved organic materials in seawater, and Meyers and Quinn [45] tried to use the effect as a method for the collection of fatty acids. As a collection technique, adsorption on calcite has several advantages. The pH of the sample is not greatly altered by the addition of small amounts of calcite; the precipitate is dense and should settle quickly; and after filtration the inorganic support can be removed by acidification. Unfortunately, the recovery of added fatty acids was inefficient; of the order of 18%. Meyers and Quinn [46] achieved a somewhat greater efficiency of collection of fatty acids with clays, but the insolubility of the clays nullified one of the advantages of this concentration technique.

6.31 Miscellaneous adsorbents

6.31.1 Miscellaneous waters

Reviews of published work on the use of solid phase adsorbents for the preconcentration of organic compounds are presented in Table 6.1.

Goto and Taguchi [66] have reviewed the use of soluble filters for the preconcentration of traces of impurities in non saline waters.

Aikin [67] compared the extraction schemes and parameters such as pH, pore size etc. which must be considered in choosing a resin for the preconcentration of organic acids from large volumes of water.

In tests on nine different 0.5–0.5μm filter units Walsh *et al.* [68] showed that both sample contamination and significant loss of analyte by adsorption occurred. This problem could be mitigated by the addition of methyl alcohol prior to filtration.

To determine chlorinated dioxins and furans down to 1ppq in seawater, Petrick *et al.* [65] took 2L water, filtered it, and extracted it using various resins to determine dioxins, chlorinated biphenyls, HCB, DDE and polyaromatic hydrocarbons. Down to 505g L $^{-1}$ of chlorobiphenyls could be determined.

Table 6.1 Review of miscellaneous adsorbents used in preconcentration of organic compounds

Compound	Type of water	Detection limit*	Analytical finish	Ref.
Jet fuel components	Non saline	–	GC	[47]
Hydrocarbons	Ground water	–	GC	[48]
Substituted benzenes	Non saline	–	–	[49]
Benzene and sulphur compounds	Non saline	–	–	[50]
Phenols	Non saline	–	HPLC fluorometric detection	[51]
1.1.1 trichloroethane	Non saline	–	Gas extraction kinetics	[52]§
Chlorophenols	Potable water	0.001–1 µg L⁻¹	GC of acetylated derivatives	[53]
Polychlorobiphenyls and chlorinated insecticides	Non saline	–	GC of hexane extract	[39]
N-methyl carbamates	Non saline	–	GC	[54]
N and P containing pesticides	Non saline	–	GC	[55]
Metribuzin, atrazine, metalochlor and estervalerate	Non saline	–	25 factorial experiment, factors, pH, elution solvent strength and organic modifiers	[56]
Propoxur, Carbofuran, propham, captan, chloroprotan, barban and butyrate	Non saline	–	HPLC	[57]
Misc. pesticides	Non saline	–	HPLC	[58,59]
N-methyl carbamates	Non saline	–	GC	[54]
Pyrethroids	Non saline	–	GC	[60]
Volatile organics	Potable	sub µg L⁻¹	–	[61]§
Misc. organics	Non saline	sub µg L⁻¹	GC and LC	[62]
Misc. organics	Non saline	–	GC	[63]§
Misc. organics	Non saline	–	GC/MS	[64]
Dibenzo-p-dioxins and polychlorodibenzo furans	Seawater	1ng L⁻¹ to pg L⁻¹	–	[65]

*µg L⁻¹ unless otherwise stated, §Hollow fibre membrane, GC = gas chromatography, HPLC = high performance liquid chromatography

Source: Own files

References

1 Lee, Y.H. *International Journal of Analytical Chemistry*, **29**, 263 (1987).
2 Lee, Y.H. and Mowrer, J. *Analytica Chimica Acta*, **221**, 203 (1989).
3 Kurtz, D.A. *Bulletin of Environmental Contamination and Toxicology*, **17**, 391 (1977).
4 Musty, P.R. and Nickless, G. *Journal of Chromatography*, **100**, 83 (1974).
5 Mantoura, R.F.C. and Lewellyn, C.A. *Analytica Chimica Acta*, **151**, 297 (1983).
6 Jenkins, T.F., Knapp, L.K. and Walsh, M.E. Report ORREL–SR–87–2 AMXTH–TE–FR 86103, 31pp. (1987).
7 Geissbuhler, H. and Gross, D. *Journal of Chromatography*, **27**, 296 (1967).
8 Musty, P.R. and Nicklin, G. *Journal of Chromatography*, **120**, 369 (1976).
9 Ahling, B. and Jensen, S. *Industrial Chemistry*, **42**, 1483 (1970).
10 Pilipenko, A.T., Yurchenko, V.V., Zhuk, P.F. and Zul'figarov, O.S. Khim. *Tekhnol. Vody*, **9**, 420 (1987).
11 Noren, K. and Sjovall, J. *Journal of Chromatography*, **414**, 55 (1987).
12 Aue, W.A., Kapila, S. and Hastings, C.R. *Journal of Chromatography*, **73**, 99 (1972).
13 Miliotis, T., Knutsson, M., Joensson, J.A. and Mathiasson, L. *International Journal of Environmental Analytical Chemistry*, **64**, 35 (1996).
14 Josefson, G.M., Johnston, J.B. and Trubey, R. *Analytical Chemistry*, **56**, 764 (1984).
15 Weil, L. *Gas–u–Wassfach*, **113**, 64 (1972).
16 Weil, L. *Analytical Abstracts*, **24**, 1259 (1973).
17 Beyermann, K. and Eckrich, W. *Zeit Analytical Chemistry*, **265**, 1 (1974).
18 Rice, M.R. and Gold, H.S. *Analytical Chemistry*, **56**, 1436 (1984).
19 Korenman, Y.L., Alymova, A.T., Medvecheka, E.I., Zhininskaya, I.J. and Lorents, K.B. *Soviet Journal of Water Chemistry and Technology*, **8**, 84 (1986).
20 Maskarinec, G.P., Manning, D.L., Harvey, R.W., Griest, W.H. and Tomkins, B.A. *Journal of Chromatography*, **302**, 51 (1984).
21 Mosesman, N.H. Betz, W.R. and Corman, S.D. Proceedings of Water Quality Technical Conference 1986. 14 (Advanced Water Analysis and Treatment) pp. 245–250 (1987).
22 O'Keefe, P., Mayer, C., Smith, R. *et al. Chemosphere*, **15**, 1127 (1988).
23 Russo, M.V., Goretti, G. and Liberti, A. *Chromatographia*, **35**, 290 (1993).
24 Gorenc, B., Goreng, D. and Rosker, A. *Vesin. Slov. Kem, Druc.*, **33**, 467 (1988).
25 Baykut, S. and Aroguz, A.Z. *Chim. Acta. Turc.*, **13**, 161 (1985).
26 Yao, C.C.D. and Zlatkis, A. *Chromatographia*, **23**, 370 (1987).
27 Rostad, C.E., Perrira, W.E. and Ratcliff, S.M. *Analytical Chemistry*, **56**, 2856 (1984).
28 Yang, Z. and Cal, Z. *Fenxi Huaxue*, **14**, 13 (1988).
29 Rossner, B. and Schwedt, G. *Fresenius Zeitschrift für Analytische Chemie*, **315**, 610 (1983).
30 Sackmauereva, M., Pal'usova, O. and Szokolay, A. *Water Research*, **11**, 551 (1977).
31 Semin, N.N., Filippov, V.S., Toliskina, N.F. *et al. Journal of Chromatography*, **364**, 315 (1986).
32 Belkin, F. and Hable, M.A. *Bulletin of Environmental Contamination and Toxicology*, **40**, 244 (1988).
33 Cannavacinolo, F., Goretti, G., Lagana, A., Petronia, B.B. and Zoccolillo, L. *Chromatographia*, **13**, 223 (1980).

34 Ogawa, I. and Fritz, J.S. *Journal of Chromatography*, **329**, 81 (1985).
35 Uchiyama, M. *Water Research*, **12**, 299 (1978).
36 Crescenzi, C., Di Corcia, A., Possariello, G., Samperi, R. and Turnes Carou, M.I. *Journal of Chromatography, A*, **733**, 41 (1996).
37 Di Corcia, A., Crescenzi, C., Samper, R. and Scappaticcio, L. *Analytical Chemistry*, **69**, 2819 (1997).
38 Ghaoui, L. Journal of *Chromatography*, **302**, 51 (1984).
39 Noroozian, E., Maris, F.A., Nieleu, M.W.F. *et al. Journal of High Resolution Chromatography and Chromatography Communications*, **10**, 17 (1987).
40 Hicks, E. and Riley, J.P. *Analytica Chimica Acta*, **116**, 137 (1980).
41 Kalb, G.W. *Atomic Absorption Newsletter*, **9**, 84 (1970).
42 Emteborg, H., Baxter, D.C., Sharp, M. and Fresh, W. *Analyst (London)*, **120**, 69 (1995).
43 Neihof, R. and Loeb, G. *Journal of Marine Research*, **32**, 5 (1974).
44 Chave, K.E. *Science*, **148**, 1723 (1965).
45 Meyers, P.A. and Quinn, J.G. *Limnology and Oceanography*, **16**, 992 (1971).
46 Meyers, P.A. and Quinn, J.G. *Geochimica and Cosmochimica Acta*, **37**, 1745 (1973).
47 Lagenfeld, J.J., Hawthorne, S.B. and Miller, D.J. *Analytical Chemistry*, **68**, 144 (1996).
48 Ritter, J., Stromquist, V.K., Mayfield, H.T., Henley, M.V. and Lavine, B.K. *Microchemical Journal*, **54**, 59 (1996).
49 Thomas, S.P., Sri Ranjan, R., Webster, G.R.B. and Sarna, L.P. *Environmental Science and Technology*, **30**, 1521 (1996).
50 Rivasseau, C. and Cande, M. *Chromatographia*, **41**, 462 (1995).
51 Takami, K., Mochizuki, K., Kamo, T., Sugimae, A. and Nakamoto, M. *Bunseki Kagaku*, **36**, 601 (1987).
52 Prath, K.F. and Pauliszyn, J. *Analytical Chemistry*, **64**, 2101 (1992).
53 Shi, M., Zhu, X., Tao, F. and Hu, Z. *Huanjina Kexue*, **8**, 67 (1987).
54 Ballesteros, E., Gallego, M. and Valcarcel, M. *Environmental Science and Technology*, **30**, 2071 (1996).
55 Choudbury, T.K., Gerhardt, K.O. and Mawhinney, T.P. *Environmental Science and Technology*, **30**, 3259 (1996).
56 Wells, M.J.M., Riemer, D.D. and Wells-Knecht, M.C. *Journal of Chromatography*, **659**, 337 (1994).
57 Jones, E.O. *Analytical Chemistry*, **63**, 580 (1991).
58 Huen, J.M., Gillard, R., Mayer, A.G., Baltensperger, B. and Keen, H. *Fresenius Zeitschrift für Analytische Chemie*, **348**, 606 (1994).
59 Boyd-Boland, A.A., Magdic, S. and Pawliszyn, J.B. *Analyst (London)*, **121**, 929 (1996).
60 Van der Hoff, G.R., Pelusio, F., Brinkman, U.A.T., Baumann, R.A. and Van Zoonen, P. *Journal of Chromatography, A*, **719**, 59 (1996).
61 Sliver, L.E., Ho, J.S. and Budde, W.L. American Chemical Society Symposium Series No. 508 (Pollution Prevention Industrial Processes), pp. 169–177 (1992).
62 Slobodnik, J., Hogenboom, A.C., Louter, A.-G.H. and Brinkman, U.A.T. *Journal of Chromatography*, **730**, 353 (1996).
63 Pratt, K.D. and Paulizzyn, J. *Analytical Chemistry*, **64**, 2107 (1992).
64 Louter, A.J.H., Van Beekvelt, C.A., Montanes, P.C. *et al. Journal of Chromatography, A*, **725**, 67 (1996).
65 Petrick, G., Schulz-Bull, D.E., Martens, V., Scholz, K. and Duinker, J.C. *Marine Chemistry*, **54**, 97 (1996).

66 Goto, K. and Taguchi, S. *Analytical Science*, **9**, 1 (1993).
67 Aiken, G.R. *Advances in Chemistry Series* (Org. Pollution, Water), **214**, 295 (1987).
68 Walsh, M.E., Knapp, L.K. and Jenkins, T.F. *Environmental Technology Letters*, **9**, 45 (1988).

Chapter 7

Organics: Preconcentration on polymers and copolymer adsorbents

7.1 1,4 di(methacrylolyoxymethyl) naphthalene–divinylbenzene copolymer

7.1.1 Non saline waters

7.1.1.1 Chlorophenols

Gawdzik et al. [1] preconcentrated chlorophenols by solid phase extraction using cartridges of this copolymer. They compared recoveries and breakthrough volumes of phenol and several mono-, di- and tri-chlorophenols for this copolymer and several other chemically bonded phases containing hexyl and octadecyl groups (alkylsilane modified silica RP–6 and RP–18). 1,4–di(methacryloyloxymethyl)naphthalene–divinyl-benzene copolymer gave yields of about 100% for aqueous solutions containing 2µg L $^{-1}$ of each study compound. This level of recovery was achieved for RP–18 only with 2,4,6–trichlorophenol and not at all for RP–6.

7.2 Styrene–divinylbenzene copolymer

7.2.1 Non saline waters

7.2.1.1 Hydrocarbons and phthalate esters

Robinson et al. [2] evaluated disposable cartridges containing styrene–divinylbenzene copolymer hydrophobic resin (Chrom–Prep RRP–1) (containing octadecyl bonded phase silica) for preconcentration of selected aromatic hydrocarbons, phthalate esters, and food, drug, and cosmetic (FD and C) dyes in aqueous solutions. Chrom–Prep cartridges were compatible with eluents and samples over the pH range 1–13 whereas Sep–Pak cartridges were limited to pH 1–7. The former had a greater absorption capacity for model aromatic hydrocarbons and phthalate esters, whereas the latter exhibited greater capacity for the FD and C dyes, which had breakthrough volumes that were a function of pH. Analysis of aqueous samples spiked with ^{14}C-labelled solute standards showed that both cartridge types leached a small amount of analyte during the trace enrichment step.

7.2.1.2 Phenols

Pocurull *et al.* [3] used tetrabutyl ammonium bromide as an ion pair reagent to extract µg L $^{-1}$ concentrations of phenols from non saline waters onto carbon-based or styrene–divinylbenzene materials.

7.2.1.3 Humic acids

Boening *et al.* [4] studied the use of styrene divinyl benzene copolymers in the preconcentration of humic substances in non saline waters.

7.2.1.4 Miscellaneous organics

Styrene–divinylbenzene copolymer cartridges have been used to preconcentrate ultratrace levels of organic compounds in waters [5].

7.3 Acrylonitrile–divinylbenzene copolymers

7.3.1 Non saline waters

7.3.1.1 Miscellaneous organics

Acrylonitrile–divinylbenzene copolymer has been used to preconcentrate methanol, acetaldehyde, hexafluoroacetone and acetic acid from non saline water [6] prior to determination by gas chromatography.

7.4 Polymethacrylic ester

7.4.1 Non saline waters

7.4.1.1 Aromatic bases

Stuber and Leenheer [7] assessed the aqueous elution, selective concentration approach for isolating aromatic bases from water, and the factors controlling the concentration process identified. The degree of concentration attainable depends on the ratio of the capacity of the neutral form of the amine to that of the ionised form. The capacity factors of ionic forms of amines on polymethylacrylic ester resins are 20–250 times lower than those of the neutral forms and increase with the hydrophobicity of the amine.

7.5 Vinyl pyridine–2–vinyl benzene copolymer

7.5.1 Non saline waters

7.5.1.1 Phenols

Kawabata and Ohira [8] have studied the removal and recovery of phenols from aquatic samples using 5–vinyl pyridine–divinylbenzene copolymer as an adsorbent. Although the analytical implications of this work were not discussed these clearly exist. Elution of the concentrated phenols from the resin column was accomplished by a treatment with acetone or methanol.

7.5.1.2 Miscellaneous organics

Sakodynskii *et al.* [9] studied polycomplexonates of copper, mercury and silver on vinyl derivatives of pyridine as potential selective adsorbents in gas chromatography. These workers studied the specificity of these sorbents and their high selectivity for halogen, sulphur and nitrogen compounds, and aromatic and unsaturated hydrocarbons.

7.6 N-vinyl–2–pyrrolidone

7.6.1 Non saline water

7.6.1.1 Phenols

Carpenter *et al.* [10] separated phenolic materials from aqueous solutions on cross-linked aqueous insoluble N-vinyl–2–pyrrolidine polymer. The pH for maximal binding of the phenolic compound to the resin was found to be dependent an the acidity of the phenolic compound. Binding to resin was particularly favourable for polyhydroxyl and extended aromatic compounds. Columns packed with this resin removed more than 95% of simple phenolic compounds from aqueous solution, and quantitative recovery of the bound phenolic compound was possible by elution with 4M urea solution.

7.7 Vinyldithiocarbamate

7.7.1 Non saline waters

7.7.1.1 Organomercury compounds

Minagawa *et al.* [11] described a technique employing chelating resins which has been applied to the determination of very low concentrations down to 0.2ng L $^{-1}$ of organic and inorganic mercury in non saline waters including rivers, lakes and rainwaters. The resin used contains dithiocarbamate groups which bind mercury but not alkali and alkaline earth metals. Both forms of mercury are collected at pH1–11 and eluted with slightly acid 4% thiourea in water. Large volumes of water can be concentrated to determine mercury by cold vapour atomic absorption spectrometry. Mercury vapour is generated from inorganic mercury with alkaline stannous chloride and from organic and inorganic mercury with a cadmium chloride–stannous chloride solution.

No interference was produced in the determination of 0.1µg of mercury(II) by the presence of at least 1000µg of each of the following ions or substances added to 5L aliquots of river water: chromium(III), magnesium, sodium, potassium, calcium, nickel(II), copper (II), lead(II), cadmium, gold(III), iron(III), aluminium, zinc, phosphate, chloride, carbonate, nitrate, sulphate, silicate, cysteine and humic acid. The accuracy of the method was tested by analysing river water samples spiked with known amounts of mercury and CH_3Hg^+. The accuracy of the method was satisfactory. The precision of the method was 9.5 ± 0.43ng L $^{-1}$ at the 10ng L $^{-1}$ level and 0.36ng L $^{-1}$ at the 15ng L $^{-1}$ level for inorganic and organic mercury respectively.

Yamagami et al. [12] also applied the chelating resins (dithiocarbamate type) to the determination of micrograms per litre of mercury in water. The samples are adjusted to pH2.3 and passed through a column packed with 5g of the resin, at a flow rate of 50ml min $^{-1}$. The resin is then digested under reflux with concentrated nitric acid and the mercury is determined by atomic absorption spectrophotometry, using the reduction aeration technique. The method is relatively simple and inexpensive and the detection limit is 10ng mercury in water samples as large as 10L.

7.8 Wofatit Y–77

7.8.1 Non saline waters

7.8.1.1 Miscellaneous insecticides and herbicides

Dedak et al. [13] showed that the organic polymeric adsorbent Y–77 exhibited recoveries exceeding 90% for hydrophillic pesticides and herbicides (such as Methamidophos, Trichlorfon, sodium trichloroacetate, Dimethoate, Propachlor, 2,4–D and Fenuron) from non saline waters in saturated sodium chloride solutions. The target compounds were subsequently desorbed from the adsorbent with methyl alcohol. It is recommended that the more hydrophobic compounds such as 2,4–D be eluted with hot solvents.

7.9 Miscellaneous adsorbents

7.9.1 Non saline waters

7.9.1.1 Miscellaneous organics

Jahangir and Samuelson [14] have discussed the sorption of cyclohexane derivatives with a hydroxyl or carbonyl group separated by one or two methylene groups from aqueous media by resins. They found that these compounds adsorbed more strongly than the corresponding aromatic compounds both on sulphonated styrene–divinylbenzene resins and on non-ionic styrene–divinylbenzene resins. These observations and the lower temperature dependence observed for the cyclohexane derivatives indicate that hydrophobic interactions have a marked influence on the absorption.

Ghaoqui [15] has compared different adsorbents for the isolation of polynuclear aromatic hydrocarbons from waters.

Gawdzik and Matynia [16] studied the use of several porous polymers, with different functional groups, for use as sorbents for nitrobenzenes.

Yang et al. [17] used solid–phase microextraction with a polymeric phase immobilised on to a hollow membrane for the determination of $1\mu g\ L^{-1}$ of polyaromatic hydrocarbons in non saline waters. The hollow fibre membrane was directly interfaced to a gas chromatograph.

Potter and Pawliszyn [18] used solid–phase microextraction with a polymeric phase immobilised onto a fused-silica filter for the determination of $1–20ng\ L^{-1}$ of polyaromatic hydrocarbons in non saline waters.

References

1 Gawdzik, J., Gawdzik, B. and Czerwinska-Bil, U. *Chromatographia*, **25**, 504 (1988).

2 Robinson, J.L., John, J., Safa, A.I., Kirkes, K.A. and Griffiths, P.E. *Journal of Chromatography*, **402**, 201 (1987).

3 Pocurull, E., Calull, M., Marce, R.M. and Borrull, F. *Journal of Chromatography*, **719**, 105 (1996).

4 Boening, P.H., Beckmann, P. and Snoeyink, V.I. *Journal of the American Water Works Association*, **72**, 54 (1980).

5 Takami, K., Okumura, T., Yamasaki, H. and Nakamoto, M. *Bunseki Kagaku*, **37**, 195 (1988).

6 Grob, R.L. and Kaiser, O. *Journal of Environmental Science and Heath*, **A11**, 623 (1976).

7 Stuber, H. and Leenheer, J.A. *Analytical Chemistry*, **55**, 111 (1983).

8 Kawabata, N. and Ohira, K. *Environmental Science and Technology*, **13**, 1396 (1979).

9 Sakodynskii, K.I., Panina, L., Reznckova, Z.A. and Kargman, V.B. *Journal of Chromatography*, **364**, 455 (1988).

10 Carpenter, A., Siggia, S. and Carter, S. *Analytical Chemistry*, **48**, 225 (1976).

11 Minagawa, K., Takizawa, Y. and Kufune, I. *Analytica Chimica Acta*, **115**, 103 (1980).
12 Yamagami, E., Tateishi, S. and Hashimoto, A. *Analyst (London)*, **105**, 491 (1980).
13 Dedak, F., Wenzel, K.D., Luft, F., Obelander, H. and Mothes, B. *Fresenius Zeitschrift für Analytische Chemie*, **328**, 484 (1987).
14 Jahangir, L.M. and Samuelson, O. *Analytica Chimica Acta*, **100**, 53 (1978).
15 Ghaoqui, L. *Journal of Chromatography*, **399**, 69 (1987).
16 Gawdzik, B. and Matynia, T. *Journal of Chromatography, A*, **733**, 491 (1996).
17 Yang, M.J., Harms, S., Luo, Y.Z. and Pawliszyn, J. *Analytical Chemistry*, **66**, 1339 (1994).
18 Potter, D.W. and Pawliszyn, J. *Environmental Science and Technology*, **28**, 298 (1994).

Chapter 8

Organics: Polyurethane foam adsorbent

Organic compounds

8.1 Non saline waters

8.1.1 Aliphatic hydrocarbons

Ahmed *et al.* [1] investigated the precision and accuracy of infrared spectroscopic methods for determining oil in seawater using two techniques of oil preconcentration, namely carbon tetrachloride extraction and polyurethane foam adsorption. The foam adsorption method of Schatzberg and Jackson [2] was used. Estuary water (10–15L) was passed through a foam disc (8cm diameter, 4cm thick, 100 pores in $^{-1}$) and then the retained oil was extracted in a Soxhlet apparatus with carbon tetrachloride. This extract was measured as described above. It was necessary to clean new discs by Soxhlet extraction with carbon tetrachloride for 4–6h to reduce blank analysis to acceptable levels. Recoveries of known concentrations of oil were greater than 85% for those concentrations above 5mg L $^{-1}$.

The results in Fig. 8.1 obtained on synthetic solutions of oil in carbon tetrachloride show that the average recovery is 101% and the average error is 13%. Therefore, the method itself seems quite reliable since the quantity of oil in the sample bottle can be determined with a relative standard deviation of less than 15% for concentrations exceeding about 0.05mg L $^{-1}$. Replication experiments on samples gave relative standard deviations for each set of replicates from 46 to 130% with the average being 75%; this variability is far in excess of the variability of the analytical procedure itself (15%). This error must, therefore, be associated with sampling from a heterogeneous system such as oil in water.

Comparative data for the extraction and foam adsorption preconcentration methods are given in Table 8.1. In all cases, the oil concentration resulting from the foam techniques is the lower of the two values; in fact, it is lower by a factor of five on the average. Although there are sampling errors associated with both techniques it is clear that

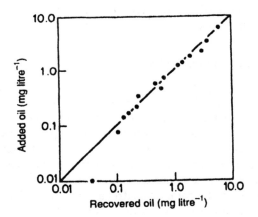

Fig. 8.1 Oil concentration as determined by infrared spectrometric technique versus known concentration
Source: Reproduced with permission from Ahmed, S.M. *et al.* [1] American Chemical Society, Washington

Table 8.1 Comparison of oil concentrations determined using different preconcentration techniques (surface sample only)

CCl₄ extraction (mg L⁻¹)	Foam adsorption (mg L⁻¹)	Ratio
84.1	7.2	11.7
0.10	0.08	1.25
0.21	0.03	7.0
0.283	0.19	1.5
0.239	0.07	3.4
		Av. 5.0

Source: Reproduced with permission from Ahmed, S.M. *et al.* [1] American Chemical Society

the foam technique has a much lower collection efficiency than liquid–liquid extraction and is, therefore, a less useful procedure.

Carsin [3] has carried out a literature review of the preconcentration of hydrocarbons from polluted water. He reviewed sampling equipment used in the determination of hydrocarbon contaminants in non saline waters. A new type of sampler incorporating a floating collector is employed, and an aspirator pump draws the contaminated water through a bed of oil-absorbent polyurethane foam. Problems of operation of the sampling apparatus, including the behaviour at sea under different weather conditions are discussed. The recovery and subsequent analysis of the oil fraction by infrared spectrometry are also described.

8.1.2 Polyaromatic hydrocarbons

Saxena et al. [4] used polyurethane foams to preconcentrate trace quantities of six representatives of polynuclear aromatic hydrocarbons (fluoranthene, benzo(k)fluoranthene, benzo(j)fluoranthene, benzo(a) pyrene, benzo(ghi) perylene, and indeno (1,2,3–cd)pyrene) prior to regular screening of these compounds in US raw and potable waters. Final purification and resolution of samples was by two-dimensional thin layer chromatography followed by fluorometric analysis and quantification.

In this method the polyaromatic hydrocarbons are collected by passing water through polyurethane foam plugs. Water is heated to $62 \pm 2°C$ prior to passage and flow rate is maintained at approximately 250ml min $^{-1}$ to obtain quantitative recoveries. The collection is followed by elution of foam plugs with organic solvent, purification by partitioning with solvents and column chromatography on Florisil. Analysis is by two-dimensional thin layer chromatography on cellulose acetate–alumina plates followed by fluorometry and gas liquid chromatography using flame ionisation detection. The latter method was less sensitive than thin layer chromatography. Employing this method and a sample volume of 60L, polyaromatic hydrocarbons were detected in all the water supplies sampled. Although the sum of the six representative polyaromatic hydrocarbons in potable waters was small (0.9–15µg L $^{-1}$) the values found for raw waters were as high as 600µg L $^{-1}$.

In further work Saxena et al. [5] and Basu and Saxena [6] showed that the polyurethane foam plug method had an extraction efficiency for polyaromatic hydrocarbons of at least 88% from treated waters and 72% from raw waters. Foam retention efficiencies of the six polyaromatic hydrocarbons from spiked laboratory potable water at 25µg L $^{-1}$ are shown in Table 8.2. The data confirm that polyurethane foam plugs under suitable conditions not only effectively concentrate benzo(a) pyrene but other polyaromatic hydrocarbons as well. Foam plugs concentrated polyaromatic hydrocarbon almost quantitatively from finished water at lower concentrations also (Table 8.3). The high polyaromatic hydrocarbon retention efficiencies were also maintained with heavily polluted surface waters (Table 8.4). It is unclear why the retention values for three of the polyaromatic hydrocarbons are well above 100%. A possible explanation for this may be the inability of cyclohexane extraction to quantitatively recover these polyaromatic hydrocarbons from the heavily polluted water, this will give rise to lower polyaromatic hydrocarbon concentration in the aqueous phase than actually present.

Table 8.2 Foam retention efficiencies for PAH from potable water

Compound[a]	Amount added to water (µg)	% retention
FL	100	100
BjF	100	
BkF	100	88[b]
BaP	100	81
IP	100	89
BghiP	100	91

[a]FL, fluoranthrene; BjF, benzo(j)fluoranthrene; BkF, benzo(k) fluoranthrene; BaP, benzo(a)pyrene; IP, indeno(1,2,3cd) pyrene; BghiP, benzo(ghi) perylene
[b]Combined value given since the compounds could not be separated on the GLC column. Water source, laboratory tap water; water volume, 4L; conc. of each PAH, 25ppb; detection method, GLC–FID

Source: Reproduced with permission from Saxena, J. et al. [4] US National Technical Information Service, Springfield, Virginia

Table 8.3 Foam retention efficiencies for PAH from treated water

Compound[a]	Conc. in aqueous phase (ng L^{-1})	Amount retained by foam from 1L of water (ng)	% retention
FL	278.6	260.4	93.5
BjF	48.3	47.4	98.1
BkF	51.7	50.6	97.6
BaP	36.4	33.6	92.3
IP	25.5	23.9	93.7
BghiP	22.6	19.8	87.6

[a]See Table 8.2
Water source, laboratory tap water; water volume, 60L; conc. of fluoranthene, 500ppt; all others, 100ppt; detection method, TLC–fluorometric

Source: Reproduced with permission from Saxena, J. et al. [4] US National Technical Information Service, Springfield, Virginia

Table 8.5 shows the detection limits of the six polyaromatic hydrocarbons using the polyurethane foam preconcentration method obtained by gas chromatography with a flame ionisation detector or thin layer chromatography fluorometric analysis. The detection of fluoranthene and benzo(a)pyrene in thin layer chromatography fluorometry is restricted by the background levels of these compounds contributed from the foam

Table 8.4 Foam retention efficiencies for PAH from raw water

Compound*	Conc. in aqueous phase (ng L^{-1})	Amount retained by foam from 1L of water (ng)	% retention
FL	289.1	343.7	118.9
BjF	77.6	94.0	121.1
BkF	66.1	55.6	84.1
BaP	74.5	59.7	80.1
IP	85.2	61.2	71.8
BghiP	23.9	28.3	118.4

*See Table 8.2
Water source, Onondaga Lake water; volume, 30L; Conc. of fluoranthene, 500ppt; all others, 100ppt; detection method, TLF–fluorometric

Source: Reproduced with permission from Saxena, J. et al. [4] US National Technical Information Service, Springfield, Virginia

Table 8.5 Limit of detection of PAH with foam preconcentration coupled with ILC–fluorometric or GLC–FID method

Compound*	TLC–fluorometric detection		GLC–FID detection	
	Absolute limit (ng)	Limit in 60L water (ng L^{-1})	Absolute limit (ng)	Limit in 60L water (ng L^{-1})
Fl	140.0	2.3	13.6	4.5
BjF	7.5	0.1	10.1	3.4
BkF	5.0	0.1		
BaP	10.0	0.2	11.9	4.0
IP	10.0	0.2	14.7	4.9
BghiP	20.0	0.3	14.9	5.0

*See Table 8.2

Source: Reproduced with permission from Saxena, J. et al. [4] US National Technical Information Service, Springfield, Virginia

plugs. Their detection limits are assumed to be twice the background fluorescence level. In the case of the gas chromatographic FID method, the detection limits for polyaromatic hydrocarbons are based on a minimum output response of five times the background noise level and a maximum volume of 5µL from a total of 100µL concentrate. Details of this procedure are given below.

During the concentration of polyaromatic hydrocarbons from water on foam plugs, several other contaminants were also concentrated and some of these were eluted during polyaromatic hydrocarbon elution. In addition several impurities originating in the foam were also leached during the elution process. The latter impurities could be partially eliminated only by precleaning of the plugs with cyclohexane and/or benzene by batch or Soxhlet extraction. The impurities interfered with the analysis of polyaromatic hydrocarbons, therefore a clean-up procedure was devised. The levels of impurities and subsequently the extent of the clean-up necessary are directly dependent upon the sample volume and the number of foam plugs employed for concentration. A sample volume of 60L was adequate for detection of polyaromatic hydrocarbon. 60L of unspiked finished water was passed through six precleaned foam columns each containing two plugs, in three successive steps maintaining the water temperature at $62 \pm 2°C$ and flow rate at $250 \pm 10ml$ min $^{-1}$. Each column was eluted with 30ml acetone and 125ml cyclohexane. The combined extract was concentrated and subjected to clean-up. At no time was the polyaromatic hydrocarbon mixture allowed to proceed to complete dryness since this has been shown to result in loss of polyaromatic hydrocarbons [7].

Additional clean-up involving a short Florisil column was found to be necessary for further separation from impurities. Chromatographic grade Florisil (60–100 mesh) was washed with methanol and 1:1 hexane–benzene and activated for at least 4h at 130°C. The Florisil was cooled to room temperature and 8g was transferred to a glass column (1.5cm × 30cm) with benzene by slurry method and washed with an addition of 100ml of benzene prior to passing sample concentrate through it. The flow chart illustrated in Fig. 8.2 depicts the complete clean up procedure. Thin layer chromatography was performed using aluminium oxide acetylated cellulose plates, 2% (w/v) of $CaSO_4.2H_2O$ (200 mesh) was added to the slurry to increase binding of the layer to the surface. The plates were developed in two dimensions, n-hexane–benzene (4:1 v/v) and methanol–ether–water (4:4:1 v/v).

The emission and excitation spectra used for identification of the suspected polyaromatic hydrocarbon spots were run directly on the plates at room temperature with a thin film scanner attached to a spectrophotofluorometer. This procedure eliminated the losses usually encountered during removal of polyaromatic hydrocarbon spots for fluorescence measurement in solvents and the interferences arising due to solvent interaction. For quantification of the spots, the excitation wavelength was fixed at 365nm and the fluorescence intensities were measured at the following wavelengths (nm): fluoranthene (458), benzo(j)fluoranthrene (427), benzo(k)fluoranthene (428), benzo(a)pyrene (427), and benzo(ghi)pyrene (416).

Navra'til et al. [8] have also investigated the use of high capacity open pore polyurethane columns for the collection and preconcentration of

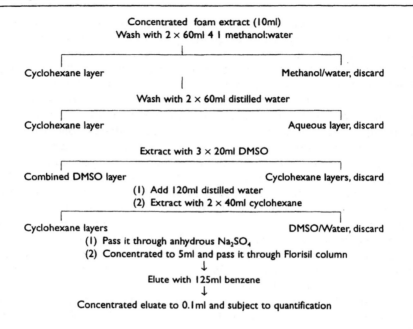

Concentrated foam extract (10ml)
Wash with 2 × 60ml 4 1 methanol:water

| Cyclohexane layer | Methanol/water, discard |

Wash with 2 × 60ml distilled water

| Cyclohexane layer | Aqueous layer, discard |

Extract with 3 × 20ml DMSO

| Combined DMSO layer | Cyclohexane layers, discard |

(1) Add 120ml distilled water
(2) Extract with 2 × 40ml cyclohexane

| Cyclohexane layers | DMSO/Water, discard |

(1) Pass it through anhydrous Na₂SO₄
(2) Concentrated to 5ml and pass it through Florisil column
↓
Elute with 125ml benzene
↓
Concentrated eluate to 0.1ml and subject to quantification

Fig. 8.2 Flow chart of clean up method
(DMSO: dimethylsulphoxide)

Source: Reproduced with permission from Saxena, J. *et al.* [4] US National Technical Information Service, Springfield, Virginia

polyaromatic hydrocarbons from water. They tested open pore polyurethane columns for ion exchange properties, solvent compatibilities, and ability to remove and concentrate polynuclear aromatic hydrocarbons from water. Their performance was compared with that of other sorbent materials using aqueous solutions of pyrene.

Afghan *et al.* [9] present results from studies on the feasibility of using polyurethane foams to extract polynuclear aromatic hydrocarbons from water prior to their determination. They also studied factors affecting adsorption. Preliminary filtration of the sample to remove particulate matter is recommended. Polyurethane foam is particularly useful when large samples are routinely analysed in that it offers high flow rates, good capacities, and low costs compared to solvent extraction. Polyaromatic hydrocarbons can be extracted at the part per trillion (pg ml $^{-1}$) from non saline waters.

8.2 Potable waters

8.2.1 Polychlorobiphenyls

Porous polyurethane foam has also been used to preconcentrate polychlorobiphenyls from water samples. Gesser *et al.* [10] found that the compounds could be adsorbed on a column composed of two polyurethane plugs (each 38mm × 22mm) inside a glass tube. The sample water was poured through at 250ml min $^{-1}$ then the plugs were removed and squeezed free from water; the polychlorobiphenyls were then extracted by treating the plugs with acetone and hexane. The concentrated extract was analysed by gas chromatography on a glass column (180cm × 3mm) packed with Chromosorb W HMDS supporting 2% SE–30 and 3% of QF–1 and operated at 200°C; the carrier gas was helium and a ^{63}Ni electron capture detector was used. In tests on 20µg of polychloro-biphenyls added to 1L of water, the recovery (based on measurements of 13 peaks on the chromatogram) ranged from 18.2 to 19.6µg. Bedford [11] also used polyurethane foam plugs to extract polychlorobiphenyls from non saline waters; these results indicated that Aroclor is probably adsorbed on to small particles in unfiltered lake water which can pass through the foam. Although this makes the foam method somewhat impractical for enriched and turbid waters, it is still a valuable technique for clear water containing low quantities of polychlorobiphenyls.

Polyurethane foam has been used as a liquid–liquid partitioning filter for the concentration of polychlorobiphenyls from water samples [10,12].

Organometallic compounds

8.3 Non saline waters

8.3.1 Organomercury compounds

Phenylmercury and methylmercury compounds have been pre-concentrated on diethylammonium polyurethane foam prior to determination by X-ray fluorescence analysis [13].

Braun *et al.* [14] showed that polyurethane foam loaded with diethyl-dithiocarbamate is suitable for the preconcentration of trace amounts of organic and inorganic mercury prior to its chemical analysis.

Braun [15] has shown that polyurethane foam loaded with diethyl-dithiocarbamate is suitable for the concentration of trace amounts of organic and inorganic mercury from non saline water samples prior to chemical analysis. Mercury(II), methylmercury and phenylmercury ions at around 1µg L $^{-1}$ levels can be almost quantitatively preconcentrated from water and eluted from the foam with acetone.

References

1 Ahmed, S.M., Beasley, M.D., Etromson, A.C. and Hites, R.A. *Analytical Chemistry*, **46**, 1858 (1974).
2 Schatzberg, P. and Jackson, D.F. In *US Coast Guard Report* No. 734209.9, Washington DC. November (1972).
3 Carsin, J.L. *Revue International d'Oceanographic Medicale*, **48**, 77 (1977).
4 Saxena, J., Basu, D.K. and Kozuchowski, J. In Report No. PB276635, US National Technical Information Service, Springfield, VA, p. 94 (1977).
5 Saxena, J., Kozuchowski, J. and Basu, D.K. *Journal of Environmental Science and Technology*, **11**, 682 (1977).
6 Basu, D.K. and Saxena, J. *Journal of Environmental Science and Technology*, **12**, 791 (1978).
7 Josefsen, C.M., Johnston, J.B. and Trubey, R. *Analytical Chemistry*, **56**, 764 (1984).
8 Navra'til, J.D., Sievers, R.E. and Walton, H.F. *Analytical Chemistry*, **49**, 2260 (1977).
9 Afghan, B.K., Wilkinson, R.J., Chow, A. *et al. Water Research*, **18**, 9 (1984).
10 Gesser, H.D., Chow, A., Davis, F.C., Uthe, J.F. and Reinke, J. *Analytical Letters (London)*, **4**, 883 (1971).
11 Bedford, J.W. *Bulletin of Environmental Contamination and Toxicology*, **12**, 662 (1974).
12 Uthe, J.F., Reinke, J. and Gesser, H. *Environmental Letters*, **3**, 117 (1972).
13 Musty, P.R. and Nickless, G. *Journal of Chromatography*, **120**, 369 (1976).
14 Braun, T., Abbas, M.N., Torak, S. and Zvakefalvi-Nagy, Z. *Analytica Chimica Acta*, **160**, 277 (1984).
15 Braun, T., Abbas, M.M., Bakos, L. and Elek, A. *Analytica Chimica Acta*, **131**, 311 (1981).

Chapter 9

Organics: Covalently bonded octadecyl silica gel adsorbent

Organic compounds

9.1 Non saline waters

9.1.1 Aromatic hydrocarbons

Nakamura et al. [1] studied the extraction of 25 aromatic compounds and 20 agricultural chemicals from non saline waters using Sep–Pak (octadecyl resin–silica) cartridges. Analyte recoveries were related to their *Kow* values.

9.1.2 Polyaromatic hydrocarbons

El'Harrak et al. [2] used C_{18} membranes to extract low ppt concentrations of polyaromatic hydrocarbons from non saline waters. These workers studied the use of Brij–35 surfactant to prevent analytes from adsorbing on the inner walls or surfaces of the extraction vessel.

Ogan et al. [3] preconcentrated polyaromatic hydrocarbons on a column of Sep–Pak prior to desorption and determination by high performance liquid chromatography using a spectrofluorometric detector.

9.1.3 Phenols

To determine low ppt amounts of phenols in non saline waters Bao et al. [4] acetylated the phenols with acetic anhydride and collected them on a C_{18} resin disc.

9.1.4 Surfactants

Scullion et al. [5] used C_{18} and SAX extraction to preconcentrate surfactants in non saline waters. Analysis of the concentrates was carried out by high performance liquid chromatography with mass spectrometric detection.

9.1.5 Volatile organic compounds

Liska *et al.* [6] have shown that polymeric PLRPs in combination with C_{18} was the best solid phase extractant for the enrichment of polar organic compounds in water.

Ho *et al.* [7] used C_{18} resin discs to extract 43 semi-volatile organic compounds from non saline waters. These compounds were then eluted by supercritical carbon dioxide.

9.1.6 Chlorophenols

C_{18}-bonded silica has been used to preconcentrate chlorophenols in non saline waters prior to gas chromatography [8].

Househam *et al.* [9] compared columns packed with C_{18}-bonded silica with columns packed with C_6 and C_8-bonded silica for the extraction of chlorophenols. C_{18}-bonded silica was the most suitable.

Ono *et al.* [10] used solid phase extraction with C_{18}-bonded silica of water samples to determine pentachlorophenol and tricholorophenol in non saline waters.

9.1.7 Aliphatic diamines

Nishikawa [11] preconcentrated aliphatic diamines (after derivitivisation with acetylacetone) on Sep–Pak C_{18} cartridges. The diamines were determined by high performance liquid chromatography in the range 1.23–14.32µg L^{-1} with relative standard deviations of 0.4–4.9%, and detection limits of 0.14–1.78µg L^{-1}. 100ml water samples were used. Analyte recovery was 88–96% in river water (relative standard deviations 1.3–4.0%), and 34–93% in seawater (relative standard deviation 1.3–9.8%).

9.1.8 Polychlorobiphenyls

Thome and Vandaele [12] preconcentrated polychlorinated biphenyls using microcolumns by quantitative adsorption on to covalently bonded silica gel with octadecyl groups. Maximal efficiency of adsorption was achieved by successive treatment with 1ml hexane, acetone, methanol, and methanol in water (1:1 v/v). A sample flow rate of 10ml min^{-1} was achieved by vacuum which was maintained to dry the column. Interfering polar compounds were eluted with 0.2ml methanol/water. Polychlorinated biphenyls were eluted with 2ml hexane, concentrated by evaporation, and analysed by gas chromatography with electron capture detection. Mean recovery was 95.2% with polychlorinated biphenyl concentrations of 0.01–10ppb.

Hermanau *et al.* [13] and Thome and Vandaale [12] used solid phase extraction of water samples using columns packed with C_{18} bonded silica packings to concentrate polychlorobiphenyls in non saline waters.

9.1.9 Chlorinated insecticides

C_{18}–silica-bonded phase columns [14,15] have been used to preconcentrate chlorinated insecticides prior to gas chromatographic [15] and thin layer chromatographic analysis [14].

Sherma [14] described a procedure for preconcentrating and determining triazine and chlorophenoxy acid herbicides from non saline waters at 10µg L $^{-1}$ concentrations involving extraction and concentration on disposable columns containing C_{18}-bonded silica gel, separation by thin layer chromatography on pre-adsorbent silica gel layers impregnated with silver nitrate, and detection by exposure to ultraviolet light with densitometric scanning. Herbicide concentrations were interpolated from calibration lines relating peak areas to weight of standards spotted on the TLC plate. Recoveries from triazines and chlorophenoxy herbicides were 70–88% and 93–100%, respectively.

9.1.10 Miscellaneous insecticides and herbicides

Sep–Pak C_{18}-bonded silica has been used for the preconcentration of various organic substances in water. Pyrazole herbicide in non saline water was adsorbed from a large volume of water, then desorbed into a small volume of methylene dichloride prior to analysis by high performance liquid chromatography [16]. Polyaromatic hydrocarbons in potable water were adsorbed on a C_{18} reversed-phase column, then desorbed with methanol [18].

Organophosphorus insecticides. Bargnoux *et al.* [18] used preconcentration at low temperatures as a means of extracting organophosphorus insecticides from water, prior to gas chromatography.

Fenithrothion derivatives. Volpe and Mallet [19] developed a method for determining down to 0.5ng of Fenitrothion and five Fenitrothion derivatives in water by adsorption on XAD–4 and XAD–7 resins, followed by solvent elution and gas–liquid chromatography of the extract.

Picloram and 2,4 dichlorophenoxyacetic acid. Wells and Michael [20] used solid phase extraction of water samples using columns packed with C_{18}-bonded silica to determine Picloram and 2,4 dichlorophenoxyacetic acid in non saline waters.

Fluridone. West and Turner [21] used Sep–Pak C_{18} cartridges to extract Fluridone from water samples while N-methylformamide passed through unretained. Water containing N-dimethylformamide was treated with methanol and glycerol before evaporation *in vacuo*. Residual N-dimethylformamide was dissolved in methanol and analysed by gas chromatography with a Hall electrolytic conductivity detector operated in the nitrogen mode. Fluridone was eluted from the cartridge with methanol and analysed by liquid chromatography with ultraviolet detection at 313nm. Limits of detection for Fluridone and N-dimethylformamide were, respectively, 1 and 2µg L $^{-1}$. Mean Fluridone recoveries were 98% at 1–400µg L $^{-1}$, and mean N-dimethylformamide recoveries were 87% at 2–50µg L $^{-1}$.

Azine type herbicides. Baker's yeast cells (*Saccharomices cerevisae*) have been successfully immobilised on to silica gel and used in on-line isolation and trace enrichment of Desisopropylatrazine, Desethyl-atrazine, hydroxy-atrazine, Simazine, Cyanazine, Atrazine, Carabaryl, Propanil, Linuron and Fenamiphos [22]. Since humic and fulvic acids were not extracted, no clean-up was necessary. The pesticides were spiked at 0.1–1µg L $^{-1}$ in tap water, ground water and seawater and were precon-centrated using on-line solid-phase extraction into a yeast immobilised on silica gel precolumn followed by liquid chromatography with diode–array detection. All the variables that affect the enrichment step, such as amount of yeast immobilised, dimensions of the precolumn, sample pH and preconcentration flow rate, were optimised. The degree of selectivity was evaluated by comparing the chromatograms obtained after on-line sample preconcentration on the yeast precolumn with those obtained by on-line solid-phase extraction using a precolumn filled with C_{18} material. The relative standard deviation for the whole procedure in the determination of the selected pesticides at the 0.3µg L $^{-1}$ concentration level ranged from 1 to 9%, depending on the pesticide and the type of water. Detection limits within the range 0.01–5µg L $^{-1}$ were obtained by percolating only 25ml of water sample without any additional clean-up step.

Chlorosulphuron, Sulphometon methyl and AC 243977. Reverse phase C_{18} columns have been used to extract herbicides such as AC 243997, Chlorosulphuron and Sulphometon methyl from environmental waters. Recoveries of the targeted compounds at the low µg L $^{-1}$ range was approximately 90% [23].

Butachlor, Oxdiazone, Chlormirophen and Chlormethoxynil. Sep–Pak C_{18} columns have been used to preconcentrate the herbicides Butachlor, Oxdiazone, Chloronitrophen and Chlormethoxynil from agricultural waters [24].

Pesticides. Carr and Harris [25] and Junk and Richard [26] used solid phase extraction of non saline water samples with C_{18}-bonded silica to determine pesticides and polycyclic aromatic hydrocarbons in non saline waters.

Hydroxytriazines. Solid phase extraction with C_{18} resin followed by liquid chromatography or GC/MS has been used to determine 100µg L $^{-1}$ hydroxytriazines in non saline waters [27].

9.1.11 Methylisoborneol

Conte [28] used C_{18} resin to sample methylisoborneol in amounts down to 10ng L $^{-1}$ in non saline waters.

9.1.12 Miscellaneous organics

Nakamura *et al.* [1] studied the extraction of 25 aromatics and 20 agricultural chemicals from non saline waters using eight Sep–Pak cartridges. Analyte recoveries were related to their Kw values.

Reighard and Olesik [29] studied the use of supercritical carbon dioxide and 'enhanced fluidity' liquid carbon dioxide as an extraction solvent for analytes adsorbed on to a C_{18} solid phase, the higher strength of the solvents containing methanol reduced the volume and time required for efficient analyte extraction.

9.2 Ground waters

9.2.1 Organochlorine insecticides

Bagnetti *et al.* [15] give details of equipment and procedure for simultaneous semi-quantitative determination of sub µg L $^{-1}$ traces of 21 pesticides in groundwater, involving rapid extraction using C_{18}–silica-bonded phase columns and capillary gas chromatography with selected ion recording mass spectrometry.

9.3 Seawater

9.3.1 Miscellaneous organics

Sauer *et al.* [30] studied the use of Sep–Pak C_{18} cartridges for the preconcentration of organic compounds from estuary and sea water.

9.4 Potable water

9.4.1 Polyaromatic hydrocarbons

Polyaromatic hydrocarbons in potable water have been preconcentrated on Sep–Pak C_{18} [17].

9.4.2 2,3,7,8 tetrachlorodibenzo-p-dioxin

O'Keefe *et al.* [31] have described a sampling cartridge containing octadecyl reverse phase adsorbent, for collecting of 2,3,7,8–tetrachloro-dibenzo-*p*-dioxin from potable water. Procedure for subsequent clean-up and analysis of the samples is also described and results of tests to assess the efficiency of the method are included. The lower limit of detection for 2,3,7,8–tetrachlorodibenzo-*p*-dioxin by this method was about 1pg L $^{-1}$.

9.5 Wastewaters

9.5.1 Phenols

Nielen *et al.* [32] used precolumns packed with C_{18}-bonded silica and sulphonic acid cation exchange materials for the on-line separation and trace enrichment of trace organics in industrial waste water samples. The precolumn fractions were analysed by automated liquid chromatography and diode array ultraviolet visible detection. Multi-signal plots and three-dimensional spectrochromatograms were used in the identification of the pollutants. The technique was optimised for the analysis of 29 organic compounds in industrial waste water. 2,5–dimethylphenol, 2,6–dichloro-phenol, and *o*-phenylphenol recoveries were all in excess of 90%, when the water sample was first acidified to pH2 and 30g sodium chloride per 100ml added before passage through the column. For phenol, however, a recovery of only 42% was achieved under these conditions.

9.6 Sewage effluents

9.6.1 Alkylbenzene sulphonates

A C_{18} Empane disc has been used to preconcentrate linear alkylbenzene sulphonates from primary sewage effluents [33]. In-vial elution with tetrabutyl ammonium hydrogen sulphate in chloroform was followed by gas chromatography with FID detection of the resulting butyl esters.

9.7 Trade effluents

9.7.1 Polyaromatic hydrocarbons

Separations on C_{18} columns have also been used to analyse PAHs in trade effluents [34,35].

Symons and Crick [34] have conducted a study of the concentration of polynuclear aromatic hydrocarbons from laboratory water and refinery effluents by means of Sep–Pak C_{18} cartridges. Reversed-phase liquid chromatography with coupled ultraviolet and fluorescence detection was applied to separate and quantify these hydrocarbons. The method was used to determine several polynuclear aromatic hydrocarbons in refinery effluents at the 0.1–50µg L $^{-1}$ level.

Organometallic compounds

9.8 Rain

9.8.1 Alkyllead compounds

To preconcentrate trialkyllead species Blaszkewicz *et al.* [36] complexed interfering metal ions in rainwater with EDTA before adjustment of the pH to 10. Samples were pumped through an extraction column of silica gel to adsorb lead compounds which were then desorbed with acetate buffer containing methanol at pH3.7. The eluate was diluted and adjusted to pH8 with borate buffer before further concentration on a C_{18} precolumn. Adsorbed trialkyllead compounds were eluted by back-flushing on to a RP–C_{18} column and separated with methanolic acetate buffer. On-line detection used a post–column chemical reaction detector. Detection limits for sample volumes of 500ml were 15µg L $^{-1}$ and 20µg L $^{-1}$ for trimethyl- and triethyllead, respectively. Standard deviation was less than 4% for a sample containing 90pg triethyllead per ml.

9.9 Non saline waters

9.9.1 Alkyltin compounds

Muller [37] has described a procedure for the preconcentration and determination of mono-, di-, tri- and tetra-substituted organotin compounds in lake water samples. The ionic compounds were extracted from diluted aqueous solutions as chlorides by using a Tropolin–C_{18} silica cartridge and from sediments and sludges by using an ethereal tropolone solution. The extracted compounds were then ethylated by a Grignard reagent, and analysed by high resolution gas chromatography with flame photometric detection.

Cai and Bayona [38] used sodium tetraethylborate to ethylate butyl, phenyl and cyclohexyltin compounds which were then collected on a C_{18} disc, eluted with supercritical carbon dioxide and measured by gas chromatography with flame photometric detection.

9.10 Sewage sludge

9.10.1 Alkyltin compounds

The procedure described by Muller [37] referred to in section 9.9.1 has also been applied to sewage sludges.

References

1 Nakamura, M., Nakamura, M. and Yamada, S. *Analyst (London)*, **121**, 469 (1996).

2 El'Harrak, R., Cabull, M., Marce, R.M. and Barrul, F. *International Journal of Environmental Analytical Chemistry*, **64**, 47 (1996).
3 Ogan, K., Katz, E. and Slavin, M. *Journal of Chromatographic Science*, **16**, 517 (1978).
4 Bao, M.L., Pantani, F., Barbieri, K., Burrini, D. and Griffini, O. *Chromatographia*, **42**, 227 (1996).
5 Scullion, S.D., Clench, M.R., Cooke, M. and Ashcroft, A.E. *Journal of Chromatography, A*, **733**, 207 (1996).
6 Liska, I., Brouer, E.R., Lingeman, H. and Brinkman, U.A.T. *Chromatographia*, **37**, 13 (1993).
7 Ho, J.S., Tang, P.H., Eichelberger, J.W. and Budde, W.L. *Journal of Chromatographic Science*, **33**, 1 (1995).
8 Fingler, S., Drevenbar, V. and Vasitic, Z. *Mikrochimica Acta*, **No. 4/6**, 163 (1987).
9 Househam, B.C., Van den Berg, C.M.G. and Riley, J.P. *Analytica Chimica Acta*, **200**, 291 (1987).
10 Ono, H., Tsuyama, A., Aoyama, T. *et al. Nagoya-Shi Eisel Fenkyushoho*, **33**, 88 (1987).
11 Nishikawa, Y. *Journal of Chromatography*, **392**, 349 (1987).
12 Thome, J.P. and Vandaele, Y. *International Journal of Environmental Analytical Chemistry*, **29**, 95 (1987).
13 Hermanau, H., Stottmeister, E. and Hendel, P. *Acta Hydrochim. Hydrobiol*, **16**, 45 (1988).
14 Sherma, J. *Journal of Liquid Chromatography*, **11**, 2121 (1988).
15 Bagnetti, R., Bentenati, E., Davoli, E. and Fanelli, R. *Chemosphere*, **17**, 59 (1988).
16 Nyagah, G. *Journal of Chromatographic Science*, **19**, 500 (1981).
17 Fisher, R. *Fresenius Zeitschrift für Analytische Chemie*, **311**, 109 (1982).
18 Bargnoux, H., Pepin, D., Chahard, J.L. *et al. Analysis*, **5**, 170 (1977).
19 Volpe, G.G. and Mallet, V.N. *International Environmental Analytical Chemistry*, **8**, 291 (1980).
20 Wells, M.G.N. and Michael, J.L. *Analytical Chemistry*, **59**, 1739 (1987).
21 West, S.D. and Turner, L.G. *Journal of the Association of Official Analytical Chemists*, **71**, 1049 (1988).
22 Martin-Esteban, A., Fernandez, P. and Camera, C. *Analytical Chemistry*, **69**, 3267 (1997).
23 Wells, M.J.M. and Michael, J.L. *Journal of Chromatographic Science*, **25**, 345 (1987).
24 Yasumasu, S. and Takata, I. *Fukuoka-shi Eisel. Shikenshoto*, **11**, 54 (1988).
25 Carr, J.W. and Harris, J.M. Government Report Announce Index. (US) Abstract No. 756, 473.87 (24) (1987).
26 Junk, G.A. and Richard, J.J. *Analytical Chemistry*, **60**, 451 (1988).
27 Farber, H., Nick, K. and Scholer, H.T. *Fresenius Journal of Analytical Chemistry*, **350**, 145 (1994).
28 Conte, E.D., Conway, S.C., Mitter, D.W. and Perschbacher, P.W. *Water Research*, **30**, 2125 (1996).
29 Reighard, T.S. and Olesik, S.V. *Journal of Chromatography, A*, **737**, 233 (1996).
30 Sauer, W.S., Jadamse, J.R., Sagar, R.W. and Kileen, T.J. *Analytical Chemistry*, **51**, 2180 (1979).
31 O'Keefe, P., Meyer, C., Smith, R. *et al. Chemosphere*, **15**, 1127 (1986).
32 Nielen, M.W.F., Brinkman, U.A.T. and Frei, R.W. *Analytical Chemistry*, **57**, 806 (1985).
33 Krueger, C.J. and Field, J.A. *Analytical Chemistry*, **67**, 3363 (1995).
34 Symons, R.E. and Crick, I. *Analytica Chimica Acta*, **151**, 237 (1983).

35 Alexander, R., Cumbers, K.M. and Kagi, R.I. *International Journal of Environmental Analytical Chemistry*, **12**, 161 (1982).
36 Blaszkewicz, M., Baumhoer, G. and Neidhart, B. *International Journal of Environmental Analytical Chemistry*, **28**, 207 (1987).
37 Muller, M.D. *Analytical Chemistry*, **59**, 617 (1987).
38 Cai, Y. and Bayona, J.M. *Journal of Chromatographic Science*, **33**, 89 (1995).

Chapter 10

Organics: Active carbon adsorbent

Activated carbon filters have been employed for adsorption of different kinds of organics in non saline waters since it was developed and introduced by the US Public Health Service [1,2]. However, the lack of adsorption and desorption control in addition to bacterial and oxidising attack on the organics does place some limitations on the method.

10.1 Active carbon filters

10.1.1 Non saline waters

10.1.1.1 Phenols

Eichelberger *et al.* [3] collected the phenols from water samples on columns of activated carbon. The phenols were stripped from the carbon with chloroform prior to gas chromatography of the cleaned up extract. Clean-up was achieved by a treble extraction of the chloroform phase with aqueous sodium hydroxide. The aqueous extracts were combined, acidified with concentrated hydrochloric acid to pH2, and extracted three times with ethyl ether, and the combined ether extracts were passed through a 10cm Florisil column topped with anhydrous sodium sulphate (2cm). The phenols were eluted with ether, and the eluate concentrated by evaporation and analysed by gas chromatography. An aluminium column (300cm × 3.1mm od) packed with 10% of Carbowax 20M terephthalic acid on HMDS-treated Chromosorb W (70–80 mesh) was used operated at 210°C for the majority of phenols with nitrogen (50ml min $^{-1}$) as carrier gas and flame ionisation detection. Recoveries were on average, 85% for alkylphenols and 94% for chlorophenols.

10.1.1.2 Alkylbenzene sulphonates

Wang *et al.* [4–7] discussed carbon absorption for the concentration of linear alkylate sulphonates.

10.1.1.3 Chlorinated insecticides

Various workers [1,2,8–14] have investigated the use of activated carbon filters to concentrate chlorinated insecticides from water samples prior to gas chromatographic analysis. Quantitative values reported by these methods must be considered minimal because of lack of control on adsorption and desorption characteristics of the carbon filter bed and also because of bacterial breakdown of adsorbent organics during the sampling period. Use of the carbon adsorption method determined that maximum organic adsorption efficiencies are achieved at a reduced flow rate (not greater than 120ml min $^{-1}$ and total throughput volume of 1500L or less). In a study designed to determine carbon adsorption–desorption efficiency and reproducibility, it was determined by Eichelberger and Lichtenberg [15] that the carbon is useful for some organochlorines giving a 70–85% recovery rate and less dependable for others, including DDT, which was determined to have an average 37% recovery rate.

Eichelberger and Lichtenberg [15] used carbon adsorption prior to gas chromatography for the determination of Methoxychlor, Lindane, Endrin, Dieldrin, Chlordane, DDT, Heptachlor epoxide and Endosulphan in water samples. The eluate from the carbon was extracted with ethyl ether–hexane (3:17) and the carbon was dried and extracted with chloroform.

Brodtmann [16] carried out a long term study on the qualitative recovery efficiency of the carbon adsorption method versus that of a continuous liquid–liquid extraction method for several chlorinated insecticides. Comparative results obtained by electron capture gas chromatography indicate that the latter method may be more efficient. Samples of river water for analysis by the carbon adsorption method were collected at a rate of approximately 120ml min $^{-1}$ for a 7 day period in each case. At the end of the weekly sampling period, the total throughput volume was recorded. The carbon chloroform extract was then obtained by chloroform extraction of the carbon in a modified Soxhlet apparatus for 36h using glass distilled, pesticide grade petroleum ether. The neutral fraction of the chloroform extract was then prepared for gas chromatography by the methods of Breidenbach [17].

Breidenbach [17] used a continuous liquid–liquid extraction apparatus as described by Kahn and Wayman [18] for the extraction of non-polar solutes from river water. Pesticide grade petroleum ether, used in all cases, was recycled internally (initial solvent charge of 350ml) thereby continuously exposing essentially fresh solvent to the river water. The throughput rate of river water was set at 33ml min $^{-1}$ for the sampling period of 7d. To further enhance solute recoveries, three extractors were connected in series and solvent charges were pooled prior to concentration and clean up. The pooled extracts were then reduced in volume over a steam bath in flasks equipped with three-ball Snyder columns. A Florisil clean-up step using sequential elutions with 6% and 15% ethyl ether–petroleum ether solutions was employed (the Florisil was activated at 130°C). A 300mm × 120mm plug of Florisil was topped by a 15mm × 20mm plug of anhydrous sodium sulphate and a wad of extracted glass wool. The Florisil was settled by tapping the tube lightly. This packing was then pre-wetted with 35ml of hexane which was then discarded. When the last of the hexane reached the top of the Florisil packing, the sample was quantitatively transferred to the column. The sample was then eluted with 200ml of 6% v/v ethyl ether–petroleum ether solution. When the last of the 6% eluate had reached the top of the packing, 200ml of 15% v/v ethyl ether–petroleum ether solution was added for the elution of Dieldrin and Endrin. Elution rates for both procedures were adjusted to approximately 5ml min $^{-1}$.

At this stage, the 6% eluate was ready for gas chromatography, but the 15% eluate required a further clean-up step on a magnesium oxide–Celite 545 column. The magnesium oxide was prepared for use by forming a slurry of 200g magnesium oxide in distilled water. The slurry was then heated on a steam bath for 30min, vacuum filtered, and dried overnight at 130°C. A blender was used to pulverise the dried filtrate which was then mixed in a 1:1 ratio (w/w) with Celite 545. A second 300mm × 20mm id chromatographic tube was attached to a 250ml vacuum flask. A plug of glass wool was placed in the bottom of this tube, then 10g of 1:1 magnesium oxide–Celite 545 was added under full vacuum (approximately 640mmHg) to pack it tightly. The vacuum line was then bled so that 35ml petroleum ether was eluted at a rate of 15–20ml min $^{-1}$. It was not necessary to discard this wash. After refluxed evaporation to 10ml, the 15% eluate was quantitatively transferred to the magnesium oxide–Celite 545 columns and eluted with 100ml petroleum ether at the 15–20ml min $^{-1}$ rate with partial vacuum. This eluate was again evaporated to 15ml by use of a Snyder column equipped flask and was then ready for gas chromatographic analysis as described below.

A dual column gas chromatograph equipped with two tritium electron capture detectors was employed by Brodtmann [16]. Both columns were acrylic glass, 1.83mm long by 0.32cm id. Column A was packed with 5% DC–260 on 80/100 DCMS Chromosorb W. Flow rate of carrier gas (5% methane–argon) through this column was 80ml min–1. Column B was packed with 1.5% OV–17/1.95% QF–1 on 80/100 Chromosorb W DCMS support. Flow rate of carrier gas (5% methane–argon) through this column was 50ml min $^{-1}$. Both pairs of injectors, columns, and detectors were maintained at, respectively, 212, 184, and 204°C.

Quantitation of unknown samples was accomplished by the comparison of peak heights for peaks in the unknown samples, confirmed by retention times on two different columns, to those of calibration mixtures of reference standard pesticides on both columns. Pesticide

Table 10.1 Overall recovery rates

Insecticide	% recovered	
	Carbon adsorption method	Liquid–liquid method
α-BHC	75.8	82.2
γ-Chlordane	78.4	95.2
p,p'-DDD	84.8	91.5
p,p'-DDE	55.6	93.8
o,p-DDT	74.0	95.9
p,p'-DDT	83.9	85.8
Dieldrin	85.2	95.6
Endrin	55.0	97.5
Heptachlor	52.6	83.8
Heptachlor epoxide	73.8	96.8
Lindane	77.5	92.2

Source: Reproduced with permission from Brodtmann, N.V. [16] American Water Works Association

recoveries obtained by Brodtmann [16] are given in Table 10.1. These values were obtained by spiking carbon samples followed by extraction and clean-up for the carbon adsorption method and by spiking aliquots of petroleum ether after a 7d sample period in the extractors. The overall conclusion of this work was that use of the liquid–liquid extractor method often finds low levels of chlorinated insecticides in water samples when the carbon adsorption method is insufficiently sensitive to do so.

10.1.1.4 Polychlorobiphenyls, polychlorodibenzo-p-dioxins, polychlorodibenzofurans and chlorinated insecticides

Berg et al. [19] separated polychlorobiphenyls from chlorinated insecticides on an activated carbon column prior to derivatisation and gas chromatographic separation on the column. Separation is based on the observation that polychlorobiphenyls adsorbed on activated charcoal cannot be removed quantitatively with hot chloroform but can be with cold benzene. Insecticides of the DDT group and a variety of others (eg γ-BHC, Aldrin, Dieldrin, Endrin and Heptachlor and its epoxide) can be eluted from the charcoal with acetone–ethyl ether (1:3). Typical recoveries from a mixture of p,p'-DDE (1,1–dichloro–2–bis–(4–chlorophenyl)ethylene), p,p'TDE, o, p-DDT, and a polychlorobiphenyl (Aroclor 1254) by successive elution with 90ml of the 1:3 acetone–ether and 60ml of benzene were 92, 92, 94 and 90% respectively. Identification and determination of the polychlorobiphenyls was effected by catalytic dechlorination to bicyclohexyl or perchlorination to decachlorobiphenyl, followed by gas chromatography. For bicyclohexyl, gas chromatography was carried out on a column (240cm × 6mm) consisting of 10% DC–710 on Chromosorb W, operated at 90°C for 2.5min then temperature programmed at 10°C min^{-1} with flame ionisation detection. For decachlorobiphenyl, the column (60cm × 3.1mm) is 5% SE–30 on Chromosorb W, operated at 215°C, with nitrogen as carrier gas, and a tritium detector. Chriswell et al. [20] have also discussed the preconcentration of polychlorobiphenyls on charcoal.

10.1.1.5 Chlorophenols

Activated carbon has been used to preconcentrate chlorophenols in non saline waters prior to gas chromatography [21].

10.1.1.6 Humic and fulvic acids

Boening et al. [22] have discussed the use of activated carbon adsorption for the removal of soil and leaf fulvic acid from water. These workers used total organic carbon and fluorescence measurements to determine the concentration of humic substances. although this work was orientated to the removal of humic substances from water on a commercial scale, it may, nevertheless, be of analytical interest.

10.1.1.7 Nitrosamines

Fine et al. [23] and Kimoto et al. [24] have preconcentrated nitrosamines on carbon and subsequently desorbed them with chloroform, ethanol or carbon disulphide prior to analysis by gas chromatography.

N-nitrosodimethylamine was detected in low ppt amounts by extraction with a carbon–based disc followed by gas chromatography with short path thermal desorber and a chemiluminescence nitrogen detector [25].

10.1.1.8 Miscellaneous organics

Bacaloni et al. [26] listed adsorption data for 51 compounds, including alcohols, acids, phenols, ethers, aldehydes, ketones, esters, pesticides, and PCBs, on graphitised carbon black at concentrations of 5–200µg L^{-1} of the compounds in water. The graphitised carbon black had a surface area of 100m^2 g^{-1}, particle size 80–100 mesh, and pH10.25. Chlorinated pesticides are among the compounds which are entirely adsorbed. Graphitised carbon black gave a larger specific retention volume and was more effective than Tenax, especially in non-potable water treatment. The best eluant for the recovery of chlorinated pesticides is 50:50 hexane: diethyl ether; recovery is generally 100% with 5ml of eluant.

Hertjies and Meijens [27] have conducted an examination of the effect of particle size, contact time, temperature, pH and other factors on the adsorption of organics by active carbon. Mangani et al. [28] examined the performance of graphitised carbon black (Carbopak B) in the preconcentration of some organic priority pollutants (pesticides, herbicides, phthalates, polyaromatic hydrocarbons) from non saline water. In most cases 100% recoveries were obtained in tests carried out by spiking either the adsorbent or the water sample. A very low eluent volume enabled high preconcentration ratios (1:1000). Coupled with gas chromatography–mass spectrometry, compounds of interest were determined at levels as low as 10ppt.

Rivera et al. [29] desorbed organics from 2000L water supply samples on to granular activated carbon. Organics were then desorbed with dichloromethane, evaporated to dryness, dissolved in diethyl ether and fractionated. Soluble fractions were analysed by GC/MS/DS. Ether insoluble compounds were analysed and fractionated by high performance liquid chromatography with diode array detection followed by FAB, and FAB–CID–MIKE characterisation. Two years' analytical data indicated that surfactants, plasticisers, ethylene glycol derivatives, phosphates and hydrocarbons could be considered chronic pollutants in the rivers studied. The occurrence of 152 compounds was reported.

Blanchard and Hardy [30] have described a method for determining benzene, toluene, ethylbenzene, dichloromethane, chloroform and carbon tetrachloride in non saline waters. Samples were collected using a silicone polycarbonate membrane and adsorbed on to activated carbon. The volatile components were desorbed with carbon disulphide and their levels determined by gas chromatography. Linearity of response and response time of the membrane were evaluated. The response time was less than 5min in all cases.

Carbon disulphide is an excellent solvent in the washing of active carbon traps. An additional advantage of carbon disulphide is the relatively low sensitively of the flame ionisation detector for this solvent [31].

Thakkar and Manes [32] have described a technique for the isolation of multiple trace components in non saline waters on activated carbon using adsorptive displacement. In this method the sorbed analytes are equilibrated with a solvent containing a large excess of a strongly sorbing solute/displacer.

Other applications of carbon adsorption are summarised in Table 10.2.

10.1.2 Seawater

10.1.2.1 Miscellaneous organics

One of the earliest choices of adsorbent for seawater organics was activated charcoal [39,40]. This technique has been refined for use in both

Table 10.2 Use of active carbon in preconcentration of organics

Substance preconcentrated	Method	Type of water	Ref.
Trihalomethanes	Mini-column	Potable	[33]
Chlorinated hydrocarbons	–	Potable, waste water	[34–36]
Polyaromatic hydrocarbons	Short column	Potable	[37]
Metal chelates or aromatic and aliphatic phosphonates	Filtered through active carbon	Non saline	[38]

Source: Own files

fresh and seawater by a number of workers [41,42]. A major problem in this technique has been the unknown efficiencies of collection and desorption. Jeffrey [43], using ¹⁴C–labelled material, found that 80% of the organic material in the seawater was adsorbed by the charcoal. Of that 80% approximately 80% again was desorbed by the solvents used. The overall efficiency of the method was therefore about 64%. This method of collection is one of the few that permits the accumulation of gram amounts of organic materials from seawater and therefore also permits the application of many of the standard techniques of organic analysis. However, it is not known how the distribution of compounds is changed in the process of adsorption and desorption; certain classes of compounds are probably entirely removed from the mixture, while others may be retained only in part. When so active a surface as carbon is used, there is also a possibility that chemical changes may take place during adsorption. The mixture as released from the charcoal may be considerably different from that originally present in seawater. This collection method, although attractive for its simplicity and speed, is limited to qualitative results.

10.1.3 Potable water

10.1.3.1 Polyaromatic hydrocarbons

Active carbon has been used to preconcentrate polyaromatic hydrocarbons [37].

10.1.3.2 Chlorinated hydrocarbons

Active carbon has been used to preconcentrate chlorinated hydrocarbons from potable and waste waters prior to analysis [34,35].

10.1.3.3 Anionic surfactants

Anionic detergents have been preconcentrated on active carbon prior to their determination [4–7,44,45].

10.1.3.4 Haloforms

Reichert and Lochtman [45] studied the appearance of haloforms in raw water and their formation during potable water treatment, as well as their elimination and the prevention of their formation. Techniques for the determination of organohalogen compounds are discussed, in particular, the extractive organic halogens and adsorbed organic halogen; methods are described and compared. A combination of these methods has been used to analyse surface waters in West Germany. The adsorbed organic halogens are first determined by adsorption on activated carbon followed by pyrolysis and microcoulometry. Parallel to these determinations the extractive organic halogens are determined according to the DIN regulation method, which involves extraction with pentane–di–isopropyl ether, pyrolysis and determination of halogens in a condenser, by coulometry or nephelometry.

Dressman and Stevens [33] and Oake and Anderson [46] have also studied the adsorption of haloforms on to carbon.

10.1.3.5 Miscellaneous organics

Rivera et al. [29] used activated carbon to preconcentrate organics from the Barcelona water supply. Organics were desorbed with methylene chloride, evaporated to dryness and dissolved in diethyl ether and fractionated. Ether soluble fractions were analysed by gas chromatography–mass spectrometry and ether insoluble fractions were analysed by high performance liquid chromatography with diode–array detection followed by fast ion bombardment.

10.2 Graphitised carbon black

10.2.1 Non saline waters

10.2.1.1 Phenols

Di Corcia et al. [47] used graphitised carbon black to extract phenols from water.

10.2.1.2 Non ionic surfactants

Preconcentration of graphitised carbon black followed by liquid chromatography–mass spectrometry with electrospray interface has been used to determine ng L^{-1} of non ionic polyethoxylate surfactants in non saline waters [48].

10.2.1.3 Chlorophenols

Turnes *et al.* [49] extracted chlorinated phenols by solid phase extraction as acetylated compounds using graphitised carbon. Sub ng L^{-1} concentrations were determined.

10.2.1.4 Insecticides and herbicides

Di Corcia *et al.* [50] and Mangani *et al.* [28] found that graphitised carbon black efficiently preconcentrated organochlorine pesticides from non saline waters.

Azine herbicides. Di Corcia *et al.* [51] preconcentrated Simazine and Atrazine on a miniaturised graphitised carbon black column (Carbopack B). A 250ml aqueous sample was passed through an adsorbent bed (50mg) of graphitised carbon black (at an optimal flow rate of 30ml min^{-1}). After washing with methanol (150µL) the herbicides were desorbed with 0.7ml of dichloromethane–methanol (3:2 v/v). Extracts were fractionated and analysed by reversed phase high performance liquid chromatography with an ultraviolet detector set at 220nm. Using this rapid method recoveries of Simazine and Atrazine added to 250ml of water at a concentration of 50ng L^{-1} were 97.2 and 95.8% with limits of detection of 0.07ng and 0.15ng, respectively.

Cai *et al.* [52] used graphiticised carbon black to extract ng L^{-1} concentrations of didealkylatrazine from water.

Shtivel [53] used graphitised carbon black cartridges to extract polar pesticides prior to high performance liquid chromatography. The cartridge was back flushed with the eluent to extract the pesticides prior to chromatography of the extract. Down to 0.01µg L^{-1} of 27 polar insecticides were determined; including Dichlorvos, Methoate, Oxamyl, Methomyl.

A multi-residue method for determining pesticides has been described [54] which uses a graphitised carbon black cartridge to remove pesticides from the water sample followed by liquid chromatographic analysis. Down to 0.003 to 0.07µg L^{-1} of 35 pesticides could be determined. Pesticides studied included Oxamyl, Methomyl, Phoxan, 2,4,5–trichlorophenoxyacetic acid (2.4–5T), 2.4–DB and MCPB.

References

1 Braus, H., Middleton, P.M. and Walton, G. *Analytical Chemistry*, **23**, 1160 (1951).
2 Sproul, O.J. and Ryckman, D.W. *Journal of Water Pollution Control Federation*, **33**, 1188 (1961).

3 Eichelberger, J.W., Dressman, R.C. and Longbottom, A. *Journal of Environmental Science and Technology*, **4**, 576 (1970).

4 Wang, L.K., Kao, S.I., Wang, W.H., Kao, J.H. and Loshkin, A.L. *Ind. Engrg Chem Prod. Res. Devel.*, **17**, 186 (1978).

5 Wang, L.K., Wang, M.H. and Kao, J.T. *Water Air Soil Pollution*, **9**, 337 (1978).

6 Wang, L.K., Yang, J.Y., Ross, R.G. and Wang, M.H. *Water Research Bulletin*, **11**, 267 (1975).

7 Wang, L.K., Wang, J.Y. and Wang, M.H. Proceedings of 28th Industrial Waste Conference, USA. pp. 76–82 (1973).

8 Goldberg, E.G. (ed.) In *A Guide to Marine Pollution*, Chapters 1, 2 and 4. Gordon and Breach, New York (1972).

9 Middleton, F.M,. Grant, W. and Rossen, A.A. *Industry Engineering Chemistry*, **48**, 268 (1956).

10 Ludzok, F.J., Middleton, F.M. and Ettinger, E.B. *Sewage Industrial Wastes*, **30**, 622 (1958).

11 Polange, R.C. and Magrigian, S. *Journal American Water Works Association*, **50**, 1214 (1958).

12 Middleton, R.M. and Lichtenberg, J.J. *Journal Industrial Engineering Chemistry*, **52**, 99A (1960).

13 Standard Methods for Examination of Water and Water APHA (13th edn). American Public Health Authority, Washington DC, p. 103 (1971).

14 Rosen, A.A. and Middleton, F.M. *Analytical Chemistry*, **31**, 1729 (1959).

15 Eichelberger, A. and Lichtenberg, J.J. *Journal of American Water Works Association*, **25**, 63 (1971).

16 Brodtmann, N.V. *Journal of American Water Works Association*, **67**, 558 (1975).

17 Breidenbach, A.W. In *Identification and Measurement of Chlorinated Hydrocarbon Pesticides in Surface Water*, 1968–0–315–842, US Government Printing Office, Washington DC (1968).

18 Kahn, L. and Wayman, C.H. *Analytical Chemistry*, **36**, 1340 (1964).

19 Berg, O.W., Diosady, P.L. and Rees, G.A.V. *Bulletin of Environmental Contamination and Toxicology*, **7**, 338 (1972).

20 Chriswell, C.D., Ericson, R.L., Junk, G.A. *et al. Journal of American Water Works Association*, **69**, 669 (1977).

21 Noll, K.E. and Gounaris, A. *Water Research*, **22**, 815 (1988).

22 Boening, P.H., Beckman, P. and Snoeyink, V.L. *Journal of American Waterworks Association*, **72**, 54 (1980).

23 Fine, D.H., Rounbehler, D.P., Huffman, F. and Epstein, S.S. *Bulletin of Environmental Contamination and Toxicology*, **14**, 404 (1975).

24 Kimoto, W.J., Dooley, C.J., Carre, J. and Fiddler, W. *Water Research*, **15**, 1099 (1981).

25 Tomkins, B.A. and Griest, W.H. *Analytical Chemistry*, **68**, 2533 (1996).

26 Bacaloni, A., Goretti, G., Lagana, A., Petronio, B.M. and Rotatori, M. *Analytical Chemistry*, **52**, 2033 (1980).

27 Hertjies, P.M. and Meijens, A.P. *Wasser Abwasser*, **111**, 61 (1970).

28 Mangani, F., Crescentini, P., Palma, P. and Bruner, F. *Journal of Chromatography*, **452**, 527 (1988).

29 Rivera, J., Ventura, F., Caixach, J., De Torres, M. and Figueras, A. *International Journal of Environmental Analytical Chemistry*, **29**, 15 (1987).

30 Blanchard, R.D. and Hardy, J.K. *Analytical Chemistry*, **50**, 1621 (1984).

31 Steven, S.F., Smith, J.F., Flego, U. and Renber, I. *Water Research*, **12**, 447 (1978).

32 Thakkar, S. and Manes, M. *Environmental Science and Technology*, **22**, 470 (1988).

33 Dressman, R.C. and Stevens, A.A. *Journal of the American Water Works Association*, **75**, 431 (1983).
34 Hahn, J. *Korrespondenz Abwasser*, **29**, 539 (1982).
35 Heckel, E. *Umwelt*, **No. 1**, 31 (1986).
36 Kleopter, R.D. and Fairless, B.T. *Environmental Science and Technology*, **6**, 1036 (1972).
37 Legana, A., Petronio, B.W. and Rotatori, M. *Journal of Chromatography*, **198**, 143 (1980).
38 Danz, O. and Jackwerth, E. *Fresenius Zeitschrift für Analytische Chemie*, **318**, 22 (1984).
39 Jeffrey, L.M. and Hood, D.W. *Journal of Marine Research*, **17**, 247 (1958).
40 Wangersky, P.J. *Science*, **115**, 665 (1952).
41 Vaccaro, R.F. *Environmental Science and Technology*, **5**, 134 (1971).
42 Grob, K. and Zuercher, F. *Journal of Chromatography*, **117**, 285 (1976).
43 Jeffrey, L.M. In Development of a Method for Isolating Gram Quantities of Dissolved Organic Matter from Sea Water and Some Chemical and Isotopic Characteristics of the Isolated Material, PhD Thesis, A&M University, Texas, USA, p.152 (1969).
44 Taylor, C.G. and Waters, J. *Analyst (London)*, **97**, 533 (1972).
45 Reichert, J.K. and Lochtman, J. *Environmental Science Letters*, **4**, 15 (1983).
46 Oake, R.J. and Anderson, I.M. Water Research Centre, Medmenham, UK. Report No. TR 217. The determination of carbon adsorbable organohalide in water (1984).
47 Di Corcia, A., Marchese, S., Samperi, R., Cecchini, G. and Cirilli, L. *Journal AOAC International*, **77**, 446 (1994).
48 Crescenzi, C., Di Corcia, A., Samperi, R. and Marcomini, A. *Analytical Chemistry*, **67**, 1797 (1995).
49 Turnes, J., Rodriguez, L., Garcia, C.M. and Cela, R. *Journal of Chromatography*, **743**, 283 (1996).
50 Di Corcia, A., Carfagnini, G. and Marchett, M. *Ann Chem. (Rome)*, **77**, 825 (1987).
51 Di Corcia, A., Marchett, M. and Samperi, R. *Journal of Chromatography*, **405**, 357 (1987).
52 Cai, Z., Gross, M.L. and Spalding, R.F. *Analytica Chimica Acta*, **304**, 67 (1995).
53 Shtivel, N.K., Gipokva, L.E. and Smirnova, Z.S. *Soviet Journal of Water Chemistry and Technology*, **7**, 66 (1985).
54 Jones, E.O. *Analytical Chemistry*, **63**, 580 (1991).

Chapter 11

Organics: Coprecipitation techniques

Organic compounds

11.1 Ferric hydroxide coprecipitation

11.1.1 Non saline waters

11.1.1.1 Phenols

Coprecipitation with ferric hydroxide has been used to preconcentrate phenols in non saline waters [1–4].

11.1.1.2 Humic acid

Hiraide *et al.* [5] applied coprecipitation with ferric hydroxide and flotation with cetyldimethylbenzylammonium chloride to preconcentrate humic acid from fresh waters.

11.1.2 Seawater

11.1.2.1 Miscellaneous organics

Ferric hydroxide is the precipitant most commonly used for the collection of organics. It is formed by the *in situ* formation of hydrated ferric oxides, usually by the addition of ferrous iron, followed by potassium hydroxide. The technique was first used for the precipitation of organic matter from an aged algal culture [1]. They recovered 79–95% of the ^{14}C-labelled material from such cultures. Williams and Zirino [2] measuring efficiencies of removal of dissolved organic carbon, found that such scavenging collected between 38 and 43% of the organic carbon measurable by wet oxidation with persulphate. Chapman and Rae [4] examined the effect of this precipitation on specific compounds. They found coprecipitation to be more complete with copper hydroxides, but still far from satisfactory. Only certain compounds were removed effectively by this treatment and the efficiency of removal varied with the water type and the organic compound involved.

11.2 Zirconium hydroxide and iron hydroxide coprecipitation

11.2.1 Seawater

11.2.1.1 Amino acids

A limited amount of work has been carried out using zirconium phosphates, compounds with well-defined coagulation and adsorption properties. The efficiency of coprecipitation was about 70% for free amino acids and albumin [5]. These methods may prove useful in the qualitative analysis of organic compounds, once the selectivities of the precipitants are understood. The metallic oxides suffer from the disadvantage of producing a precipitate which is difficult to filter, while calcite and zirconium phosphates produce relatively well-mannered precipitates. Even when the efficiencies of collection of various model compounds in seawater are known, the immense variety of organic compounds in seawater will keep this technique largely qualitative.

Husser et al. [6] determined 18 amino acids in seawater by a procedure based on coprecipitation with ferric hydroxide. Iron and cations are removed by ion exchange and amino acids determined by paper and ion exchange chromatography.

11.3 Barium sulphate coprecipitation

Organic compounds

11.3.1 Non saline waters

11.3.1.1 Polyaromatic hydrocarbons

Kusada et al. [7] have shown that small amounts of polycyclic aromatic hydrocarbons in aqueous solution are almost completely adsorbed on barium salts of copper(II) sulphophthalocyanines and cobalt(II) phthalocynaine, which were precipitated from the solution. Recoveries of the PAHs from the precipitates by thermal desorption gas chromatography were in the range 71–95%. The method is useful for the concentration and analysis of medium molecular weight, thermally stable PAHs.

Organometallic compounds

11.3.1.2 Organolead compounds

Mikac and Branica [8] preconcentrated dissolved dialkyllead and inorganic lead species by coprecipitation with barium sulphate. Alkyllead was then determined in the concentrate by differential pulse anodic-stripping voltammetry.

11.4 Magnesium hydroxide

11.4.1 Non saline waters

11.4.1.1 Polyaromatic compounds and polychlorobiphenyls

Hess *et al.* [9] used *in situ* precipitation with magnesium hydroxide to selectively preconcentrate polynuclear aromatic hydrocarbons and polychlorinated biphenyls. Magnesium sulphate and ammonium hydroxide were added to a sample aqueous solution (1L) spiked with polychlorinated biphenyls and polynuclear aromatic hydrocarbons. Centrifugation produced a precipitate which was then dissolved in ammonium chloride and sulphuric acid. The solution was extracted three times with methylene chloride and the combined extracts analysed by gas chromatography with flame ionisation detection. This method of *in situ* magnesium hydroxide precipitation was selective for PAHs and polychlorinated biphenyls of high molecular weight. The method discriminated against acidic molecules (phenols), basic molecules (amines) and other neutral molecules such as phthalate esters. Results from samples of non saline waters were more variable and showed lower recoveries than for pure water samples.

References

1 Jeffrey, L.M. and Hood, D.W. *Journal of Marine Research*, **17**, 247 (1958).
2 Williams, P.M. and Zirino, A. *Nature (London)*, **204**, 462 (1964).
3 Tatsumota, M., Williams, W.T., Prescott, J.M. and Hood, D.W. *Journal of Marine Research*, **19**, 89 (1961).
4 Chapman, G. and Rae, A.C. *Nature (London)*, **214**, 627 (1967).
5 Hiraide, M., Ren, F.N., Tamura, R. and Mizuike, A. *Mikrochimica Acta*, **416**, 137 (1987).
6 Husser, E.R., Stehl, R.H., Price, D.R. and De Lap, R.A. *Analytical Chemistry*, **49**, 154 (1977).
7 Kusada, K., Shiraki, K. and Miva, T. *Analytica Chimica Acta*, **224**, 1 (1989).
8 Mikac, X. and Branica, M. *Analytica Chimica Acta*, **212**, 349 (1988).
9 Hess, G.G., McKenzie, D.E. and Hughes, B.M. *Journal of Chromatography*, **366**, 197 (1988).

Chapter 12

Organics: On-line and off-line preconcentration methods

The advantage of on-line liquid–solid extraction procedures for the preconcentration of organic compounds prior to analysis is that a large volume of sample is passed through a column containing the adsorbent material that adsorbs the organic compounds of interest. Subsequently the organics in the adsorbent cartridge are desorbed by a relatively small volume of a reagent or a solvent which contains an appreciably enriched concentration of the organic compounds. This is then passed into the detection instrument. The detection limit can thereby be improved by one to three orders of magnitude. Also, the reduction in sampling handling in this type of procedure can considerably reduce analytical errors.

There is a general trend to use liquid chromatography instead of gas chromatography combined with liquid–solid extraction in environmental water analysis of polar analytes [1,2]. Trace enrichment of organic pollutants from water matrixes by off-line liquid–solid extraction has some disadvantages caused by sample manipulation and losses of the most volatile analytes, particularly phenols, during the concentration of the extract [3]. For this reason, automated on-line liquid–solid extraction methods have been developed to minimise sample manipulation and analyte losses [4,5].

Liquid chromatography–mass spectrometry using thermospray [5] and particle beam [6] interfaces have been reported for the analysis of phenols, although problems associated with the lack of structure information were shown. The advent of liquid chromatography–mass spectrometry with atmospheric pressure ionisation interfaces, mainly atmospheric pressure chemical ionisation, electrospray, and ion spray, has overcome those disadvantages because of higher sensitivity [7–9]. Atmospheric ionisation techniques have been less frequently used in environmental analysis, especially in the negative ion mode. Applications using electrospray, ion spray and/or atmospheric pressure chemical ionisation for the analysis of organophosphorus, chlorinated phenoxy acids, triazines and carbamates have been reported. Pentachlorophenol and a few chloronitrophenols were analysed by conventional electrospray methods and polar pesticides by thermospray methods.

Organic compounds

12.1 On-line methods

12.1.1 Non saline waters

12.1.1.1 Aromatic sulphonates

Lange *et al.* [7] used on-line solid phase extraction followed by liquid chromatography to determine aromatic sulphonates in non saline waters.

12.1.1.2 Aminophenols and chloroanilines

Guenn and Hennion [8] have discussed an on-line method for preconcentrating aminophenols and chloroanilines using porous graphite carbon as the adsorbent prior to high performance liquid chromatography.

12.1.1.3 US Environmental Protection Agency priority pollutants

Puig *et al.* [9] have described a method for determining ng L^{-1} levels of methylphenol, nitrophenol and chlorophenols in river water samples by automated on-line liquid–solid extraction followed by liquid chromatography–mass spectrometry using atmospheric pressure chemical ionisation and ion-spray interfaces.

12.1.1.4 Insecticides and herbicides

Haman and Kettrup [10] developed two high performance liquid chromatographic methods for trace analysis of phenoxyherbicides in range 1mg L^{-1} to 1ng L^{-1} using spiked water samples. The sample consisted of 1L extraction with methylene dichloride followed by on-line enrichment, solid phase preconcentration and on-line enrichment. The detection of targeted compounds was accomplished by ultraviolet detection at 230 and 280nm.

Other methods for solid phase on-line preconcentration of insecticides and herbicides are reviewed in Table 12.1.

12.1.1.5 Miscellaneous organics

Hankmeier *et al.* [21] used solid phase extraction followed by gas chromatography–AED. This has been used to determine sub ng L^{-1} amounts of pollutants in non saline waters (100mL sample).

Table 12.1 Solid phase on-line preconcentration methods for insecticides and herbicides in non saline waters

Compound	Adsorbent	Detection limit*	Solvent elution	Analytical detection	Ref
Organochlorine insecticides	–	–	–	LG–GLC combination	[11]
Organochlorine insecticides and phase adsorbent polychlorobiphenyls	Reversed phase	–	Hexane	LC	[12]
Organophosphorus insecticides	–	–	–	Electrospray ion spray atmospheric chemical ionisation LC–mass spectrometry	[13]
Chlorotriazines	–	–	–	LC	[14]
Triazine type	–	ng L^{-1}	–	Gas chromatography–mass spectrometry	[15,16]
30 pesticides including atrazine, simazine, alachlor and molinate	–	0.01–0.5µg L^{-1}	–	HPLC	[2]
Carbamate type	–	–	–	–	[17]
Misc. pesticides	Membrane disc	–	–	Gas chromatography	[18]
Acidic herbicides	–	sub µg L^{-1}	–	Electrospray mass spectrometry with chemical ionisation	[19]
Polar pesticides LC mass spectrometry	–	100ng L^{-1}	–	Thermospray	[20]

LC = liquid chromatography GC = gas chromatography
*µg L^{-1} unless otherwise stated

Source: Own files

Brinkman [22] has reviewed on-line extraction coupled to gas chromatography or liquid chromatography for the detection of organic compounds in non saline waters.

Organometallic compounds

12.1.2 Non saline waters

12.1.2.1 Organomercury compounds

Falter and Schoeler [23] used on-line preconcentration combined with liquid chromatographic separation and cold vapour atomic absorption spectrometry to determine down to 0.5ng L^{-1} methylmercury in non saline waters.

12.2 Off-line methods

12.2.1 Non saline waters

12.2.1.1 Herbicides

Honing et al. [17] used off-line liquid chromatography–mass spectrometry with a thermospray interface to determine sub µg L^{-1} concentrations of carbamate pesticides in non saline waters.

Lacorte et al. [24] used off-line liquid chromatography–mass spectrometry with a thermospray interface to determine sub µg L^{-1} concentrations of Temephos and its degradation products in non saline waters.

References

1 Puig, D. and Barceló, D. Trends Anal. Chem., 15, 362 (1996).
2 Chiron, S., Fernandez Alba, A. and Barceló, D. Environmental Science and Technology, 27, 2352 (1993).
3 Liska, I. Journal of Chromatography, A, 665, 163 (1993).
4 Barceló, D. and Hennion, M.C. Analytica Chimica Acta, 318, 1 (1995).
5 Puig, D and Barceló, D. Chromatographia, 40, 435 (1995).
6 Cappiello, A., Famiglini, G., Palma, P., Berioni, A. and Bruner, F. Environmental Science and Technology, 29, 2295 (1995).
7 Lange, F.T., Wenz, M. and Branch, H.J. Journal of High Resolution Chromatography, 18, 243 (1995).
8 Guenn, S. and Hennion, M.C. Journal of Chromatography, 665, 243 (1994).
9 Puig, D., Silgoner, M., Grasserbauer, M. and Barceló, D. Analytical Chemistry, 69, 2756 (1997).
10 Haman, R. and Kettrup, A. Chemosphere, 18, 527 (1987).
11 Jones, B. Journal of High Resolution Chromatography, Chromatography Communications, 11, 181 (1988).
12 Noroozian, E., Moris, F.A., Nielen, M.W.F. et al. Journal of High Resolution Chromatography, Chromatography Communications, 10, 17 (1987).

13 Lacorte, S. and Barceló, D. *Analytical Chemistry*, **68**, 2464 (1996).
14 Onnerfjord, P., Barceló, D., Emneus, J., Gorton, L. and Marko Varga, G. *Journal of Chromatography*, **737**, 35 (1996).
15 Vreuls, J.J., Bultermann, A.J., Ghijsen, R.T. and Brinkman, U.A.T. *Analyst (London)*, **117**, 1701 (1992).
16 Molina, C., Durand, G. and Barceló, D. *Journal of Chromatography, A,* **712**, 113 (1995).
17 Honing, M., Riu, J., Barceló, D., Van Barr, B.L.M. and Brinkman, U.A.T. *Journal of Chromatography, A,* **733**, 283 (1996).
18 Louter, A.J.H., Brinkman, U.A.T. and Ghijsen, R.T.J. *Microcolumn Separation,* **5**, 303 (1993).
19 Chiron, S., Papilloud, S., Haerdi, W. and Barceló, D. *Analytical Chemistry*, **67**, 1637 (1995).
20 Sennert, S., Volmer, D., Levsen, K. and Wuensch, G. *Fresenius Journal of Analytical Chemistry*, **351**, 642 (1995).
21 Hankemeier, T., Louter, A.J.H., Rinkema, F.D. and Brinkman, U.A.T. *Chromatographia*, **40**, 119 (1995).
22 Brinkman, U.A.T. *Environmental Science and Technology*, **29**, 79A (1995).
23 Falter, F. and Schoeler, H.F. *Fresenius Journal of Analytical Chemistry*, **353**, 34 (1995).
24 Lacorte, S., Ehresmann, N. and Barceló, D. *Environmental Science and Technology*, **30**, 917 (1996).

Chapter 13

Organics: Solvent extraction methods

13.1 Non saline waters

Organic compounds

13.1.1 General discussion

Solvent extraction is an attractive method for concentrating a particular fraction of the dissolved organic matter, the fraction concentrated being determined by the choice of solvent. The most obvious limitation on the method is that set by solvent choice, the solvent should have only limited solubility in water, which limits the materials removed to the less polar compounds. Another limitation is that set by contamination. The solvent used must be purified carefully, since the amounts of the various organic compounds collected from water could be about as large as the trace impurities in the solvents. Once the separation into the organic solvent has been accomplished, any of a number of techniques of fractionation and analysis can be applied. The first step in the solvent extraction is the actual sampling of the water. All of the problems associated with sampling can occur in this step and may be aggravated by the large volumes of water customarily employed.

Solvent extraction has been applied both as a batch and a continuous process [1–5] and to non saline waters and seawaters. Since a large mass of water is to be extracted, the batch methods are limited although Kawahara et al. [6] constructed a semi-automatic device to speed up the procedure and reduce the organic solvent volume. Batch methods are usually constructed without continuous refreshing systems. Instead a repeated shaking with a fresh portion of solvent is used.

Various workers [7–9] have described an *in situ* solvent extraction equipment suitable for on-site operation on river banks etc. Ahnoff and Josefsson [7] point out that to avoid sample loss and contamination it is desirable to carry out sample enrichment for the analysis of trace organic compounds in water as near to the sampling point as possible. Even sampling points on adjacent land stations or ships can lead to contamination from tubing walls and the inner surfaces of the pump and also deposition of particles can occur in the tubing. The *in situ* apparatus they describe is designed to perform solvent extraction of large amounts of water, using the apparatus continuously while situated at the sampling point at a desired depth.

Peters [2] has compared continuous extractors for the extraction and concentration of trace organics from water. Baker *et al.* [3] evaluated a Teflon helix liquid–liquid extraction for concentration of trace organics in water into methylene chloride. The apparatus utilises Teflon coils for phase contact and gravity–phase separation. Gomella *et al.* [4] devised a method of extracting and concentrating micropollutants in water with cyclohexane. The cyclohexane extracts are frozen and the cyclohexane is then allowed to sublime. Fritz [5] has compared various methods for concentrating trace organics from water including resin sorption, solvent extraction, water gas distribution, gas stripping, direct injection into a gas chromatograph and resin sorption with thermal desorption. Burchill *et al.* [10] give details of equipment and procedure for the routine monitoring of water for trace organic compounds down to the $0.01\mu g\ L^{-1}$ level involving solvent extraction with dichloromethane followed by gas chromatography with flame ionisation detection and mass spectrometry.

A liquid–liquid extraction procedure for the simultaneous recovery of a wide range of basic, neutral and phenolic organic compounds in a single container has been described [11]. Samples were extracted with a mixture of dichloromethane and ether, using a specially designed strirrer and modified micro Snyder evaporator. Analysis was possible by various chromatographic methods and recoveries varied between 72 and 109% with a detection limit of $0.01\mu g\ L^{-1}$ for each compound.

Eichelberger *et al.* [12] studied the effect of sample pH on precision and accuracy in the determination of organics in water by methylene dichloride extraction followed by fused silica capillary column gas chromatography–mass spectrometry and packed column gas chromatography–mass spectrometry.

Murray [13] has described a rapid micro-extraction procedure for preconcentration and analysis of trace amounts of organic compounds in water and compared the results obtained with those obtained by macro-extraction methods. Final analysis of the solvent extract was carried out by gas chromatography. The method using a flask with a side arm and fitted with a capillary tube involved shaking 980ml of water with 200μL of hexane. By carefully adding water through the side arm the solvent layer was displaced into the capillary tube after tilting the flask. About 50μL were recovered and was suitable for direct analysis by gas chromatography. This method gave as good a recovery and was much more rapid than macro-extraction or continuous extraction using three inverted flasks with a vibratory mixer and siphon system.

In the micro method an extraction flask (Fig. 13.1) similar to the rapid liquid extraction used by Grob *et al.* [14] was used. The flask, which contained 980ml of water and 200µL of hexane, was manually shaken for 2min. By tilting the extraction flask and carefully adding water through the side arm, the solvent layer could be held in the centre portion and finally displaced into the capillary tube. About 50µL could be recovered and was suitable for direct analysis by gas chromatography.

For the macro method a steam distillation solvent extraction head similar in design to No. 6555, manufactured by Ace Glass Co. (as used by Veith and Kiwus [15]) was made with a reduced solvent capacity of 10ml. Each 10L water sample was passed through this system (Fig. 13.1(b)) in about 8h. The hexane layer was removed and analysed by gas chromatography.

A continuous extraction apparatus was developed (Fig. 13.1(c)) which consisted of three inverted 250ml volumetric flasks clamped to a retort stand, from which a bar was connected to a vibratory mixer. This was adjusted to give a vigorous shaking of all three flasks. Each 10L sample was siphoned through the apparatus in about 4h and extracted with a total solvent volume of 10ml. The hexane was recovered in a separatory funnel for analysis by gas chromatography. Quantitative analyses are based on the use of internal standards and integration of peak areas. Glass distilled hexane was used throughout.

Recovery of hydrocarbons with boiling points up to 250°C was low (from 8–16%) with both the micro Snyder column and the rotary evaporator. The recovery of phthalate esters and pesticides using the rotary evaporator was slightly better than the micro Snyder column but both techniques showed losses of 10% or more for pesticides (Table 13.1).

A Hewlett Packard 5750 gas chromatograph was used with an Inctronics C.R.S. 208 integrator for quantitative analysis. A 2m × 6mm od glass column was packed with 10% Dexsil 400 coated on Chromosorb W AW, 80–100 mesh.

Potable water samples spiked with ng L^{-1} concentrations of selected pesticides and µg L^{-1} concentrations of hydrocarbons (C_{10}, C_{12}, C_{14} and C_{16}) were extracted successively with 200µL of hexane and analysed by gas chromatography using electron capture and flame ionisation detectors, respectively. There was a mean recovery of 58.3% in the first 200µL extract and 94.3% in three 200µL extracts. The standard deviations (calculated from 16 values, four analyses of four extracts) for the first extract range from 2.9 to 10.3% with the higher values at the µg L^{-1} level (Table 13.2).

When 1L of potable water was extracted with 200µL of solvent, only about 50µL were recovered because of losses due to solubility and evaporation. Thus, a concentration factor of ×20000 was achieved. But since the extraction efficiency was only about 50% in the first extract the true concentration factor was ×10000. When 10L of potable water were

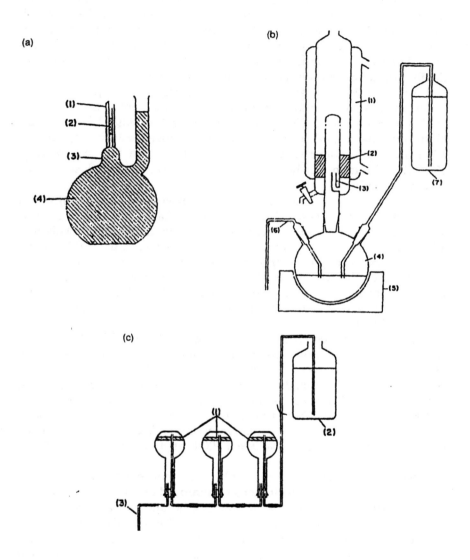

Fig. 13.1 (a) Micro extraction flask: 1 = capillary tube; 2 = solvent layer; 3 = modified 1L volumetric flask; 4 = water sample. (b) Continuous steam distillation and extraction apparatus (modified Nielsen–Kryger): 1 = Condenser; 2 = solvent layer; 3 = water overflow; 4 = boiling flask; 5 = heating mantle; 6 = overflow; 7 = sample reservoir. (c) Continuous solvent extraction apparatus: 1 = solvent layer (hexane); 2 = sample reservoir; 3 = overflow

Source: Reproduced with permission from Murray, D.A.J. [13] Kluwer/Plenum Academic Publishers, New York

Table 13.1 Recoveries on concentrating from 10ml to 1ml in hexane

Substance	Boiling point (°C)	Recovery (%) Micro Snyder column (90°C)	Recovery (%) Rotary evaporator (50°C)
Lindane	–	81	92
Aldrin	–	77	84
Heptachlor epoxide	–	78	90
α-cis–Chlordane	–	78	87
Dieldrin	–	79	85
n-Octane	126	37	8
n-Decane	174	57	25
n-Dodecane	216	59	39
n-Tetradecane	254	61	49
Dipropyl phthalate	305	81	100
Diisobutyl phthalate	298	82	96
Dibutyl phthalate	340	82	100
Butyl glycolyl butyl phthalate	219 at 5mmHg	84	100

Source: Reproduced with permission from Murray, D.A.J. [13] Kluwer/Plenum Publishing

Table 13.2 Recoveries from tap water using micro-extraction procedure

Substance	Concen-tration	Recovery in extract (%)	Rel. stand. dev.[a] (%)	Total recovery in extractions (%)
Aldrin	10ng L⁻¹	47.9	5.6	89.4
Heptachlor epoxide		58.5	8.2	91.3
α-cis–Chlordane		59.2	10.2	92.0
Dieldrin		62.2	10.0	92.6
n-Decane	50µg L⁻¹	69.3	3.7	98.6
n-Dodecane		60.3	3.3	97.3
n-Tetradecane		56.0	3.9	96.7
n-Hexadecane		52.9	2.9	96.4
Dibutyl phthalate	2.5µg L⁻¹	65.5	5.9	96.6
Butyl glycolyl butyl phthalate	25µg L⁻¹	43.6	4.0	89.7

[a]Four analyses of four extracts

Source: Reproduced with permission from Murray, D.A.J. [13] Kluwer/Plenum Publishing

extracted with 10ml of solvent, losses due to solubility and evaporation are proportionally less significant than with the micro-procedure and about 8ml of solvent were recovered. Thus the concentration factor was about ×1250. The extraction efficiency will be higher than with the microprocedure but the concentration factor cannot exceed ×1250 without using a solvent concentration step.

The superior concentration factor of the micro method, the absence of a concentration step, and speed of analysis are advantages in routine quantitative analyses of water samples. Preconcentration factors of 20000 for traces of organics in potable water samples have been achieved by extraction using diethyl ether and petroleum ether (bp 30–40°C) [15–16]. One portion of the extract was reduced to a volume suitable for analysis by gas chromatography–mass spectrometry. The other portion was reacted with diazomethane, which converted some components to methyl derivatives which were then amenable to analysis by a computerised gas chromatography–mass spectrometry system. In order to check for possible procedural contamination two solvent blanks (one for each solvent) were processed in the same way as the extracts.

High resolution glass capillary (porous layer open tubular, PLOT) columns were used for the gas chromatographic separation and each concentrated solvent extraction and blank was run on two different columns, one of which had OV–1 (a methyl silicone gum) as a stationary phase, while the other had FFAP (Carbowax 20M treated with 2-nitroterephthalic acid) as stationary phase.

Sampling: Collect 20L samples in four pre-cleaned 5L glass stoppered bottles and store in the dark during transportation to the laboratory. Before taking a sample, the bottle is rinsed once with the water to be sampled.

Extraction: Obtain solvent extracts for gas chromatography–mass spectrometry from the samples collected in a specially designed rig (Fig 13.2). After the sample has been transferred to the rig (previously filled with nitrogen) seal the apparatus under a flow of nitrogen. Then acidify the sample to pH2 by addition of sulphuric acid (50% v/v usually about 30ml). Stir the sample (6000rev min^{-1}) for a few minutes, check the pH and add purified petroleum ether (500ml, bp 30–40°C). Stop the stirrer after about 5min and allow the solvent layer to separate (about 20min). Then pressurise the solvent layer (using nitrogen via the adjustable take off head and remove it into the separating funnel. Add a second aliquot of petroleum ether (500ml, bp 30–40°C) to the sample and repeat the extraction process. Combine the two extracts and dry (pre-treated anhydrous sodium sulphate). For the extraction with diethyl ether, add two aliquots (each of 1L) with stirring (5min) after each addition. Allow the solvent layer to separate and remove from the rig. Due to the solubility of diethyl ether in water, the volume of this first extract (about 500ml) is considerably less than total amount (2L) of diethyl ether added. Add a further aliquot (500ml) of diethyl ether and extract the sample.

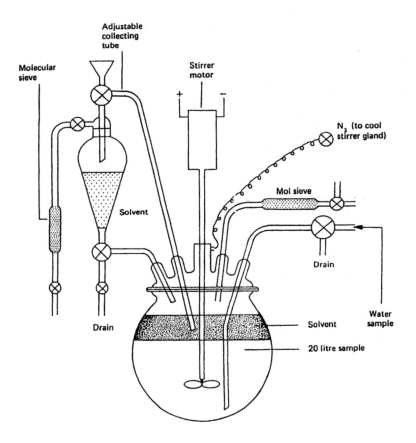

Fig. 13.2 Solvent extraction apparatus
Source: Reproduced with permission from Murray, D.A.J. [13] Kluwer/Plenum Academic
Publishers, New York

Remove the extract from the rig and combine with the first diethyl ether
extract prior to drying (pre-treated anhydrous sodium sulphate).

Concentrating: Concentrate the dried solvent extracts (to 10ml) on a
rotary evaporator, and transfer to precleaned 10ml centrifuge tubes. Carry
out further concentration (to 1ml) under a stream of dry clean air. Split the
extracts equally into two fractions (500µL), store one fraction (at −18°C)
prior to derivatisation, and concentrate the other further (to 100µL) prior
to examination by gas chromatography and mass spectrometry.

Derivatisation: Add a solution (10ml) of N-methyl-N-nitrosotoluene-*p*-
sulphonamide (100g L^{-1}) in diethyl ether:methanol (2:3) to a solution
(50% v/v) of potassium hydroxide (1ml) and distil the diazomethane

formed on to the sample to be methylated. Stopper the sample tube and allow to stand for 30min. Remove the stopper and allow the excess diazomethane to evaporate. Reduce the methylated extract in volume to 100µL in a stream of dry, clean air.

Gas chromatography–mass spectrometry operating conditions:

Hewlett Packard 5710 A (VG I6F).

Carrier gas: helium at 4–5ml min $^{-1}$.

Capillary columns: 50m PLOT columns containing OVI or FFAP (Phase Separations Ltd or Perkin–Elmer Ltd).

Injector: SGE splitless injection system SCI–B operated at 250°C.

Column temperature: Initial 80°C held for 4min, then programmed at 8°C min $^{-1}$ to 250°C (OV–1) or 220°C (FFAP) and held at top temperature for at least 10min.

All spectra were obtained under electron impact (EI) conditions.

Ionisation voltage: 70 eV.

Trap current: 100µA.

Resolution: 10000.

Source temperature: 220°C.

Transfer line temperature: 260°C.

Electron multiplier gain: 2 kV.

Mass range: 20–460 amu.

Scan speed: 1 decade $^{-1}$.

Interscan delay: 1.

Data system: The data system used throughout was a VG 2040 system. Perfluorokerosene was used as the calibration compound for all the gas chromatography–mass spectrometry runs. The calibration step was always followed by the acquisition of several mass spectra of perfluoro-kerosene in order to check the calibration.

Generally, 1000 to 1250 mass spectra were acquired during each gas chromatography–mass spectrometry run. The initial data interpretation step involved an inspection, usually on a visual display unit, of the computer reconstructed total ion current chromatogram and the mass spectra obtained for each obvious peak in the chromatogram was examined. If necessary a background subtraction was performed. If the mass spectrum of a compound of interest was not visually recognised as a known compound, reference was made either to the Eight Peak Index of Mass Spectra [17] or the EPA/NTH Mass Spectral Data Base [18]. If a suitable spectral match was not obtained from either of these sources, the fragmentation pattern of the spectrum of interest was examined and an attempt made to deduce the structure of the compound giving rise to the spectrum. Identifications made either from reference data or from an examination of the mass spectrum produced were considered to be tentative until a pure sample of a known compound gave a suitable match for the mass spectrum and gas chromatographic retention time.

Mass chromatography was used to search for compounds of interest which did not appear as obvious peaks on the TLC chromatogram.

Rooney et al. [19] extracted miscellaneous organics from non saline water using diethyl ether, carbon disulphide or methylene dichloride and achieving preconcentration factors of 10^5. Gas chromatography was used in the analytical finish. Gomez-Taylor et al. [20] similarly achieved a preconcentration factor of 2.5×10^5 in diethyl ether extractions of miscellaneous organics from non saline water. These workers used Fourier transform infrared spectroscopy for the analytical finish.

Ibrahim and Suffet [21] used Freon 113 (trichlorotrifluoroethane) as an alternative to methylene dichloride for the preconcentration of organics from chlorinated potable water. Freon minimised the interference from solvent artifacts arising from the reaction of cyclohexane (preservative in methylene chloride) with residual free chlorine (present in chlorine treated water samples). The extraction efficiency of Freon FC–113 was comparable with that of methylene chloride when 100–300ng non-polar organics per litre was extracted and concentrated 1000-fold. No by-products were observed when Freon was used to extract chlorinated water.

Yriezx et al. [22] used a large volume (10L) extractor using dichloro-methane in a pulsed column system to extract ng L^{-1} amounts of polar and non-polar organics from non saline waters.

Bisiuk [23] has described a device for the isolation of organic compounds from an aqueous phase which utilises the spray technique to increase the surface in the solvent:water interface exchange.

Umano et al. [24] carried out large-scale extractions of trace organics in non saline waters by continuous liquid–liquid extraction with dichloro-methane.

The preconcentration by solvent extraction of various individual types of organic compounds is now discussed. See Table 13.3 for a review of the following types of organic compounds: nitrosamines, non ionic detergents, chloroanisoles, pyrethrins, amines, siloxanes, vegetable oils, linear alkyl benzene sulphonates, geosmin, squoxin, EDTA and carbon disulphide.

13.1.2 Aliphatic hydrocarbons

Fitzpatrick and Tan [63] showed that the use of Freon solvent for the determination of total petroleum hydrocarbons in non saline waters can also be eliminated by using gas chromatography–FID measurements.

13.1.3 Aromatic hydrocarbons

Stoffmeister et al. [68] preconcentrated monocyclic aromatic hydro-carbons from potable water by micro-extraction with n-pentane and addition of di-isobutylene as an internal standard before determination by gas chromatography with a flame ionisation detector. The detection limit was in the range 0.1–0.5µg L $^{-1}$. See also Table 13.3.

13.1.4 Polyaromatic hydrocarbons

Concawe [195] recommends methods described by the Environmental Protection Agency for the determination of polyaromatic hydrocarbons in oil refinery effluents. The method involves extractive preconcentration of the effluent with methylene dichloride followed by clean-up procedures followed by gas chromatography or liquid chromatography.

For the gas chromatographic method no detector is specified; flame ionisation seems to be the best choice. Lack of selectivity in the method can lead to interference by compounds that are not completely removed by the clean-up. Higher selectivity can be achieved by use of a photoionisation detector. In addition some pairs of the polyaromatic hydrocarbon isomers are incompletely separated by the 1.8m column used in this method while the heavier polyaromatic hydrocarbons often show tailing peaks. Almost all polyaromatic hydrocarbon isomers can be separated using a longer column (9.0m) with the same packing material, or by using a glass capillary column; only benzo(b)fluoranthrene–benzo(k)fluoranthene are not resolvable even by the long packed, or capillary column approach, both of which are recommended as a final step in preference to a conventional column. The liquid chromatographic procedure uses reversed phase liquid chromatography with fluorescence detection to separate all 16 polyaromatic hydrocarbons completely. The method is sensitive and so selective as often to allow the method to be applied without a clean-up procedure.

For gas chromatographic methods detection limits are about 1µg L $^{-1}$ whereas for the liquid chromatographic methods limits are between 1 and 100µg L $^{-1}$ for 2- and 3-ring aromatics and below 1µg L $^{-1}$ for the 4-, 5- and 6-ring compounds. For polyaromatic hydrocarbon analysis the extraction with methylene chloride may successfully be substituted by cyclohexane allowing high recoveries by one single extraction. Isolation of polyaromatic hydrocarbons from the water matrix may also conveniently be performed via adsorption using prepacked small high performance liquid chromatography columns circumventing the problems of emulsification often arising during liquid–liquid extraction. This technique has been used in a study of polynuclear aromatic hydrocarbons in aqueous effluents from refineries as it was shown to differentiate between certain compounds which were not resolved by gas chromatography–mass spectrometry.

Yu et al. [82] have described a method for the enrichment of trace levels of polycyclic aromatic hydrocarbons in non saline waters involving the use of ultrasonic stirring to enhance the extraction capability of the organic solvent for the polycyclic aromatic hydrocarbons and heating to break down the emulsion. See also Table 13.3.

13.1.5 Phenols

Czuczwa *et al.* [196] preconcentrated traces of phenols and cresols from rain water by a method based on continuous liquid–liquid extraction (solvents heavier than water) and normal phase and high performance liquid chromatography.

Onuska and Terry [89] used micro-extraction using true miscable phases (water and 2-propanol) followed by demixing of these phases, caused by the addition of ammonium sulphate, to extract µg L^{-1} concentrations of phenols from 50ml of water. See also Table 13.3.

13.1.6 Anionic surfactants

Inaba [193] described improvements in the tetrathiocyanatocobalt(II) dipotassium salt method to achieve greater sensitivity and simplicity in the determination of polyoxyethlene-type non ionic surfactants. The amount of cobalt(II) extracted into the organic phase (toluene) was determined by using 4-(2-pyridylazo)–resorcinol as a spectrophotometric reagent. The molecular absorptivity of cobalt(II)–PAR complex was 33 times larger than that of the thiocyanate complex. Octaethyleneglycol mono-*n*-dodecylether was used as a standard polyoxyethylene-type non ionic surfactant. Absorbance of the organic phase increased with increasing thiocyanate and/or cobalt(II) concentration, when the aqueous phase contained no potassium chloride salt. Interferences from coexisting substances such as linear alkylbenzene sulphonates and humic acid were eliminated during the toluene preconcentration step. Preliminary ion exchange procedures were recommended for samples with high salinity (seawater) and/or high concentrations of interfering substances. The applicability of the method was demonstrated by the analysis of several highly polluted Japanese lakes and marshes. See also Table 13.3.

13.1.7 Carboxylic acids

Takati and Vernon [96] used chloroform extraction followed by capillary column gas chromatography to estimate extremely low levels of fatty acids in river waters and sewage effluents. See also Table 13.3.

13.1.8 Chlorinated hydrocarbons

See Table 13.3.

Table 13.3 Preconcentration solvents for organics

Substance preconcentrated	Type of water	Solvent	Finish technique	Ref.
Alkanes	Non saline	Pet. ether	GLC	[25]
		C_5H_{12}	GLC	[26,27]
		C_6H_{14}	GLC	[28,29]
		C_6H_{14}	UV	[29]
		C_8H_{18}	GLC	[30,31]
		iso–C_8H_{18}	GLC	[32]
		$PhCH_3$	GLC	[27]
		$PhCH_3$–PhC_2H_5	GLC	[27]
		$(C_2H_5)_2O$	GLC	[33,34]
		$(C_2H_5)_2O$	Fluorescence	[35–38]
		CH_3OH–C_6H_6	GLC	[39,40]
		$C_6H_5NO_2$	GLC	[41–43]
		CH_2Cl_2	IR	[44]
		$CHCl_3$	GLC	[26,45]
		CCl_4	GLC	[26,27,41, 46–48]
		CCl_4	UV	[49,50]
		CCl_4	IR	[27,41,48, 51–61]
		CCl_4	NMR	[62]
		$C_2Cl_3F_3$	NMR	[62]
		$C_2Cl_3F_3$	IR	[232]
		Freon	GC, FID	[63]
Aromatic hydrocarbons	Non saline	CS_2	GLC	[64]
	Rain	$C_6H_5NO_2$	GLC	[65]
	Rain	C_6H_6	Spectrophoto.	[66]
	Non saline	$CH_3COOC_2H_5$	GLC–MS	[67]
	Non saline	n-C_5H_{12}	GC–FID	[68]
Polyaromatic hydrocarbons	Non saline	n-C_5H_{12}	GLC	[69]
	Non saline	cyclo C_5H_{10}	TLC	[70,71]
	Potable	cyclo C_5H_{10}	Spectrofluoro	[72]
	Potable	cyclo C_5H_{10}	GLC	[73]
	Non saline	C_6H_6	TLC	[74]
	Non saline	CH_2Cl_2	TLC	[75]
	Non saline	CH_2Cl_2	GLC	[69,75–79]
	Non saline	CCl_4	TLC	[80]
	Non saline	$(C_2H_5)_2O$	GLC	[69]
	Non saline	$(CH_3)_2CO$	GLC	[69]
	Non saline	Acetonitrile	GLC	[69]
	Non saline	Dimethyl sulphoxide	GLC	[75,81]
	Non saline	Ultrasonic stripping	–	[82]
Phenols	Non saline	CH_2Cl_2	GLC	[83]
	Potable	$(C_2H_5)_2O$	Spectrophoto	[84]
	Non saline	$(C_2H_5)_2O$	GLC	[85]
	Non saline	$CH_3COOC_2H_5$	GLC–MS	[86]

Table 13.3 continued

Substance preconcentrated	Type of water	Solvent	Finish technique	Ref.
	Non saline	$CH_3COOC_2H_5$	GLC	[87]
	Rain	$C_6H_5CH_3$	GLC	[88]
	Non saline	water C_3H_7OH Micro-extraction and $(NH_4)_2SO_4$ demixing	–	[89]
Carboxylic acids	Non saline	CH_2Cl_2	GLC	[90]
		$CHCl_3$	TLC	[91]
		$(C_2H_5)_2O$	Spectrophoto	[92]
		$CH_3OH–CHCl_3$	GLC	[93]
		$C_6H_6–CH_3OH$	GLC	[94,95]
	Non saline	$CHCl_3$	GLC	[96]
Chlorinated hydrocarbons	Non saline	C_5H_{12}	GLC	[97–99]
	Lake	C_5H_{12}	GLC	[100]
	Non saline	$C_6H_6–C_6H_{14}$	GLC–MS	[101]
	Non saline	CH_2Cl_2	GLC	[102]
	Non saline	CCl_4	NMR	[62]
	Non saline	$C_2F_3Cl_3$	NMR	[62]
	Waste	$n\text{-}C_5H_{12}$	GLC	[103]
Chlorophenols	Non saline	$n\text{-}C_6H_{14}$	GLC	[104]
	Rain	$C_6H_5CH_3$	GLC	[88]
	Non saline	$C_6–C_8$ alkanes	–	[105]
	Non saline	Acetylation–solvent extraction	–	[106]
Polychlorobiphenyls	Non saline	C_6H_{14}	GLC	[107,108]
Polychlorobiphenyl/ DDT	Non saline	C_6H_{14}	–	[109]
Haloforms	Potable	Pet. ether	GLC	[110]
		Pet. ether	HPLC	[111]
		C_5H_{12}	GLC	[112–118]
		C_6H_{14}	GLC	[112,119]
		C_7H_{16}	GLC	[112]
		$iso\text{-}C_8H_{18}$	GLC	[112–114]
		$n\text{-}C_9H_20$	GLC	[112]
		Me cyclohexane	GLC	[112,120, 121]
		$C_6H_4(CH_3)_2$	GLC	[122]
Chlorinated insecticides	Non saline	Pet. ether	GLC	[123–127]
	Non saline	$n\text{-}C_6H_{14}$	GLC	[123–129]
	Rain	C_5H_{12}	GLC	[130,131]
	Non saline	$(C_2H_5)_2O$	GLC	[132–134]
	Non saline	C_6H_6	GLC	[127,135]
	Non saline	$CH_2CCl_2\text{:}C_6H_{14}$ (15:85)	GLC	[136]
	Non saline	CH_2C_{l2}	GLC	[102,136]
	Non saline	$CHCl_3$	TLC	[137]

Table 13.3 continued

Substance preconcentrated	Type of water	Solvent	Finish technique	Ref.
	Non saline	Acetonitrile	GLC	[127]
	Non saline	$C_2F_3Cl_3$	NMR	[62]
Polychlorobiphenyl,	Non saline	C_6H_{14}	GLC–EC	[138]
DDT, HCH, HCB chlorinated insect-icides, polyaromatic hydrocarbons, polychlorobiphenyls, phthalates and organophosphorus insecticides	Non saline	Microwave	–	[139]
Misc pesticides	Non saline	n-C_7H_{16}	–	[140]
Misc pesticides	Non saline	Continuous on-line solvent extraction	–	[141]
α-Bendo–sulphon	Non saline	C_6H_{14}	GLC	[142]
Vapona	Lake	$CHCl_3$	GLC	[143]
Trichlorpyr	Non saline	$(C_2H_5)_2O$	GLC	[144]
Dichlorvos	Non saline	C_6H_{14} $C_6H_5CH_3$ C_6H_{14}–C_6H_6	GLC	[145]
Dieldrin	Non saline	Pet. ether	GLC	[146]
Phosphoric acid esters (after hydrolysis)	Non saline	$CHCl_3$	GLC	[147]
	Non saline	CH_2Cl_2	GLC–MS	[148]
Organophosphorus insecticides	Lake	n-C_6–H_{14}	TLC	[149]
	Non saline	C_6H_6	GLC Spectrophoto	[150,151]
	Non saline	CH_2Cl_2	GLC	[102,152]
	Non saline	$CHCl_3$	TLC	[153]
	Non saline	$CHCl_3$	GLC	[143]
Herbicides	Non saline	iso-C_8H_{18}	GLC	[154–159]
Herbicides carbamate type	Non saline	C_6H_6	GLC	[154–159]
	Lake	CH_2Cl_2	Spectrophoto	[160]
	Non saline	CH_2Cl_2	GLC	[102,161]
	Non saline	CH_2Cl_2	TLC	[158,162–164]
m-s-Butyl phenyl methyl (phenylthio) carbamate	Non saline	$CHCl_3$	GLC	[161]
	Non saline	Acetonitrile	GLC	[161]
Phenoxyacetic acid type	Non saline	C_6H_{14}	GLC	[165]
		C_6H_6	GLC	[166]
		$(C_2H_5)_2O$	Spectrophoto	[167]
		$(C_2H_5)_2O$	TLC	[168]
		$(C_2H_5)_2O$	GLC	[169]
		$CH_3COOC_2H_5$	GLC	[169]

Table 13.3 continued

Substance preconcentrated	Type of water	Solvent	Finish technique	Ref.
Triazine type	Non saline	CH_2Cl_2	GLC	[170–172]
Urea type	Non saline	$C_6H_5CH_3$	GLC	[173]
		CH_2Cl_2	GLC	[174,175]
		CH_2Cl_2	TLC	[162]
		$CHCl_3$	GLC	[173]
		CCl_4	GLC	[169]
		$(C_2H_5)_2O$	GLC	[173]
		CH_3NO_2	GLC	[173]
Misc. types	Non saline	CH_2Cl_2	GLC	[176]
Dalapon	Non saline	$(C_2H_5)_2O$	TLC	[177]
Paraquat, diquat	Non saline	CH_2Cl_2	GLC	[178]
Dicamba	Non saline	$(C_2H_5)_2O$	GLC	[179]
Nitrosamines	Non saline	CH_2Cl_2	GLC	[180,181]
Non ionic detergents as tetrathiocyanato cobalt(II) complex	Lakes	$C_6H_5CH_3$	AAS	[182]
Chloroanisoles	Non saline	CH_2Cl_2	GLC	[183]
Pyrethrins	Potable	C_6H_{14}	GLC	[184]
Amines	Potable	C_6H_{14}	GLC–MS	[185]
Siloxanes	Non saline	Pet. ether	ICPAES	[186]
Vegetable oil	Non saline	CCl_4	IR	[187]
Linear alkyl benzene sulphonates	Non saline	C_6H_{14}	GLC	[188]
Geosmin	Non saline	CH_2Cl_2	GLC–MS	[189]
		CH_2Cl_2-n-tridecane	GLC	[190]
Squoxin	Non saline	n-C_3H_7OH–CCl_4 (2:1)	Spectrophoto	[233]
ETDA	Non saline	$CHCl_3$	GLC	[191,192]
Nitrophenols	Non saline	$C_6H_5CH_3$	GLC	[88]
Anionic surfactants	Non saline	Toluene extraction of 4(–2-pyridylazo)resorcinol complex	Spectrophoto	[193]
Volatile organic compounds	Non saline	Thermospray solvent extraction	–	[194]

Source: Own files

13.1.9 Chlorophenols

Fingler et al. [105] carried out a comparative study of trace enrichment of chlorophenols from ground and potable waters by extraction with C_6 and C_8 alkanes and by octadecyl carbon C_{18} reversed phase adsorption. All three procedures could efficiently enrich traces of chlorophenols, the choice of procedure depending on the required analytical sensitivity. Extraction procedures could be applied to concentrations of chlorophenols at or above the $1\mu g$ L^{-1} level. However, the C_{18} reversed-phase adsorption procedure achieved a detection limit of $10ng$ L^{-1}.

Dano *et al*. [106] used acetic analydride and pentafluorobenzoyl chloride to derivitivise chlorophenols *in situ* in non saline waters. The derivatives were extracted from water and determined by gas chromatography. Acetic anhydride was the more suitable reagent for the determination of poly-chlorophenols while pentafluorobenzoyl chloride was appropriate for less chlorinated phenolic compounds. See also Table 13.3.

13.1.10 Polychlorobiphenyls

Kahn and Wayman [197] reported a continuous multi-chamber liquid–liquid extractor with internal solvent refreshing for the extraction of non-polar contaminants, especially polychlorobiphenyls and chlorinated pesticides from non saline waters. With this apparatus, it was possible to extract 135L of water at rates of 0.5 to $1.0L$ h^{-1} and in that way isolate and identify μg L^{-1} concentrations of these compounds with a recovery of 97%. Goldberg *et al*. [9] presented a similar apparatus which was capable of handling flow rates up to $2L$ h^{-1} designed to operate with heavier than water solvents. In a later work Goldberg *et al*. [1] described modified extractors for both heavier than water and lighter than water solvents. Three solvent heavier than water extractors were set up in one series and four solvent lighter than water extractors were set up in a second series.

Spiked water was pumped at a rate of 7 to $8L$ h^{-1} through each set of extractors and the extraction efficiencies were determined. A concentration factor of up to 105 was obtained and the dipole moment difference between the solute and solvent was demonstrated to be an index of the extraction efficiency.

Ahnoff and Josefsson [198] modified the solvent extraction apparatus of Kahn and Wayman [197] to fit field conditions by using one chamber and enlarging the outlet tube. The phase separation became more efficient and the stirring rate could be increased without losing drops of the organic solvent. This arrangement is similar to that described by Kerner *et al*. [199] who used a bulb for better phase separation. However, a system where the organic solvent is recycled is too complicated for routine field application, especially when immersed in the water. The changes of the outer temperature will disturb the heating of the solvent. The few theoretical plates in the column between the evaporation flask and the reflux condenser did not prevent the organic compounds cycling together with the solvent. Water is dissolved in the vaporising flask. The boiling of the solvent together with water causes, for example, uncontrollable heat changes, hydrolysis reactions, and steam distillation. In the system devised by Ahnoff and Josefsson [198] the organic refreshing system normally used was eliminated. Chlorinated biphenyls were extracted from some hundred litres of water at the Gota river with different multi-chamber arrangements. Finally, the content of polychlorinated biphenyl was determined with ECD gas chromatography in the range 0.1–1.0ng L^{-1} of water.

Apparatus: The apparatus is shown in Fig. 13.3. In principle, it is a mixing settler although the mixing and settling chambers are not completely separated but combined in one cylinder. The water is continuously drawn through the cylinder under vigorous mixing with a lighter than water organic solvent, thereby forming a vortex. The mixing is performed by a magnetic stirrer (A) located under the cylinder. The organic solvent is trapped by a flange in the cylinder and is not replenished during the procedure. The mixing of the water and solvent takes place in the upper part of the cylinder and the emulsion is separated successively in the lower part. The lighter solvent droplets are allowed to rise and return into the mixing zone. In this way, very little of the extracting solvents is lost during the procedure. Use of a solvent volume of 100–300ml with the 800ml apparatus (shown in Fig. 13.3(a)) is recommended.

The apparatus consists of three main parts; the extraction unit, the magnetic stirring device (A), and the pump (B) (ProMinent Electronic 0304T, CFG, West Germany). The extraction unit is designed in four parts; top (C), bottom (D), bowl (E) all in PTFE and a borosilicate glass cylinder (F), 80mm id by 190mm, and a wall thickness of 5mm. The top and bottom parts are pressed water tight to the glass cylinder by plates of acid proof stainless steel (G) and four screwbolts (H). The bottom steel plate has an opening in the middle (65mm in diameter) to actuate the stirring bar inside (I). The magnetic stirring bar (28mm × 18.5mm) rotates at about 900rpm. In the top part are two channels (3mm in diameter) one for the water inlet and one, equipped with a stopcock, for the filling and the draining of the apparatus. The bottom part (D) has a circular hollow to fit the bottom of the bowl (E). Three symmetrically placed holes (K) at the edge of the hollow are in contact with a circular channel (L) which leads to the outlet tube (B). In this way, the extracted water is sucked out from the calm circular compartment (O) with a symmetrical flow. The bowl (E) is designed to keep the mixing process above and inside whereas the outside of the bowl acts to separate the emulsion and return the organic solvent into the mixing zone. The bowl has a double flange (M) to quell the vertical emulsion movement. The circular cross section area between the flange and the glass wall is important because the flow rate acts to bring the emulsion downward while the lighter solvent is striving upward. The rotating movement in the upper compartment is definitely

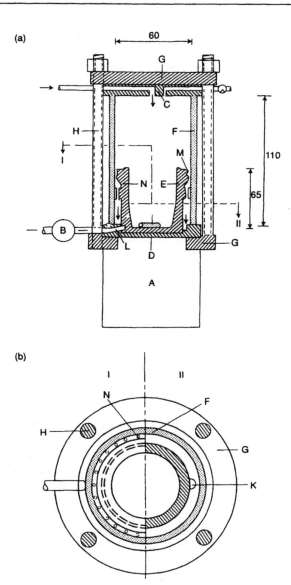

Fig. 13.3 (a) Schematic diagram of the mixed settler (dimensions given in mm). (b) Cross-section at I and II from the apparatus shown above
A = magnetic stirrer; B = pump; D–E = top, middle and bottom of extraction unit; F = borosilicate glass cylinder; G = acid proof stainless steel plate; H = screwbolts (bottom plate with stirring bar I); K = three symetrically placed holes; L = circular channel

Source: Reproduced with permission from Ahnoff, M. and Josefsson, B. [198] American Chemical Society, Washington

Table 13.4 Concentration of PCB found in Gota River water at *c.* 1 m

Arrangement	Date	Water vol. (L)	Pump rate (L h⁻¹)	PCB (L⁻¹)
Parallel	8 Sept. 1972	290	5.0	0.31
	8 Sept. 1972	305	5.2	0.29
Parallel	17 Jan. 1973	170	3.5	0.21
	17 Jan. 1973	185	3.8	0.22
Series 1				
Extractor 1	29 Jan. 1973	190	3.8	0.32
Extractor 2	29 Jan. 1973	190	3.8	0.16
Series 2				
Extractor 1	31 Jan. 1973	170	3.4	0.20
Extractor 2	31 Jan. 1973	170	3.4	0.12
Series 3				
Extractor 1	3 June 1973	240	3.0	0.28
Extractor 2	3 June 1973	240	3.0	0.17
Extractor 3	3 June 1973	240	3.0	0.09

Source: Reproduced with permission from Ahnoff, M. and Josefsson, B. [198] American Chemical Society

quelled at N (Figs. 13.3(a),(b)) by letting the water pass 36 holes (3.6mm in diameter). The final phase separation occurs in the cavity (O) below the holes where the mixing procedure has no influence. By this arrangement of the settling zones, most of the lighter solvent is trapped and returns to the mixing compartment, provided the flow rate is not too high. Since there is no pressure drop in the apparatus, the pump can be placed at the outlet. Thus contamination caused by the pump is eliminated.

Table 13.4 presents the amounts of polychlorinated biphenyls found in the river water extracts from different field experiments at the Gota river. Three different extraction arrangements were set up in five experiments; two parallel extractor experiments and three experiments with the extractors arranged serially. The water flow varied as well as the total amount of water pumped through the system.

Petrick *et al.* [200] used hexane to preconcentrate polychlorobiphenyls from non saline water prior to analysis by gas chromatography. Aliphatic and polyaromatic compounds were first separated from the extract by high performance liquid chromatography. See also Table 13.3.

13.1.11 Haloforms

Vartiainen *et al.* [201] used diethyl ether liquid–liquid extraction to pre-concentrate mutagenic organic compounds from chlorinated potable water with a high humus content. See also Table 13.3.

13.1.12 Chlorinated insecticides

Mohnke *et al.* [138] have described a method for determining polychlorobiphenyls, DDT, HCHs and HCB at the low ng L^{-1} level in non saline waters involving extraction with hexane, clean-up with sulphuric acid and analysis by gas chromatography with electron capture detection.

Schoeler and Brodesser [140] used liquid–liquid extraction using *n*-heptane on a light phase rotation perforator to preconcentrate pesticides in non saline waters.

Bourgeois *et al.* [109] carried out micro-extractions of 40ng L^{-1} poly-chlorobiphenyls and DDT using 1ml aliquots of hexane. Analyte recoveries of about 90% were achieved.

Continuous liquid–liquid extraction with on-line gas chromatography–AED provided detection of µg L^{-1} concentrations of pesticides [141].

Chee *et al.* [139] have discussed the use of microwave-assisted solvent elution to remove analytes, including organochlorine pesticides, polychlorinated biphenyls, polycyclic aromatic hydrocarbons, phthalates and organophosphorus pesticides from C$_{18}$ discs obtained in preconcentration studies on non saline waters. See also Table 13.3.

13.1.13 α Bendo–sulphan

See Table 13.3.

13.1.14 Vapona

See Table 13.3.

13.1.15 Trichlorpyr

See Table 13.3.

13.1.16 Dichlorvos

See Table 13.3.

13.1.17 Dieldrin .

See Table 13.3.

13.1.18 Phosphoric acid esters

See Table 13.3.

13.1.19 Organophosphorus insecticides

Leoy [25] preconcentrated Azinphos methyl, Fenthion and Diazinon by extraction from aqueous solution with n-heptane, isoamyl alcohol, cyclohexane, diethyl ether, dichloromethane or octanol. Diazinon was not efficiently extracted. Gas chromatography calibration graphs were linear for 0.5–0.7 and 8–20mg L $^{-1}$ for Azinphos methyl (extractions were 90 and 70%, respectively), and up to 4mg L $^{-1}$ for Fenthion (extraction efficiency 33%). Detection limits and relative standard deviation were 0.04 and 0.09mg L $^{-1}$ and 3.4 and 5.3% for Azinphos methyl and Fenthion respectively. Other pesticides and related compounds did not interfere. See also Table 13.3.

13.1.20 Carbamate type herbicides

See Table 13.3.

13.1.21 Phenoxyacetic acid type herbicides

See Table 13.3.

13.1.22 Triazine type herbicides

See Table 13.3.

13.1.23 Substituted urea type herbicides

See Table 13.3.

13.1.24 Miscellaneous herbicides

See Table 13.3.

13.1.25 Nitrosamines

See Table 13.3.

13.1.26 Non ionic surfactants

See Table 13.3.

13.1.27 Chloroanisoles

See Table 13.3.

13.1.28 Pyrethrins

See Table 13.3.

13.1.29 Amines

See Table 13.3.

13.1.30 Siloxanes

See Table 13.3.

13.1.31 Vegetable oils

See Table 13.3.

13.1.32 Linear alkylbenzene sulphonates

See Table 13.3.

13.1.33 Geosmin

See Table 13.3.

13.1.34 Squoxin

See Table 13.3.

13.1.35 Ethylene diamine tetraacetic acid

See Table 13.3.

13.1.36 Nitrophenols

See Table 13.3.

13.1.37 Volatile organics

Farrell and Pacey [194] used a thermospray extractor to remove semi-volatile organic compounds from 300ml samples of non saline waters.

Organometallic compounds

13.1.38 Organomercury compounds

Non saline waters: Methyl-, ethyl- and phenyl–mercury compounds have been preconcentrated from non saline waters by extraction with benzene, followed by conversion to their cysteine complexes and gas chromatography [202,203]. Pentane:diethyl ether (1:1) has also been used to preconcentrate dialkylmercury compounds prior to analysis by gas chromatography [204,205].

Stary et al. [206] developed a preconcentration radioanalytical method for determining down to 0.01µg L^{-1} of methyl- and phenylmercury and inorganic mercury using 100–500mL samples of potable or river water. Extraction chromatography and dithizone extraction were the most promising methods for the concentration of organomercurials in the concentration range 0.01–2µg L^{-1}. The dithizone extraction was used for the preconcentration of inorganic mercury.

Schintu et al. [207] preconcentrated organic and inorganic mercury by extraction with dithizone solution. Organic mercury was then back-extracted with aqueous sodium thiosulphate prior to determination by cold-trap cold–vapour atomic absorption spectrometry.

13.1.39 Organotin compounds

Various solvents have been used to preconcentrate organotin compounds from non saline waters including benzene–tropolone [208,209], n-hexane–tropolone [210], chloroform–tropolone [209], methylene dichloride [209] and benzene [211]. The concentrated organotin compounds are analysed either by direct gas chromatography [208–210,212] or are converted to tin hydrides and gas chromatographed [211].

13.1.40 Organolead compounds

Non saline waters: Chau et al. [213] extracted organolead compounds from water into benzene, prior to conversion to their dithiocarbamates and subsequently to butyl–lead derivatives prior to gas chromatography. Hexane has also been used [214] to preconcentrate alkyl–lead compounds from non saline water prior to analysis by gas chromatography using an atomic absorption detector. Down to 0.5µg L^{-1} organolead compounds can be determined by this procedure.

13.2 Seawater

Organic compounds

13.2.1 Sampling techniques

Because of the relatively large amount of time and equipment required for the processing of each sample, these methods must be used to characterise the organic compounds at a few selected stations and depths, rather than in mass surveys. Once the separation into the organic solvent has been accomplished, any of a number of techniques of fractionation and analysis can be applied.

The first step in the solvent extraction is the actual sampling of the water column. All of the problems associated with sampling can occur in this step, and may be aggravated by the large volumes of seawater customarily employed. The cleverest approach to this problem is to avoid sampling in the normal manner altogether. Ahnoff and Josefsson [7] have described an *in situ* apparatus for solvent extraction. This apparatus is buoyed at the sampling depth, anywhere between the surface and 50m, and water is pumped through a series of extraction chambers. The capacity of the unit is 50L per 48h. The use of an *in situ* pumping system on the far side of the extraction chambers eliminates the pump and hose contamination, as well as much of the contamination coming from the passage through the surface film. Since the apparatus is battery powered, the unit may be suspended from a free-floating surface buoy; no ship time is required, except for placement and recovery of the samplers. Thus while each sample may require up to 48h to collect, a number of depths and areas may be sampled in the same time period.

Since the non-polar organic content of seawater is fairly low, identification of unknown compounds requires the processing of large quantities of seawater. Where possible, continuous extraction should be favoured over batch extraction. Several workers have developed such systems: Goldberg *et al.* [1] designed and evaluated a system for fresh water. This apparatus seems fragile, and would require considerable redesign before it could be used routinely on shipboard. Ahnoff and Josefsson [198] built a solvent extraction apparatus for river work which was later modified into their *in situ* extractor [7]. The unit as described in the earlier work could easily be adapted for seawater analysis. A unit based on a Teflon helix liquid–liquid extractor, some 101.5m in length, was constructed by Wu and Suffet [215]. The extractor was optimised for the removal of organophosphorus compounds, specifically pesticides, with an efficiency of around 80%. For some compounds, these continuous extraction methods should be the methods of choice, and should be further explored. These compounds are the ones that are not seriously affected by hose or pump contamination, are present in low concentrations, and are important enough, for either scientific or practical reasons to warrant the extra time spent on station.

The conventional approach to solvent extraction is the batch method. Early work with this method was hampered by the low concentrations of the compounds present and the relative insensitivity of the methods of characterisation. Thus lipids and hydrocarbons have been separated from seawater by extraction with petroleum ether and with ethyl acetate. The fractionation techniques included column and thin layer chromatography with final characterisation by thin layer chromatography, infrared and ultraviolet spectroscopy, and gas chromatography. Of these techniques, only gas chromatography is really useful at the levels of organic matter present in seawater. With techniques available today such as glass capillary gas chromatography and mass spectrometry, much more information could be extracted from such samples [216].

The information could be restricted to a tractable amount by performing some preliminary fractionation before the gas chromatography–mass spectrometry step. This type of separation and fractionation has been proposed by Copin and Barbier [217]. During the investigation of pollution in coastal seawaters, Werner and Waldichuk [218] pointed out the need for concentrating and isolating trace amounts of certain substances with a continuous solvent extractor. They constructed a modified Scheibel apparatus by changing the organic solvent cycle system.

13.2.2 Aliphatic hydrocarbons

Zsonay and Kiel [219] have used flow calorimetry to determine total hydrocarbons in seawater. In this method the seawater (1L) was extracted with trichlorotrifluorethane (10ml) and the extract was concentrated, first in a vacuum desiccator, then with a stream of nitrogen to 100μL. A 50μL portion of this solution was injected into a stainless steel column (5cm × 1.8mm) packed with silica gel (0.063–0.2mm) deactivated with 10% of water. Elution was effected, under pressure of helium, with trichloro-trifluorethane at 5.2ml $^{-1}$ and the eluate passed through the calorimeter. In this the solution flowed over a reference thermistor and thence over a detector thermistor. The latter was embedded in porous glass beads on which the solutes were adsorbed with evolutions of heat. The difference in temperature between the two thermistors was recorded. Each solute first displayed the adsorbed solvent molecules, giving a desorption peak (negative); this was followed by an adsorption peak (positive). The area of the desorption peak was proportional to the amount of solute present. All hydrocarbons were eluted as one peak, the silica gel column removing the non-hydrocarbon solutes. Responses to unsaturated hydrocarbons were less than those to saturated ones; if the peak area was calculated as nonocane, an error of 25% could be introduced. The limit of detection was about 5μg L $^{-1}$.

Law et al. [220] used pentane, hexane and methylene chloride to extract hydrocarbons from coastal seawaters. The use of extraction solvents of similar polarity (ie pentane or hexane) give comparable results (average variation 22–23%) for samples with total hydrocarbon concentrations in the range 1–2μg Ekofisk crude oil equivalents per litre. Results obtained using dichloromethane for extraction were higher by a factor of around 2. Probable reasons were the greater extraction efficiency of dichloromethane owing to its higher polarity; more efficient scavenging of adsorbed hydrocarbons by dichloromethane owing to its higher water miscibility.

13.2.3 Halogenated hydrocarbons

Eklund *et al.* [221,222] have developed a capillary column method [223, 224] for the determination of down to 1µg L $^{-1}$ volatile organohalides in waters which combine the resolving power of the glass capillary column with the sensitivity of the electron capture detector. The eluate from the column is mixed with purge gas of the detector to minimise band broadening due to dead volumes. This and low column bleeding give enhanced sensitivity. Ten different organohalides were quantified in seawater. Using this technique these workers detected bromoform in seawater for the first time. Halogenated hydrocarbons in different waters were identified by comparison with a standard solution. Retention times were measured on two columns with different stationary phases, ie SE–52 and Carbowax 400. Extraction of 100ml water with various amounts of pentane ranging from 1 to 15ml showed that the extraction efficiency was increased when using a lower water to pentane ratio.

13.2.4 Chlorinated insecticides

Since petroleum ether was the solvent used in earlier studies for extracting the DDT from seawater, Wilson and Forester [225] initiated further studies to evaluate the extraction efficiencies of other solvent systems. The recovery rates of *o,p'*-DDE in all tests were greater than 89% with petroleum ether, 15% diethyl ether in hexane followed by hexane or methylene dichloride, indicating no significant loss during analyses. The average percentage recovery of *p,o'*-DDT extracted from seawater containing 3.0µg L $^{-1}$ of the DDT isomer (salinity 16ng L $^{-1}$) or distilled water samples up to 14d after initiation of the experiment was of a similar order.

13.2.5 Aminoacids

Palmark [226] determined glycine, threonine, valine and phenyl alanine by conversion to their DNPPA derivatives and solvent extraction from seawater.

13.2.6 Miscellaneous organics

The Oil Companies International Study Group for Conservation of Clean Air and Water–Europe (Concawe) [227] have made a detailed study of the application of solvent extraction to the determination of organics in water. In this procedure, one portion of the aqueous sample is adjusted to pH11 and extracted with pure methylene dichloride. The methylene dichloride extract containing basic and neutral substances is examined by gas chromatography coupled to a mass spectrometer. The alkaline aqueous phase is acidified and extracted with methylene dichloride to provide an extract containing phenols which, again is examined by gas chromatography–mass spectrometry. A second portion of the water sample is acidified and extracted with methylene dichloride to provide a further phenol–containing extract for examination by high performance liquid chromatography. A third portion of the water sample is extracted with methylene dichloride and examined for polynuclear aromatic hydrocarbons by high performance liquid chromatography using an ultraviolet–fluorescence detector and by gas chromatography using a flame ionisation detector.

Solvent extraction has proved to be most useful when applied to the concentration of particular compounds for which there exists an analytical method of great sensitivity. The major application of the method has been for the determination of hydrocarbons in seawater. In general, solvent extraction is an excellent method for the concentration and determination of specific compounds, chiefly non-polar, in seawater [4]. Special precautions must be taken to prevent contamination from trace materials in the solvents used. *In situ* methods offer many advantages, not the least being the elimination of lengthy processing in the ship-board laboratory; these methods should be investigated more thoroughly and, if possible, extended to greater depths. When coupled to modern separation and detection systems, the methods may offer the simplest and most direct approach to the measurement of certain classes of compounds. In most cases, we have little or no estimate of the efficiency of the solvent extraction techniques. Because of the great variety of compounds present in any one sample, a true efficiency of extraction may be impossible to obtain. Working efficiencies, using model compounds, may be the only approach in trying to make the analysis truly quantitative.

Hon-Nami and Hanya [228,229] used chloroform extraction to extract linear alkyl benzene sulphates from seawater prior to analysis by gas chromatography–mass spectrometry.

Organometallic compounds

13.2.7 Organomercury compounds

Organomercury compounds have been preconcentrated from seawater by forming the dithizonates and extracting these with a small volume of carbon tetrachloride. Mercury is then back-extracted from the organic phase with a small volume of hydrochloric acid and determined by stannous chloride reduction and cold vapour atomic absorption spectrometry [205].

13.2.8 Organotin compounds

Triphenyltin compounds in water at concentrations of 0.004 to 2mg L $^{-1}$ are readily extracted into toluene and can be determined by spectrofluorometric measurements of the triphenyltin–3–hydroxyflavone complex.

13.3 Potable waters

13.3.1 Miscellaneous organics

The procedure described by Murray [230] discussed in section 13.1.1 has also been applied to the determination of pesticides and hydrocarbons in non saline waters and has also been applied to potable waters.

13.4 Waste waters

13.4.1 Halogenated aliphatic hydrocarbons

Pfannhauser and Thaller [103] have described a gas chromatographic method for quantitatively estimating traces of 16 different halogenated solvents in waste water. The solvent residues were extracted using *n*-pentane on a column containing a mixture of deactivated Florasil with the ground sample. The elute was injected into a fused silica capillary column and the peaks recorded by electron-capture detection (Ni–63). The method could detect as little as 0.01µg L $^{-1}$ of most halogenated short-chain aliphatic hydrocarbons.

13.4.2 Miscellaneous organics

Kuhelka *et al.* [231] used a preconcentration technique involving extraction of the sample with organic solvent followed by gas–liquid chromatographic determination on a packed column or by mass fragmentography. Using the second technique the detection limit could be lowered by a factor of 10–100. Detailed procedures for analysis of chemical works waste waters by gas chromatograph–mass spectrometry techniques are outlined.

References

1 Goldberg, M.C., LeLong, L. and Sinclair, M. *Analytical Chemistry*, **45**, 89 (1973).
2 Peters, T.L. *Analytical Chemistry*, **54**, 1913 (1982).
3 Baker, R.J., Gibs, J., Meny, A.S. and Suffet, L.H. *Water Research*, **21**, 179 (1987).
4 Gomella, C., Belle, J.P. and Auvray, J. *Techniques et Sciences Municipals*, **71**, 439 (1976).
5 Fritz, J.S. *Accounts of Chemical Research*, **10**, 67 (1977).

6   Kawahara, F.K., Eichelberger, J.W., Reid, B.H. and Stierly, H. *Journal of the Water Pollution Control Federation*, **39**, 572 (1967).
7   Ahnoff, M. and Josefsson, B. *Analytical Chemistry*, **48**, 1268 (1976).
8   Stepan, S.F., Smith, J.F., Flego, U. and Renkers, L. *Water Research*, **12**, 447 (1978).
9   Goldberg, M.C., LeLong, L. and Kahn, L. *Environmental Science and Technology*, **5**, 161 (1971).
10 Burchill, P., Herod, A.A., March, K.M., Pirt, C.A. and Pritchard, E. *Water Research*, **17**, 1891 (1983).
11 Theron, S.J. and Hassett, D.W. *Water South Africa*, **12**, 31 (1986).
12 Eichelberger, J.W., Kerns, K.H., Olynk, P. and Budde, J.L. *Analytical Chemistry*, **55**, 1471 (1983).
13 Murray, D.A.J. *Journal of Oceanography*, **117**, 135 (1979).
14 Grob, K., Grob, K. and Grob, G. *Journal of Chromatography*, **106**, 299 (1975).
15 Veith, G.D. and Kiwus, L.M. *Bulletin of Environmental Contamination and Toxicology*, **17**, 631 (1977).
16 Fielding, M., Gibson, T.M., James, H.A., McLoughlin, K. and Steel, C.P. In Water Research Centre Technical Report TR 159 Organic Micropollutants in Drinking Water. February (1981).
17 Mass Spectrometry Data Centre. *Eight Peak Index of Mass Spectra*, 2nd edn. MSDC, Automic Weapons Research Establishment, Altermaston, Reading (1974).
18 Heller, S.R. and Milne, G.W.A. In *EPA/NIH Mass Spectral Data Base*, US Department of Commerce/National Bureau of Standards (1978).
19 Rooney, T.A., Freeman, R.R., Hewlett Packard Ltd., Avondale, US. In Technical Paper No. 69. Analysis of organic contaminants in surface water using high resolution gas chromatography and selective detectors, 174th Meeting of the American Chemical Association National Meeting, Chicago, Illinois, 28 August–2 September (1977).
20 Gomez-Taylor, M.M., Kyehl, D. and Griffiths, P.R. *International Journal of Environmental Analytical Chemistry*, **5**, 102 (1978).
21 Ibrahim, E.A. and Suffet, I.H. *Journal of Chromatography*, **454**, 217 (1988).
22 Yrieux, C., Gonzalez, C., Deroux, J.M., Lacoste, C. and Leybros, J. *Water Research*, **30**, 1791 (1996).
23 Bisiuk, N., Namiesnik, J. and Torres, L. *Analusis*, **15**, 560 (1987).
24 Umano, K., Reece, C.A. and Shibamoto, T. *Bulletin of Environmental Contamination and Toxicology*, **56**, 558 (1996).
25 Leoy, E.M. *Water Research*, **5**, 723 (1971).
26 Bridie, A.L., Bos, J. and Henzberg, S.C. *Journal of Institute of Petroleum*, **59**, 263 (1973).
27 Ruebelt, C. *Helgolander wiss Meersunters*, **16**, 306 (1967).
28 Przybylski, A. *Chemia Analit.*, **8**, 601 (1963).
29 Reisus, K. *Fortschr Wasserchem, ihrer, grenzgeb*, **10**, 43 (1968).
30 Osipov, V.M. and Belova, T.D. *Khimiya Tekhnol. Tolp. Masel.*, **13**, 56 (1968).
31 Lurje Yu Yu. Nauch. Soobschch; Vses. Nauchno–issled, Inst. Vodosnabzh, Kanaliz Gidrotekhn. Soor zuz henii. i Inzh Gidrogoel. Ochiska Prom. Stock n Vod., Moscow, 34 (1963).
32 Jeltes, R. and Veldink, R. *Journal of Chromatography*, **27**, 242 (1967).
33 Skotnikova, L.A. *Izy Vyssk Ucheb. Zavod Neft Gas.*, **12**, 111 (1969).
34 Jeltes, R. *Water Research*, **3**, 931 (1969).
35 Danyl, F. and Nietsch, N. *Microchemie, Mikrochem Acta*, **39**, 333 (1952).
36 Nietsch, B. *Angew. Chem.*, **66**, 571 (1954).
37 Nietsch, B. *Gass Wass Warme*, **10**, 66 (1956).

38 Nietsch, B. *Mikrochemica Acta*, 171 (1956).
39 Peake, E. and Hodgson, G.W. *Journal of the America Oil Chemists Society*, **43**, 215 (1966).
40 Peake, E. and Hogdson, G.W. *Journal of the American Oil Chemists Society*, **44**, 696 (1967).
41 Golden, J. *Techniques et Sciences Minicipales*, **71**, 17 (1967).
42 Davies, A.W. *Bulletin of the Belgian Centre for Water Studies*, **330**, 252 (1971).
43 Caruso, S.C. *Developments in Applied Spectroscopy*, **6**, 323 (1967).
44 Kawahara, F.K. *Journal of Chromatographic Science*, **10**, 629 (1972).
45 Kawahara, F.K. In Laboratory *Guide for the Identification of Petroleum Products*. 1014 Broadway, Cincinnati, Ohio. US Dept. of the Interior, Federal Water Pollution Central Administration, Analytical Quality Control Laboratory, 41 pp. (1969).
46 Jeltes, R. and Van Tonkelaar, W.A.M. *Water Research*, **6**, 271 (1972).
47 Desbaumes, E. and Imhoff, C. *Water Research*, **6**, 885 (1972).
48 Benyon, L.R. In *Methods for the Analysis of Oil in Water and Soil*, The Hague, Stickting, CONCAWE, pp. 40 (1968).
49 Nadzhafova, K.N. Trady Vses. Nanchro-issled. Inst. Vodosnabzh, Kanaliz, Gidrotekhn, Sooruzhenii, Inzh. Gidrogeol, No. 23, 107 (1970).
50 Harva, O. and Somersalo, A. *Acta Chem. Fenn.*, **31**, 384 (1958).
51 Simard, R.C., Hasecawa, W., Bondaruk, W. and Headington, C.E. *Analytical Chemistry*, **23**, 1384 (1951).
52 Hellman, H. *Deutsch gewaesserck Mitt.*, **13**, 19 (1969).
53 Fastabend, W. *Chemie-Ingr-Tech.*, **37**, 728 (1965).
54 Ruebelt, C. *Gas–u–Wass Fach.*, **108**, 893 (1967).
55 Ruebelt, C. *Zeito Analytical Chemistry*, **221**, 299 (1966).
56 Osipov, V.M. *Khimiya Tekhnol Topl Masel*, **17**, 52 (1971).
57 Golubeva, M.T. *Lab Delo*, **11**, 665 (1966).
58 Lindgreen, C.G. *Journal of American Water Works Association*, **49**, 55 (1971).
59 Mallevialle, J. *Water Research*, **8**, 1071 (1974).
60 Martin, P. and Geyer, D. *Korrespondez Abwasser*, **21**, 202 (1974).
61 Shtivel, N.K., Gipokova, L.I. and Smirnova, Z.S. *Soviet Journal of Water Chemistry and Technology*, **7**, 66 (1985).
62 Becconsall, J.K. *Analyst (London)*, **103**, 1233 (1978).
63 Fitzpatrick, M.G. and Tan, S.S. *Chemistry New Zealand*, **57**, 22 (1993).
64 Mirzayanov, V.S. and Bugrooyn, F. *Zavod Lab.*, **B8**, 656 (1972).
65 Mel'kanovitskaya, C.G. *Gidrokhim Mater.*, **53**, 153 (1972).
66 Dudova, M. Ya and Diterikhas, O.D. *Didrokhim Mater.*, **50**, 115 (1969).
67 Matsumoto, G. and Hanya, T. *Journal of Chromatography*, **194**, 199 (1980).
68 Stoffmeister, E., Hendel, P. and Hermanan, H. *Chemosphere*, **17**, 801 (1988).
69 Constable, D.J.T., Smith, S.R. and Tomaka, J. *Environmental Science and Technology*, **18**, 895 (1984).
70 Schloz, L. and Altmann, H.J.Z. *Analytical Chemistry*, **240**, 81 (1968).
71 Schnossner, H., Falkenberg, W. and Althaus, H. *Zeitschrift für Wasser und Abwasser Forschung*, **16**, 132 (1963).
72 Monorca, S., Causey, B.S. and Kirkbright, G.C. *Water Research*, **13**, 503 (1979).
73 Grimmer, G., Dettbarn, G. and Schreider, D. *Zeitschfirt für wasser und Abwasser Forshung*, **14**, 100 (1981).
74 Borneff, J. and Kunte, H. *Archives Hygiene Balt.*, **153**, 220 (1969).
75 Acheson, M.A., Harrison, R.M., Perry, R. and Wellings, R.A. *Water Research*, **10**, 207 (1976).
76 Wedgewood, R. and Cooper, R.C. *Analyst (London)*, **78**, 170 (1953).

77 Wedgewood, P. and Cooper, R.C. *Analyst (London)*, **79**, 163 (1954).
78 Wedgewood, P. and Cooper, R.C. *Analyst (London)*, **81**, 42 (1956).
79 Hess, G.G., Mckenzie, D.E. and Hughes, B.M. *Journal of Chromatography*, **366**, 197 (1986).
80 Koppe, P. and Muhle, P. *Vom Wasser*, **35**, 42 (1968).
81 Natusch, D.F.S. and Tomkins, B.A. *Analytical Chemistry*, **50**, 1429 (1978).
82 Yu, W., Chen, Z. and Sha, G. *Huanjing Huaxue*, **7**, 457 (1988).
83 Goldberg, M.C. and Weiner, E.R. *Analytica Chimica Acta*, **115**, 373 (1980).
84 Environmental Protection Agency. In *Quality Criteria for Water, Superintendent of Documents*, US Government Printing Office, No. 005–09–01049–4, Washington, DC (1976).
85 Semenchenko, L.L. and Kaplin, V.T. *Zhur Analit Khim.*, **23**, 1257 (1968).
86 Matsumoto, G., Ishiwateri, O. and Hanya, T. *Water Research*, **11**, 693 (1977).
87 Yorkshire Water Authority, UK. In *Determination of phenols in potable waters by gas chromatography*. Report, Method 655–01 (1977).
88 Bengtsson, G. *Journal of Chromatographic Science*, **23**, 397 (1985).
89 Onuska, F.I. and Terry, K.A. *Journal of High Resolution Chromatography*, **18**, 564 (1995).
90 Thompson, S. and Eglington, G. *Geochim, Cosmochim Acta*, **42**, 199 (1978).
91 Maktaz, E.D., Batvinova, L.E. and Kruchinina, A.A. *Soviet Journal of Water Chemistry and Technology*, **6**, 59 (1984).
92 Goncharova, I.A. and Khomenko, A.N. *Didrokhim Mater.*, **47**, 161 (1968).
93 Johnson, R.W. and Calder, J.A. *Geochimica Cosmochim Acta*, **37**, 264 (1973).
94 Van Hoevan, W., Maxwell, J.R. and Calvin, M. *Geochim Cosmochimica Acta*, **33**, 877 (1969).
95 Mishimura, M. *Geochim Cosmochimica Acta*, **41**, 1817 (1977).
96 Takati, O.S. and Vernon, F. *Water Research*, **23**, 123 (1989).
97 Oliver, B.G. and Bothen, K.D. *Analytical Chemistry*, **52**, 2066 (1980).
98 Dietz, F. and Trund, J. *Vom Vasser*, **41**, 137 (1973).
99 Deetman, A.A., Demeuleemeester, P., Garcia, M. and Hanck, G. *Analytica Chimica Acta*, **82**, 1 (1976).
100 Hagenaier, H., Werner, G. and Jager, W. *Zeitschrift für Wasser und Abwasser Forshing*, **15**, 195 (1982).
101 Bungasser, A.J. and Calarmotolo, J.F. *Analytical Chemistry*, **49**, 1588 (1977).
102 Thompson, J.F., Reid, S.J. and Kantor, E.J. *Archives of Environmental Contamination and Toxicology*, **6**, 143 (1977).
103 Pfannhauser, W. and Thaller, A. *Fresenius Zeitschrift für Analytische Chemie*, **322**, 220 (1985).
104 Haben, H.J., Ching, S.A., Caserette, L.J. and Young, R.A. *Bulletin of Environmental Contamination and Toxicology*, **15**, 78 (1976).
105 Fingler, S., Porevenkar, U. and Vasilie, Z. *Mikrochimica Acta*, **No. 416**, 163 (1987).
106 Dano, S.D., Chambon, P., Chambon, R. and Sanou, A. *Analysis*, **14**, 538 (1988).
107 Bauer, U. *Gas–u Wasserfach Wasser und Abwasser*, **113**, 58 (1972).
108 Girenko, D.B., Klisenko, M.A. and Dishchalka, Y.K. *Hydrobiological Journal*, **11**, 60 (1975).
109 Bourgeois, D.J., Devaux, P.H. and Mallet, V.N. *International Journal of Environmental Analytical Chemistry*, **59**, 15 (1995).
110 Fielding, M., McLoughlin, K. and Steel, C. In *Water Research Centre Enquiry Report ER 532, August 1977*, Water Research Centre, Stevenage Laboratory, Elder Way, Stevenage, Herts, UK (1977).

111 Reunanen, M. and Kronfeld, R. *Journal of Chromatographic Science*, **20**, 449 (1982).
112 Bush, B., Norang, R.S. and Syrotynstu, S. *Bulletin of Environmental Contamination and Toxicology*, **18**, 436 (1977).
113 Dressman, R.C., Stevens, A.A., Fair, J. and Sinch, B. *Journal of American Waterworks Association*, **71**, 392 (1979).
114 Richard, J.J. and Junk, G.A. *Journal of American Water Works Association*, **79**, 62 (1977).
115 Henderson, J.E., Peyton, G.R. and Glaze, W.H. A convenient liquid–liquid extraction method for the determination of halomethanes in water at the parts per billion level. In *Identification and Analysis of Organic Pollutants in Water*. (L.H. Keith ed.), Ann Arbor Science, Ann Arbor Michigan, pp. 105, 195 (1976).
116 USEPA Part III, Appendix C. Analysis of trihalomethanes in drinking water. Federal Register, **44**, No. 231 68672, 29 November (1979).
117 Method 501.1. In *The Analysis of Trihalomethanes in Finished Waters by the Purge and Trap Method*. ESEPA, EMSL, Cincinnati, Ohio, 45628, 6 November (1979).
118 Kirschen, N.A. *Varian Instrument Applications*, **14**, 10 (1980).
119 Von Rensberg, J.F.T., Von Higssteon, J.T. and Hassett, A.J. *Water Research*, **12**, 127 (1978).
120 Nicolson, B.C., Bursill, D.B. and Couche, D.J. *Journal of Chromatography*, **325**, 221 (1985).
121 Mieure, J.P. *Journal of American Waterworks Association*, **69**, 62 (1977).
122 Department of the Environment (Natural Water Council Standing Committee of Analysts). In *Methods for the Examination of Waters and Associated Materials Chloro and Dromtrihalo Methanes in Water*. HMSO, London (1981).
123 Saekmanreva, M., Pal'usova, O. and Szokalay, A. *Water Research*, **11**, 551 (1977).
124 Taylor, R., Bogacka, R. and Krasnicki, K. *Chemica Analitica*, **19**, 73 (1974).
125 Taylor, R. and Bogacka, T. *Analytical Abstracts*, **17**, 1206 (1969).
126 Weil, L. and Ernest, K.E. *Gas–u–Wassfach*, **112**, 184 (1971).
127 Leoni, V. *Journal of Chromatography*, **62**, 63 (1971).
128 American Public Health Association. In *Standard Methods for the Examination of Water and Wastewater*, 15th edn., Method 509 A. p. 493 (1978).
129 American Public Health Association. Method 573. In *Supplement to the 15th Edition of Standard Methods for the Examination of Water and Wastewater; Selected Analytical Methods Approved and Cited by the United States Environmental Protection Agency, American Public Health Association, American Water Works Association, Water Pollution Control Federation* (1978).
130 Engst, R. and Knoll, R. *Nahrung*, **17**, 837 (1973).
131 Schafer, M.L., Peeler, J.T., Gardner, W.D. and Campbell, J.E. *Environmental Science and Technology*, **3**, 1261 (1969).
132 Erney, D.R. *Analytical Letters*, **12**, 501 (1979).
133 Weil, L. and Quentin, K.E. *Zeit für Wasser und Abwasser Forschung*, **7**, 147 (1974).
134 Laurin, G. *Bulletin of Environmental Contamination and Toxicology*, **5**, 542 (1970).
135 Konrad, J.G., Pionke, H.B. and Chesters, G. *Analyst (London)*, **94**, 490 (1969).
136 Millar, J.D., Thomas, R.E. and Schattenberg, H.I. *Analytical Chemistry*, **53**, 214 (1981).
137 Mosinska, K. Proceedings of the Institute Przem Org., 1971, 253 (1972).

138 Mohnke, M., Rohde, K.H., Bruegmann, L. and Franz, P. *Journal of Chromatography*, **364**, 323 (1986).
139 Chee, K.K., Wong, M.K. and Lee, M.K. *Analytica Chimica Acta*, **330**, 217 (1996).
140 Schoeler, H.F. and Brodesser, J. Committee of European Communities (Ref) EUR.EUR 11350. Organic Micropollutants in the Aquatic Environment, pp. 69–74 (1988).
141 Goosens, E.C., de Jong, D., de Jong, G.J. Rinkema, F.D. and Brinkman, U.A.T. *Journal of High Resolution Chromatography*, **18**, 38 (1995).
142 Gorback, S., Haarring, R., Knauf, W. and Werner, H. *Bulletin of Environmental Contamination and Toxicology*, **6**, 40 (1971).
143 Blankit, P.F. *Journal of Chromatography*, **179**, 123 (1979).
144 Tuskioka, T., Talishita, R. and Murakami, T. *Analyst (London)*, **111**, 145 (1986).
145 Shevchuk, I.A., Dubchenko, Y,.G. and Naidenova, T.S. *Soviet Journal of Water Chemistry and Technology*, **7**, 73 (1985).
146 Simal, J., Crous Vidal, J., Maria-Charro Arial Boado, M.A., Diaz, R. and Vilas, D. *An Bromat (Spain)*, **23**, 1 (1971).
147 Murray, D.A.J. *Journal of the Fisheries Research Board, Canada*, **132**, 457 (1975).
148 Ishikawa, S., Taketami, M. and Shimohara, R. *Water Research*, **19**, 119 (1985).
149 Mallet, V. and Brun, G.L. *Bulletin of Environmental Contamination and Toxicology*, **12**, 739 (1974).
150 Blanchet, P.F. *Journal of Chromatography*, **179**, 123 (1979).
151 Venkataraman, S. and Sathyamurthy, V. *Journal of the Indian Water Works Association*, **11**, 351 (1979).
152 Rice, J.R. and Dishberger, H.J. *Journal of Agriculture and Food Chemistry*, **16**, 67 (1968).
153 Zycinski, D. *Roczn. Panst. Zahl. Hig.*, **22**, 189 (1971).
154 Rollo, J.W. and Cortes, A. *Journal of Gas Chromatography*, **2**, 132 (1964).
155 Holden, E.R., Jones, W.M. and Beroza, M. *Journal of Agriculture and Food Chemistry*, **17**, 56 (1969).
156 Cohen, I.C., Norcup, J., Ruzicka, J.H.A. and Wheals, B.R. *Journal of Chromatography*, **49**, 215 (1970).
157 Gutermann, W.H. and Listi, D.J. *Journal of Agriculture and Food Chemistry*, **3**, 48 (1965).
158 Coburn, J.A., Riplay, B.D. and Chan, A.S.Y. *Journal of the Association of Official Analytical Chemists*, **59**, 188 (1976).
159 Lewis, D.L. and Paris, D.F.J. *Journal of Agriculture and Food Chemistry*, **22**, 148 (1974).
160 Handa, S.K. and Dikshit, A.K. *Analyst (London)*, **104**, 1185 (1979).
161 Westlake, W.E., Monika, I. and Gunther, F.A. *Bulletin of Environmental Contamination and Toxicology*, **8**, 109 (1972).
162 Frei, R.W., Lawrence, J.F. and Le Gay, D.S. *Analyst (London)*, **98**, 9 (1973).
163 Reeves, E.G. and Woodham, D.W. *Journal of Agriculture and Food Chemistry*, **22**, 76 (1974).
164 Frei, R.W., Larwence, J.F. and Belliveau, P.F. *Zeit Analytical Chemistry*, **254**, 271 (1971).
165 Fredeen, F.J.L.I., Saha, J.G. and Balba, M.H. *Pesticides Monitoring Journal*, **8**, 241 (1975).
166 Devine, J.N. and Zweig, G. *Journal of Association of Official Analytical Chemists*, **52**, 187 (1969).
167 Bogacka, T. *Chemical Analysis*, **16**, 59 (1971).
168 Bogacka, T. and Taylor, R. *Chemie Analit.*, **15**, 143 (1970).
169 Suffet, I.H. *Journal of Agriculture and Food Chemistry*, **21**, 591 (1973).

170 Hermann, W.D., Tourayre, J.C. and Engli, H. *Pesticide Monitoring Journal*, **13**, 128 (1979).
171 Method S69. In *Method for Triazine Pesticides in Water and Wastewater*. Interim pending issuance of methods for organic analysis of water and wastes. US Environmental Protection Agency, Environmental and Monitoring Support Laboratory, September (1978).
172 Lee, H.B. and Stokker, Y.D. *Journal of the Association of Official Analytical Chemists*, **69**, 568 (1986).
173 Deleu, R., Barthelemy, J.P. and Copin, A. *Journal of Chromatography*, **134**, 483 (1977).
174 McKane, C.E. and Hance, R.J. *Bulletin of Environmental Contamination and Toxicology*, **4**, 31 (1969).
175 McKane, C.E. and Hance, R.J. *Analytical Abstracts*, **17**, 3849 (1969).
176 McKane, C.E., Byast, T.H. and Hance, R.J. *Analyst (London)*, **97**, 653 (1972).
177 Frank, P.A. and Demint, R.J. *Environmental Science and Technology*, **3**, 69 (1969).
178 Payne, W.R., Pope, J.D. and Benner, J.E. *Journal of Agriculture and Food Chemistry*, **22**, 79 (1974).
179 Norris, L.A. and Montgomery, M.L. *Bulletin Environmental Contamination and Toxicology*, **13**, 1 (1975).
180 Richardson, M.L., Webb, K.S. and Gough, T.A. *Ecotoxical Environmental Safety*, **4**, 207 (1980).
181 Fine, D.H., Raunbehler, D.B., Huffman, F. and Epstein, S.S. *Bulletin Environmental Contamination and Toxicology*, **14**, 464 (1975).
182 Inaba, K. *International Journal of Environmental Analytical Chemistry*, **31**, 63 (1987).
183 Lee, H.B. *Journal of Association of Official Analytical Chemists*, **71**, 803 (1988).
184 Kawano, Y. and Bevenue, A. *Journal of Chromatography*, **72**, 51 (1972).
185 Avery, M.J. and Junk. G.A. *Analytical Chemistry*, **57**, 790 (1985).
186 Wanatabe, N., Yasuda, Y., Katok Nakamura, T. *et al. Science of the Total Environment*, **34**, 169 (1984).
187 Uchiyama, M. *Water Research*, **12**, 299 (1978).
188 Waters, J. and Garrigan, J.T. *Water Research*, **17**, 1549 (1983).
189 Yasuhara, A. and Fuwa, F. *Journal of Chromatography*, **172**, 453 (1979).
190 MacDonald, J.C., Bock, C.A. and Slater, G.D. *Applied Microbiology and Biotechnology*, **25**, 392 (1987).
191 Gardiner, J. *Analyst (London)*, **102**, 120 (1977).
192 Gardiner, J. In Technical Memorandum, TM101, Water Research Centre, Stevenage Laboratory, Herts, UK (1975).
193 Inaba, K. *International Journal of Environmental Analytical Chemistry*, **31**, 63 (1987).
194 Farrell, E.S. and Pacey, G.E. *Analytical Chemistry*, **68**, 93 (1996).
195 The Oil Companies International study group for conservation of clean air and water. In Europe CONCAWE Report No. 6/82. Analysis of Trace Substances in Aqueous Effluents from Petroleum Refineries (1982).
196 Czuczwa, J., Levenberg, C., Tromp, J., Giger, W. and Ahlel, M. *Journal of Chromatography*, **403**, 233 (1987).
197 Kahn, L. and Wayman, C.H. *Analytical Chemistry*, **36**, 1340 (1964).
198 Ahnoff, M. and Josefsson, B. *Analytical Chemistry*, **46**, 658 (1974).
199 Kerner, I., Goto, M. and Korte, F. *International Journal of Environmental Analytical Chemistry*, **2**, 57 (1972).
200 Petrick, G., Schulz, D.E. and Duiker, J.C. *Journal of Chromatography*, **435**, 241 (1988).

201 Vartiainen, T., Lizmatainen, A., Jaaskelainen, S. and Kauranen, P. *Water Research*, **21**, 773 (1987).
202 Nishi, S. and Horimoto, H. *Japan Analyst*, **17**, 1247 (1968).
203 Nishi, S. and Horimoto, J. *Japan Analyst*, **19**, 1646 (1970).
204 Pressman, R.C. *Journal of Chromatographic Science*, **10**, 472 (1972).
205 Department of the Environment and National Water Council (UK). In *Determination of Organic and Inorganic Mercury in Seawater*. HMSO, London 23 pp. (1978).
206 Stary, J., Havlik, B., Prasilova, J., Kratzer, K. and Haunasova, J. *International Journal of Environmental Chemistry*, **5**, 89 (1978).
207 Schintu, M., Kauri, T., Contu, A. and Kudo, A. *Ecotoxicology and Environmental Safety*, **14**, 208 (1987).
208 Chau, Y.K., Wong, P.T.S. and Bengert, G.A. *Analytical Chemistry*, **54**, 246 (1982).
209 Meinema, H.A., Burger, W.T., Verslins-Dehaar, G. and Geners, E.C. *Environmental Science and Technology*, **12**, 288 (1978).
210 Unger, M.A., MacIntyre, W.C., Greaves, J. and Huggett, R.J. *Chemosphere*, **15**, 461 (1986).
211 Hattori, Y., Kobayashi, A., Takemoto, S. *et al. Journal of Chromatography*, **315**, 341 (1984).
212 Soderquist, C.J. and Crosby, D.C. *Analytical Chemistry*, **50**, 1435 (1978).
213 Chau, Y.K., Wong, P.T.S. and Kramer, O. *Analytica Chimica Acta*, **146**, 211 (1983).
214 Chau, Y.K., Wong, P.T.S., Bengent, G.A. and Kramer, O. *Analytical Chemistry*, **186**, 51 (1979).
215 Wu, C. and Suffet, I.H. *Analytical Chemistry*, **49**, 231 (1977).
216 Fielding, W., Gibson, T.M., James, H.A., McLoughlin, K. and Steep, C.P. In *Organic Micropollutants in Drinking Water*, Technical Report TR 159, Water Research Centre, Medmenham, UK.
217 Copin, G. and Barbier, M. *Oceanography*, **23**, 455 (1971).
218 Werner, A.E. and Waldichuk, M. *Analytical Chemistry*, **34**, 1674 (1962).
219 Zsonay, A. and Kiel, W.J. *Journal of Chromatography*, **90**, 74 (1974).
220 Law, R.J., Marchand, M., Dahlmann, G. and Fileman, T.W. *Marine Pollution Bulletin*, **18**, 486 (1987).
221 Eklund, G., Josefsson, B. and Roos, C. *Journal of High Resolution Chromatography*, **1**, 34 (1978).
222 Eklund, G., Josefsson, B. and Roos, C. *Journal of Chromatography*, **142**, 575 (1977).
223 Grob, K. and Grob, G. *Journal of Chromatography*, **125**, 471 (1970).
224 Grob, K. *Chromatographie*, **10**, 181 (1977).
225 Wilson, A.J. and Forester, J. In UK Environmental Protection Agency Report No. EPA 600–7–74, 108 (1974).
226 Palmark, K.H. *Anal. Chem. Scandinavia*, **17**, 1456 (1963).
227 The Oil Companies International Study Group for Conservation of Clean Air and Water, Europe. In *Analysis of Trace Substances in Aqueous Effluents from Petroleum Refineries*. Concawe Report No. 6/82 (1982).
228 Hon-Nami, H. and Hanya, T. *Water Research*, **14**, 1251 (1980).
229 Hon-Nami, H. and Hanya, T. *Journal of Chromatography*, **161**, 205 (1978).
230 Murray, A.D.J. *Journal of Chromatography*, **177**, 1797 (1979).
231 Kuhelka, V., Mitera, J., Novak, J. and Mostecky, J. Scientific Paper, Prague Institute of Chemical Technology, F22, 151 (1978).
232 Gruenfeild, M. *Environmental Science and Technology*, **7**, 636 (1973).
233 Kugemagi, U., Burnard, J. and Terrier, L.D. *Journal of Agriculture and Food Chemistry*, **23**, 717 (1975).

Chapter 14

Organics: Supercritical fluid extraction

The advantages of this recently developed technique are two-fold. By passing large volumes of water samples through solid phase traps and desorbing organics from the solid phase with a relatively small volume of supercritical fluid very large preconcentration factors can be achieved. Also, the use of carbon dioxide or methanol modified carbon dioxide or one of the other supercritical fluids now available, eg Freons, avoid the introduction into the extract of organic solvents and contaminants therein which could interfere with subsequent analysis of the extracts. Recent work on the applications of supercritical fluid extractions to water and soil samples is discussed below.

14.1 Non saline waters

14.1.1 General discussion

Dirksen et al. [1] addressed the problem of using supercritical fluid extraction in water samples with high particle concentrations. These workers used glass heads to inhibit plugging of the supercritical fluid disc.

Liska et al. [2] have studied the effectiveness of various solid phase extraction materials for the extraction of organic compounds from non saline waters. They showed that polymeric PLRP,s in combination with C_{18} was the best material for the enrichment of polar organic compounds.

Senseman et al. [3] have shown that various pesticides show a greater stability when stored or frozen on solid phase extraction discs than when stored in water at 4°C.

Games et al. [4] used high performance liquid chromatography for solute focusing prior to superfluid chromatography–mass spectrometry in the analysis of waste streams.

Reighard and Olesik [5] used supercritical carbon dioxide and 'enhanced fluidity' liquid carbon dioxide as extraction solvents for analytes adsorbed on to C_{18} resin discs, the higher strength of the solvents containing methanol reduced the volume and time required for efficient analyte extraction.

Wells *et al.* [6] have discussed the coordination of supercritical fluid, and solid phase extractions with separation and detection methods. The applications of solid phase extraction–supercritical fluid extraction to various specific organic compounds is discussed below.

14.1.2 Petroleum hydrocarbons

Fourteen laboratories participated in a round robin study of proposed EPA methods 3560 and 8440 which involve supercritical fluid extraction followed by infrared spectrometry for the determination of petroleum hydrocarbons. Good recoveries were obtained and overall accuracy was 82.9% [7].

Eckert-Tilotta *et al.* [8] obtained good recoveries of total petroleum in hydrocarbons from non saline waters by supercritical fluid extraction. These were better than those obtained by conventional extraction methods but with supercritical fluid extraction the use of Freon–113 solvent was reduced by more than a factor of 10.

14.1.3 Non ionic surfactants

Kane *et al.* [9] used supercritical fluid extraction of solid phase materials to determine non ionic surfactants in aqueous samples.

14.1.4 Polychlorobiphenyls

Bowadt *et al.* [10] used supercritical fluid extraction of solid phase materials to determine down to 10µg L^{-1} polychlorobiphenyls in non saline waters.

14.1.5 Polychlorobiphenyls, polychloro dibenzo-p-dioxins and polychlorodibenzofurans

Van Bavel *et al.* [11] have developed a solid-phase carbon trap used in conjunction with higher performance supercritical carbon liquid chromatography for the simultaneous determination of polychlorobiphenyls, polychlorodibenzofurans and polychlorodibenzodioxins and pesticides in environmental waters. The purpose of their study was to find materials that can be used as a solid-phase trap in a commercial supercritical fluid extraction instrument. PX–21 active carbon was particularly suitable. The polar fraction containing polychlorodibenzofurans and polychlorodibenzodioxins were successfully separated on a PX–21 solid–phase trap from the non-polar fraction containing polychlorobiphenyls and organochlorine pesticides.

Direct injection of the concentrated fractions on to a gas

chromatographic column equipped with a mass spectrometric detector was possible without the need for sample clean-up.

14.1.6 Organochlorine insecticides

Direct supercritical fluid extraction has been used to determine low levels of organochlorine pesticides in water [12].

14.1.7 Explosives

Slack *et al.* [13] used supercritical fluid extraction of solid phase materials to determine explosives in aqueous samples.

14.1.8 Miscellaneous organics

Tang *et al.* [14] used supercritical fluid extraction of solid phase materials to determine polyaromatic hydrocarbons, polychlorobiphenyls, organochlorine pesticides and phthalate esters in non saline waters.

Ehntoff *et al.* [15] used supercritical carbon dioxide to extract low levels of volatile and relatively insoluble compounds from water.

Organometallic compounds

14.1.9 Organotin compounds

Liu [16,17] has used a combination of supercritical fluid extraction and gas chromatography–atomic emission spectrometry to speciate 13 organotin compounds in non saline waters. After extraction the organotin compounds were treated with pentyl magnesium bromide to convert inorganic tin compounds into their neutral derivatives.

14.2 Soils

Supercritical fluid extraction is one of the few preconcentration techniques that have been applied to solids such as soils and sediments. As this application is of interest to the water chemist it is briefly discussed below.

14.2.1 General discussion

Dankers *et al.* [18] have shown that by using dichloromethane as a static modifier, 20–30M supercritical fluid extractions from soil gave results comparable to those obtained in 4h conventional sample preparation methods.

Meyer *et al.* [19] showed that supercritical fluid extraction can give recoveries comparable to Soxhlet extraction methods, even for soils with a high carbon content.

14.2.2 Gasoline hydrocarbons

Burford et al. [20] have described a coupled supercritical fluid extraction gas chromatographic method for the extraction and determination of gasoline and diesel range organics from contaminated soils. The direct transfer of the extract to a gas chromatography reduced analysis to ~80min compared to the 18h required for conventional sonication analysis.

14.2.3 Polyaromatic hydrocarbons

In the supercritical fluid extraction of polyaromatic hydrocarbons in soil Lee et al. [21] showed that while carbon dioxide is less efficient for the extraction of heavier polyaromatic hydrocarbons than other extractants such as nitrous oxide and Freon–22, its deficiency was remedied by using a mixture of water, methanol and methylene dichloride as modifiers and by adjusting other experimental conditions.

Meyer and Kleihoehmer [22] showed that better polyaromatic recoveries from soil were obtained after supercritical fluid extractions by using liquid–solid traps rather than analyte trapping in pure organic solvents.

Reindl and Hoefler [23] showed that by optimising supercritical fluid extraction parameters, followed by high performance liquid chromatography detection a precision of about 10% RSD was achieved for 15 polyaromatic hydrocarbons in soils.

14.2.4 Polychlorobiphenyls

Microwave-assisted extraction with gas chromatography–ECD was compared to ELISA for the determination of polychlorobiphenyls in soils and sediments. Both techniques were amenable to field screening and monitoring applications [24].

14.2.5 Miscellaneous organics

Preconcentration procedures based on supercritical fluid extraction have also been applied to a range of other organic compounds in soils including organic acids and ketones [25], phenols and cresols [26], aromatic amines [27], trialkyl and triaryl phosphates [28], herbicides [29] and pyridine [30].

14.3 Sediments

14.3.1 Fatty acids

Hordijk et al. [31] determined lower volatile fatty acids in river sediments using chemically bonded FFAP GLC columns.

14.3.2 Polychlorobiphenyls

Onuska and Terry [32] used supercritical fluid extraction in tandem with high resolution gas chromatography for the determination of polychloro-biphenyls in sediments.

Methods have been described [33] for the determination of polychlorobiphenyls in sulphur containing sediments. A supercritical fluid extraction method did not involve any manual clean-up or sample pretreatment. Lee and Peart [34] optimised supercritical fluid extraction procedures for the determination of polychlorobiphenyls in sediments.

14.3.3 Miscellaneous organics

Wells [35] has discussed the application of supercritical fluid extraction to the determination of organic compounds in aquatic sediments and biota.

14.4 Sewage

14.4.1 Hydrocarbons and phenols

A combination of steam distillation with high performance liquid chromatography fractionation and gas chromatography with flame ionisation detection has been used to determine n-alkanes, linear alkyl-benzenes, polynuclear aromatic hydrocarbons and 4–nonylphenol in digested sewage sludges [36].

14.4.2 Cationic surfactants

Fernandez et al. [37] used solid phase extraction and high performance liquid chromatography to determine cationic surfactants in sludges.

References

1 Dirksen, T.A., Price, S.M. and Mary, S.J. American Laboratory, 25, 24 (1993).
2 Liska, I., Brouwer, E.R., Lingeman, H. and Brinkman, U.A.T. Chromatographia, 37, 13 (1993).
3 Senseman, S.A., Lavy, T.L., Mattice, J.D., Myers, B.M. and Skulman, B.W. Environmental Science and Technology, 27, 516 (1993).
4 Games, D.E., Rontree, J.H. and Fowlis, I.A. Journal of High Resolution Chromatography, 17, 68 (1994).
5 Reighard, T.S. and Olesik, S.V. Journal of Chromatography, A, 737, 233 (1996).
6 Wells, M.J.M. and Stearman, G.K. ACS Symposium Series No. 630, pp. 18–33 (1996).
7 Lopez-Avila, V., Young, R., Kim, R. and Beckert, W.F. Journal AOAC International, 76, 555 (1993).
8 Eckert-Tilotta, S.E., Hawthorne, S.B. and Miller, D.J. Fuel, 72, 1015 (1993).
9 Kane, M., Dean, J.R., Hitchen, S.M., Dowle, C.J. and Tranter, R.L. Analytical Proceedings, 30, 399 (1993).

10 Bowadt, S., Johansson, B., Pelusio, F., Larcen, B.R. and Rowda, C. *Journal of Chromatography*, **662**, 424 (1994).
11 Van Bavel, B., Jaremoto, M., Karisson, L. and Linstrom, G. *Analytical Chemistry*, **68**, 1279 (1996).
12 Barnabas, I.J., Dean, J.R., Hitchen, S.M. and Owen, S.P. *Journal of Chromatography, A*, **665**, 307 (1994).
13 Slack, C.G., McNair, H.M., Hawthorne, S.B. and Miller, D.J. *Journal of High Resolution Chromatography*, **16**, 473 (1993).
14 Tang, P.H., Ho, J.S. and Eichberger, J.W. *Journal AOAC International*, **76**, 72 (1993).
15 Ehntoff, D.J., Eppig, C. and Thrun, K.E. Advanced Chemistry Series (Org. Pollut. Water), **214**, 483 (1987).
16 Liu, Y., Lopez-Avila, V., Alcarez, M. and Beckert, W.F. *Journal of High Resolution Chromatography*, **16**, 106 (1993).
17 Liu, Y., Lopez-Avila, V., Alcaraz, M. and Beckert, W.F. *Analytical Chemistry*, **66**, 3788 (1994).
18 Dankers, J., Groenenboom, M., Scholtis, L.H.A. and Van der Heiden, C. *Journal of Chromatography*, **641**, 357 (1993).
19 Meyer, A., Kleiboekmer, W. and Cammann, K. *Journal of High Resolution Chromatography*, **16**, 491 (1993).
20 Burford, M.D., Hawthorne, S.B. and Miller, D.J. *Journal of American Environmental Laboratory*, **8**, 1 (1996).
21 Lee, H.B., Peart, T.E., Hong-You, R.L. and Gere, D.R. *Journal of Chromatography*, **653**, 83 (1993).
22 Meyer, A. and Kleihoehmer, W. *Journal of Chromatography*, **657**, 327 (1993).
23 Reindl, S. and Hoefler, F. *Analytical Chemistry*, **66**, 1808 (1994).
24 Lopez-Avila, V., Benedicto, J., Charcan, C., Young, R. and Beckert, W.F. *Environmental Science and Technology*, **29**, 2709 (1995).
25 Longbehn, A. and Steinart, H. *Journal of High Resolution Chromatography*, **17**, 293 (1994).
26 Futter, J.E. and Wall, P. *Journal of Planar Chromatography, Mod TLC* **6**, 372 (1993).
27 Oostdyk, T.S., Grob, R.L., Snyder, J.L. and McNally, M.E. *Analytical Chemistry*, **65**, 596 (1993).
28 DeGens, H., Zegars, B.N., Lingeman, H. and Brinkman, U.A.T. *International Journal of Environmental Analytical Chemistry*, **56**, 119 (1994).
29 Lopez-Avila, V., Dodhiwals, N.S. and Beckert, W.F. *Journal of Agriculture and Food Chemistry*, **41**, 2038 (1993).
30 Peters, R.J.B., Van Renesse, E. and Von Duivenbode, J.A.D. *Fresenius Journal of Analytical Chemistry*, **348**, 249 (1994).
31 Hordijk, C.A., Burgers, I., Phylipsen, G.J.M. and Cappenberg, T.E. *Journal of Chromatography*, **511**, 317 (1990).
32 Onuska, F.I. and Terry, K.A. *Journal of High Resolution Chromatography*, **12**, 527 (1989).
33 Bowadt, S. and Johansson, B. *Analytical Chemistry*, **66**, 667 (1994).
34 Lee, H.B. and Peart, T.E. *Journal of Chromatography*, **663**, 87 (1994).
35 Wells, D.E. Tech. Indust. Anal. Chem. 13 (Environmental Analysis), 79 (1993).
36 Sweetman, A.J. *Water Research*, **28**, 343 (1994).
37 Fernandez, P., Alder, A.C. and Giger, W. *National Meeting of the American Chemical Society, Division of Environmental Chemistry*, **33**, 303 (1993).

Chapter 15

Organics: Static headspace analysis

This technique differs from purge and trap techniques (see Chapter 16) in the following respect. In purge and trap, the water sample is purged with a gas and the organics removed and adsorbed on to a solid such as Tenax or active carbon. The adsorbed organics are then desorbed in a small volume of an organic solvent, usually chloroform or carbon disulphide which is subsequently analysed by gas chromatography. Alternatively the organics are released thermally from the solid adsorbent. Although capable of achieving very high concentration factors a disadvantage of this technique is that the organic solvent and impurities therein can cause difficulties at the gas chromatographic state of the analysis. Headspace analysis is solventless in that the aqueous sample is heated say to 30°C in the presence, usually, of inorganic salts which help to drive the organics into the inert gas filled headspace above the sample. Portions of the headspace gas are then withdrawn by syringe for gas chromatographic analysis. Thus headspace analysis unlike purge and trap, is a static method. By attention to detail in the design of the apparatus and operating conditions, it is possible by this technique to obtain concentration factors of up to 100 for the organics in the original water sample.

Thus, Friant and Suffet [1] obtained 66-fold enrichment factors by:

1   increasing the temperature of the water sample from 30 to 50°C
2   saturating the water sample with sodium sulphate
3   adjusting the pH of the original sample to 7.1.

The detection limits ranged from 50μg L^{-1} for methyl ether ketone to 740μg L^{-1} for dioxane.

By definition headspace analysis is only applicable to organics which are sufficiently volatile to be released from the water sample by sweeping with an inert gas. Compounds which come in this category include aliphatic and aromatic hydrocarbons, low boiling chloro and fluoro hydrocarbons, haloforms and alcohols, ketones and aldehydes etc.

This static headspace system can operate in a closed loop or open loop mode.

Definitions

Open-loop system. The water sample is placed in a sealed vessel and, if required, an inorganic 'salting out' agent is added. An inert gas, eg helium is injected into the headspace and the vessel sealed and maintained at a particular temperature for a fixed period of time. An equilibrium is then set up between the concentration of organics in the water phase and the overlying gas phase. At the end of the time interval a portion of the headspace gas is withdrawn with a gas syringe and this gas is analysed by gas chromatography or another appropriate technique.

Closed-loop system. This is the same as the above system except that the headspace gas is continuously recycled through the water sample into the headspace for a fixed period of time. Generally, higher concentrations of the organics will be obtained in the headspace under these conditions than will be obtained in the open loop system leading to greater sensitivity in the analysis.

There are two types of headspace analysis, static and dynamic.

The *static headspace* method discussed in this chapter is based on the fact that when a water sample that contains organic compounds is sealed in a vial, organics will equilibrate between the water and vial headspace. Distribution of compounds between the two phases depends on temperature, vapour pressure for each compound, sample matrix influences on compound activity coefficients and ratio of headspace to liquid volume in the vial. A major advantage of this method is that only relatively volatile water-insoluble compounds tend to partition into the headspace; therefore, a form of sample clean-up is provided. Also, since only gaseous samples are injected into the gas chromatograph, column and detector contamination are prevented and chromatographic interferences are minimised.

15.1 Non saline waters

Wyllie [2] has observed that purge and trap (ie dynamic headspace) methods are less readily automated, liable to loss of light volatiles, readily contaminated by samples containing a very high concentration of some volatiles, subject to contamination by the ambient air, expensive on glassware and/or glassware washing, and have a low sample throughput. Static headspace methods are more readily automated, usable on other matrices besides drinking water, portable (and hence available for field use), and labour-saving to the extent that all glassware was disposable; but it too was liable to contamination by the ambient air. Under suitable conditions, the techniques were equally sensitive, but headspace had the advantage in precision.

Munz and Roberts [3] elucidated the interactions of Henry's constant, solubility, and activity coefficients by the multiple equilibration method of a closed system with analysis of the aqueous phase. Methanol and isopropanol were used as model co-solvents in a study of eight common chlorinated solvents. 100ml gas tight syringes were used as containers. Solvents were spiked with solute and equilibrated for 25min before expelling the gas phase and analysing the liquid phase by gas chromatography. Experimental results were compared with predictions from a semi-empirical thermodynamic model and from vapour pressure and solubility date (UNIFAC). No effect of solute concentration on the solute's Henry's constant was found up to solute–liquid mole fractions of 0.001. Very high co-solvent concentrations in excess of 10g L^{-1} were required to reduce the Henry's constant. No effect was observed in multisolute systems at low concentrations. The Henry's constant of the organic compounds increased by a factor of approximately 1.6 for each 10°C rise in temperature and was considered to be the most important factor in determining the removal of volatile solutes in gas–liquid contacting systems. The UNIFAC model predicted relative effects of concentration and co-solvent but could only be applied with caution to predict temperature effects.

Malten and Vreden [4] applied the gas chromatographic headspace analysis technique to the detection of organic volatiles in very small volumes of aqueous samples. Cowen *et al.* [5] have described the construction of a septumless injection device for delivering headspace gases into a gas chromatograph.

Distillation techniques have been combined with headspace analysis as a means of improving preconcentration factors of up to 100 [5–7]. The distillation technique has been used [6,7] to concentrate volatile polar water-soluble organic compounds in the distillate for later headspace gas chromatographic analysis. A fast and reliable headspace gas injection system equipped with an evacuated gas sampling valve was used. Methanol, ethanol, acetone, 2–propanol, and methyl ethyl ketone and other volatile polar organics can be determined at concentrations ranging from 0.008 to 13mg L^{-1}. Combining the headspace gas chromatographic method of analysis with sample preconcentration by the distillation technique made it possible to determine volatile polar organics at the µg L^{-1} level.

The distillation procedure was as follows: 100ml of each sample solution was spiked with *n*-butanol to give a 0.8mg L^{-1} level and saturated with 21.3g of dried sodium sulphate in a 250ml round bottomed flask. A 7.75 in Vigreux column was connected to the flask. A small distilling head with a thermometer and a small condenser attached was connected to the upper end of the Vigreux column. During the distillation process, the vapour temperature at the distilling head was controlled at 100°C by regulating the voltage supply to the heating mantle. Distillation continued until the first 1.5ml of distillate was collected. From that amount, 1ml was injected with a 2ml syringe into a 15ml headspace serum bottle containing 1.2g dried sodium sulphate. Distillations were done in duplicate for each sample. The standard solutions and distilled water were distilled in the same way.

Headspace bottles were heated in a 70°C water bath for at least 70min. The headspace gas was injected via an eight-port gas sampling valve into a gas chromatograph equipped with flame ionisation detectors. The gas chromatograph was fitted with a 183cm 2mm id glass column packed with 0.4% Carbowax 1500 on 60/80 mesh Carbopack A (Supelco. Inco., Bellefonte, Pa.). After injection the column temperature was held at 60°C for 2min, then increased at 8°C min $^{-1}$ to 150°C and held for 5min. The carrier helium flow rate was 10ml min $^{-1}$. An electronic integrator was interfaced to the flame ionisation detector for acquisition of retention time and peak area data. The ratios of the peak areas of the resulting chromatograms to internal standard peak area were used for standard curves.

Drodz and Novak [8] have shown that low concentrations of volatile organics can be determined in water by the double sampling method of headspace gas chromatography using a closed loop strip/trap technique. Fielding *et al.* [9] have developed the following procedure of dynamic headspace analysis.

Carry out the extraction in the apparatus shown in Fig. 15.1. Place the sample (2L) in a precleaned quinol glass stoppered bottle in the extraction flask and add a solution (2µL) of the internal standard (100µg L $^{-1}$) in acetone. Use deuterated internal standards to provide quantitative information on some of the compounds identified, chlorobenzene–d_5, *p*-xylene–d_{10}, phenol–d_5, naphthalene–d_8, hexadecane–d_{34} and phenanthrene–d_{10} added to each sample immediately before extractions are suitable. Place the flask in the water bath (thermostatted at 28°C) and leave for 30min. Then switch on the pump, and continue the purging (1.5L min $^{-1}$) for 3h. Switch off the pump, remove the filter tube from the filter holder and connect to an eluate collection tube (Fig. 15.1) with PTFE tubing. Connect a short length (2cm) of capillary tubing (5.5mm od, 1.0mm id) to the other end of the filter tube to prevent evaporation of the solvent while extracting the filter. Elute the carbon filter with purified methylene chloride (2 × 15µL, 1 × 20µL) to give an extract suitable for gas chromatography–mass spectrometric analysis.

A further refinement of headspace analysis is called closed loop stripping analysis. In this technique the sample at a controlled temperature is purged with inert stripping gas in a closed systems whereby the gas is recycled through the sample continuously for a period of 0.5–2h. In one version of the technique a sample is withdrawn from the cycling gas periodically and analysed directly for volatile organics by gas chromatography or gas chromatography–mass spectrometry. In another

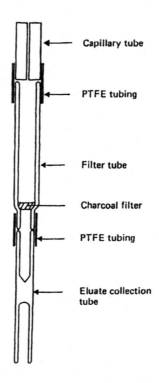

Fig. 15.1 Apparatus for elution of organic compounds from carbon adsorption filter after dynamic headspace extraction
Source: Reproduced with permission from Fielding, M. et al. [9] Water Research Centre, UK

version of the apparatus the recycling gas passes continuously through a short column of a material such as Tenax GC which adsorbs organics or through a cryogenic trap. The column or trap is removed periodically and its contents analysed by gas chromatographic techniques.

McNally and Grob [10] determined the solubility limits of organic priority pollutants by gas chromatographic headspace analysis. A simple and rapid method is described which accurately determines these limits. Headspace, or vapour equilibration analysis, may be used to determine the solubilities of the compound with even moderate volatility. As long as the compound shows some degree of volatility (vapour pressure), the method may be used to determine solubility. The method has been applied to chlorinated aromatics on the US Environmental Protection Agency's pollutants' lists. Otsen and Williams [11] obtained detection limits of 1µg L $^{-1}$ and linear plots of peak area versus concentration in the range 0.25–16µg L^{-1} for 32 organic compounds examined by the headspace analysis technique. At concentrations between 4 and 16µg L^{-1} peak area precisions of 10% RSD were obtained.

Polak and Lu [12] have described a gas stripping method for the determination of the total amount of volatile but slightly soluble organic materials dissolved in water from oil and oil products. Helium is bubbled through a sample of the aqueous liquid and the gas carries the organic vapours directly to a flame ionisation detector. The detector response plotted against time gives an exponential curve, from which the amount of organic material is derived with the aid of an electrical digital integrator. A detector–response factor is required and this is determined with samples prepared by saturating water with hexane or with benzene. Some applications of static headspace analysis are listed in Table 15.1 and further details of particular types of organic compounds in waters are discussed in sections 15.1 to 15.7.

15.1.1 Aliphatic hydrocarbons

Khazal et al. [14] and Drodz and Novak [15] examined and compared the methods of headspace–gas and liquid–extraction analysis, comprising the gas chromatography of samples of the gaseous or liquid–extract phases withdrawn from closed equilibrated systems and involving standard-addition quantitation, for the determination of trace amounts of hydrocarbons in water. The liquid–extraction method [62] is more accurate but it yields chromatograms with an interfering background due to the liquid extractant. The sensitivity of determination of volatile hydrocarbons in water is roughly the same for each method, ie micrograms per litre.

These workers showed that the standard-addition technique is suitable for quantitative determination of trace amounts of hydrocarbons in water, using both headspace analysis and liquid–extraction techniques.

Drodz et al. [16] examined the reliability and reproducibility of qualitative and quantitative headspace analysis of parts per billion of various aliphatic and aromatic hydrocarbons in water using capillary column gas chromatography utilising a simple all-glass splitless sample injection system. They examined the suitability of the standard-addition method for quantitative headspace gas analysis for concentrations in the condensed phase varying from units to hundreds of parts per billion.

With both methods, the chromatographic analysis were carried out at a sensitivity attenuation 1/16; glass column (180cm × 3mm id) packed with 8.14g bf 10% (w/w) Apiezon K on Chromaton N (0.2–0.25mm); column temperature 80°C; nitrogen carrier gas; flow rated of 26, 29.4 and 200ml min^{-1}, for nitrogen, hydrogen and air, respectively. The Chromaton N and Apiezon K were products of Lachema and AEI (Manchester, UK) respectively.

Table 15.1 Preconcentration of organics from various types of water using static headspace analysis

Type of organic	Type of water sample	Detection limit*	Analytical finish	Ref.
Aliphatic hydrocarbons	Sea	–	GLC	[13]
Aliphatic hydrocarbons	Non saline	–	GLC	[14–17]
Naphtha	Sea	–	GLC	[18–21]
Aromatic hydrocarbons	Non saline	sub µg L^{-1}	GLC	[16,22,23]
Aromatic hydrocarbons	Non saline	–	GLC	[24]
Alcohols, ketones, aldehydes	Sea	µg L^{-1}	GLC–MS	[7,25]
Trichloroethylene tetrachloroethane	Non saline	–	GLC	[26]
Chloroaliphatic hydrocarbons	Ground waters	–	–	[27]
Chloroform, dichlorobromo ethane, dibromochloro ethane, bromoform and carbon tetrachloride	Non saline	0.1ng L^{-1}	GLC electron capture detector	[28]
Chlorinated hydrocarbons	Industrial effluents and potable waters	–	GLC–MS	[23,29]
Chlorinated hydrocarbons	Waste waters	0.5mg L^{-1}	GLC	[30]
Halogenated hydrocarbons	Non saline	–	GLC	[22,31–33]
Octafluorobutylene	Non saline	0.01mg L^{-1}	–	[34]
Trichlorofluoro methane and dichloro difluoromethane	Sea	0.05×10^{-12}	GLC	[35]
Haloforms	Potable	–	–	[29,36–45, 50]
Haloforms	Rain	0.001µg L^{-1}	GLC	[46]

Table 15.1 continued

Type of organic	Type of water sample	Detection limit*	Analytical finish	Ref.
Haloforms	Potable	–	GLC	[22,29,36–39, 47–52]
Haloform	Potable	1µg L^{-1}	GLC electron capture	[51]
Chloroform	Potable	–	GLC–MS	[53]
Chloroform, carbon tetrachloride and and ethylene chloride	Potable	–	–	[54]
Chloroform and bis(2) chloroethyl ether	Potable	–	–	[49]
Vinyl chloride	Non saline	–	Photo ionisation detector	[55]
Methyl bromide	Potable	5ng L^{-1}	GLC	[56]
Methyl bromide	Non saline	–	GLC	[57]
Trifluoroacetic acid	Rain	10ng L^{-1}	GLC–MS	[58]
Tetrahydrothiophen	Non saline	–	GLC–MS	[59]
Volatile organics	Wastewater	–	GLC	[12,19–21,57, 60,61]
Misc organics	Non saline	–	GLC	[62]
Misc. organics	Potable	–	GLC	[1,61]
Haloform	Potable	–	GLC	[29]
Haloform	Trade	–	GLC	[29]

*µg L^{-1} unless otherwise stated

Source: Own files

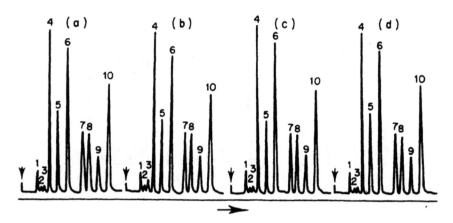

Fig. 15.2 Chromatograms of 1ml headspace gas samples, after an equilibration time of (a) 10min, (b) 30min, (c) 2h and (d) 2.5h at 40°C. Peaks: 1, acetone (10ppm in the liquid phase); 2, n-hexane (25ppb); 3, benzene (46ppb); 4, n-heptane (27ppb); 5, toluene (57ppb); 6, n-octane (42ppb); 7, ethylbenzene (79ppb); 8, m–xylene (78ppb); 9, o-xylene (69ppb); 10, n-nonane (44ppb). Column 1, carrier gas N_2 inlet pressure 0.1 1atm, temperature 70°C Source: Reproduced with permission from Drodz, J. et al. [16] Elsevier Science UK

The headspace method of analysis is less accurate but more sensitive than methods based on liquid extraction. The results in Fig. 15.2 show that with the Drodz et al. [16] method an equilibration time of 10min is adequate for equilibrium between the water sample and the headspace to be achieved.

Various other workers [17,22] have studied the application of headspace analysis to the determination of hydrocarbons in water. McAucliffe [63] determined dissolved individual hydrocarbons in 5ml aqueous samples by injecting up to 5ml of the headspace. For petroleum oils which contain numerous hydrocarbons, very much larger aqueous samples are required. The percentage of hydrocarbons in the gaseous phase, after water containing the hydrocarbons in solution was equilibrated with an equal volume of gas, was found to be 96.7–99.2% for most C_3–C_8 alkenes. In the case of benzene and toluene the values were 18.5 and 21.0%, respectively, indicating that the lower aromatic hydrocarbons may be less amenable to the technique.

Majid [13] determined naphtha in aqueous effluents by headspace gas chromatography. There was minimal interference from high boiling point compounds.

Colenutt and Thorburn [19] applied a gas stripping technique to various synthetic and actual samples of hydrocarbons in water. Synthetic solutions of 10µg L $^{-1}$ n-alkanes from n-octane to n-hexadecane prepared by adding

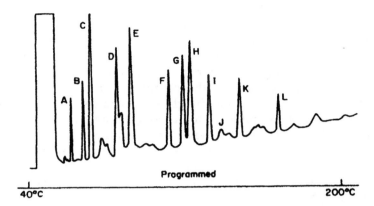

Fig. 15.3 Chromatogram of Welsh Harp water sample. Analytical conditions: 10% Carbowax 20m coated on acid-washed silanised Chromosorb W. Nitrogen carrier gas at a flow rate of 30ml min $^{-1}$. A, n-octane; B, n-nonane; C, benzene; D, n-decane; E, toluene; F, n-undecane; G, ethyl benzene; H, xylenes; I, n-dodecane; J, n-propyl benzene; K, n-tridecane; L, n-tetradecane
Source: Reproduced with permission from Colenutt, B.A. and Thorburn, S. [19] Gordon AC Breach, Amsterdam

acetone solutions of the hydrocarbon to distilled water were put through the procedure. Gas chromatograms were prepared of carbon disulphide extracts. if the solution was analysed almost immediately after preparation a value close to the nominal 10µg L $^{-1}$ for each component was obtained. However, if the aqueous sample was left exposed in an open laboratory for any length of time, the concentration of the lower molecule weight compounds decreased. Thus the concentrations of the lower alkanes are somewhat suspect in that the evaporation effects prior to sampling are unknown.

Fig. 15.3 shows a typical gas chromatogram obtained by Colenutt and Thorburn [19] for the extract of a river water sample showing the presence of aliphatic and aromatic hydrocarbons at the 1–10µg L $^{-1}$ level. These workers found up to 10 times greater concentrations of hydrocarbons in rainwater to that found in river water.

See also Table 15.1.

15.1.2 Aromatic hydrocarbons

Roe *et al.* [24] determined volatile aromatic compounds in soil or ground water using a headspace method. A preservative was needed to prevent microbial degradation of analytes and refrigeration at 4°C was not sufficient to preserve sample integrity.

Desbaumes and Imhoff [20] swept volatile hydrocarbons and their halogenated derivatives from water contained in a heated metallic column by a current of purified air and their concentrations were determined with a flame ionisation detector. The condensed vapours are analysed quantitatively by gas chromatography. Details and diagrams of the equipment are given and the operating procedure is described. Samples must be stored only in glass or stainless steel containers. Substantial losses may still occur if the storage time is more than 10h.

Kaiser [21] has described a sensitive degassing technique for trace hydrocarbons in which volatile hydrocarbons up to C_{12} (probably higher if the aqueous sample is warmed) are removed from aqueous solution at 20°C by a stream of dry nitrogen during 2–10min, and passed into a gas chromatographic column cooled in liquid nitrogen. After the degassing period was completed, the column temperature was programmed at a rate of 7.5°C min^{-1}, and the hydrocarbons eluted and detected in the usual manner. Substances can be removed from water, other liquids, and solids without any further preparation. The detection limit achieved for individual hydrocarbons in water was 10^2ppb (10^{-9}wt %).

Bruckner et al. [64] used a headspace cell in their apparatus for the preconcentration and analysis of aromatic hydrocarbons such as toluene and o-xylene in non saline waters. They used a flame ionisation detector to detect volatile aromatic compounds that have been separated by water–reversed-phase liquid chromatography. The mobile phase is 100% water at room temperature, without the use of organic solvent modifiers. An interface between the liquid chromatograph and the detector is presented, whereby a helium stream samples the vapour of volatile components of individual drops of the liquid chromatograph eluent, and the vapour enriched gas stream is sent to the flame ionisation detector. The design of the headspace cell is simple because the water-only nature of the liquid chromatographic separation obviates the need to do any ultrasonic solvent removal prior to gas phase detection. Despite the absence of organic modifier, hydrophobic compounds can be separated in a reasonable time due to the low phase volume ratio of the water–reversed-phase liquid chromatography columns. The headspace interface easily handles liquid chromatography flows of 1ml min^{-1} and, in fact, compound detection limits are improved at faster liquid flow rates. The transfer efficiency of the headspace interface was estimated at 10% for toluene in water at 1ml min^{-1} but varies depending on the volatility of each analyte. The detection system is linear over more than 5 orders of 1–butanol concentration in water and is able to detect sub-ppb amounts of o-xylene and other aromatic compounds in water. In order to analyse volatile and non-volatile analytes simultaneously, the flame ionisation detector is coupled in series to a water–reversed-phase liquid chromatographic system with ultraviolet absorbance detection.

Water–reversed-phase liquid chromatography improves ultraviolet absorbance detection limits because the absence of organic modifier allowed the detector to be operated in the short-wavelength ultraviolet region, where analytes generally have significantly larger molar absorbtivities. The selectivity the headspace interface provides for flame ionisation detection of volatiles is demonstrated with a separation of 1–butanol, 1,1,2–trichloroethane, and chlorobenzene in a mixture of benzoic acid in water. Despite coelution of butanol and 1,1,2 trichloroethane with the benzoate anion, the non-volatile benzoate ion does not appear in the flame ionisation detector signal, allowing the analytes of interest to be readily detected. The complementary selectivity of ultraviolet-visible absorbance detection and implementation of flame ionisation detection allows for the analysis of volatile and non-volatile components of complex samples using water–reversed-phase liquid chromatography without the requirement that all the components of interest be fully resolved, thus simplifying the sample preparation and chromatographic requirements. This instrument should be applicable to routine automated water monitoring, in which repetitive injection of water samples onto a gas chromatograph is not recommended. See also Table 15.1.

15.1.3 Halogenated aliphatic compounds

Kaiser and Oliver [22] have determined volatile halogenated hydro-carbons at the 0.1–10µg L $^{-1}$ level in water by headspace and gas chromatography. Hrivnak et al. [65] determined chlorinated C_1–C_4 hydrocarbons in water using capillary gas chromatography. For the isolation of chlorinated hydrocarbons (n-butyl chloride, di-, tri- and tetra-chloromethane, 1, 2–dichloroethane, 1, 2–dichloropropane and tri-chloethylene), a stripping technique was used. The hydrocarbons were analysed in a capillary stainless steel column at 80°C. Using electron capture it is possible to determine down to 0.1µg L $^{-1}$ of these substances.

Hellmann [26] has applied the headspace technique to the determination of tetrachloromethane and trichloroethylene in river water.

Biebier-Dammaan et al. [66] compared headspace and solvent extraction methods for the determination of halogenated hydrocarbons in non saline waters. Yurteri et al. [67] studied the effects of salts, surfactants and humic material in clean and polluted water on the Henry's law constants governing headspace analysis.

Headspace analysis has been applied to the determination of volatile chlorinated hydrocarbons in water. Hellman [31] determined chloroform, carbon tetrachloride, trichloroethylene and perchloroethylene. Studies of the operating variables on the headspace technique are described including the effects of filling volume, bath temperature and duration of heating in the thermoblock. The method gave satisfactory and reproducible results, with detection limits of 0.05µg L $^{-1}$ in all cases.

Mehran et al. [32] evaluated various gas chromatographic methods employing direct headspace and water injection into fused silica capillary columns for their ability to separate a model system composed of deionised water and trace amounts of 16 halocarbons. For headspace injection, both separation and detection sensitivity were affected by the lengths of solute bands and enhanced by focusing. Phase ratio focusing and distribution constant focusing were considered. Aqueous injection techniques required compromises when choosing detector temperatures and gas flow rates, with optimal values being system dependent.

Static headspace techniques for determining trihalomethanes have been studied by Keith [38], Bush et al. [49], Morris and Johnson [50] and Gomella and Belle [40].

Friant and Suffet [1] have described a direct head gas analysis procedure for the isolation of chloroform from aqueous environmental samples. The technique included gas chromatography and mass spectrometry. This worker carried out fundamental studies of the partitioning of organic compounds between the aqueous and vapour phases and systematically examined effects of variations in operating parameters. Possible errors and limitations of this method are discussed.

Varma et al. [37] carried out a comparative study of the determination of trihalomethanes in water. They compared the results of chloroform extraction using six liquid–liquid extraction solvents (pentane, methyl–cyclohexane, iso–octane, hexane, n–heptane and n–nonane) with the vapour space extraction method. The vapour space method yielded the poorer results.

Headspace gas chromatography has been used to determine 0.1ppb to 1ppm chloroform carbon tetrachloride, trichloroethylene and tetra-chloroethylene in drinking, non saline and industrial waters [29]. These workers carried out a systematic study of the parameters affecting the accuracy and precision of results obtained by this technique.

Headspace analyses were conducted using a Tracor 222 gas chromatograph equipped with a ^{63}Ni electron-capture detector which was operated in the pulsed linearised mode. Separations were effected with a 3m × 4mm id glass column containing 20% SP–2100–0.1% Carbowax 1500 on 100–200 mesh Supelcoport. The column was operated at 85°C using 90:10 argon–methane carrier gas (65ml min $^{-1}$) which was prepurified with a molecular sieve filter and an oxygen trap. The detector was maintained at 275°C and purged with 40ml min $^{-1}$ of 90:10 argon–methane. The injection port was maintained at 130°C and was fitted with Microsep–138A septa. Chromatographic data were recorded using a Hewlett Packard 3380A printer–plotter integrator using a chart speed of 1cm min $^{-1}$ and a slope sensitivity of 1.0. Under these conditions, respective retention times for air, $CHCl_3$, CCl_4, C_2HCl_3 and C_2Cl_4 were 0.88, 3.64, 5.12, 6.15 and 12.88min.

Headspace samples were injected into the gas chromatograph using Precision Sampling Corp. syringes (Pressure–Lok, series D) of 5, 2 and 1ml capacity and Hamilton syringes (1000 series) of 500, 250 and 100μL capacity.

Purge and trap analyses were conducted using the above chromatograph and column; however, a Tracor 700 Hall electrolytic conductivity detector was used. The detector was operated at 900°C with a hydrogen reaction gas flow of 40cm³ min⁻¹. The electrolytic fluid flow was set a 1.2ml min⁻¹; detector conductivity range was 10. The chromatographic column employed a nitrogen flow of 60ml min⁻¹ and was operated at 80°C. The injector was 90°C and the column to detector transfer line was maintained at 120°C. Liquid sample concentration was provided by a Tekman LSC–1 (all Tenax trap). The Tekmar trap effluent port was directly interfaced to the gas chromatograph injection port with 0.8mm id (1.6mm od) stainless steel capillary tubing. The desorption heater was a modified replacement heater (P/N 12082) which provides a rapid temperature ramp (180°C in 40s). The Tekmar unit was operated using a 15min sample purge of 20ml min⁻¹ of nitrogen, a 4min desorption at 200°C, and 5ml samples. The purge and trap method was calibrated using aqueous standards (and dilutions thereof). Integration of detector signals was conducted as described above. Observed retention times (from initiation of desorption) for the respective components $CHCl_3$, CCl_4, C_2HCl_3 and C_2Cl_4 were 4.5, 6.2, 7.2 and 14.6min. Single or multiple component primary aqueous standards were prepared. The primary aqueous standard contained the following halocarbon concentrations: 50μg L⁻¹ of chloroform, 2μg L⁻¹ of carbon tetrachloride, 20μg L⁻¹ of trichloroethylene and 20μg L⁻¹ of tetrachloroethylene.

Dietz and Singley [29] also showed that equilibration of the aqueous sample with the headspace with no agitation is very slow. Even after 2h, equilibration is far from complete when vials are not shaken. One minute of agitation is sufficient for sample phase equilibration. Equilibrium is completely attained by rapid hand agitation which is comparable to a long-term agitation on a mechanical shaker.

When very carefully conducted, analytical precision of approximately 3% relative standard deviation (r.s.d.) can be achieved by this method. For routine analyses of many samples, a precision of from 5 to 1.0% (r.s.d.) is readily attained. Table 15.2 presents a study in which 10 lake water samples were collected during a 1h period. These samples were collected, transported and stored for 30 days at 4°C, then analysed. The first five samples (group A) were analysed very carefully; a precision of –3% (r.s.d.) was observed. The second set of samples (group B) was analysed using less stringent attention to detail. In this case, a precision of about

Table 15.2 Results from analyses of 10 lake water samples

Sample	Results (µg L^{-1})			
	CHCl$_3$	CCl$_4$	C$_2$HCl$_3$	C$_2$Cl$_4$
Group A				
1	56.1	11.8	11.4	7.9
2	54.8	12.0	11.0	7.8
3	54.6	12.2	11.0	7.9
4	55.1	12.0	11.2	7.9
5	53.3	11.5	11.7	8.2
Average	55.2	11.9	11.3	7.9
R.s.d.	1.1%	2.2%	2.7%	2.0%
Group B				
1	56.2	13.0	12.7	10.8
2	55.8	13.1	11.8	10.1
3	59.1	14.2	13.0	11.4
4	59.1	14.3	12.9	11.3
5	57.6	13.5	12.8	11.4
Average	57.6	13.6	12.6	11.0
R.s.d.	2.7%	4.5%	2.3%	5.1%

Source: Reproduced with permission from Deitz, E.A. and Singley, K.F. [29] American Chemical Society

5% (r.s.d.) was noted. The analytical method thus provides excellent precision even for rapid screening of samples.

A purging method has been described for the specific determination of low levels of methyl bromide fumigant in water [57]. The analysis was performed with a Hewlett Packard Purgatrator, mounted on a PTC (packed trap capillary) module, with two electron-capture detectors. The module was used in a Hewlett Packard Model 433 gas chromatograph.

Using headspace analysis, Jin et al. [34] determined octaflurobutylene in amounts down to 0.01mg L^{-1} in non saline waters. The presence of salts in the samples increased the detection limit [55].

Headspace analysis with a photoionisation detector [55] has been used to determine vinyl chloride in a non saline water.

See also Table 15.1.

15.1.4 Tetrahydrothiophen

See Table 15.1

15.1.5 Miscellaneous organics

Vitenberg and Kostkina [68] compared three variants of splitless introduction of headspace vapours into a gas chromatograph by

(i)   direct introduction
(ii) direct introduction with cryogenic preconcentration and
(iii) after preconcentration.

Cryogenic preconcentration was found to decrease the detection limit by an order of magnitude.

15.2 Seawater

15.2.1 Aliphatic hydrocarbons

While the gases used in stripping are usually air, nitrogen or helium, electrolytically evolved hydrogen has been used as a collector for hydrocarbons [18]. In this technique the gas is not passed through a column of adsorbent, but instead collects in the headspace of the container. Since the volume of seawater and of hydrogen are known, the hydrocarbon concentration in the headspace can be used to calculate the partition coefficients and the concentrations of hydrocarbon in the seawater. See also Table 15.1.

15.2.2 Aromatic hydrocarbons

See Table 15.1.

15.2.3 Trichlorofluoromethane and dichlorodifluoromethane

Bullister and Weiss [35] used headspace analysis and gas chromatography with electron capture gas chromatography to determine trichlorofluoromethane and dichlorodifluoromethane in seawater. The reported detection limit in 30mL samples was 0.005×10^{-12}mol kg^{-1}. See also Table 15.1.

15.2.4 Alcohols, ketones and aldehydes

Chian et al. [7] pointed out that the Bellar and Lichtenberg [25] procedure of gas stripping followed by adsorption on to a suitable medium and subsequent thermal desorption on to a gas chromatograph–mass spectrometer is not very successful for trace determinations of volatile polar organic compounds such as the low molecular weight alcohols, ketones, and aldehydes. To achieve their required sensitivity of parts per billion Chian et al. [7] carried out a simple distillation of several hundred ml of sample to produce a few ml of distillate. This achieved a concentration factor of between 10 and 100. The headspace gas injection–gas chromatographic method was then applied to the concentrate obtained by distillation. See also Table 15.1.

15.2.5 Miscellaneous organics

There have been many other applications of the technique [1,5,6,25,54, 69–72]. The major advantage of the headspace method is simplicity in handling the materials. At most, only one chemical, the salt used in the salting-out procedure, needs to be added and in most cases the headspace gas can be injected directly into a gas chromatograph or carbon analyser. On the other hand, the concentration of organic materials present is limited by the volume of seawater in the sample bottle. This is very much a batch process. Equilibration between the headspace gas and the solution can take a considerable time. This is not a problem when the salting-out material is added at sea, and the samples are then brought into the laboratory for analysis some time later. When the salting-out is done in the laboratory, equilibration can be hastened by recirculating the headspace gas through the solution. A system could be devised which would permit the accumulation of volatiles from a large volume of water into a relatively small headspace, perhaps by recirculating both water and headspace gas through a bubbling and collection chamber, but much of the simplicity and freedom from possible contamination would be lost in the process.

Volatile organic materials can also be removed from solution by distillation, either at normal or at elevated pressures. While the amounts to be collected in this fashion are small, if headspace samples are taken at elevated temperatures and pressures trace quantities of organics can be detected [82]. It should be emphasised however, that whenever extreme conditions are employed to free an organic fraction, that fraction is defined by the conditions of the separation and cannot profitably be compared with fractions defined by different sampling conditions. The use of elevated temperatures and pressures may also alter the compounds separated, limiting the amount of information that can be extracted from the analyses.

15.3 Rainwater

15.3.1 Haloforms

See Table 15.1.

15.3.2 Trifluoroacetic acid

Zehavi and Seiber [58] have discussed a headspace, gas chromatographic method for the determination of microgram per litre levels of trifluoroacetic acid in fog and rain water and surface waters. The described method determines trace levels of trifluoroacetic acid, an atmospheric breakdown product of several of the hydrofluorocarbon and hydrochlorofluorocarbon replacements for the chlorofluorocarbon refrigerants, in water and air. Trifluoroacetic acid is derivatised to the volatile methyl trifluoroacetate and determined by automated headspace gas chromatography with electron-capture detection or manual headspace gas chromatography using gas chromatography–mass spectrometry in the selected ion monitoring mode. The method is based on the reaction of an aqueous sample containing trifluoroacetic acid with dimethyl sulphate in concentrated sulphuric acid in a sealed headspace vial under conditions favouring distribution of methyltrifluoroacetate to the vapour phase. Water samples are prepared by evaporative concentration, during which trifluoroacetic acid is retained as the anion, followed by extraction with diethyl ether of the acidified sample and then back-extraction of trifluoroacetic acid (as the anion) in aqueous bicarbonate solution. The extraction step is required for samples with a relatively high background of other salts and organic materials. Air samples are collected in sodium bicarbonate–glycerin-coated glass denuder tubes and prepared by rinsing the denuder contents with water to form an aqueous sample for derivatisation and analysis. Recoveries of trifluoroacetic acid from spiked water, with and without evaporative concentration, and from spiked air were quantitative, with estimated detection limits of $10\mu g\ L^{-1}$ (unconcentrated) and $25ng\ L^{-1}$ (concentrated 250mL:1mL) for water. Several environmental air, fog water, rain water and surface water samples were successfully analysed; many showed the presence of trifluoroacetic acid.

15.3.3 Haloforms

See Table 15.1.

15.4 Groundwater

15.4.1 Chloroaliphatic hydrocarbons and haloforms

McLary and Barker [27] used headspace analysis to monitor levels of trichloromethane, 1:1:1 trichloroethylene and tetrachloroethylene in groundwater samples. See also Table 15.1.

15.5 Potable water

15.5.1 Haloforms and aliphatic halogen compounds

Static headspace techniques have been studied by Keith [38], Bush et al. [49], Morris and Johnson [50], Gomella and Belle [40] and Suffet and Radziul [41].

The earliest work on the application of headspace methods to the determination of haloforms is probably that of Rook [42] and Kaiser and Oliver [22]. Kaiser and Oliver [22] have described a headspace method for the determination of chloroform, dichlorobromomethane, dibromochloromethane, bromoform, and carbon tetrachloride in water. This method is based on the equilibration of the dissolved compounds in water with a small volume of gaseous headspace under reduced pressure at elevated temperature. Headspace samples, so equilibrated, are directly injected into conventional gas chromatograph inlets for rapid quantification of the volatile compounds present. With a ^{63}Ni electron-capture detector, quantitative determinations of chloroform and similar compounds in the 0.1–10µg L^{-1} range in water samples of less than 60ml are performed in approximately 0.75h.

Suffet and Radziul [41] applied various techniques including headspace analysis and solvent extraction to the screening of volatile organics including chloroform and bis(2–chloroethyl) ether in water supplies. Bush et al. [49] developed a method for the determination of halogen-containing organic compounds using measurement of peaks in headspace vapour with an electron-capture detector and a chromatograph system. The method was used in a screening survey of New York State potable water with the objectives of determining significant seasonal variations in halo-organic concentrations of the compounds in chlorinated ground water and surface water, and the frequency of occurrence of halo-organic compounds in chlorinated water. It is shown that chloroform and bromodichloromethane occur most frequently in chlorinated water.

Friant [43] described a direct head gas analysis procedure for the isolation of chloroform from aqueous environmental samples. The technique included gas chromatography and mass spectrometry. This worker carried out fundamental studies of the partitioning of organic compounds between the aqueous and vapour phases and systematically examined effects of variation in operating parameters. Possible errors and limitations of this method are discussed.

Varma et al. [37] carried out a comparative study of the determination of trihalomethanes in water. They compared the results of chloroform extraction using six liquid–liquid extraction solvents (pentane, methylcyclohexane, isooctane, hexane, n-heptane and n-nonane) with the vapour space extraction method. The vapour space method yielded the poorer results.

Montiel [28] applied the headspace analysis technique to the determination of the halomethane content of chlorinated water samples. He considered the effects of operating variables on the sensitivity of the method and the impact of a number of interfering substances (surface active agents and soluble salts). Particular attention is paid to the operation of the electron-capture detector system.

Headspace gas chromatography has been used to determine 0.1µg to 1mg per litre chloroform, carbon tetrachloride, trichloroethylene and tetrachloroethylene in potable waters [44]. These workers carried out a systematic study of the parameters affecting the accuracy and precision of results obtained by this technique.

Simmonds and Kerns [53] used a permaselective membrane of perfluorosulphonic acid polymeric material to remove water from aqueous samples by injecting them into a length of tubing made from the material, positioned ahead of a gas chromatographic column. The method allows direct injection of aqueous samples without altering column performance. The method was demonstrated in the analysis of trace amounts of halocarbons in potable water samples by direct electron-capture gas chromatography. The water removal technique was also used in conjunction with headspace analysis.

Deitz and Singley [29] observed that analytical accuracy is influenced by the effects of sample matrices on chromatographic responses. Several matrix conditions were examined experimentally. The presence of organics in the water sample might upset halocarbon phase equilibration. However, no effects (2% r.s.d.) from either methanol or acetone were observed up to a 2% (v/v) concentration. Methanol is observed in the gas chromatograms at a 2min retention time but does not interfere in the halocarbon analyses. Sample pH also does not affect (2% r.s.d.) chromatographic results. Sodium chloride concentrations greater than 1% will significantly increase chromatographic responses. These increased responses reflect an increase in halocarbon activity coefficients (salting out).

Otson et al. [36] have compared dynamic headspace, solvent (hexane) extraction, and static headspace techniques utilising Tenax gas chromatographic columns, a ^{63}Ni electron capture detector and a Hall electrolytic conductivity detector for the determination of trihalomethanes in water. The static headspace technique, involving equilibration of trihalomethanes between the water sample and air space in a closed vessel, allows an aliquot of the air space to be analysed for trihalomethanes. The relative standard deviation between trihalomethane values obtained by the three techniques ranged 9–10% for chloroform, 3–24% for dichlorobromomethane and 13–61% for dibromochloromethane.

Otson et al. [36] found that improved sensitivity for trihalomethanes was achieved by operating the Hall detector in the pyrolytic mode. As expected, the electron-capture detector gave considerably lower method detection limits than the Hall detector. The precision of the gas sparging and solvent extraction techniques was comparable to the static headspace technique.

Otson *et al.* [36] also studied the effects on reported results of delays of up to 30h between taking the sample and carrying out the analysis. Bulk aqueous solutions containing 20–120µg L^{-1} chloroform, 1.6–9.9µg L^{-1} dichlorobromomethane and 0.4–2.4µg L^{-1} bromochloromethane showed respectively concentration losses of 12 ± 3, 16 ± 5 and 21 ± 4%. Otson *et al.* [36] concluded that the three analytical techniques gave comparable trihalomethane values for potable water samples. Although the gas sparging technique was the most sensitive, the solvent extraction technique gave comparable precision. The static headspace technique showed relatively poor precision and inferior sensitivity. The sensitivity of the static headspace and solvent extraction techniques can be improved by increasing the volume of the aliquot injected into the gas chromatograph, but overloading of the electron-capture detector by organohalides must be avoided.

Castello *et al.* [39] determine trihalomethanes using a polar and non-polar column mounted in series connected to a ^{63}Ni electron-capture detector to separate the trihalomethanes and any other halocarbons present, using temperature programming. The series arrangements enabled the analysis of each headspace sample to be completed in less than 30min.

Castello *et al.* [39] evaluated the gas chromatographic sensitivity and linearity of the ^{63}Ni asymmetric type electron-capture detector for trihalomethanes and other haloalkanes using headspace and liquid–liquid extraction techniques for mixed column gas chromatographic separation. Each extraction method showed a wide linearity range, allowing simplified calibration techniques and automation.

Croll *et al.* [51] determined trihalomethanes in potable water using gas syringe injection of headspace vapours and electron-capture gas chromatography. Headspace vapours were withdrawn from a sample container, which had been equilibrated at ambient temperature, by a gas-tight syringe with a valved needle. The samples were injected into a gas chromatograph equipped with an electron-capture detector. The use of sealed vials minimised contamination. The method was simple and insensitive to sample volume, natural salt concentration and minor vapour loss during sampling. The relative standard deviation was less than 2% and the results were comparable with liquid–liquid extraction. The limit of detection of trihalomethane was less than 1µg L^{-1}.

Static headspace analysis has been used to determine methyl bromide in potable waters [56]. The technique used to determine traces of methyl bromide involved addition of sodium iodide to convert methyl bromide to methyl iodide, which was then determined by gas chromatography. The detection limit was approximately 5ng L^{-1} See also Table 15.1.

15.5.2 Miscellaneous organics

Friant and Suffet [1] obtained 66-fold enrichment factors by:

(1) increasing the temperature of the water sample from 30 to 50°C;
(2) saturating the water sample with sodium sulphate; and
(3) adjusting the pH of the original sample to 7.1.

Their method used a Tracor MT–550 gas chromatograph equipped with dual flame ionisation detectors or equivalent, and a chromatographic column 2.4m × 3mm id stainless steel packed with 20% SE–30 on 80/100 mesh Gas Chrom Q. The column temperature was maintained isothermally at 130°C. Inlet, outlet, and detector temperatures were 180, 200, and 200°C respectively. Glassware in contact with both water and vapour phases was silanised to minimise surface adsorption. First the glassware was washed with detergent. This was followed by rinsing with distilled water and air drying. The dry surface was treated with Glass–treet (Alltech Associates) and rinsed with anhydrous methanol.

The sampling bottle used for laboratory test conditions was a modified 2L Pyrex reagent bottle. The neck was reformed by using 30 glass O–ring joints and the top was rounded and sealed. The two joints were sealed by a Buna rubber O–ring and a compression clamp. To allow syringe sampling of the vapour two (3mm diameter) glass side arms were attached. A stainless steel (3–6mm) Swagelok reducing union was attached to the glass sidearms by 6mm rubber O–rings and a Swagelok nut. The sampling port was sealed by a chromatographic silicon septum. The gas syringe was a Precision Scientific (Baton Rouge, LA) pressurisable Series A–2 10.0ml gas syringe equipped with sideport needle and stop/go valve. The valve permitted the sampling of larger vapour volumes by allowing compression of the sample prior to injection, thereby reducing peak broadening. The syringe also permit easier injections since the pressure in the syringe can be equal to the GLC column head pressure.

All experiments were run isothermally in a constant temperature air bath controlled to ±0.5°C. All samples were stirred on a magnetic stirrer. Anhydrous sodium sulphate 475g L^{-1} (3.35M) dissolved in 0.20M orthophosphate buffer) was placed in a clean sample bottle at the required experimental temperature 16h prior to the analysis to reduce the time required to dissolve the salt. A sampling temperature of 50°C was used. Vapour samples were withdrawn through the sidearm sampling port of the sample bottle with the gas syringe. The syringe was flushed with vapour phase prior to withdrawing the sample. After sampling the vapour, the syringe valve was closed and the volume reduced to 10% of the sample volume followed by direct injection into the gas chromatographic column. The general experimental conditions were:

Table 15.3 Three-dimensional presentation of the experimental results for the analysis of variance of methyl ethyl ketone, $K \times 10^{-3}$

| | Temperature (°C) | | | | | |
| | 30 | | Salt concentration (M) | | 50 | |
	0.00	1.41	3.35	0.00	1.41	3.35
pH4.5	4.19	21.3	118	21.3	39.8	234
pH7.1	3.90	20.0	109	20.0	37.6	260
pH9.1	4.56	18.7	105	18.7	35.0	229

Source: Reproduced with permission from Friant, S.L. and Suffet, J.H. [1] American Chemical Society

Table 15.4 Partition coefficient of model compounds at the optimum and reference states

| | Compound $K \times 10^{-3}$ | |
	Optimum conditions pH7.1, 50°C, 3.35M[a]	Reference condition, pH7.1, 30°C, no salt[a]
Methyl ethyl ketone	260	3.90
Nitroethane	72.5	2.89
Butanol	44.3	0.746
Dioxane	13.7	0.278

[a]0.2 Morthophosphate buffer

Source: Reproduced with permission from Friant, S.L. and Suffet, J.H. [1] American Chemical Society

Equilibration time	5h
Vapour volume	1100–1200ml
Liquid volume	1000–1040ml
Volume of vapour samples	5.0ml
pH	7.1
Sodium sulphate concentration	0.00–3.35M
Ionic strength	0.00–10.02
Orthophosphate buffer	0.20M

Table 15.3 is a three-dimensional presentation of the results obtained from the $3 \times 3 \times 2$ statistical experiments. Methyl ethyl ketone data are shown as an example. It is seen that the highest partition coefficient (K) for methyl ethyl ketone is obtained with 3.35M sodium sulphate, pH7.1, and

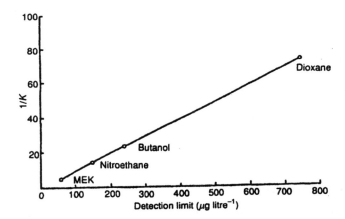

Fig. 15.4 Theoretical detection limits for model compounds plotted versus the reciprocal of the partition coefficients; the nominal FID detection limit is 50ng and the amount is 5.0ml of injected vapour
Source: Reproduced with permission from Friant, S.L. and Suffet, J.H. [1] American Chemical Society, Washington

50°C. In Table 15.4 are shown partition coefficients for four different organic compounds under these conditions, relative to conditions where sodium sulphate is not added and the temperature is reduced to 30°C.

The calculated theoretical detection limits for the four organic compounds examined by Friant and Suffet [1] are shown plotted against the reciprocal of the partition coefficient ($1/K$) in Fig. 15.4. The theoretical limits were determined by assuming a nominal flame ionisation detector limit of 50ng absolute amount injected. The detection limits ranged from $50\mu g\ L^{-1}$ for methyl ethyl ketone to $740\mu g\ L^{-1}$ for dioxane. The smaller $1/K$ is the lower the detection limit. Fig. 15.4 can be used to determine the theoretical detection limits for any solute from the measured partition coefficient under comparable conditions.

The general 'screening' procedure for quantitative analysis with semi-quantitative evaluation has been utilised for taste and odour profiling of water. A primary reason for the investigation of headspace analysis was to subsequently develop a method that would isolate volatile organics causing taste and odour in potable water. Isolation of possible taste and odour organics under the enhanced conditions allows determination of the initial water phase concentrations. These concentrations can then be used to determine the vapour phase concentrations of the organic compounds presented to the consumer of drinking water. All that is needed is the partition coefficients of the organic compound at the enhanced and potable water conditions, respectively.

Application of the method of headspace analysis as described was utilised for gas chromatography–mass spectrometry of a potable water sample in Philadelphia. The results of mass spectral identification of the compounds found in the potable water showed the presence of toluene, two C_{-2} benzene isomers, chloroform, chlorodibromomethane, dichloro-bromoethane, and 1,1,2,2–tetrachloroethane. The analysis was completed in 125ml bottle containing 100ml of potable water. A 50μL gas sample was injected on to the gas chromatograph using a [63]Ni electron capture detector.

See also Table 15.1.

15.6 Wastewaters

15.6.1 Aliphatic hydrocarbons

CONCAWE [61] have applied the gas stripping analysis technique developed by the Environmental Protection Agency [73–75] for the preconcentration of organics in oil refinery wastes prior to their identification and determination by gas chromatography–mass spectrometry. The method has been used to determine down to 1μg L^{-1} of benzene, toluene, ethylbenzene and mono- and di-chlorobenzenes in aqueous refinery wastes [74] and up to 29 purgeable chlorine- and bromine-containing organics in aqueous industrial wastes. For the determination of purgeable trace aromatic and aliphatic hydrocarbons a flame ionisation detector should be used. For the determination of purgeable halogenated organics a halide selective detector should be used (the Hall 700A electrolytic conductivity detector is recommended, but the alkali flame ionisation or electron capture devices could be used).

15.6.2 Chlorinated hydrocarbons

Lukacovic et al. [45] applied headspace gas chromatography to the determination down to 0.5mg L^{-1} of chlorinated hydrocarbons in waste waters. Techniques for enrichment and clean-up of the extract were devised, the preferred method consisting of liquid–liquid extraction with methylene chloride, followed by flash evaporation, clean-up of the concentrate by column chromatography on Florasil, a further evaporation step and subsequent gas chromatography with an electron-capture detector. Two gas chromatographic packings were compared for use with these haloethers, 3% SP–1000 on Supelcoport giving better results than Tenax–GC on account of better peak shape and resolution. The validity of the method was confirmed by application to samples of either municipal or industrial effluents spiked with known amounts of particular haloethers. See also Table 15.1.

15.6.3 Miscellaneous volatiles

Leinster *et al.* [83] analysed aqueous volatile organic compounds in waste water by solvent extraction, dynamic headspace analysis (gas sparging) and the closed-loop stripping technique and the methods were evaluated. If the volatile organic content is of most importance the purging technique combined with thermal desorption gas–liquid chromatography is recommended.

Malten and Vreden [4] applied the gas chromatographic headspace analysis technique to the detection of organic volatiles in small volumes of aqueous samples. Cowen *et al.* [5] have described the construction of a septumless injection device for delivering headspace gases into a gas chromatograph.

Distillation techniques have been combined with headspace analysis as a means of improving preconcentration factors of up to 100 [5–7]. Heyndrickx and Peteghem [54] identified volatile components in waste waters by combined headspace analysis and gas chromatography. Drodz and Novak [8] have shown that low concentrations of volatile organics can be determined in water by the double sampling method of headspace gas chromatography using a closed-loop strip–trap technique.

15.7 Trade effluents

15.7.1 Chlorinated hydrocarbons

See Table 15.1

15.7.2 Haloforms

The headspace method described by Deitz and Singley [29] and discussed in section 15.5.1 has been applied to the determination of haloforms in trade effluents.

15.8 Soils

15.8.1 Phenols

Direct acetylation followed by gas chromatography headspace analysis has been used to determine phenols and cresols in soils [30].

15.8.2 Miscellaneous volatiles

Gas chromatography with flame ionisation and photoionisation detection can be used on-site by adopting a jar headspace procedure for the analysis of soils [60].

15.9 Closed loop headspace analysis

15.9.1 Aliphatic and aromatic hydrocarbons

Gomez-Belinchon and Albaiges [76] optimised the conditions for the closed-loop stripping analysis method for the quantitative determination of volatile organics in water to reduce analysis time. Extraction efficiencies with respect to volatility, water temperature and time were calculated for a series of 21 aliphatic and aromatic hydrocarbons and their chlorinated derivatives. Optimal recoveries were obtained by stripping at 45°C for 0.5h instead of the conventional 35°C for 2.0h. Operational procedures for overcoming background contamination were discussed.

Drodz et al. [77] used water–air systems with low µg L^{-1} levels of benzene, toluene, n-decane, n-undecane, and n-dodecane to evaluate the analytical method of repetitive stripping and trapping of analytes. Closed-circuit and open arrangements were investigated to determine the reliability of the method. In a closed circuit, the stripping–trapping process was accomplished under a conservation or an equilibrium regime, whereas in an open arrangement, conservation or pseudoequilibration models of trapping were possible. All the above were used for quantitative analysis. Conservation trapping gave better results when working in an open arrangement. Systematic negative errors of 20 and 40% were obtained for the higher aliphatic hydrocarbons and were attributed to varying matrix effects associated with the adsorption of analytes at the air–water interface.

Closed loop stripping analysis is a further refinement of the technique in which preconcentration factors of up to one million-fold are possible. Westendorf [78] has described a commercially available apparatus, the Tekman CLS–1 instrument, for the semi-automatic determination of traces of organics in water. This instrument can accommodate any sample size up to 4L. Analysis of fuels in water by closed loop stripping analysis can provide an excellent fingerprint of the sample. The detection limit obtainable by closed loop stripping analysis is even less than 1µg L^{-1} which is 3–4 orders of magnitude more sensitive than any other technique. Detection limits for other fuels, such as gasoline, kerosene or jet fuel are even lower.

15.9.2 Bromoalkanes and chloroalkanes

Janda et al. [79] used closed loop stripping analysis in conjunction with capillary column gas chromatography to identify unexpected substances appearing in the gas chromatograms of potable water following the addition of bromoalkanes as internal standards to the water sample. Due to the presence of free chlorine in the water sample bromoalkanes were converted to chloroalkanes which did not occur in unchlorinated water samples.

15.9.3 Miscellaneous organics

Boren *et al.* [80] optimised the open stripping system for the analysis of trace organics in water. They examined advantages of the open system compared to the closed loop stripping analysis. Stripping temperatures between 30 and 90°C were optimised for the recovery of polar compounds, low volatile, semi-volatile and volatile compounds, and phenols. They investigated the effect of extraction solvent (*n*-hexane, benzene, carbon disulphide, methylene dichloride, diethyl ether) on recovery. Extraction efficiency was generally low. A solvent mixture of carbon disulphide–benzene–methanol (63:30:5) was recommended.

Smith [81] has discussed the design and features of an on-line purge and cold-trap preconcentration device for the rapid analysis of volatile organic compounds in aqueous samples. A theoretical model describing ice formation and growth in the capillary trap was evaluated. Synthetic mixtures covering concentrations in the range 0.1–10μg L^{-1} and different chemical classes were used to study the effects of various process factors on the efficiency and selectivity of water removal as well on the purging recovery. Either a condenser or a Nafion water permeable membrane could be used to remove water selectively from non-polar compounds. Polar and high boiling non-polar solutes were partly or completely lost by both methods. The importance of the concentration of solutes, flow rate relative to the volume of the purge gas, and the temperatures of the condenser, cold trap and sample was emphasised. Non-polar volatile organic compounds could be enriched to give a detection limit of 1μg L^{-1}. Process time could be reduced by purging at elevated sample temperatures and/or at large purge gas flow rates.

References

1 Friant, S.L. and Suffet, J.H. *Analytical Chemistry*, **51**, 2161 (1979).
2 Wyllie, P.L. Proceedings of Water Quality Technical Conference 1986. (Adv. Water Anal. Treatment), 14, pp.185–202 (1987).
3 Munz, C. and Roberts, P.V. *Journal of the American Water Works Association*, **79**, 62 (1987).
4 Malten, L. and Vreden, N. *Forum Stadte Hygiene*, **29**, 37 (1978).
5 Cowen, W.F., Cooper, W.J. and Highfill, J.W. *Analytical Chemistry*, **47**, 2483 (1975).
6 Gjavotchanoff, S., Luessem, H. and Schlimme, K. *Gas–u–Wasser Fach*, **112**, 448 (1971).
7 Chian, E.S.K., Kuo, P.P.K., Cooper, W.J., Cowen, W.F. and Fuentes, R.C. *Environmental Science and Technology*, **11**, 282 (1977).
8 Drodz, J. and Novak, J. *International Journal of Environmental Analytical Chemistry*, **11**, 241 (1982).
9 Fielding, M., Gibson, T.M., James, H.A., McLoughlin, K. and Steel, C.P. In Technical Report TR 159G. Organic micropollutants in drinking water. Water Research Centre, Medmenham, UK, February (1981).
10 McNally, M.F. and Grob, R.L. *Journal of Chromatography*, **260**, 23 (1983).

11 Otsen, R. and Williams, D.T. *Analytical Chemistry*, **54**, 942 (1982).
12 Polak, J. and Lu, B. *Analytica Chimica Acta*, **63**, 231 (1974).
13 Majid, A. *ADSTRA Journal of Research*, **2**, 241 (1986).
14 Khazal, W.J., Vejrostra, J. and Novak, J. *Journal of Chromatography*, **157**, 125 (1978).
15 Drodz, J. and Novak, J. *Journal of Chromatography*, **152**, 55 (1978).
16 Drodz, J., Novak, J. and Rijks, J. *Journal of Chromatography*, **158**, 471 (1978).
17 McAuclliffe, C.J. *Chemical Technology*, **1**, 46 (1971).
18 Wasik, S.P. *Journal of Chromatographic Science*, **12**, 845 (1974).
19 Colenutt, B.A. and Thorburn, S. *International Journal of Environmental Studies*, **15**, 25 (1980).
20 Desbaumes, E. and Imhoff, C. *Water Research*, **6**, 885 (1972).
21 Kaiser, R. *Journal of Chromatographic Science*, **9**, 227 (1971).
22 Kaiser, L.E. and Oliver, B.G. *Analytical Chemistry*, **48**, 2207 (1976).
23 Schulz, J. *Vom Wasser*, **69**, 49 (1987).
24 Roe, U.D., Lacy, M.J., Stuart, J.D. and Robbins, G.A. *Analytical Chemistry*, **61**, 2584 (1989).
25 Bellar, T.A. and Lichtenberg, J.J. *Journal of the American Water Works Association*, **566**, 739 (1974).
26 Hellmann, H. *Zeitschrift für Wasser und Abwasser Forschung*, **21**, 67 (1988).
27 McLary, T.A. and Barker, J.F. *Groundwater Monitoring Review*, **7**, 63 (1988).
28 Montiel, A. *Tribune de Cebedeau*, **32**, 422 (1979).
29 Deitz, E.A. and Singley, K.F. *Analytical Chemistry*, **51**, 1809 (1979).
30 Llopart-Vizoso, M.P., Lorenzo-Ferreira, R.A. and Cela-Torrijos, R. *Journal of High Resolution Chromatography*, **19**, 207 (1996).
31 Hellman, Z. *Zeitschrift für Wasser und Abwasser Furschung*, **18**, 92 (1985).
32 Mehran, M., Cooper, M. and Jennings, W. *Journal of Chromatographic Science*, **24**, 142 (1986).
33 Kirshen, N. *American Laboratory*, **16**, 60 (1984).
34 Jin, X., Shen, S., He, P. and Cheng, W. *Gaoedeng Xuaxiao Huaxue Xuebao*, **9**, 195 (1988).
35 Bullister, J.L. and Weiss, R.F. *Deep Sea Research, Part A*, **35**, 839 (1988).
36 Otson, R., William, D.T. and Bothwell, P.D. *Environmental Science and Technology*, **13**, 936 (1979).
37 Varma, M.M., Siddigue, M.R., Doty, K.T. and Machis, H. *Journal of American Water Works Association*, **71**, 389 (1979).
38 Keith, L.H. In *Identification and Analysis of Organic Pollutants in Water*. Ann Arbor Science Publishers Ltd., Ann Arbor, Michigan (1976).
39 Castello, G., Gerbino, T.C. and Kanitz, S. *Journal of Chromatography*, **351**, 165 (1986).
40 Gomella, C. and Belle, J.P. *Techniques et Sciences Municipales*, **73**, 125 (1978).
41 Suffet, L.H. and Radziul, O. *Journal of the American Water Works Association*, **68**, 520 (1976).
42 Rook, J.J. *Water Treatment and Examination*, **21**, 259 (1972).
43 Friant, S.L. Thesis, Drexel University. University Microfilms Ltd., London 468pp. 29162 (1972).
44 Glaze, W.H. and Rowley, R. *Journal of the American Water Works Association*, **71**, 509 (1979).
45 Lukacovic, L., Mikulas, M., Vanko, A. and Kiss, G. *Journal of Chromatography*, **207**, 373 (1981).
46 Comba, M.E. and Kaiser, K.L. *International Journal of Environmental Analytical Chemistry*, **16**, 17 (1983).

47 Symons, J.M., Bell, T.A., Carswell, J.K. *et al. Journal of American Water Works Association*, **67**, 643 (1975).
48 In *National Survey for Halomethanes in Drinking Water*. Health and Welfare, Canada, 77–EHD–9 (1977).
49 Bush, B., Narang, R.S. and Syrotynski, S. *Bulletin Environmental Contamination and Toxicology*, **13**, 436 (1977).
50 Morris, R.J. and Johnson, L.J. *Journal of American Water Works Association*, **68**, 492 (1976).
51 Croll, B.T., Summer, M.E. and Leathard, D.A. *Analyst (London)*, **111**, 73 (1986).
52 Castello, G., Gerbino, T.C. and Kanitz, S. *Journal of Chromatography*, **247**, 263 (1982).
53 Simmonds, P.G. and Kerns, E. *Journal of Chromatography*, **186**, 785 (1979).
54 Heyndrickx, A. and Peteghem, C. *European Journal of Toxicology*, **8**, 275 (1975).
55 Stein, U.B., Narang, R.S. *Bulletin of Environmental Contamination and Toxicology*, **27**, 583 (1981).
56 Cirilli, L. and Borgioli, A. *Water Research*, **20**, 273 (1986).
57 Packard Technical News. Hewlett Packard 1. No. 7258 February (1981).
58 Zehavi, D. and Seiber, J.N. *Analytical Chemistry*, **68**, 3450 (1996).
59 Gorlucci, G., Airoldi, L. and Farelli, R. *Journal of Chromatography*, **287**, 425 (1984).
60 Fitzgerald, J. In *Principles and Practices for Petroleum Contaminated Soils*. Eds. E.J. Calabrese and P.T. Kostecki. Lewis Boca Raton, Florida, pp.49–66 (1993).
61 The Oil Companies International Study Group of Conservation of Clean Air and Water, Europe CONCAWE Report No. 6/82, Analysis of trace substances in aqueous effluents from petroleum refineries (1982).
62 Drodz, J. and Novak, J. *Journal of Chromatography*, **136**, 37 (1977).
63 McAucliffe, C. *Journal of Chemical Geology*, **4**, 225 (1969).
64 Bruckner, C.A., Ecker, S.T. and Synovec, R.E. *Analytical Chemistry*, **69**, 3465 (1997).
65 Hrivnak, F., Siskupic, P. and Hassler, J. *Vodni Hospodarstivi, Series B*, **28**, 195 (1978).
66 Biebier-Dammann, V., Funk, W., Von Mareard, G. and Rinne, D. *Zeitschrift für Wasser und Abwasser Forschung*, **20**, 22 (1987).
67 Yurteri, C., Ryan, D.F., Callow, J.J. and Gurol, M.D. *Journal of Water Pollution Contr. Fed.*, **59**, 950 (1987).
68 Vitenberg, A.G. and Kostkina, M.I. *Zhur. Anal., Khim*, **43**, 318 (1988).
69 Chesler, S.N., Gump, B.H. and Hertz, H.P. *Analytical Chemistry*, **50**, 805 (1978).
70 Dowty, B., Green, L. and Laseter, J.L. *Journal of Chromatographic Science*, **14**, 187 (1976).
71 Waggott, A. In *Proceedings of the Analytical Division of the Chemical Society, London*, **15**, 232 (1978).
72 Malter, L. and Vreden, V. *Forum Stadte–Hygiene*, **29**, 37 (1978).
73 Environmental Protection Agency Report EPA 624, Environmental Protection Agency, USA.
74 Environmental Protection Agency Report EPA 602, Environmental Protection Agency, USA.
75 Environmental Protection Agency Report EPA 601, Environmental Protection Agency, USA.
76 Gomez-Belinchon, J.I. and Albaiges, J. *International Journal of Environmental Analytical Chemistry*, **30**, 183 (1987).
77 Drodz, J. Vodakova, Z. and Novak, J. *Journal of Chromatography*, **354**, 47 (1986).
78 Westendorf, R.G. *International Laboratory*, **32**, September (1982).

79 Janda, V., Marha, K. and Mitera, J. *Journal of High Resolution Chromatography and Chromatography Communications*, **11**, 541 (1988).
80 Boren, H., Grimvall, A., Palmborg, J., Sauerhed, R. and Wigilius, B. *Journal of Chromatography*, **348**, 67 (1985).
81 Smith, R. *Journal of High Resolution Chromatography and Chromatography Communications*, **10**, 60 (1987).
82 Luessen, H. Detection of trace materials by concentration by pressure Distillation. Hans. Tech. Essen, Vortrages Verhoff, 62 (1972).
83 Leinster, P. McIntyre, J.N. and Perry, R. *Chemosphere*, **10**, 291 (1981).

Chapter 16

Organics: Dynamic headspace purge and trap analysis

As explained at the beginning of Chapter 15, this technique differs from static headspace analysis in that in dynamic headspace analysis the water sample is purged with an inert gas and the volatile organics removed and adsorbed on to a solid, such as Tenax GC or active carbon. The adsorbed organics are then desorbed in a small volume of organic solvent, usually chloroform or carbon disulphide which is subsequently analysed by gas chromatography. Although capable of achieving very high concentration factors, a disadvantage of this technique is that the organic solvent and impurities therein can cause difficulties at the gas chromatographic stage of the analysis. Static headspace analysis is solventless in that the aqueous sample is heated to say 30°C in the presence, usually, of inorganic salts which help to drive the organics into the inert gas-filled headspace above the sample. Portions of the headspace gas are then withdrawn by syringe for gas chromatographic analysis. This headspace analysis, unlike purge and trap, is a static method.

As an alternative to adsorption of the volatile organics on a solid followed by extraction with an organic solvent, the adsorbed organics can be released from the adsorbent by heating and the released solids swept directly into a gas chromatograph, ie solventless analysis.

Very large preconcentrations of organics in the sample can be achieved by these methods. The technique was first described in 1973 [15–17,68].

16.1 Non saline waters

16.1.1 General discussion

Colenutt and Thorburn [1,2] carried out more work to establish the practical limits of the gas stripping preconcentration technique when applied to standard mixtures and to real water samples containing mg L^{-1} quantities of organics and to optimise the various factors affecting the efficiency of the technique. Important variables were the flow rate of stripping gas, the time period for which stripping is carried out, the particle size of the adsorbent, and the choice of solvent used for

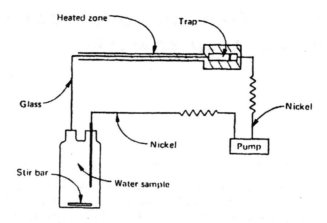

Fig. 16.1 **Flow scheme used in gas stripping preconcentration**
Source: Reproduced with permission from Colenutt, B.A. and Thorburn, S. [1] Gordon
AC Breach, Amsterdam

desorption. It was shown that sample contamination is an even more serious problem than low recoveries. Problems of contamination of aqueous samples by adsorption of atmospheric vapours prior to analysis were also considered.

The stripping system used is illustrated diagrammatically in Fig. 16.1. Nitrogen from a gas cylinder was purified by passing through an active carbon trap and then a flow controller. The gas was then bubbled through the aqueous sample which was contained in a flat bottomed flask. The gas outlet into the sample was passed through a coarse frit and in all cases the volume of the flask was selected to be as near to the sample volume as possible so reducing any contamination by laboratory air entering the system. After passing through the water the gas was passed through a splash trap to remove any drops of water and then through the active carbon trap.

The active carbon trap was formed from a piece of 6mm od glass tube. This was drawn to a fine jet at one end and had an overall length of 8cm. Active carbon (between 5 and 25mg) was held in the tube by two pieces of glass fibre yarn. The combined length of the plugs of glass fibre yarn and the active carbon was about 5cm, leaving an empty portion of tube at the wider end. Flow of gas through the trap was always from the wider end to the narrower end, eliminating any possibility of the adsorbent being blown out of the trap by the gas. The flow control valve ensured constant flow throughout a particular experiment. A soap bubble flowmeter was connected to the outlet end of the trap in order to measure the gas flow at the start and end of each stripping operation. All the glass joints in the system were lined with PTFE sleeves. Connections between glass tubing were made with Swagelok stainless steel fittings incorporating combined PTFE ferrules.

Desorption from the adsorbent was carried out in the following manner, a 5cm length of PTFE tube was connected to the wider end of the trap. The solvent was then dropped on to the adsorbent by means of a dropping pipette. The active carbon provided a restriction to the flow of the solvent so that when a dropping pipette with a rubber teat was connected to the trap by means of the PTFE tube it was possible either to force the solvent through the adsorbent or suck the solvent up through the solid. Thus by careful manipulation of the pipette bulb the solvent was drawn back and forth over the adsorbent several times before the solvent eventually reached the drawn end of the trap and was collected in a calibrated tube. The volume of solvent used for desorption was variable but was routinely between 50µL and 1ml.

Initially the active carbon traps were spiked with known amounts of selected compounds using a microlitre syringe. The percentage recovery of the adsorbate from the adsorbent was measured. Desorption was carried out with a range of solvents in the manner described above. The solutions were then made up to standard volume and analysed by gas chromatography. All the analyses were carried out on a Perkin–Elmer Model F11 gas chromatograph fitted with a flame ionisation detector and an electron capture detector. Column effluent was split in a 1:1 ratio between the two detectors. The columns used for the analysis were either 2m 6mm od glass column packed with 5% OV 17 coated on acid washed Chromosorb W (80–100 mesh) or a 2m long, 6mm od glass column packed with 5% Carbowax 20M coated on silanised acid washed Chromosorb W. The oven temperature used varied depending upon the compound being analysed. Nitrogen was the carrier gas at a flow rate of 25ml min^{-1}.

For each compound in each solvent the size of the peak on the chromatogram was compared with the peak obtained by injection of a solution of known comparable concentration under the same conditions. It was thus possible to calculate the percentage of the adsorbate recovered. Typical recoveries of compounds from 100–200 mesh active carbon achieved with particular solvents are listed in Table 16.1. Recoveries obtained with different particle size adsorbent with carbon disulphide are given in Table 16.2. A similar pattern of results was obtained using n-hexane solvent.

The possibility of contamination from external sources at these exceedingly low concentrations was high. The solutions were prepared in freshly distilled water which had itself been stripped by a flow of nitrogen for 2h immediately prior to the preparation of the solution. The particle size of the active carbon used in the traps for this work was 100–200 mesh. To ensure that none of the compounds was passing

Table 16.1 Recovery of compounds from 100/120 mesh active carbon traps with various solvents

Adsorbate	Solvent					
	n-Hexane	Toluene	Carbon disulphide	Diethyl ether	Carbon tetrachloride	Methanol
n-Octane	99	97	99	97	96	90
n-Decane	98	97	99	95	96	90
Toluene	97	–	100	97	96	89
Ethyl benzene	98	99	100	98	96	90
1,2–Dichlorobenzene	96	96	100	95	98	85
Ethanol	90	90	95	92	88	95
n–Butanol	91	88	96	91	92	98
Phenol	94	91	96	92	90	96
Methyl ethyl ketone	91	90	94	93	91	92
Ethyl benzoate	92	91	92	92	90	90
Acetic acid	90	90	95	95	88	92

Source: Reproduced with permission from Colenutt, B.A. and Thorburn, S. [1] Gordon AC Breach

Table 16.2 Effect of adsorbent particle size on recovery efficiency (percentage of adsorbate recovered after extraction with carbon disulphide)

Adsorbate	Mesh range and corresponding particle size				
	44/60 0.251– 0.353mm	60/85 0.178– 0.251mm	85/100 0.150– 0.178mm	100/120 0.124– 0.150mm	120/200 0.076– 0.124mm
n-Octane	92	95	98	99	99
n-Decane	92	96	98	99	99
Toluene	91	95	99	100	100
Ethyl benzene	92	97	99	100	100
1,2–Dichlorobenzene	94	96	99	100	100
Ethanol	86	90	93	95	96
n-Butanol	86	91	92	96	97
Phenol	90	93	95	96	97
Methyl ethyl ketone	86	90	92	94	94
Ethyl benzoate	83	86	90	92	93
Acetic acid	86	90	93	95	95

Source: Reproduced with permission from Colenutt, B.A. and Thorburn, S. [1] Gordon AC Breach

through the traps and failing to be adsorbed in some experiments a second identical trap was connected in series with the first trap. This was analysed at the end of the stripping procedures in the same way as the first trap. In no case was there breakthrough of sufficient quantities to be detected in the chromatographic analysis.

Blank analyses were performed in two ways. If a sample having once been stripped was subjected to a second stripping operation a zero blank was obtained, with the chromatogram showing no peaks. However, if distilled water from the laboratory stock was stripped in the same manner the resulting chromatogram frequently showed traces of some compounds. This was suspected to be as a result of contamination by these compounds present in the laboratory atmosphere. This was confirmed when it was found that distilled water samples left standing in the laboratory overnight showed higher concentrations of trace pollutants than a sample of freshly distilled water. Problems could arise if water samples with very low concentrations of organic solutes were left exposed in the laboratory since adsorption of further traces of organic compounds could occur. For this reason prior to analysis all samples were kept in securely sealed bottles filled to capacity.

Table 16.3 shows the satisfactory recoveries obtained by the technique for synthetic solutions of polychlorinated biphenyls and chlorinated insecticides.

Table 16.3 Recovery of pesticides and polychlorinated biphenyls by stripping

Solute	n-Pentane	Carbon disulphide	Carbon tetrachloride	Diethyl ether
Aldrin	95	96	96	88
Dieldrin	92	94	95	85
Lindane	94	94	95	80
p,p'-DDT	95	92	95	85
Endrin	90	92	95	85
Heptachlor	92	94	98	90
Methoxychlor	92	90	94	88
Aroclor 1221	94	92	92	90
Aroclor 1248	96	95	94	89
Aroclor 1254	96	98	96	92

Source: Reproduced with permission from Colenutt, B.A. and Thorburn, S. [1] Gordon AC Breach

Haberer and Schredlskar [3] designed an apparatus for predicting the ease of removal of volatile organics from water by air stripping. The method involves passage of a stream of nitrogen through 2L of the aqueous liquid for measured intervals of time, and adsorption of the volatilised organic molecules on to activated carbon. The residual organic content was determined by gas chromatography. The method can be used to determine the rate constant for the stripping effect, from which the time required to reduce the concentration by half (or any other desired amount) can be calculated.

Friant and Suffet [4] have investigated in detail the interactive effects of temperature, salt concentration, and pH on headspace analysis for isolating volatile trace organics in aqueous samples. Optimal conditions were derived from a statistical evaluation of the effect of parameter variation on the partition coefficient. These were a pH of 7.1, a sample temperature of 50°C, and a salt concentration equivalent to 3.35M sodium sulphate. Dowty et al. [5] passed the headspace purge gas through a column of Tenax GC (poly(p-2, 6–diphenylphenylene)oxide) adsorbent to trap the organics. The organics are then released from the Tenax GC and swept into a gas chromatograph for analysis in the μg L^{-1} range. Bellar and Lichtenberg [6] also used the principle of adsorption of the organic components of the purge gas on a solid adsorbent material.

Kuo et al. [7] investigated gas stripping and thermal desorption procedures for preconcentrating volatile polar water soluble organics from water samples for analysis by gas chromatography. They investigated the recovery efficiency of the process for many model compounds in aqueous solutions. Detection limits were at the μg L^{-1} level or lower for most of the compounds studied.

Yoshioka *et al.* [8] evaluated purge and trap parameters such as purge gas flow rate, purge gas volume and sample volume in the analysis of volatile organic compounds.

Dynamic headspace analysis has been used to measure volatile organic compounds [9]. The system comprised a flow cell with a constant headspace which was sampled automatically and analysed with a portable gas chromatograph.

Some further applications of the purge and trap technique with carbon adsorption are listed in Table 16.4.

Individual types of organic compounds are discussed below.

16.1.2 Aliphatic hydrocarbons

Swinnerton and Linnenbom [13] were the first to examine the applicability of gas stripping methods to the determination of hydrocarbons in water. They determined C_1–C_6 hydrocarbons by stripping them from water with a stream of helium.

After gas stripping, the hydrocarbons can either be passed direct to a gas chromatograph or, to increase sensitivity, trapped in a cold trap and then released into the gas chromatograph. Alternatively, the stripped hydrocarbons can be trapped in, for example, active carbon, then released into the gas chromatograph. This method offers the possibility of determining trace amounts of organic compounds in water even below the ng L^{-1} level (1 part in 10^{12}, w/w), particularly for the most volatile compounds [14]. Many factors, such as interference by artefacts because of impurities in the stripping gas, the large amount of water passing the trap, adsorption of less volatile compounds in drying filters, the selection of sorbents and the adsorption and desorption efficiency, are serious drawbacks of the method, particularly for quantitative analysis. However, Grob and co-workers [15–18,68] reported an impressive improvement of the method by using a closed-loop system, provided with a small-volume effective charcoal filter, but several precautions are necessary when working at such low concentrations. The complicated procedure and the sophisticated equipment required result in many more or less unknown factors and a semi-quantitative analysis. In view of the absolute amounts of pollutants involved, their overall results were excellent.

It is possible to also collect the volatile in a cold trap [19]. A more favoured technique is the collection of the gases by adsorption on some support such as one of the Chromosorbs, or Tenax GC [19,20]. The volatiles are then desorbed by heating and injected into a gas chromatograph.

Hammers and Bosman [21] quantitatively evaluated a simple dynamic technique for non-polar pollutants in aqueous samples at the µg L^{-1} level.

Table 16.4 Applications of dynamic headspace purge and trap analysis to the determination of organics in water

Compound	Type of water	Solid phase adsorbent	Method of desorption	Analytical finish	Detection limit*	Ref.
Alkanes	Non saline	–	CS₂	GLC	1µg L⁻¹	[2,10–12]
C₁–C₆ alkanes	Non saline	Active carbon or cold trapping	–	GLC	ng L⁻¹	[13–18]
Alkanes and aromatic hydrocarbons	Non saline	Chromosorb or Tenax GC	Thermal	GLC	1–5µg L⁻¹	[19–21]
Gasoline, (aromatic aliphatics)	Non saline	Glass bead Tenox TA Ambersorb X–340 Charcoal composite column	Thermal	GLC	10–500µg L⁻¹	[22]
Aromatic hydrocarbons	Non saline	–	CS₂	GLC	–	[12,23]
Alicyclic hydrocarbons	Non saline	–	CS₂	GLC	–	[12]
Dioxane, dibutanone, 4-methylpentanone and butoxylethanol	Non saline	–	CS₂/MeOH	GLC	1µg L⁻¹ (4L sample)	[24]
Acetic acid	Non saline	–	–	GLC	0.02M	[25]
Dioxane	Non saline	–	–	GLC–MS	–	[24]
Aldehydes	Non saline	Tenox TA	Thermal	GLC	–	[26]
Odour components (2-methylisoborned, 2,3,6-trichloroanisole and geosmin)	Potable	–	–	GLC	–	[27]
Haloforms	Non saline	Tenox GC	Thermal	GLC	–	[28]
Haloforms	Non saline	–	–	GLC	–	[29]
Haloforms	Non saline	Tenox GC	Thermal	GLC	–	[30–34]

Table 16.4 continued

Compound	Type of water	Solid phase adsorbent	Method of desorption	Analytical finish	Detection limit*	Ref.
Haloforms	Non saline	Tenox GC	Thermal	Atmospheric microwave emission detector	–	[35–37]
Haloforms	Non saline	–	–	–	1–500μg L⁻¹	[38]
Haloforms	Non saline	–	–	Conductivity detector	–	[39]
Haloforms	Non saline	–	–	GLC	–	[40–47]
1,1 dichloroethane	Non saline	–	–	GLC	–	[48]
Chlorobenzene, chlorotoluene	Non saline	–	CS₂	GLC	–	[49]
Chloroaliphatic compounds	Non saline	–	CS2	GLC	–	[50]
Chloroaliphatic compounds	Non saline	–	–	GLC–MS	–	[51]
Halogenated compounds	Non saline	–	–	GLC	–	[52]
Chlorinated solvents	Non saline	–	–	GLC	μg L⁻¹	[53]
Chloroalkanes	Non saline	–	CS₂	GLC	1 μg L⁻¹	[46]
Halogenated organic compounds	Non saline	–	–	Capillary GLC	–	[54]
Chloro and bromo alkanes	Non saline	–	Thermal	GLC halogen specific detector	>0.5μg L⁻¹	[15–17,30, 55–58]

*μg L⁻¹ unless otherwise stated

Table 16.4 continued

Compound	Type of water	Solid phase adsorbent	Method of desorption	Analytical finish	Detection limit*	Ref.
Chloroalkanes	Non saline	Tenox GC	Thermal	GLC	–	[30]
Chloroalkanes	Potable	–	Thermal	GLC	–	[59]
Dichloromethane and aromatic hydrocarbons	Non saline	–	–	Photoionisation and Hall Electrolytic conductivity GLC	–	[60]
Vinyl chloride, bromomethane, chloroethane, dichlorodifluoromethane and dichloromethane and 36 volatile organics	Non saline	–	–	Wide bore capillary GLC	–	[61]
22 Volatile organics	Non saline	–	–	GLC-MS	–	[62]
Chloroform, bromodichloromethane, dichlorobromomethane and bromoform	Non saline	–	–	GLC	10–50µg L^{-1}	[63]
Vinyl chloride	Non saline	–	CS$_2$	GLC	–	[64]
Vinyl chloride	Non saline	Silica gel and Carbosieve B	–	GLC halogen specific detector	4µg L^{-1}	[46]
Ethane thiol, carbon disulphide and dimethyl sulphide	Non saline	Tenax TA	Cyrotrap trapping	Capillary GLC	–	[65]
Dimethyl sulphide	Non saline	–	–	GLC	0.8ng L^{-1}	[66]

Table 16.4 continued

Compound	Type of water	Solid phase adsorbent	Method of desorption	Analytical finish	Detection limit*	Ref.
Volatile organosulphur compounds	Non saline	–	–	Microwave induced plasma atomic emission spectrometry	ppt	[67]
Misc. organics	Non saline	Active carbon	n-hexane, toluene, CS$_2$, EtOH, CCl$_4$, MeOH	GLC	–	[1,2]
Misc. organics	Non saline	Active carbon	–	GLC	–	[3]
Misc. organics	Non saline	–	–	GLC	–	[4]
Misc. organics	Non saline	Thermal	–	GLC	μg L^{-1}	[5]
Misc. organics	Non saline	–	–	GLC	–	[6]
Misc. organics	Various	–	–	GLC	–	[7]
Misc. organics	Various	–	–	GLC	–	[8,9]
Misc. organics	Non saline	Active carbon	–	GLC	–	[15,17,68]
Misc. organics	Non saline	–	–	GLC	–	[69]
Misc. organics	Non saline	Carbon	CS$_2$	GLC	–	[70]
Misc. organics	Non saline	Thermal	–	GLC	–	[71]
Misc. organics	Non saline	Tenox GC	Thermal	GLC	0.9μg L^{-1}	[72]
Misc. organics	Non saline	–	–	GLC	–	[73]
Misc. organics	Non saline	–	–	GLC	–	[74–79]
Organomercury compounds	Non saline	–	–	GLC–FTIR	50–100ppb	[80]

*μg L^{-1} unless otherwise stated

Table 16.4 continued

Compound	Type of water	Solid phase adsorbent	Method of desorption	Analytical finish	Detection limit*	Ref.
Aliphatic hydrocarbons	Sea	Cryogenic trapping	–	GLC	–	[14]
Aliphatic hydrocarbons	Sea	Cold trapping or adsorption on solid	–	GLC	–	[81]
Aliphatic hydrocarbons	Sea	Active carbon	Misc	GLC–MS	–	[58,82]
Aliphatic hydrocarbons	Sea	Thermal and Tenax GC (cold trapping)	–	GLC	–	[19,20 83–85]
Aliphatic hydrocarbons	Sea	Active carbon	–	GLC–MS	ppb	[86]
Aliphatic hydrocarbons	Sea	Tenax GC	–	GLC or GLC–MS	–	[85]
Aliphatic hydrocarbons	Sea	–	–	GLC	–	[2,10,19, 23,87–89]
Aldehydes and ketones	Sea	–	–	GLC	–	[90]
Halocarbons	Sea	–	–	GLC	–	[83]
Vinylchloride	Sea	Silica gel or Carbosieve B	–	GLC–MS	μg L^{-1}	[46]
Chlorinated pesticides and polychlorobiphenyls	Sea	Active carbon	CS$_2$	GLC	–	[1,2]
Misc. organics	Sea	Active carbon	Cl$_2$ or Cl$_2$CH$_2$C (closed loop analysis)	GLC–MS	–	[15–17, 91,92]
Chloroalkanes	Sea	Active carbon	–	GLC	μg L^{-1}	[15–17,91]
Chloroalkanes and hydrocarbons	Sea	Active carbon	–	HPLC	–	[93]

Table 16.4 continued

Compound	Type of water	Solid phase adsorbent	Method of desorption	Analytical finish	Detection limit*	Ref.
Isobutyraldehyde isovaleraldehyde and 2-methyl butyraldehyde	Potable	–	–	GLC	0.5–100µg L⁻¹	[26]
Chloroaliphatic compounds and hydrocarbons	Potable	–	Closed loop analysis	GLC	<1µg L⁻¹	[92]
Haloforms	Potable	Misc.	Thermal	GLC	<0.5µg L⁻¹	[28]
Haloforms and chloroaliphatics	Potable	–	–	GLC	1µg L⁻¹	[95,96]
Haloforms	Potable	Tenox GC	–	GLC atmospheric pressure microwave emission detector	0.1µg L⁻¹	[37]
Haloforms	Potable	–	–	GLC	–	[34]
Haloforms	Potable	SP100 on Carbopack–B	Thermal	GLC	–	[47]
Haloforms	Potable	Tenox GC	–	GLC–atmospheric pressure helium microwave induced plasma emission spectrometry	–	[37]
Haloforms	Potable	–	–	GLC (closed loop)	–	[97]

*µg L⁻¹ unless otherwise stated

Table 16.4 continued

Compound	Type of water	Solid phase adsorbent	Method of desorption	Analytical finish	Detection limit*	Ref.
Haloforms	Potable	Tenax GC or Chromosorb	Thermal	GLC	µg L^{-1}	[5,98]
Haloforms	Potable	Active carbon	Misc. solvents	GLC–MS	–	[99]
Chloroaliphatics	Groundwater	–	–	GLC photo-ionisation and Hall electrolytic detectors	<0.1–1µg L^{-1}	[100]
Misc. organics (EPA priority pollutants)	Groundwater	–	–	–	<20µg L^{-1}	[73]
Aromatic hydrocarbons and chlorobenzenes	Waste	–	–	GLC–MS	1µg L^{-1}	[86,101, 102]
Misc. organics	Waste	Active carbon	CS$_2$ or thermal	GLC	–	[94]
Aromatic hydrocarbons and chlorinated hydrocarbons	Trade effluents	–	–	GLC electrolytic conductivity and photoion-isation detectors	–	[8]
Misc. organics	Landfill waters	–	–	GLC	1–100µg L^{-1}	[38]
Gasoline	Soil	–	–	GLC	5µg L^{-1}	[103]
Misc. organics	Non saline	Carbon	CS$_2$	GLC	–	[104]

*µg L^{-1} unless otherwise stated

Source: Own files

The aqueous sample was purged with nitrogen and *n*-alkanes, benzene, and alkylbenzenes collected in a Tenax trap. The trap tube was mounted at the injection port of a gas chromatograph, where analytes were trapped at the head of the capillary column by cooling with liquid nitrogen. Quantitative data was obtained using flame ionisation detection. Distilled water, natural ditch water and a soil water suspension, each spiked with known amounts of solutes, were analysed. Recoveries of alkylbenzenes were higher than expected and might have been caused by transfer of solute molecules to the gas above the water sample when they were adsorbed to the gas–water interface of the gas bubbles. Quantitative recoveries were obtained for alkanes and alkylbenzenes from distilled water and ditch water but not for chlorinated ethenes. The applicability of headspace sampling under these conditions was best restricted to analytes with Henry's coefficients similar to that of benzene or larger. Detection limits for alkanes and alkylbenzenes and for the examined chlorinated ethenes were 1 and 5µg L $^{-1}$, respectively. The use of a number of strip vessels and traps would reduce analysis time to less than 2h.

Belkin and Hable [22] have described the measurement of low levels (10, 100, 50µg L $^{-1}$) of gasoline in water using a stripping thermal desorption procedure. A multicomponent collection tube containing glass beads, Tenax TA, Ambersorb XE–340 and charcoal was used in place of the more common Tenax collection tube. Gasoline recovery from spiked samples was 95–104% with coefficients of variation between 9.44 and 10.6%. In the case of gasoline components (3–methylpentane, *n*-hexane, benzene, isoctane, *n*-heptane, toluene, *n*-octane, 1–chlorohexane, ethylbenzene, *p*-plus *m*-xylene, *o*-xylene, *n*-nonane, *n*-undecane) each at a concentration of 100µg L $^{-1}$ in water, recoveries ranged from 77 to 128% and coefficients of variation from 4.3 to 33%. The procedure used small water samples (15mL) and was relatively rapid, as three samples could be sparged simultaneously and thermal desorption gas chromatography time was less than 30min. See also Table 16.4.

16.1.3 Aromatic hydrocarbons

See Table 16.4.

16.1.4 Alicyclic hydrocarbons

See Table 16.4.

16.1.5 Dioxane, 2–butanone, 4–methyl pentanone and butoxyethanol

See Table 16.4.

16.1.6 Acetic acid

See Table 16.4.

16.1.7 Dioxane

Epstein *et al.* [24] compared two methods for the determination of μg L $^{-1}$ levels of 1,4 dioxane in non saline waters. The first method is based on purge and trap gas chromatography–mass spectrometry and the second is based on sorption/desorption of the analyte followed by determination by gas chromatography with a flame ionisation detector.

16.1.8 Aldehydes

Daignault *et al.* [26] determined low molecular weight aldehydes in non saline waters by a purge and trap technique using Tenax TA as adsorbent. The aldehydes were then thermally desorbed from the Tenax onto cryogenically cooled gas chromatography column and quantified.

16.1.9 Odour components (2 methyl isoborneol, 2,3,6 trichloroanisole and geosmin)

See Table 16.4.

16.1.10 Haloforms

Dressman *et al.* [28] compared determinations of haloforms by methods involving extraction with methylcyclohexane, iso–octane and pentane with results obtained by a purge and trap method involving purging the sample with nitrogen and adsorption on a Tenax–GC trap, followed by thermal desorption from the trap and gas chromatography.

For comparison the average percentage recovery of the trihalomethanes obtained by the purge and trap method in the 1–200mg L $^{-1}$ range was 75–95% much the same as in the liquid–liquid extraction methods.

Dressman *et al.* [28] are of the opinion that purge and trap methods are very amenable to gas chromatographic–mass spectrometric confirmation of the identity of volatile compounds and at levels lower than can be obtained with liquid–liquid extraction methods. The lower minimum detectable concentrations are attainable because virtually all of the trihalomethanes purged from the sample are transferred to the gas chromatographic column and the detector – without the relatively non-volatile interferences coextracted by liquid–liquid extraction methods and without introducing solvent-related interference.

The liquid–liquid extraction methods rarely provide a compound sufficiently concentrated for gas chromatography–mass spectrometry and it may be necessary to resort to a solvent enrichment technique. Enriching the solvent by evaporation commonly results in interference from concentrated solvent impurities to the extent that mass–spectral analysis cannot be performed, or in loss of the more volatile components such that actual enrichment is not achieved.

Ammonia has been shown to be a major interferent in the gas chromatographic analysis of trihalomethanes and other organohalogens in water using the purge and trap technique [29]. The ammonia must be removed prior to analysis.

Haloforms have been preconcentrated by adsorption on Tenax GC followed by thermal desorption on to a gas chromatograph [30–34] or an atmospheric pressure microwave emission detector [35–37].

Chichester-Constable *et al.* [38] have developed an improved sparger for a purge and trap concentrator for the analysis of halocarbons in the 1–500µg L $^{-1}$ range.

Pierce *et al.* [39] describe the use of the purge and trap method with an electrolytic conductivity detector for the determination of volatile halogenated organics including the trihalomethanes.

Other workers who have studied the application of dynamic headspace analysis to the determination of haloforms include Symons [40], Keith [41], workers at the Health and Welfare Departments, Canada [42], and the US Environmental Protection Agency [43–46] and Kirschen [47]. See also Table 16.4.

16.1.11 Chloroalkanes

This section deals with gas purging methods which have been shown to be applicable to determining the following halocarbons in water:

Bromoform	1,2–Dichloroethane
Bromodichloromethane	1,1–Dichloroethene
Bromomethane	*trans*-1,2–Dichloroethene
Carbon tetrachloride	1,2–Dichloropropene
Chlorobenzene	*cis*-1,3–Dichloropropene
2–Chloroethylvinyl ether	*trans*-1,3–Dichloropropene
Chloroform	Methylene chloride
Chloromethane	1,1,2,2–Tetrachloroethane
Dibromochloromethane	Tetrachloroethane
1,2–Dichlorobenzene	1,1,1–Trichloroethane
1,3–Dichlorobenzene	1,1,2–Trichloroethane
1,4–Dichlorobenzene	Trichloroethene
Dichlorodifluoromethane	Trichlorofluoromethane

1,1–Dichloroethane	Vinyl chloride

Chloroethane

In one gas purging method the volatile compounds are extracted from the water sample by passing pure gas (eg nitrogen) through the water sample and collecting the volatile compounds on a small adsorption column. The compounds are introduced into a chromatograph by heating the adsorption column. The determination is carried out by temperature-programmed gas chromatography using a halogen-specific detector (electrical conductivity detector) or a more generalised detector (flame ionisation or mass spectrometry) [15–17,30,55–59]. This technique is quite sensitive (>0.5µg L $^{-1}$).

Bellar and Lichtenburg [30] used a purge and trap method in connection with temperature-programmed gas chromatography to resolve 23 volatile organohalides all of which have been identified at one time or another in various water samples. This method entails pumping of the water sample with an inert gas, collection of the purged trihalomethane on an adsorbent (eg Tenax GC), followed by the thermal desorption.

Lopez-Avila et al. [60] determined dichloromethane and aromatic compounds using photoionisation and Hall electrolytic conductivity gas chromatography in which the gas chromatograph is connected in series to a purge and trap analyser.

Mosesman et al. [61] investigated factors influencing the analysis of volatile pollutants by wide-bore capillary chromatography and a purge and trap system using five volatile gas mixtures: dichlorodifluorometh-ane, chloromethane, vinyl chloride, bromomethane and chloroethane. The factors studied were initial column temperature (10 or 35°C), carrier gas (helium), flow rate (5, 10, 15ml min $^{-1}$), speed with which pollutants were desorbed from the trap, and the type of detector used (flame ionisation detector or electrolytic conductivity detector). After optimising the analytical conditions for the gases, a mixture of 36 volatile pollutants was analysed. The remaining 30 compounds were refocused at the column inlet, producing sharp, well-resolved peaks. A carrier gas flow rate of 10ml min $^{-1}$ and an initial column temperature of 10°C were the optimal conditions for purge and trap analysis of volatile priority pollutants from a VOCOL wide-bore capillary column. A new experimental trap improved the chromatography of the volatile pollutants. The choice of detectors for the analysis was critical.

Coleman et al. [62] optimised purging efficiency and quantification of organic contaminants from water using a 1L closed loop stripping apparatus and computerised capillary column gas chromatography–mass spectrometry. These workers studied four ways of improving the recovery and precision of the 4.5L closed loop stripping apparatus: by purging a smaller volume; by better control of the temperature of the gas entering the filter and of the filter itself, by increasing the temperature of the sample water bath; and by eliminating as much metal as possible in the closed loop. The 1L system as developed and optimised for maximal purging efficiency of 1–chloroalkane standards spiked in water. The optimal conditions for efficient recovery were established and more than 95% recovery was achieved for several chlorinated paraffins. Recovery studies of a 22 compound standard mixture of organics, using two 1L closed loop stripping devices of different design, showed that both systems had similar recovery efficiencies, about twice as good as those for 4.5L devices. Relative standard deviations showed similar improvement.

Leppine and Archambault [63] have carried out determination of trihalomethanes ($CHCl_3$, $BrCl_2CH_2$, Br_2ClCH_2, $CHBr_3$) at the 10–50µg L^{-1} level in non saline waters utilising purge and trap analyses and a gas chromatograph equipped with an electron-capture detector.

Workers at the National Environment Research Centre, US Environmental Protection Agency [45] have described a purge and trap method for determining vinyl chloride at the microgram per litre level in water. An inert gas is bubbled through the sample to transfer vinyl chloride to the gas phase, and the vinyl chloride is then concentrated on silica gel or Carbosieve B under non-cryogenic conditions, and determined by gas chromatography with a halogen-specific detector. Gas chromatography–mass spectrometric methods were used to provide confirmatory identification of vinyl chloride.

Bellar et al. [44] used a computer to scan the data and construct a selected ion current profile consisting of peaks that produce an m/e 62 ion. Other compounds likely to be present in the water sample which produce m/e 62 ions are easily resolved using the gas chromatographic conditions recommended by these workers, so, in this sense, the method is specific for vinyl chloride.

Data obtained by Bellar et al. [44] showed that a quantitative recovery of vinyl chloride is obtained on silica gel and Carbosieve B with purge volumes of 150–400mL at 20mL min^{-1}.

To determine the effect of sample collection and storage on the accuracy of the method, a 1L sample of river water contained in a IL separatory funnel was dosed with vinyl chloride at 20µg L^{-1}. This mixture was then used to fill several 50ml glass-stoppered bottles. The bottles were then stored under ambient conditions. Seven of the samples, having no headspace, were randomly selected and analysed over a period of 93h. The data shows that the recoveries were constant over the period of study. The average recovery was 15.1 ± 0.4µg L^{-1}. The initial 25% loss is attributed to the headspace above the dosed sample while it was contained in the separatory funnel. Losses due to headspace or exposure to the atmosphere are further illustrated below.

The time zero sample from the above experiment, now containing 5ml of headspace, was reanalysed at 15min and again at four additional times

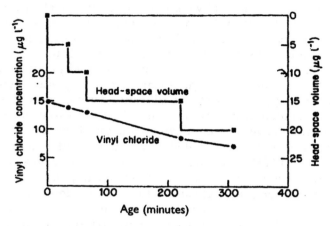

Fig. 16.2 Recovery of vinyl chloride from dosed Ohio River water stored with variable headspace at ambient temperature
Source: Reproduced with permission from Bellar, T.A. *et al.* [44] American Chemical Society, Washington

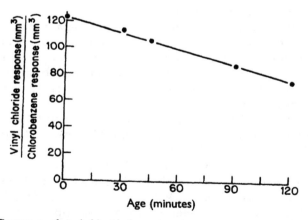

Fig. 16.3 Recovery of vinyl chloride from dosed tap water stored unstoppered at ambient temperature
Source: Reproduced with permission from Bellar, T.A. *et al.* [44] American Chemical Society, Washington

over a period of 300min (Fig. 16.2) Each time 5ml of sample was withdrawn leaving an additional 5ml of headspace. Care was taken not to agitate the sample during the storage period. The results show that as

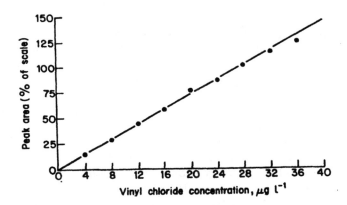

Fig. 16.4 Response curve for vinyl chloride using microcoulometric detector
Source: Reproduced with permission from Bellar, T.A. *et al.* [44] American Chemical Society, Washington

the headspace increases, the recovery of vinyl chloride decreases. The total loss over the time period was about 50% or about 10% h $^{-1}$. These observations indicate the extreme care that is essential when dealing with the analysis of volatile organics in water samples.

The loss of vinyl chloride from water in an open narrow neck container at ambient temperature was observed by dosing 50ml of tap water in a 50ml volumetric flask with 10mg L $^{-1}$ of vinyl chloride and 20mg L $^{-1}$ chlorobenzene. Chlorobenzene is relatively non-volatile and was used as an internal standard. These analyses were done by direct aqueous injection gas chromatography, not by the purge and trap technique. The recovery of vinyl chloride relative to the chlorobenzene is shown in Fig. 16.3. The loss of vinyl chloride was linear throughout the time period with a total loss of 35% or about 17% h $^{-1}$. The recovery of chlorobenzene was constant throughout the study. To test the procedure over a wide concentration range, a standard curve was prepared by injecting known amounts of a 10ng μL $^{-1}$ vinyl chloride-in-acetone solution into the purging device containing 5.0ml of organic-free water. Each mixture was then purged and analysed. The response obtained by microcoulometric titration–gas chromatography was linear over a concentration range of 4–40μg L $^{-1}$ (Fig. 16.4). Based on data collected for similar halogenated hydrocarbons, the method may be useful up to 2500μg L $^{-1}$. See also Table 16.4.

16.1.12 Organosulphur compounds

Henatsch and Juttner [65] used cryoadsorption to preconcentrate low boiling organic sulphur compounds from anoxic lake water samples. The method was based on a stripping system with a cyrotrap containing odour trapping Tenax TA cooled with liquid nitrogen. Heat desorption was used to transfer the compounds from the cyrotrap on to UCON-coated glass capillary gas chromatography columns without modifying the injection system. Sulphur compounds were easily detected by a variety of detectors including flame photometric or mass-selective detectors. Ethanethiol, carbon disulphide and dimethyl sulphide were found in anoxic freshwater lake samples. Hydrogen sulphide interferences were overcome by treatment with iron(III) chloride and excessive amounts of methane were removed by using solid carbon dioxide instead of liquid nitrogen for cyrotrap cooling.

Holdway and Nriagu [66] have described a purge and trap method for the determination of dimethyl sulphide in freshwaters. The dimethyl sulphide pumped from the sample in this way can be stored in gas sample vials for weeks once the sample has been collected. The detection limit was 0.8ng L $^{-1}$.

Gebersmann et al. [67] used purge and trap analysis followed by gas chromatography with microwave induced plasma atomic emission spectrometric detection to determine ng L $^{-1}$ levels of volatile organosulphur compounds.

16.1.13 Miscellaneous organics

Non saline waters Otson and Chau [69] evaluated a purge and trap technique for the preconcentration and determination of µg L $^{-1}$ levels of 51 target organics in water samples. The 51 compounds from the US EPA purgeable priority pollutants list were each spiked at 1, 10, and 50µg L $^{-1}$ into purified water. Detection limits were below 1µg L $^{-1}$, recoveries were greater than 70% and precision for recoveries over all three concentrations were less than 20% RDS. Improved detection limits were obtained for several water soluble and halogenated compounds when concentrator trap composition was altered, transfer line temperature decreased, and sparger vessel temperature increased. The effects of sample transport and storage on the levels of the organics in spiked water samples were investigated.

Blanchard and Hardy [70] collected volatile compounds which had been nitrogen purged from water by permeation through a silicone polycarbonate membrane and adsorption on to charcoal contained within the sampling device. Analysis was performed by desorption with carbon disulphide followed by separation and quantification by capillary gas chromatography. The method provided time-weighted average concentrations not possible with the purge and trap method. A linear relationship existed between the amount of volatile organic component collected and the product of sampler exposure time and component concentration in the environment. Results showed that the method compared favourably with the purge and trap method for precision and accuracy. Advantages of the method included the inexpensive nature of the sampling device, its simplicity, that fact that it could be reused over extended periods of time and that no power supply was required for sampling.

Melcher and Caldecourt [71] described a gas chromatographic injection and concentration technique, based on the delayed injection principle for the direct detection of trace organic compounds in water. The organic samples are retained for extended periods on a collection precolumn while the air or water is vented and then injected into an analytical column with application of a high electric current (the heat source) across the precolumn. The optimal desorption temperature is 150°C and the optimal heating rate 20°C s^{-1}. Application of the technique for the determination of glycol ether mixtures, phenol, and soluble aromatic compounds in water is described. The technique is considered particularly useful for determination of compounds which are difficult to extract or purge from water.

An alternate method is to sweep the sample volatiles into a trap containing Tenax GC adsorbent immersed in liquid nitrogen. Cailleux *et al.* [72] preconcentrated trihalomethanes, benzene and toluene from non saline waters by bubbling inert gas through the liquid or by passing the gas over the surface, followed by trapping on a polymer absorbent cooled by liquid nitrogen. For injection, the trap was simultaneously heated and purged with carrier gas. Detection limits of approximately 0.5µg L^{-1} trihalomethane for 20ml sample volumes were achieved with a relative standard deviation of 20%.

Lopez-Avila *et al.* [73] compared two wide-bore capillary columns VOCOL (diphenyl dimethyl cross-linked polysiloxane) and 007 (95% methyl and 5% phenylsilicone) in purge and trap gas chromatography–mass spectrometry techniques for determining volatile organic compounds in non saline waters. Of the two the VOCOL column resolved more of the test volatile organic compounds in the same length of time.

Various other workers have discussed the applications of dynamic purge and trap analysis to the determination of miscellaneous organic compounds in water [74–79]. See also Table 16.4.

16.1.14 Organomercury compounds

Purge and trap gas chromatography–FTIR has been used to determine in non saline waters 50–100pg methylmercury and dimethylmercury following conversion of analyte to hydrides by reaction with sodium tetraborohydride [80].

16.2 Seawater

16.2.1 Aliphatic hydrocarbons

Perras [84] has pointed out that when the concentration of hydrocarbons in seawater is great enough, as, perhaps, after a petroleum spill, the emergent gas stream can be sampled directly. This is seldom the case in true oceanic samples, however, and some form of concentration is needed.

It is possible to collect the volatiles in a cold trap [19]. A more favoured technique is the collection of the gases by adsorption on some support such as one of the Chromosorbs or Tenax GC [19,20,85].

The volatiles are then desorbed by heating and injected into a gas chromatograph.

While much of the literature on this subject is concerned with non saline water samples, it is believed that many of these procedures will also work satisfactorily with seawater; indeed, the presence of salts in the sample may assist in the removal of volatiles.

In the collection of organic materials, separation of the more volatile materials can be achieved by transferring them into the vapour phase for collection and concentration. The usual method for effecting such a transfer is to bubble some inert gas through the liquid. The effluent gas is then passed through an adsorbent such as carbon to collect the organic materials [58,82]. The solvent desorbed material is then analysed by gas chromatography or by linked gas chromatography–mass spectroscopy.

The Oil Companies International Study Group for Conservation of Clean Air and Water, Europe (Concawe) has made a detailed study [86] of the application of gas stripping to the determination of hydrocarbons in amounts down to parts per billion in water. In this procedure the water sample is purged with nitrogen and helium and the volatiles trapped on a solid adsorbent such as activated carbon. The organics are then released from the carbon by heating and purging directly into a gas chromatograph linked to a mass spectrometer. For chlorine and bromine containing impurities a halide selective detector such as the Hall electrolytic conductivity detector is used on the gas chromatograph. Alternatively, an alkali flame ionisation detector or an electron capture detector could be used.

Material removed from seawater by stripping can be concentrated by trapping in a loop immersed in a cooling bath. The usual cooling baths are liquid nitrogen or solid carbon dioxide with or without an organic solvent. The major disadvantage of the liquid nitrogen bath, along with the cost and the limited availability, particularly aboard ship, is that it condenses carbon dioxide and water, along with the organic materials actually desired. These trapping techniques have been used by many workers and are usually described in conjunction with a gas chromatographic determination of some fraction of the organic materials. An example of this kind of separation is the work of Novák *et al.* [14]. They improved the yield of volatile organics by salting out the volatiles with sodium sulphate.

Vacuum removal of volatiles has also been employed in the preconcentration of organics from seawater. This system, involving the removal of volatiles under vacuum can be set up in two ways; either as a flow-through or as a batch process. As a flow-through process, the sample is drawn continuously through the system, and the gases taken off by the vacuum pass through a sampling loop. Periodically, the material in the loop is injected into the gas chromatograph. In this manner it is possible to derive almost continuous profiles of volatile hydrocarbon concentrations [81]. In the batch mode, a larger sample can be treated over a longer period, and the volatiles collected by cold-trapping or adsorption. These techniques are not as fast as flow-through sampling, nor do they permit semi-continuous profiling, but they result in a greater concentration of the hydrocarbons, and thus in greater sensitivity [79].

May *et al.* [85] have described a gas chromatographic method for analysing hydrocarbons in marine sediments and seawater which is sensitive at the submicrogram per kilogram level. Dynamic headspace sampling for volatile hydrocarbon components, followed by coupled-column chromatography for analysing the non-volatile components, requires minimal sample handling, thus reducing the risk of sample component loss and/or sample contamination. The volatile components are concentrated on a Tenax gas chromatographic precolumn and determined by gas chromatography or gas chromatography–mass spectrometry.

Other workers have discussed the application of dynamic headspace analysis to the determination of aliphatic hydrocarbons in seawater [2,10, 19,23,87–89]. See also Table 16.4.

16.2.2 Aldehydes and ketones

Cerwen [90] used dynamic purge and trap analysis to determine ketones and aldehydes (acetone, butyaldehyde and 2–butanone) in seawater.

16.2.3 Aliphatic chloro compounds

Kaiser and Oliver [83] applied dynamic purge and trap analysis to determine halocarbons in seawater.

The Bellar *et al.* [44] purge and trap method has been applied to the determination of vinyl chloride in seawater. Fig. 16.5 represents the chromatogram obtained from a seawater sample dosed with vinyl chloride and other organohalides. Using the Hall electrolytic conductivity detector, no response was obtained for the acetone used to prepare the vinyl chloride standard solution. See also Table 16.4.

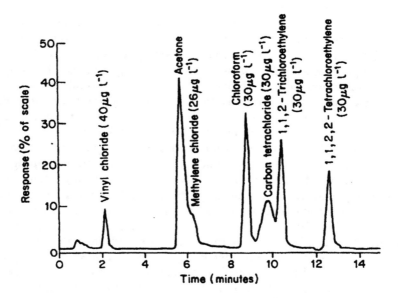

Fig. 16.5 Electrolytic conductivity gas chromatogram of organohalides recovered from seawater (full-scale response, 160μmhos)
Source: Reproduced with permission from Bellar, T.A. et al. [44] American Chemical Society, Washington

16.2.4 Chlorinated insecticides and polychlorobiphenyls

Colenutt and Thorburn [1,2] have described the procedure using gas stripping of the aqueous sample followed by adsorption to active carbon from which surface they are taken up in an organic solvent for gas chromatographic analysis. They optimised conditions for the determination of parts per billion of insecticides and polychlorinated biphenyls.

16.2.5 Closed loop analysis

The closed loop gas stripping system has been discussed by various workers [15–17,91,92]. In this technique organic compounds are removed from water by purging with a gas saturated with water vapour. Volatile and semi-volatile compounds will partition out into the headspace and are swept to an activated charcoal trap. The charcoal will retain organics while allowing the purge gas to pass through. The purge gas is then returned to re-purge the sample via a pump. At the end of the purge time, typically 2h, the trap is removed and fitted with a glass collection vial.

Organic compounds are extracted from the charcoal with a small volume of a suitable solvent such as carbon disulphide or dichloromethane which is then collected and injected into a capillary gas chromatograph or a capillary gas chromatograph coupled with a mass spectrometer.

Grob and Zurcher [15–17,91] have carried out very detailed and systematic studies of the closed loop gas stripping procedure and applied it to the determination of µg L $^{-1}$ of 1–chloroalkanes in water. Westerdorf [92] applied the technique to chlorinated organics and aromatic and aliphatic hydrocarbons. Waggott and Reid [93] reported that a factor of major concern in adapting the technique to more polluted samples is the capacity of the carbon filter which usually contains only 1.5–2mg carbon. They showed that the absolute capacity of such a filter for a homologous series of 1–chloro-n-alkanes was 6µg for complete recovery. Maximum recovery was dependent on carbon number being at a maximum between C_8 and C_{12} for the 1–chloro-n-alkane series. It is important, therefore, to balance the amount of sample stripped with the capacity of the carbon filter to obtain better than 90% recoveries.

16.3 Potable waters

16.3.1 Aldehydes

Daignault *et al.* [26] have described a purge and trap method for the determination of isobutyraldehyde, isovaleraldehyde and 2–methyl butyraldehyde in potable water in the concentration range 0.5–100µg L $^{-1}$.

16.3.2 Chloroaliphatic hydrocarbons

Westerdorf [92] has described a commercially available gas stripping apparatus, the Tekman CLS–1 instrument, for the semiautomatic determination of traces of organics in water. This instrument can accommodate any sample size up to 4L. More than 90 compounds are present in each sample.

Many of the compounds present in potable water are chlorinated organics, very few of which were present in the raw water. Raw water, however, includes a much higher degree of substituted aromatic compounds.

Analysis of fuels in water by this technique can provide an excellent fingerprint of the sample.

The Bellar *et al.* [44] purge and trap method has been applied to the determination of vinyl chloride in potable water. Fig. 16.6 represents a typical gas chromatogram obtained from chlorinated tap water which has been dosed with vinyl chloride. The chloroform, bromodichloromethane and dibromochloromethane are common to chlorinated drinking waters

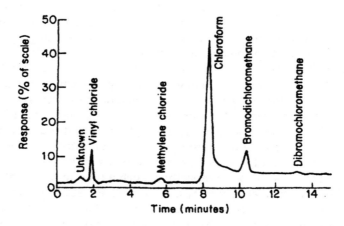

Fig. 16.6 Microcoulometric gas chromatogram of organohalides recovered from tap water dosed with vinyl chloride (sensitivity 250 ohms)
Source: Reproduced with permission from Bellar, T.A. et al. [44] American Chemical Society, Washington

and result from the chlorination process. Low levels of methylene chloride are often observed in samples analysed by this technique. These are attributed to method background. See Table 16.4.

16.3.3 Haloforms

In one gas purging method [28] the volatile compounds are extracted from the water sample by passing pure gas (eg nitrogen) through the water sample and collecting the volatile compounds on a small adsorption column. The compounds are introduced into a chromatograph by heating the adsorption column. The determination is carried out by temperature-programmed gas chromatography using a halogen-specific detector (electrical conductivity detector) or a more generalised detector (flame ionisation or mass spectrometry). This technique is quite sensitive ($>0.5\mu g\ L^{-1}$).

Dressman et al. [28] are of the opinion that purge and trap methods are very amenable to gas chromatographic–mass spectrometric confirmation of the identity of volatile compounds and at levels lower than can be obtained with liquid–liquid extraction methods. The lower minimum detectable concentrations are attainable because virtually all of the trihalomethanes purged from the sample are transferred to the gas chromatographic column and the detector – without the relatively non-volatile interference – coextracted by liquid–liquid extraction methods and without introducing solvent-related interference.

Nicolson *et al.* [95,96] have described a convenient easily automated method for the analysis of haloforms and some other volatile organohalides in potable water. This direct aqueous injection method has a detection limit at or below 1µg L $^{-1}$ for haloforms. Simultaneous analysis of finished water samples with direct aqueous injection and the gas sparging method revealed hitherto unknown aspects of water treatment chemistry. While the gas sparging technique measures only the free haloforms present in the potable water, they showed that the direct aqueous injection method quantitates the total potential haloforms that can form after chlorination.

Comparison of determinations of haloforms, particularly chloroform in potable water revealed that direct aqueous injection results were consistently higher than those obtained by gas sparging. A similar trend was shown for chlorodibromomethane. In general, the direct aqueous injection method indicated chloroform and bromodichloromethane concentrations 1.5 and 2.2 times higher respectively, than the values obtained by gas sparging.

These results suggested to Nicolson *et al.* [96] that actual chlorinated non saline water samples, as opposed to synthetic standards, contain non-volatile haloform precursors which are injected in determinations by direct injection but not by gas sparging. If the additional quantities of haloforms observed by the direct aqueous injection method are produced during analysis from non-volatile halogenated organic compounds, then in a given sample the difference should be measurable by analysing pre-purged samples by direct aqueous injection.

Earlier studies by Nicolson and Meresz [95] had indicated that a gas chromatograph equipped with a Poropak Q column and an electron-capture detector could be used for the determination of haloforms in dilute aqueous solution by direct aqueous injection. It was shown that the determination of the haloforms (chloroform, bromodichloromethane and chlorodibromomethane) as well as carbon tetrachloride, trichloroethylene and tetrachloroethylene could be achieved near the 1µg L $^{-1}$ level using 9µL injection volumes. Since no preconcentration was required, this technique was easily automated, allowing the analysis of up to 60 samples per day.

However, before adopting this method for routine monitoring, it was essential to compare it with the published gas sparging procedure. Analysis of standard aqueous haloform solutions showed excellent agreement between the two methods. However, when the two methods were compared on samples of chlorinated water obtained from treatment plants a different picture emerged. The direct aqueous injection method gave haloform values consistently higher than those obtained by gas sparging. The differences in data were consistent with the location of sampling sites. A study of this problem revealed that the two methods are measuring different parameters. Namely, the gas sparging technique quantitates only the free haloforms, while the direct aqueous injection method measures the total potential haloform concentration. The direct aqueous injection technique will produce a value which is the maximum haloform concentration which can be reached while the water is in the distribution system. When the free haloform content of the sample has been determined using both methods, the difference in the free and total potential haloform results gives an indication of the amounts of higher molecular weight haloform intermediates which cannot be detected by the gas sparging method.

Quimby and Delaney [35] and Quimby et al. [36] determined trihalomethanes in potable water by gas chromatography with an atmospheric pressure microwave-emission detector. The organics are isolated by a purge and trap technique. This detector, in addition to being very sensitive, is also element selective, distinguishing between chlorine, bromine and iodine.

The water sample is contained in a glass tube 15cm long. Helium or nitrogen purge gas, which has passed through a trap containing silver oxide, is heated at 200°C to oxidise any organic contaminants bubbled into the water sample.

The organics purged from the water sample are desorbed from the Tenax GC trap into the gas chromatograph by attaching the end of the trap to a helium line and inserting the entire length of the trap into the injection port of the gas chromatograph. Carrier gas is then purged through the heated Tenax GC trap on to the gas chromatographic column. For 5ml water samples, detection limits of ca. $0.1\mu g\ L^{-1}$ are obtained for the four bromine- and chlorine-containing trihalomethanes, and the linear range extends to above the $100\mu g\ L^{-1}$ level. Similar results are also obtained for the other organohalides. Chlorine-, bromine- and iodine-selective chromatograms for a standard aqueous solution containing several organohalides at the $25–30\mu g\ L^{-1}$ level are given in Fig. 16.7.

Pfaender et al. [32] have compared purge and trap and direct injection gas chromatographic techniques for the determination of chloroform in potable water and found evidence for the presence of non-volatile precursors. The direct aqueous injection technique, employing a bypass valve to vent water and an electron-capture detector, gave consistently higher values for chloroform than the purge method. Comparable results were obtained if the direct injection value after a 30min purge was subtracted from the before-purged value. The nature of the residual measured by direct injection after purging was investigated and shown to be due to non-purgeable intermediates that decompose within the injection port of the gas chromatograph to chloroform. The residual varied depending on the source of the water sample examined and the specific configuration of the chromatograph employed. The results

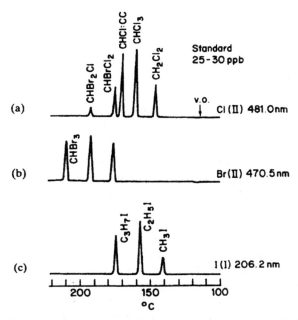

Fig. 16.7 Element-selective chromatograms from sparged standard solution. (a) Cl channel, (b) Br channel, (c) I channel
Source: Reproduced with permission from Quimby, B.D. and Delaney, M.F. [35] American Chemical Society, Washington

indicated the need for caution in the interpretation of chloroform and other trihalomethane values, especially haloform potentials, generated by direct aqueous injection.

Pfaender *et al.* [32] were unable to identify the chloroform precursor in potable water samples. He did conclude that the size of the residual measured by direct aqueous injection depends on the materials and configuration of the injection port of the chromatograph used. This raises doubts as to the usefulness of the residual haloforms measured by direct aqueous injection as an estimate of haloform formation potential. The relationship of the potential as determined by direct aqueous injection and the haloforms formed during distribution of a finished drinking water is still largely unknown.

Kirschen [47] has investigated the Environmental Protection Agency standard purge and trap method [46] for determining trihalomethanes and other halogenated volatiles in water. Kirschen [47] used a Varian Model 3700 gas chromatograph with Model 700A Hall electrolytic conductivity detector VISTA 401 chromatography data system. The 2m × 2mm glass column contained 1% Sp–1000 on Carbopack–B (60–80 mesh).

The instrument parameters were:

- Reactor temperature 750°C
- Hydrogen flow rate 30ml min $^{-1}$
- n-Propanol flow rate 0.6ml min $^{-1}$
- Detector base temperature 230°C
- Range 100
- Gas chromatography conditions: carrier gas helium, 30ml min $^{-1}$. Temperatures: column oven 45°C for 3min, 8°C min $^{-1}$ to 220°C, hold for 2min; injector 150°C.

This method was used to check the concentrations of four halomethanes, chloroform, bromoform, dichlorobromomethane and dibromochloromethane in potable water samples. Relative percentage standard deviations were normally below 5%.

Chiba and Haraguchi [37] determined trihalomethanes in microgram per litre amounts in potable water by gas chromatography–atmospheric pressure helium–microwave-induced plasma-emission spectrometry with a heated discharge tube for pyrolysis. The trihalomethanes are collected on a Tenax GC column by a gas purge and trap method.

Purge closed-loop gas chromatography has been used [97] to determine trihalomethane in potable water. This method combines the technique of gas stripping and static headspace sampling. Recovery efficiencies of between 81 and 100% are obtained. Better quality chromatograms with faster handling were achieved using the purge closed-loop method than by the purge and trap technique. See also Table 16.4.

16.3.4 Miscellaneous organics

Dowty *et al.* [5,98] used a modified headspace technique involving gas purging with subsequent adsorption of organics on poly(p-2,6–diphenylphenylene) oxide adsorbent (Tenax GC) for the monitoring of low µg L $^{-1}$ amounts of low molecular weight organics in potable water in very low concentrations. In this method volatile organics were thermally desorbed from the water sample and displaced into the headspace by purging the sample with ultrapure helium. These displaced organics were subsequently adsorbed on to Tenax GC. A glass liner which fits into a modified injection port of the gas chromatograph was packed with this adsorbent. The liner was then inserted into the injection port where the organics were thermally desorbed. These organics were then swept on to the precolumn of a gas chromatograph maintained at dry ice/methanol temperatures, where they condensed in order to ensure a 'plug' injection on to the chromatographic column. After condensation in this manner for a fixed period of time, the temperature of the precolumn was quickly raised and the organics were 'flashed' on to the chromatographic column

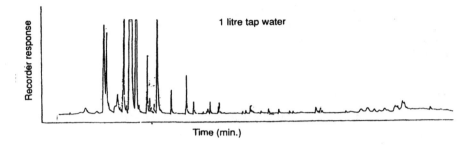

Fig. 16.8 Gas chromatogram of volatile organics present in 1L of potable water. Resolution achieved on a 91.0m × 0.05cm stainless steel capillary column coated with 10% SF–96 and 1% Igepal CO–880. Chromatographic conditions described in the text
Source: Reproduced with permission from Dowty, B. *et al.* [5] Preston Publications Inc., US

(91m × 0.05cm stainless steel capillary coated with 10% SF–96 and 1% Igepal – CO–880) where resolution of the components was achieved. The procedure was automated. Fig. 16.8 illustrates a chromatogram of organics obtained from 1L of New Orleans potable water using this procedure.

Piet [99] described a gas stripping technique for concentrating organic substances in potable water samples and followed this by adsorption on to carbon, desorption from carbon by a solvent and examination of the extract by high-resolution gas chromatography on stainless steel capillary columns coupled with an on-line mass spectrometer computer system. The procedure was applied to the identification of micro-organic constituents in potable water before and after ionisation. See also Table 16.4.

16.4 Ground waters

16.4.1 Chloroaliphatics

Lopez-Avila *et al.* [100] studied the applicability of a purge and trap technique and the use of a photoionisation detector and a Hall electrolytic conductivity detector connected in series for determining US EPA SW–846 Methods 8010 and 8020 compounds. Method precision was determined using ground water samples spiked with selected test compounds. The highest relative standard deviation value was 63% for chloromethane. Other values ranged from 1.5 to 30% with a mean value of 12%. Average recoveries ranged from 82 to 104% for six determinations. Method detection limits were determined using the SP–1000 column and the two detectors connected in series. Method detection limits for water samples ranged from 0.1–1µg L^{-1}. Lower detection limits might be achieved by using larger samples. Investigations with wide-bore open tubular capillary columns showed good results for the VOCOL or DB–624 columns for the primary analysis, although neither column could completely resolve all compounds.

16.4.2 Miscellaneous organics

Lopez-Avila et al. [73] examined the feasibility of using the purge and trap technique, with a wide-bore column directly connected to a mass spectrometer without an open slit interface for the analysis of 34 volatile compounds listed in the US EPA priority pollutant list. Two wide-bore capillary columns were evaluated, VOCOL and 007, neither of which was able to resolve all 34 compounds completely. However, the analysis was faster than for a packed column and resolutions of 20 compounds at 50µg L^{-1} per component were achieved with the VOCOL column (diphenyl-dimethyl polysiloxane with cross-linking moieties) and of 16 compounds with the 007 column (95% methyl, 5% phenyl silicone). The VOCOL column was further tested for performance with nine ground water samples containing unknown concentrations of organics and spiked with three internal standards. The instrument response was linear from 20 to 200µg L^{-1} for most compounds.

16.5 Waste waters

16.5.1 Aliphatic hydrocarbons

Concawe [86] have applied the gas stripping analysis technique developed by the Environmental Protection Agency [101,102] for the preconcentration of organics in oil refinery wastes prior to their identification and determination of organic by gas chromatography–mass spectrometry. The method has been used to determine down to 1µ.g L^{-1} of benzene, toluene, ethylbenzene, and mono- and di-chlorobenzenes in aqueous refinery wastes and up to 29 purgeable chlorine and bromine containing organics in aqueous industrial wastes. For the determination of purgeable trace aromatic and aliphatic hydrocarbons a flame ionisation detector should be used. For the determination of purgeable halogenated organics a halide selective detector should be used (the Hall 700 A electrolytic conductivity detector is recommended, but the alkali flame ionisation or electron capture devices could be used).

Graydon et al. [94] sought to combine the advantages of closed loop stripping analysis using activated carbon filters with those of thermal desorption and have evaluated the operating parameters for quantitative applications. The advantages of closed loop stripping analysis include a large enrichment factor, minimal contamination, and closed circuit recycling of substances that break through the carbon filter. One of the limitations is an inability to determine compounds that elute under the gas chromatographic peak of the solvent (usually carbon disulphide). To remedy this, thermal decomposition closed loop stripping analysis was tested. Sixteen volatile compounds with boiling points in the range 30 to 120°C were determined with recoveries of 12–52% in good agreement with the results of other methods. The filters were desorbed by direct insertion into the injection port of the gas chromatograph. The method is applicable to highly contaminated waste waters but care must be taken to ensure that the adsorbent is not overloaded.

16.6 Trade effluents

16.6.1 Aromatic hydrocarbons and chlorinated hydrocarbons

Duffy et al. [79] used a purge and trap technique to preconcentrate chlorinated hydrocarbons, benzene and toluene in industrial effluents. The purge and trap device was connected to a capillary column gas chromatograph equipped with electrolytic conductivity and photo-ionisation detectors.

16.7 Landfill waters

16.7.1 Miscellaneous organics

Chichester-Constable et al. [38] have described an improved sparger unit for the preconcentration of organics from landfill water samples containing a high proportion of solids. Volatile organic solvents in the 1–100μg L^{-1} range could be determined.

16.8 Soil

16.8.1 Gasoline hydrocarbons

Parr et al. [103] used a purge and trap gas chromatographic method to determine gasoline range organics in soil in amounts down to 0.5μg g^{-1}. Chlorinated solvents, ketones and ethers interfered.

16.9 Closed loop dynamic headspace analysis

16.9.1 Miscellaneous organics

Applications of closed loop dynamic headspace analysis include the determination of aliphatic hydrocarbons in waste waters [94] (section

16.5.1), haloforms in potable water [97] (section 15.3.3) and miscellaneous organics in seawater [15–17,91,92] (section 16.2.5).

Grob and Zurcher [16] have given a detailed description of equipment and procedures for a closed loop method of stripping trace organic constituents from water in order to obtain preconcentration. These workers point out that an important limitation is the volatility range, which includes low and medium molecular weight organic compounds up to about eicosane in the *n*-alkane series. Another limitation is that the most volatile substances, ie the substances that are eluted with the solvent used for extraction from the active carbon, cannot be determined due to volatility losses. It is feasible to analyse these substances by direct transfer of adsorbed material from the charcoal on to the gas chromatographic column, ie elimination of the solvent extraction state. One of the parameters that govern the extraction step is the amount of moisture on the charcoal. For several reasons, non-polar or weakly polar solvents with low miscibility in water have to be used and moist carbon particles are poorly wetted by such solvents. For this reason, a drying step is used prior to addition of the solvent. No wettability problems occur, even with completely non-polar solvents, when the filter temperature is kept at least 10°C above the water temperature (30°C). This temperature difference reduces the relative humidity of the gas passing the filter to about 40%. Extraction can then be started without prior drying and using any solvent.

The higher the extraction efficiency of the solvent, the smaller is the volume of solvent needed for satisfactory extraction. High efficiencies are especially important with relatively pure water samples because all available concentration procedures cause severe losses of extracted substances, not just of the most volatile substances. The most efficient solvent, carbon disulphide, unfortunately, presents problems with purification [104]. If a larger volume can be tolerated, methylene chloride, which is easier to purify, is recommended. The extraction of the most volatile substances is surprisingly difficult. The procedure can be applied quantitatively using a system of internal standards. Standard substances that do not occur in the water sample are added so as to give concentrations comparable with those of the substances to be determined.

References

1 Colenutt, B.A. and Thorburn, S. *International Journal of Environmental Analytical Chemistry*, **7**, 231 (1980).
2 Colenutt, B.A. and Thorburn, S. *International Journal of Environmental Studies*, **15**, 25 (1980).
3 Haberer, K. and Schredlskar, F. *Zeitschrift für Wasser und Abwasser Forschung*, **17**, 206 (1984).

4 Friant, S.L. and Suffet, I.H. *Analytical Chemistry*, **51**, 2161 (1979).
5 Dowty, B., Green, L. and Laseter, J.L. *Journal of Chromatographic Science*, **14**, 187 (1976).
6 Bellar, T.A. and Lichtenberg, J.J. *Journal of American Water Works Association*, **566**, 739 (1974).
7 Kuo, P.P.K., Chian, E.S.K., DeWalle, F.B. and Kim, J.H. *Analytical Chemistry*, **49**, 1023 (1977).
8 Yoshioka, M., Tsuji, M., Yamasaki, T. and Okuno, T. *Hyogo-ken. Kogal Kenkyusho Kenkyu Hakaku*, **18**, 60 (1988).
9 Lesarge, S. and Brown, S. *Analytical Chemistry*, **66**, 572 (1994).
10 Kaiser, R. *Journal of Chromatographic Science*, **9**, 227 (1971).
11 Bellar, T.A. and Lichtenberg, J.J. In *Environmental Monitoring* Series EPA–670/4–74–009. The determination of volatile organic compounds at the µg L^{-1} level in water by gas chromatography. US Environmental Protection Agency (1974).
12 Hrivnak, J. and Hassler, J. *Vodni Hospodarstvi*, **26**, 193 (1976).
13 Swinnerton, J.W. and Linnenbom, V.J. *Journal of Gas Chromatography*, **5**, 570 (1967).
14 Novák, J., Zlutick, J., Kubelka, V. and Mosteck, J. *Journal of Chromatography*, **76**, 45 (1973).
15 Grob, K. *Journal of Chromatography*, **84**, 255 (1973).
16 Grob, K. and Zurcher, F. *Journal of Chromatography*, **117**, 285 (1976).
17 Grob, K. and Grob, G. *Journal of Chromatography*, **90**, 303 (1974).
18 Grob, K., Grob, G. and Grob, K. *Journal of Chromatography*, **106**, 303 (1975).
19 Swinnerton, J.W. and Linnenbom, V.J. *Science*, **156**, 1119 (1967).
20 Bellar, T.A. and Lichtenberg, J.J. In The Determination of Voltaile Organic Compounds at the µg L^{-1} level in water by gas chromatography. US National Technical Information Service Report No. 237973/3GA, 30 pp. (1975).
21 Hammers, W.D. and Bosman, H.F.P.M. *Journal of Chromatography*, **360**, 425 (1986).
22 Belkin, F. and Hable, M.A. *Bulletin of Environmental Contamination and Toxicology*, **40**, 244 (1988).
23 Drodz, J., Vadokova, Z. and Novak, J. *Journal of Chromatography*, **354**, 47 (1986).
24 Epstein, P.S., Hauer, T., Wagner, M., Chase, S. and Giles, B. *Analytical Chemistry*, **59**, 1987 (1987).
25 Mehran, M.F., Golkar, N., Mehran, M. and Cooper, W.J. *Journal of High Resolution Chromatography and Chromatography Communications*, **11**, 610 (1988).
26 Daignault, S.A., Gac, A. and Hrydey, S.E. *Environmental Science and Technology*, **9**, 583 (1988).
27 Krasner, S.W. *Water Quality Bulletin*, **13**, 78 (1988).
28 Dressman, R.C., Steven, A.A., Fair, J. and Smith, B.J. *Journal of the American Water Works Association*, **71**, 392 (1979).
29 Department of the Environment/National Water Council Standing Committee of Analysts. *Methods for the Examination of Waters and Associated Materials. Chloro and Bromo- Tri Halogenated Methanes in Water*. HMSO, London (1981).
30 Bellar, T.A. and Lichtenburg, J.J. *Journal of American Water Works Association*, **66**, 739 (1974).
31 Federation Register (US) 69464,44, No. 223, 3 December (1979).
32 Pfaender, F.E., Jones, R.B., Stevens, A.M., Moore, L. and Hass, J.R. *Environmental Science and Technology*, **12**, 438 (1978).

33 Brass, H.J. *American Laboratory*, **12**, 23 (1980).
34 Kirshen, N. *American Laboratory*, **16**, 60 (1984).
35 Quimby, B.D. and Delaney, M.F. *Analytical Chemistry*, **51**, 875, (1979).
36 Quimby, B.D., Uden, P.C. and Barnes, P.C. *Analytical Chemistry*, **50**, 2112 (1978).
37 Chiba, K. and Haraguchi, H. *Analytical Chemistry*, **55**, 1504 (1983).
38 Chichester-Constable, D.J., Barbeau, M.E., Liu, S.L., Smith, S.R. and Stuart, J.D. *Analytical Letters (London)*, **20**, 403 (1987).
39 Pierce, C.T., Grochewzski, R.J., Kongavi, R., Narangajvanaca, K. and Brock, G.L. *American Laboratory*, April, 34 (1981).
40 Symons, J.M., Bell, T.A., Carswell, J.K. *et al. Journal of the American Water Works Association*, **67**, 643 (1975).
41 Keith, L.H. *Identification and Analysis of Organic Pollutants in Water*, Ann Arbor Science Publishers Inc., Ann Arbor, MI (1976).
42 National Survey for Halomethanes in Drinking Water, Health and Welfare Canada, 77–EHD–9 (1977).
43 US Environmental Protection Agency. Part III. Appendix C. Analysis of Trihalomethanes in Drinking Water. Federal Register, 44, No. 231 to 68672. Nov 29 (1979).
44 Bellar, T.A., Lichtenberg, J.J. and Eichelberger, J.W. *Environmental Science and Technology*, **10**, 926 (1976).
45 Method 501.1. The Analysis of Triahlomethanes in Finished Waters by the Purge and Trap Method. US Environmental Protection Agency, EMSL, Cincinnati, Ohio, 45268. November 6 (1979).
46 Method 60. US Environmental Protection Agency. Federal Register 69464, 44, No. 223. Dec 3 (1979).
47 Kirschen, N.A. *Various Instrument Applications*, **15**, 2 (1981).
48 Cooper, W.J. *Journal of High Resolution Chromatography and Chromatography Communications*, **11**, 610 (1988).
49 Thiela, D.R., Foreman, P.S., Davis, A. and Wyeth, P. *Environmental Science and Technology*, **21**, 145 (1987).
50 Leggett, D.C. *Analytical Chemistry*, **49**, 880 (1977).
51 Barber, L.B., Thurman, E.M., Takahaski, Y. and Noriega, M.C. *Groundwater*, **30**, 836 (1992).
52 Kirchen, N. *American Laboratory*, **16**, 60 (1984).
53 Buszka, P.M., Rose, D.L., Ozuna, G.B. and Groschen, G.E. *Analytical Chemistry*, **67**, 3659 (1995).
54 Kirschen, K.F. and Pauliszyn, J. *American Laboratory*, **16**, 60 (1984).
55 Zlatkis, A., Lichtenstein, A. and Tuchbee, A. *Chromagraphia*, **8**, 67 (1973).
56 Mieure, J.P. and Dietrich, M.M. *Journal of Chromatographic Science*, **11**, 559 (1973).
57 Dowty, B. and Laseter, J.L. *Analytical Letters*, **8**, 25 (1975).
58 Bertsch, W., Andersson, E. and Holzer, G. *Journal of Chromatography*, **112**, 701 (1975).
59 Stevens, A.A. and Symons, J.M. *Environmental Impact of Water Chlorination*, Oak Ridge National Laboratory, Oak Ridge, Tennessee (1975).
60 Lopez-Avila, V., Heath, N. and Hu, A.J. *Journal of Chromatographic Science*, **25**, 351 (1987).
61 Mosesman, N.H., Sidisky, L.M. and Corman, S.D. *Journal of Chromatographic Science*, **25**, 351 (1987).
62 Coleman, W.E., Munch, J.W., Slater, R.W., Melton, R.G. and Kopfler, F.C. *Environmental Science and Technology*, **17**, 571 (1983).

63 Leppine, L. and Archambault, J.F. *Analytical Chemistry*, **64**, 810 (1992).
64 Narang, R.S. and Bush, B. *Analytical Chemistry*, **52**, 2076 (1980).
65 Henatsch, J.J. and Juttner, F. *Journal of Chromatography*, **445**, 97 (1988).
66 Holdway, D.A. and Nriagu, J.O. *International Journal of Environmental Analytical Chemistry*, **32**, 177 (1988).
67 Gebersmann, C., Lobinski, R. and Adams, F.C. *Analytica Chimica Acta*, **316**, 93 (1995).
68 Grob, K. and Grob, G. *Journal of Chromatography*, **106**, 299 (1975).
69 Otson, R. and Chau, C. *International Journal of Environmental Analytical Chemistry*, **30**, 275 (1987).
70 Blanchard, R.D. and Hardy, J.K. *Analytical Chemistry*, **57**, 2349 (1985).
71 Melcher, R.G. and Caldecourt, V.J. *Analytical Chemistry*, **52**, 875 (1980).
72 Cailleux, A., Turcant, A., Allain, P. *et al. Journal of Chromatography*, **391**, 280 (1987).
73 Lopez-Avila, V., Wood, R., Flanagan, M. and Scott, R. *Journal of Chromatographic Science*, **25**, 286 (1987).
74 Rook, J.J. *Water Treatment and Examination*, **21**, 259 (1972).
75 Rook, J.J. *Water Treatment and Examination*, **23**, 234 (1974).
76 Lovelock, J.E. *Nature (London)*, **256**, 193 (1975).
77 Lovelock, J.E., Maggo, R.J. and Wade, R.J. *Nature (London)*, **241**, 194 (1973).
78 Michael, L.C., Pellizzari, E.D. and Wiseman, R.W. *Environmental Science and Technology*, **22**, 565 (1988).
79 Duffy, M., Driscoll, J.N. and Pappas Sandford, W. *Journal of Chromatography*, **441**, 73 (1988).
80 Filippelli, M. *Applied Organometallic Chemistry*, **8**, 687 (1994).
81 Brooks, J.M. and Sackett, W.M. *Journal of Geophysical Research*, **78**, 5248 (1973).
82 Dravnieks, A., Krotoszynski, S.K., Whitfield, J., O'Donnell, A. and Burgwald, J. *Environmental Science and Technology*, **5**, 1220 (1971).
83 Kaiser, K.L.E. and Oliver, B.G. *Analytical Chemistry*, **48**, 2207 (1976).
84 Perras, J.C. In *Portable Gas Chromatographic Technique to Measure Dissolved Hydrocarbons in Seawater*, USNTIS, AD Report No. 786583/5GA (1973).
85 May, W.E., Chesler, S.N., Cram, S.P. *et al. Journal of Chromatographic Science*, **13**, 535 (1975).
86 The Oil Companies International Study Group for Conservation of Clean Air and Water, Europe. In Analysis of Trace Substances in aqueous Effluents from Petroleum Refineries, Concawe Report, No. 6/82 (1982).
87 Desbaumes, E. and Imhoff, C. *Water Research*, **6**, 885 (1972).
88 Polak, J. and Lu, B. *Analytica Chimica Acta*, **63**, 231 (1974).
89 Schönmann, M. and Kern, H. *Varian Instrument News*, **15**, 6 (1981).
90 Cerwen, J.P. Volatile Organic Materials in Seawater. In *Organic Matter in Natural Waters*, ed. W. Hood. Institute of Marine Science, University of Alaska Publication No. 1, pp. 169–180 (1970).
91 Grob, K. and Grob, G. *Journal of Chromatography*, **106**, 249 (1975).
92 Westerdorf, R.G. *International Laboratory*, September, 32 (1982).
93 Waggot, A. and Reid, W.J. In *Separation of Non Volatile Organic Compounds in Sewage Effluent Using High Performance Liquid Chromatography*. Part 1, Technical Report No. 52. Water Research Centre, Medmenham, UK (1977).
94 Graydon, J.W., Grob, K., Zurcher, F. and Giger, W. *Journal of Chromatography*, **285**, 307 (1984).
95 Nicolson, A.A. and Meresz, O. *Bulletin of Environmental Contamination and Toxicology*, **14**, 453 (1975).
96 Nicolson, A.A., Meresz, O. and Lemyk, B. *Analytical Chemistry*, **49**, 814 (1977).

97 Wong, T. and Lenahan, B. *Bulletin of Environmental Contamination and Toxicology*, **32**, 429 (1984).
98 Dowty, B. and Laseter, J.L. *Analytical Letters (London)*, **8**, 25 (1975).
99 Piet, S.J. Quarterly Report. National Institute for Water Supply. The Netherlands. No. 2, 4 (1975).
100 Lopez-Avila, V., Heath, N. and Hu, A. *Journal of Chromatographic Science*, **25**, 356 (1987).
101 Environmental Protection Agency Report EPA 601. Enviromental Protection Agency, USA (1980).
102 Environmental Protection Agency Report EPA 602. Enviromental Protection Agency, USA (1980).
103 Parr, J.L., Walters, G. and Hoffman, M. *Groundwater*, **1**, 105 (1991).
104 Obach, E. In *Organic Solvents*, eds J.A. Riddick and W.B. Burger, Wiley Interscience, New York (1970).

Chapter 17

Organics: Other preconcentration techniques

Organic compounds

17.1 Electrolytic methods

17.1.1 Seawater

17.1.1.1 Hydrocarbons

Wasik [1] has used an electrolytic stripping cell to determine hydrocarbons in seawater. Dissolved hydrocarbons in a known quantity of seawater were equilibrated with hydrogen bubbles, evolved electrolytically from a gold electrode, rising through a cylindrical cell. In an upper headspace compartment of the cell, the hydrocarbon concentration is determined by gas chromatography. The major advantages of this cell are that the hydrocarbons in the upper compartment are in equilibrium with the hydrocarbons in solution and that the hydrogen used as an extracting solvent does not introduce impurities into samples.

Wasik [1] used this method to successfully determine μg L^{-1} of gasoline in seawater. He found that a convenient method for concentration of the hydrocarbons is to recycle the hydrogen stream containing the hydrocarbons many times over a small amount of charcoal (2.3mg). The charcoal, while still in the filter tube, was extracted three times with 5μL carbon disulphide. A 2–3μL aliquot of this solution was then injected into a SCOT capillary column.

A 10ml portion of the seawater extracted was diluted to 1000ml with seawater. To this solution was added toluene–d, as an internal standard to give a concentration of 0.05ppm. The diluted extract was poured into a 1000ml stripping cell located in a constant temperature water bath. The temperature of the bath was raised to 80°C and a current of 0.3A was passed through the cell. The upper chromatogram shown in Fig. 17.1 was obtained by sampling 1ml of the headspace after 30ml hydrogen had bubbled through the stripping cell. The chromatogram shows the large number of aliphatic and olefinic hydrocarbon peaks that are eluted at the

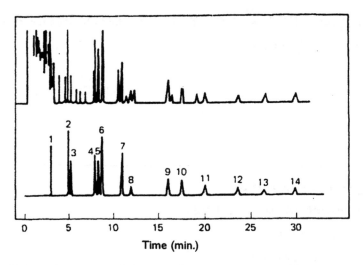

Fig. 17.1 Hydrocarbons stripped from artificial seawater after 10ml hydrogen had bubbled through the stripping cell (upper chromatogram) and after 100ml hydrogen had bubbled through the stripping cell (lower chromatogram). A 1.0ml volume of headspace gas was injected into a Scot column, 15m × 0.5mm id, coated with *m*-bis(*m*-phenoxyphenoxy)benzene; carrier gas He, 4ml min⁻¹; temperature 80°C; attenuation 1 × 8. Peak identification from retention times of known compounds. 1 = benzene; 2 = toluene; 3 = toluene–*d₈*; 4 = ethylbenzene; 5 = *para*-xylene; 6 = *meta*-xylene; 7 = *ortho*-xylene; 8 = *n*-propylbenzene; 9 = 1-methyl–4–ethylbenzene; 10 = 1,3,5–trimethylbenzene; 11 = 1-methyl–2–ethylbenzene; 12 = 1,2,4–trimethylbenzene; 13 = *n*-butylbenzene; 14 = 1,2,3–trimethylbenzene

Source: Reproduced with permission from Wasik, S.P. [1] Preston Publications Inc., US

same time as the early aromatic hydrocarbon peaks. The lower chromatogram in Fig. 17.1 was obtained after 46ml (V/V_L = 0.046) of hydrogen had bubbled through the stripping cell. All the aliphatic and olefinic hydrocarbons peaks are absent from the chromatogram. The aromatic hydrocarbon peaks were identified from retention times of known compounds.

17.2 Cryogenic methods

17.2.1 Non saline waters

17.2.1.1 Miscellaneous organics

Freeze drying. A limited amount of work has been carried out on freeze concentration methods for organics in non saline and saline waters. The method has been applied to the preconcentration of amino acids, carbohydrates, citric acid and phenylalanine [2] giving virtually complete recovery of these substances at concentrations up to 0.2mg L $^{-1}$. The technique has been used in conjunction with Sephadex gel chromatography to characterise organic components in the molecular weight range 200 to 5000 [3] in secondary sewage effluents. The filtered samples were concentrated using a freeze drying unit with methyl alcohol as the bath fluid. The organic carbon recovery in the freeze dried samples was between 92 and 96%.

Pankow [4] showed that quantitative freeze trapping of mixtures of certain volatile chlorinated hydrocarbons can be done with a 30m fused silica capillary column kept at between –60 and –100°C throughout. The effect of using different trapping temperatures on the chromatograph peak profiles was examined for 1,1-dichloroethane, dichloromethane, chloroform, carbon tetrachloride, trichloroethene, 1,2-dichloropropane, 2-chloroethylvinylether, 1,1,2-trichloroethane, dibromochloromethane, tetrachloroethane and chlorobenzene. Trapping at –80°C appeared best, giving peak breadths of only 4–7s at 8°C min $^{-1}$ and 110cm s $^{-1}$ carrier gas speed.

Cryogenic trapping. Rayer and Hennequin [5] described a purge and trap method for the separation of volatile organic compounds from the aqueous phase. Samples are degassed into the trap at –70°C for 15min under a flow of nitrogen; the trap is then closed by metal discs and stored at 4°C before the gases are desorbed at 40–45°C directly into the injection port for gas chromatographic–mass spectrometric analysis. The system was used to determine about 15 compounds, in the concentration range 0.05–10µg L $^{-1}$ in a wide range of waters.

Bodings *et al.* [6] showed that the temperature of the first cold trap was important and high concentrations of volatile compounds gave irregularities. The efficiency of different types of cryogenic trap depended on purge flow rate, total purge volume, and temperature of trapping. Hydrocarbons have been preconcentrated by cryogenic trapping of volatiles released by purging the sample with an inert gas prior to examination of the trappings by gas chromatography [7] or gas chromatography–mass spectrometry [8]. Down to 1µg L $^{-1}$ hydrocarbons could be determined by this procedure.

Cochrane [9] used a Nafion tube dryer in a purge–whole column cryotrapping method for selectively removing water from the analyte containing purge stream during the gas chromatography analysis of volatile aromatics in aqueous samples. The addition of the tube drier meant that water transferred to the column during sample purge could neither extinguish the flame of the flame ionisation detector nor distort chromatographic peak shape. The method was applied to the analysis of many volatile chlorinated and fluorinated hydrocarbons and purgeable aromatics found in contaminated aqueous samples. Gryder-Boutet and Kennish [10] described equipment for the integration of an automated headspace sampler, with whole column cryotrapping and mass selective detection. Cryogenic trapping (–80°C) gave excellent reproducibility in the µg L $^{-1}$ range. Coated fused silica capillary columns were used for gas chromatographic separation, and mass spectral data were obtained by selective ion monitoring. This method was recommended as an alternative to the US EPA method 624 which used purge and trap technology with packed columns.

Freezing out. Slow freezing, with constant stirring, results in a concentration of organic materials in the solution remaining. The technique is most effective in water of low salinity [2,11–13].

17.2.2 Seawater

17.2.2.1 Organophosphorus compounds

Freeze drying. The soluble organic phosphorus compounds arising from natural sources have been examined by a number of techniques, including concentration by freeze drying [14,15]. The most familiar of these compounds, adenosine triphosphate, is normally considered to be found in the particulate fraction, and is measured by the luciferin–luciferinase reaction. Approximately 100% recovery of glucose and lindane at the 0.1 and 0.15µg L $^{-1}$ level have been obtained from water by freeze drying.

17.2.2.2 Amino acids

Pocklington [16] has separated amino acids in seawater using freeze drying. The first step in his concentration of free acids from seawater was the freeze drying of the seawater samples. To reduce interferences in the later steps of the procedure, the sea salts were packed into a chromatographic column and washed with diethyl ether to remove non-polar compounds. The diethyl ether extract, particularly from surface water samples, quite often contained coloured materials as well as other organics. If a series of solvents of graduated polarity were passed through a sea salt column, a fractionation by polarity should be obtained. With the proper choice of solvents, a form of gradient elution could be devised which would result in a continuous, rather than a batch fractionation. Bohling [17,18] and Garrasi and Degens [19] used freeze drying followed by extraction of the dried salts with organic solvents to preconcentrate amino acids from seawater.

17.3 Froth flotation methods

17.3.1 Non saline methods

In its passage through a water column, a bubble acts as an interface between the liquid and vapour phases, and, as such, collects surface–active dissolved materials as well as colloidal micelles on its surface. Thus, in a well-aerated layer of water, the upper levels will become progressively enriched in surface–active materials. It is possible to use this effect to enrich the foamy surface layer with organic material. The foam can be as much as 200 times as concentrated in organic material as the body of the solution [20,21] and is therefore a method of precon-centrating surface–active organic materials from large volumes of water.

MacIntyre [22] suggested that adjustment of the bubble size and the depth at which the bubbles were formed might be used to control the thickness of the surface layer sampled. The collection of the charged particles ejected by bursting bubbles can be enhanced by the use of a charged glass plate as a collector [23]. A surface microlayer sampler using this principle has been devised by Fasching et al. [24]. The literature on bubble ejection and collection has been reviewed by Blanchard [25].

Hiraide et al. [26] have described a method for the rapid preconcentration of humic acid from fresh waters by coprecipitation and flotation. Flotation of suspended matter with a cationic surfactant (cetyldimethyl–benzylammonium chloride) was followed, first, by coprecipitation with iron(III) hydroxide at pH7 and, second, by flotation with anionic surfactants (sodium oleate and sodium dodecyl sulphate). This removed suspended solids and separated humic and fulvic acid. Iron(III) hydroxide was dissolved in 2M hydrochloric acid leaving acid–insoluble humic acid to be collected by ultrafiltration. Absorption spectra, molecular weight distribution, and complexing ability were determined, after dissolving humic acid in potassium hydroxide solution. Flotation decreased the time needed for separation to a fifth of that required with normal filtration procedures.

When the concentrations of surface–active materials are high, the injection of bubbles into the solution from well below the surface may result in the formation of a foam at the surface. The foam can be as much as 200 times as concentrated in organic material as the body of the solution. Natural foams of this kind can often be seen along beaches during periods of strong winds and violent wave action. Even when the surface–active materials are not present at levels high enough for foam formation, a considerable enrichment of organic material can be found in the upper part of the water column [27,28].

When bubbles break at the sea surface, some portion of the organic materials collected on the bubble surface is ejected into the air. The phen-omenon is well known [21] and can be used as a method for the collection of surface–active organic materials. While most of the bubble ejection work has been done in the laboratory, a surface microlayer sampler using this principle has been devised for work in protected inlets [24].

17.4 Evaporation/distillation methods

17.4.1 Non saline waters

Distillation. These have been studied by Kuo *et al.* [29] and Kozloski [30]. The latter worker condensed the distillate of a 1L water sample in a liquid nitrogen trap over 13min. The purgeable components were transferred to a gas chromatograph by conventional gas purging. The recoveries and purge ratios achieved by vacuum distillation of hexane, carbon tetrachloride, chloroform, dichloromethane, benzene, 1,2-dichloroethane, diethyl ether, methyl isobutyl ketone, methyl acetate, and tetrahydrofuran were determined. Misharina *et al.* [31] have reviewed the literature between 1975 and 1985 on evaporation and other methods for preconcentrating volatile organics in non saline waters.

Steam distillation. Dix and Fritz [32] examined various distillation parameters and boiling modifiers in relation to a simple steam distillation method for the extraction of organic compounds from aqueous samples. A survey mixture of 10 organic compounds, containing various functional groups, was added to water and the distillate collected for capillary gas chromatographic analysis. Recoveries of these 10 compounds collected in cold acetone, at fast-boil or with the addition of salt were tabulated. Almost all types of organic compounds with boiling points ranging from 77–238°C showed recoveries above 90%. Many azeotropes with boiling points below that of water distilled over. The addition of salt to the boiling mixture permitted the formation of an azeotrope richer in phenol.

Pervaporation. The method for separation of chlorinated hydrocarbons from dilute aqueous solutions using pervaporation membranes, has been demonstrated experimentally [33]. The mechanism of pervaporation is described as three steps: absorption of the permeating molecules at the liquid–membrane interface; diffusion of the molecule through the membranes; removal of the molecules from downstream surface of the membrane by a vacuum or carrier gas. The results indicate a viable separation can be achieved and that the solubility parameter is the most important factor for selectivity. A membrane system consisting of polyvinyl acetate as the active member on a polysulphone support has been developed and one of polyvinyl acetate sandwiched between layers of PTFE has been studied.

17.5 Osmosis

17.5.1 Non saline water

Reverse osmosis is a membrane process developed almost entirely in the past 20–25 years. The process is based on the fact that if two solutions of different concentrations are separated by a semipermeable membrane, water will move across the membrane in the direction of higher concentration. The driving force is provided by this difference in concentration. Since many of the dissolved substances will not pass through the membrane, the passage of water will eventually cease when the solutions on both sides reach the same concentration. This process of osmosis can be reversed by applying pressure to the more concentrated solution. The concentration of dissolved substances will then increase on that side of the membrane.

Although the process was initially developed for the purposes of desalination of water, the rejection of organics has also been extensively studied. Edwards and Schubert [34] show that reverse osmosis was one of the most promising techniques for removing refractory (non-volatile) organics of intermediate to higher molecular weight from water. More study is needed to predict rejections of organics by reverse osmosis; it appears, however, that rejection is related to the distribution coefficient of a given compound between the aqueous solution and the hydrated membrane. Obviously, this is complicated by the concentration of inorganic ions that continuously change the solubility of the organic compounds in solution and the fact that a high salt concentration can affect the degree of hydration of the membrane.

The scheme used by Kopfler et al. [35] for isolating organics from water is shown in Fig. 17.2. Typically, 1514L of potable water is collected into a covered 2200L stainless steel tank over a 2d period. Soon after collection is begun, 140mg of silver nitrate is added to retard bacterial growth. The collected water is adjusted to pH5.5 with hydrochloric acid. This pH is in the optimum range for stability of cellulose acetate membranes. The water is subjected to reverse osmosis with an Osmonics Model 3319–558°C unit containing a spiral wound cellulose acetate membrane at a pressure of 190lb in $^{-1}$ and a 50% conversion rate.

Immediately before entering the reverse osmosis unit, the water passes through a copper coil immersed in a refrigerated bath maintained at 6 to 10°C. The permeate (water forced through the membrane) is collected in a second stainless steel tank while the reject is recycled into the feed tank. When the volume of the feed has been reduced to 90–180L, reverse osmosis is terminated. The average total organic carbon retention in this concentrate is 85%. The concentrate is divided into 5L aliquots that are each placed in a shallow stainless steel tray. These are frozen and subsequently lyophilised. The temperature of the trays is electronically controlled and is never allowed to rise above −10°C or reach complete dryness while in the lyophiliser. When the volume of a group of trays is sufficiently reduced, the contents are allowed to thaw and are then combined into a single tray, refrozen, and lyophilised. This procedure is continued until the total volume remaining is about 3L of liquid and solids.

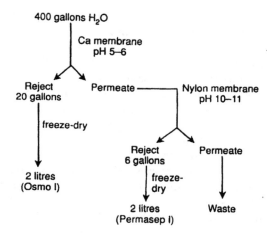

Fig. 17.2 Schematic for concentration of organics
Source: Reproduced with permission from Kopfler, F.C. *et al.* [35] J. Wiley & Sons, New York

The permeate collected from the cellulose acetate unit is again subjected to reverse osmosis in a similar manner using a Continental Water Conditioning Model 881–1 unit containing a DuPont B–9 Permasep permeator. This is an asymmetric aromatic polyamide membrane of hollow fibre configuration. It is easily degraded by the chlorine concentrations normally found in finished potable water but the chlorine is removed by the cellulose acetate membrane. This membrane can tolerate alkaline conditions and since most salts have been removed by the cellulose acetate membrane, the sample can be brought to pH10.5 with sodium hydroxide. This will ionise phenols and acids to increase rejection by the membrane. To maintain a 50% conversion rate, a pressure of 8.8kg cm^{-2} is maintained. The permeate is discarded and the feed solution is recycled until the volume is reduced to 9–13L. This concentrate is frozen and lyophilised as before.

The final product obtained by lyophilising the aqueous concentrate obtained by reverse osmosis of potable water with the cellulose acetate membrane is filtered through a coarse, sintered-glass filter to remove the precipitated salts. These salts are again lyophilised to dryness and extracted first with pentane and then with methylene chloride. The filtrate is extracted three times with pentane using 75ml pentane per litre of solution for the first extraction and 50ml pentene per litre for the other two extractions. The extracts are combined, dried with sodium sulphate, and concentrated in a Kuderna–Danish evaporator [36]. The solution is extracted again in the same manner with methylene chloride. It is then

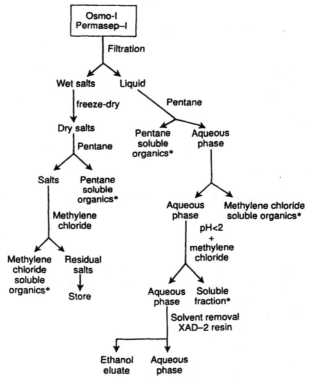

*Fractions chemically characterized and tested for toxicty.

Fig. 17.3 Schematic for extraction of organics from reverse osmosis concentrates of tap water
Source: Reproduced with permission from Kopfler, F.C. et al. [35] J. Wiley & Sons, New York

acidified with hydrochloric acid to pH2 and again extracted with methylene chloride to extract acidic and phenolic compounds.

After extraction, the solvent is removed from the concentrate in a rotary evaporator at 40°C. More organics can then be obtained by passing the concentrate through a column of XAD–2 macroreticular resin and subsequent elution with ethanol. This extraction scheme is shown in Fig. 17.3. The concentrate obtained with the DuPont Permasep membrane is extracted in the same manner. However, since the cellulose acetate membrane removes most of the salts, little if any precipitation occurs on lyophilisation and there is no need to extract salts separately.

Table 17.1 Organic compounds identified in reverse osmosis concentrates of Cincinnati drinking water

Barbital	Methyl palmitate
Benzene sulphonic acid	Methyl stearate
Benzyl butyl phthalate	Methyl tetracosanoate
2-Chloroethyl methyl ether	Octadecane
Dibutyl phthalate	Pentachlorobiphenyl
Diethyl phthalate	Phenyl benzoate
Di(2-ethyl bexyl) phthalate	Phthalic anhydride
Docosane	1,1,3,3-Tetrachloroacetone
Eicosane	Tetrachlorobiphenyl
Hexadecane	Trichlorobiphenyl
Methyl docosanoate	2,4,5-Trichlorophenol

Source: Reproduced with permission from Kopfler, F.C. *et al.* [35] American Chemical Society

Aliquots of each fraction are retained for chemical analysis. The remaining portions are combined for toxicity assays. Analysis of these fractions by gas chromatography–mass spectrometry has proved the presence of some compounds not previously identified in potable water such as those shown in Table 17.1. Because of tar-like residues remaining in the inlet of the gas chromatographic columns, it is apparent that much of the organic residue is non-volatile and, as yet, remains unidentified.

The most volatile compounds present in the water, which are lost during the concentration procedure, were collected by stripping with helium from a water sample on to a trap containing a porous polymer. This is accomplished with the use of the system developed by Bellar and Lichtenberg [37,38] modified to contain 500ml of samples. The trap is subsequently heated, and the volatile organics back-flushed into a gas chromatograph–mass spectrometer. Typical compounds identified in potable water are shown in Table 17.2.

Reverse osmosis has been used for the determination in potable water of traces of organochlorine and organophosphorus insecticides [39,40], and pentaerythritol, bis (2-chloroethyl)ether, bis(2-chloro–isopropyl) ether, urea, triethylene glycol, dipropylamine, chlorobenzene, propylbenzene and benzonitrile [41], as well as various ethers, glycols, amines, nitriles, hydrocarbons and chlorinated hydrocarbons. Cellulose acetate [42], ethyl cellulose acetate [40], and cross linked polyethyleneimine [41] were used as semi-permeable membranes. Chian *et al.* [39] studied two types of reverse osmosis membranes–cellulose acetate and cross-linked polyethylenimine. With each membrane the rejection of pesticides was better than 99%. Deinzer *et al.* [41] used cellulose acetate membranes.

Table 17.2 Volatile organic compounds identified in Cincinnati drinking water

Chloromethane	Dichloroacetylene
Bromomethane	Trichloroethylene
Bromodichloromethane	Tetrachloroethylene
Dibromochloromethane	Bromotrichloroethylene
Chloroform	Cyanogen chloride
Bromoform	Methanol
Carbon tetrachloride	Ethanol
Chloroethane	2-Methyl propanal
Chloropropane	2-Butanone
1,2-Dichloroethane	3-Methyl butanal
Acetyl chloride	Toluene

Source: Reproduced with permission from Kopfler, F.C. *et al.* [35] American Chemical Society

Klein *et al.* [40] used cellulose acetate and ethylcellulose membranes and showed that the retention of a particular organic solute in potable water by a polymer membrane is related not only to the physical interactions such as size but also is dependent upon attractive forces which are normally included in the solubility of the solute in the membrane phase. These workers used solubility parameters for projections of class and specific separations with the two membranes studied.

17.5.2 Seawater

Reverse osmosis has been applied to the preconcentration of organo-chlorine and organophosphorus insecticides [39,40] and various ethers, glycols, amines, nitriles, hydrocarbons, and chlorinated hydrocarbons in seawater and potable water. Cellulose acetate [42], ethyl cellulose acetate [40] and cross-linked polyethyleneimine [41] were used as semi-permeable membranes.

17.6 Dialysis

17.6.1 Non saline waters

Sodergren [43] studied the uptake of organic pollutants from non saline water by solvent filled dialysis membranes in static and continuous-flow systems. The pattern of uptake of DDE, DDT and a polychlorobiphenyl mixture was similar for both systems under continuous-flow conditions. Some results obtained from field experiments in which the solvent-filled membranes were exposed to actual industrial waste waters are tabulated.

It was concluded that such membranes could be used to screen and predict the fate of pollutants in industrial waste waters and aquatic ecosystems.

17.7 Semipermeable membranes

17.7.1 Non saline water

Huckins *et al.* [44] have developed a semipermeable membrane, consisting of a thin film of neutral lipid for passive *in situ* monitoring of organic compounds in non saline waters.

Bennet *et al.* [45] used semipermeable membrane devices to collect polyaromatic hydrocarbons and polychlorobiphenyls from non saline waters.

Lebo *et al.* [46] used semipermeable membrane devices to passively collect traces of chlorinated dioxins and furans from non saline waters.

Ellis *et al.* [47] used semipermeable membrane devices to passively collect traces of chlorinated insecticides from non saline waters.

Leth and Lauritsen [48] preconcentrated analytes inside a water-cooled membrane prior to being thermally released into an MS source.

Organometallic compounds

17.8 Electrolyte methods

17.8.1 Non saline waters

17.8.1.1 Organomercury compounds

Doherty and Dorsett [49] used a copper coil preconcentrator for the electrodeposition of mercury from organomercury compounds prior to its determination by flameless atomic absorption spectrometry.

References

1 Wasik, S.P. *Journal of Chromatographic Science*, **12**, 845 (1974).
2 Kammeren, P.A. and Lee, G.F. *Environmental Science and Technology*, **3**, 276 (1969).
3 Sachdev, D.R., Ferris, J.J. and Cleseri, N.L. *Journal of Water Pollution Control Federation*, **48**, 570 (1976).
4 Pankow, J.F. *Journal of High Resolution Chromatography and Chromatography Communications*, **6**, 292 (1983).
5 Rayer, J. and Hennequin, C. *Techniques et Sciences Municipales*, **77**, 25 (1982).
6 Bodings, H.T., De Jong, C. and Dooper, R.P.M. *Journal of High Resolution Chromatography and Chromatography Communications*, **8**, 755 (1985).
7 Swinnerton, J.W. and Linnenbom, V.J. *Journal of Gas Chromatography*, **5**, 570 (1976).
8 Novak, J., Zluticky, J., Kubela, V. and Mostecky, J. *Journal of Chromatography*, **76**, 45 (1973).
9 Cochrane, J.W. *Journal of High Resolution Chromatography*, **10**, 573 (1987).

10 Gryder-Boutet, D.E. and Kennish, J.M. *Journal of American Water Works Association*, **80**, 52 (1988).
11 Shapiro, J. *Science*, **133**, 2063 (1961).
12 Shapiro, J. *Analytical Chemistry*, **39**, 280 (1967).
13 Habermann, H.M. *Science*, **140**, 292 (1963).
14 Minear, R.A. *Environmental Science and Technology*, **6**, 431 (1972).
15 Minear, R.A. and Walanski, K.A. In *Investigation of the Chemical Identity of Soluble Organophosphorus Compounds found in Natural Waters*. Research Report UILUWRC–74–0086 (1974).
16 Pocklington, R. In A New Method for the Determination of Amino-Acids in Sea Water and an Investigation of the Dissolved Free Amino-Acids of North Atlantic Ocean Waters, PhD thesis, Dalhousie University, pp. 1–102 (1970).
17 Bohling, H. *Marine Biology*, **6**, 213 (1970).
18 Bohling, H. *Marine Biology*, **16**, 281 (1972).
19 Garrasi, C. and Degens, E.T. In Analytische Methoden zur saulenchromatographischen bestimmung von Aminosauren und Suckern im Meereswasser und Sediment. Berichte aus dem Projekt DFGDE 74/e: 'Litoralforschung–Abwasser in Kustennahe'. DFGAbschulbkolloquim, Bremerhaven (1976).
20 Wallace, G.T. Jr. and Wilson, D.F. In *Foam Separations as a Tool in Chemical Oceanography*, US Naval Research Laboratory Report, I, 6958 (1969).
21 Blanchard, F.C. *Science*, **146**, 396 (1964).
22 MacIntyre, P. *Journal Chemical Physics*, **72**, 589 (1968).
23 Blanchard, D.C. and Syzdek, L.C. *Limnology and Oceanography*, **20**, 762 (1975).
24 Fasching, J.L., Courant, R.A., Duce, R.A. and Piotrowicz, S.R. *Journal Rech. Atmosphere*, **8**, 649 (1974).
25 Blanchard, D.C. In *Applied Chemistry at Protein Interfaces*. Advances in Chemistry Series No. 145 ed. R.E. Baier, American Chemical Society, Washington DC, pp. 360–87 (1975).
26 Hiraide, M., Ren, F.L., Tamura, R. and Mizuike, A. *Mikrochimica Acta*, **416**, 137 (1987).
27 Dorman, D.C. and Lemlich, R. *Nature (London)*, **207**, 145 (1965).
28 Karger, B.L. and DeVivo, D.G. *Separation Science*, **1**, 393 (1968).
29 Kuo, P.P.T., Chian, E.S.K. and De Walle, F.B. *Water Research*, **11**, 1005 (1977).
30 Kozloski, R.P. *Journal of Chromatography*, **346**, 408 (1985).
31 Misharina, T.A., Zhuravleva, I.L. and Golovaya, R.V. *Journal of Analytical Chemistry, USSR*, **42**, 459 (1987).
32 Dix, K.D. and Fritz, J.S. *Journal of Chromatography*, **408**, 201 (1987).
33 Zhu, C.L., Yuang, C.N., Fried, J.R. and Greenberg, D.B. *Environmental Progress*, **2**, 132 (1983).
34 Edwards, V.H. and Schubert, P.F. *Journal of the American Water Works Association*, **66**, 610 (1974).
35 Kopfler, F.C., Melton, R.G., Mullaney, S.L. and Tardiff, R.G. *Advances in Environmental Science and Technology*, **81**, 419 (1977).
36 Gunther, F.A., Blinn, R.C., Kalbezen, H.J. et al. *Analytical Chemistry*, **23**, 1835 (1951).
37 Bellar, T.A. and Lichtenberg, J.J. *Journal of the American Water Works Association*, **66**, 739 (1974).
38 Bellar, T.A., Litchenberg, J.J. and Kroner, R.C. *Journal of the American Water Works Association*, **66**, 703 (1974).
39 Chian, E.S.K., Bruce, W.N. and Fang, H.H.P. *Environmental Science and Technology*, **9**, 52 (1975).
40 Klein, E., Eichelberger, J., Eyer, C. and Smith, J. *Water Research*, **9**, 807 (1975).

41 Deinzer, H., Melton, R. and Mitchell, D. *Water Research*, **9**, 799 (1975).
42 Kammerer, P.A. and Lee, G.F. *Environmental Science and Technology*, **1**, 276(1969).
43 Sodergren, A. *Environmental Science and Technology*, **21**, 855 (1987).
44 Huckins, J.N., Manuweera, G.K., Petty, J.D., Mackay, D. and Lebo, J.A. *Environmental Science and Technology*, **27**, 2489 (1993).
45 Bennett, E.R., Metcalfe, T.L. and Metcalfe, C.D. *Chemosphere*, **33**, 363 (1996).
46 Lebo, J.A., Gale, R.W., Petty, J.D. *et al. Environmental Science and Technology*, **29**, 2886 (1995).
47 Ellis, G.S., Huckins, J.N., Rostad, C.E. *et al. Environmental and Toxicological Chemistry*, **14**, 1875 (1995).
48 Leth, M. and Lauritsen, F.R. *Rapid Communications Mass Spectrometry*, **9**, 591 (1995).
49 Doherty, P.E. and Dorsett, R.S. *Analytical Abstracts*, **43**, 1887 (1971).

Chapter 18

Rationale, preconcentration of organic and organometallic compounds

To enable the reader to quickly locate in the text suitable preconcentration procedures for particular compounds, methods used for the preconcentration of various types of organic and organometallic compounds are reviewed in Tables 18.1 to 18.11. These include

- hydrocarbons (Table 18.1)
- oxygen containing organic compounds (Table 18.2)
- halogen containing organic compounds (Table 18.3)
- nitrogen containing organic compounds (Table 18.4)
- humic substances (Table 18.5)
- polychlorobiphenyls, polychlorodibenzo-*p*-dioxins and polychloro-dibenzofurans (Table 18.6)
- sulphur containing organic compounds (Table 18.7)
- insecticides and herbicides (Table 18.8)
- miscellaneous types of organic compounds (Table 18.9)
- mixtures of organic compounds (Table 18.10) and
- organometallic compounds (Table 18.11).

These tables list the type of organic or organometallic compounds which are required to preconcentrate, the type of water sample, the analytical finish employed and where available the detection limits achieved by various workers. The section of the book in which further information can be obtained is also listed.

Keynotes to the abbreviations used in the above tables are given below.

Abbreviations, analytical finishes

AAS:	atomic absorption spectrometry
ASV:	anodic stripping voltammetry
FTIR:	Fourier transform infrared spectroscopy
GC:	gas chromatography
GC–MS:	gas chromatography–mass spectrometry

GFAAS: graphite furnace atomic absorption spectrometry
HPLC: high performance liquid chromatography
ICPAES: inductively coupled plasma atomic emission spectrometry
IR: infrared spectroscopy
LC: liquid chromatography
LC–MS: liquid chromatography–mass spectrometry
MRNP: adsorption of macroreticular non polar solids
NAA: neutron activation analysis
NMR: nuclear magnetic resonance spectroscopy
Spec.phot.:   spectrophotometry
Specfluoro: spectrofluorometry
TLC: thin layer chromatography
UV: ultraviolet spectroscopy

Abbreviations, preconcentration methods

AER: anion exchange resins
AS: adsorption on organic and inorganic solids
C: adsorption on active carbon
CER: cation exchange resins
COP: coprecipitation techniques
CRY: cryogenic preconcentration
DHA: dynamic headspace analysis
E: electroanalytical preconcentration
OD: adsorption on octadecyl resin
P: adsorption on solid polymers
PU: adsorption on polyurethane foam
S: semipermeable membranes
SE: solvent extraction
SFC: supercritical fluid extraction
SHA: static headspace analysis

Further information on detection limits achieved by various preconcentration procedures is given in Chapter 19.

Table 18.1 Preconcentration of hydrocarbons (aliphatic, aromatic, polyaromatic, alicyclic)

Compound type	Type of water	Preconcentration method	Analytical finish	Limit of detection	Section	Chapter [Ref.]
Aliphatic hydrocarbons	Non saline	MRNP	–	–	3.2.1.11	3 [58,59]
					3.2.5	3 [13]
Aliphatic hydrocarbons	Non saline	AS	IR	–	6.22.1.2	6 [32,35]
Aliphatic hydrocarbons	Non saline	AS	GLC	–	6.31.1	6 [47]
Aliphatic hydrocarbons	Non saline	PU	IR	$<50\mu g\ L^{-1}$	8.1.1	8 [1–3]
Aliphatic hydrocarbons	Non saline	P	–	–	7.2.1.1	7 [2]
Aliphatic hydrocarbons	Non saline	SE	Fluorescence	–	13.1.1	13 [25–63,232]
Aliphatic hydrocarbons	Non saline	SE	NMR	–	13.1.2	13 [63]
Aliphatic hydrocarbons	Non saline	SFC	–	–	14.1.2	14 [8]
Aliphatic hydrocarbons	Non saline	SFC	Spec.photo	–	14.1.2	14 [7]
Aliphatic hydrocarbons	Non saline	DHA	GC	$ng\ L^{-1}$	16.1.1	16 [13–18]
Aliphatic hydrocarbons	Non saline	DHA	GC	$1\mu g\ L^{-1}$	16.1.1	16 [2,10,12]
Aliphatic hydrocarbons	Non saline	DHA	GC	$1–5\mu g\ L^{-1}$	16.1.1	16 [19–21]
Aliphatic hydrocarbons	Non saline	DHA	GC	$ng\ L^{-1}$	16.1.2	16 [13,14]
Aliphatic hydrocarbons	Non saline	DHA	GC	$10\mu g\ L^{-1}$	16.1.2	16 [22]
Aliphatic hydrocarbons	Non saline	DHA	–	–	16.1.2	16 [15–19,68]
Aliphatic hydrocarbons	Non saline	MRNP	–	–	3.2.1.1	3 [8,9,11]
Aliphatic hydrocarbons	Non saline	MRNP	–	–	3.2.5	3 [13]
Aliphatic hydrocarbons	Non saline	MRNP	–	–	3.2.5	3 [9]
Aliphatic hydrocarbons	Non saline	MRNP	–	–	3.3.1.2	3 [109]
Aliphatic hydrocarbons	Non saline	SHA	GC	–	15.1.1	15 [14,15,62]
Aliphatic hydrocarbons	Non saline	SHA	GC	–	15.1	15 [14–17]
Aliphatic hydrocarbons	Non saline	SHA	GC	$1\mu g\ L^{-1}$	15.1.1	15 [16]
Aliphatic hydrocarbons	Non saline	SHA	GC	–	15.1.2	15 [17,22]

Table 18.1 continued

Compound type	Type of water	Preconcentration method	Analytical finish	Limit of detection	Section	Chapter	[Ref.]
Aliphatic hydrocarbons	Non saline	SHA	GC	<1µg L⁻¹	15.1.1	15	[19]
Aliphatic hydrocarbons	Non saline	C	GC, GC-MS	1µg L⁻¹	17.2.2.1	17	[7-9]
Aliphatic hydrocarbons	Sea	SHA	-	-	15.2.1	15	[18]
Aliphatic hydrocarbons	Sea	SE	-	1µg	13.2.2	13	[220]
Aliphatic hydrocarbons	Sea	SE	Flow calorimetry	5µg L⁻¹	13.2.2	13	[219]
Aliphatic hydrocarbons	Sea	DHA	GC	µg L⁻¹	16.2.1	16	[2,10,19,20, 23,58,82,84, 85,87,89]
Aliphatic hydrocarbons	Sea	DHA	GC-MS	ng L⁻¹	16.1.1	16	[86]
Aliphatic hydrocarbons	Sea	SHA	GC	-	15.1	15	[13,18-21]
Aliphatic hydrocarbons	Sea	E	GC	50µg L⁻¹	17.1.1.1	17	[1]
Aliphatic hydrocarbons	Groundwater	AS	GC	-	6.31.1	6	[48]
Aliphatic hydrocarbons	Waste water	DHA	GC-MS	1µg L⁻¹	16.5.1	16	[86,94]
Aliphatic hydrocarbons	Waste water	SHA	GC-MS	1µg L⁻¹	15.6.1	15	[61,73-75]
Aliphatic hydrocarbons	Potable	SE	GC	ng L⁻¹ to µg L⁻¹	13.1.1	13	[13]
Aromatic hydrocarbons	Non saline	AS	-	-	6.31.1	6	[49]
Benzene	Non saline	AS	-	-	6.31.1	6	[50]
Dimethylbenzene	Non saline	OD	Pyrolysis GC	-	6.19.1.1	6	[28]
Aromatic hydrocarbons	Non saline	C	-	-	9.1.1	9	[1]
Aromatic hydrocarbons	Non saline	C	GC	-	10.1.1.8	10	[30]
Aromatic hydrocarbons	Non saline	SE	GC	0.1-0.5µg L⁻¹	13.1.1, 13.1.3	13	[64-68]
Aromatic hydrocarbons	Non saline	SHA	GC	sub ng L⁻¹	15.1.2	15	[16,22,23]

Table 18.1 continued

Compound type	Type of water	Preconcentration method	Analytical finish	Limit of detection	Section	Chapter [Ref.]
Aromatic hydrocarbons	Non saline	SHA	MRNP	—	3.2.1.1	3 [8,9]
Aromatic hydrocarbons	Non saline	SHA	GC	—	15.1	15 [24]
Aromatic hydrocarbons	Non saline	SHA	GC	—	15.1.1	15 [13]
Aromatic hydrocarbons	Non saline	SHA	GC	—	15.1.1	15 [20,21,24,64]
Aromatic hydrocarbons	Non saline	DHA	GC	—	16.1.1	16 [12,23]
Polyaromatic compounds	Non saline	AS	—	—	6.24.1.1	6 [38]
Polyaromatic compounds	Non saline	CER	—	—	4.1.2	4 [25]
Polyaromatic compounds	Non saline	P	—	—	7.9.1.1	7 [18]
Polyaromatic compounds	Non saline	P	GC	$1\mu g\ L^{-1}$	7.9.1.1	7 [15,17]
Polyaromatic compounds	Non saline	PU	—	$ng\ L^{-1}$	8.1.2	8 [4–8]
Polyaromatic compounds	Non saline	OD	—	—	9.1.10	9 [18]
Polyaromatic hydrocarbons	Non saline	OD	HPLC	—	9.1.2.	9 [3]
Polyaromatic hydrocarbons	Non saline	OD	—	$ng\ L^{-1}$	9.1.2	9 [2]
Polyaromatic compounds	Non saline	COP	GLC	—	11.13.1.1	11 [7]
Polyaromatic compounds	Non saline	COP	GLC	—	11.4.1.1	11 [9]
Polyaromatic hydrocarbons	Non saline	SE	GLC	—	13.1.1	13 [69–82]
Polyaromatic compounds	Non saline	—	—	—	13.1.4	13 [82,185]
Polyaromatic hydrocarbons	Non saline	MRNP	—	$0.05ng\ L^{-1}$	3.2.1.1	3 [9–11]
Polyaromatic hydrocarbons	Non saline	MRNP	GC-MS	$<0.05ng\ L^{-1}$	3.2.1.2	3 [14–16]
Polyaromatic hydrocarbons	Potable	MRNP	—	$<5\mu g\ L^{-1}$	3.2.1.2	3 [14]
Polyaromatic hydrocarbons	Potable	AS	HPLC	—	6.21.2.1	6 [33]
Polyaromatic hydrocarbons	Potable	OD	—	—	9.4.1	9 [17]
Polyaromatic hydrocarbons	Potable	—	—	—	10.1.3.1	10 [37]

Table 18.1 continued

Compound type	Type of water	Preconcentration method	Analytical finish	Limit of detection	Section	Chapter [Ref.]
Polyaromatic hydrocarbons	Potable	SE	GC	–	13.1.1	13 [13,14]
Polyaromatic hydrocarbons	Waste	MRNP	GC	<10ng L⁻¹	3.2.3.1	3 [85]
Polyaromatic hydrocarbons	Waste	MRNP	TLC–GC	<10µg L⁻¹	3.3.4.1	3 [135]
Polyaromatic hydrocarbons	Trade	OD	LC	0.1–50µg L⁻¹	9.7.1	9 [34–36]
Alicyclic hydrocarbons	Non saline	DHA	GC	–	16.1.1	16 [1,2]
Animal oils		AS	–	–	6.2l.1.2	6 [32]
Mineral, vegetable	Non saline	SE	IR	–	13.1.1	13 [187]
Mineral, vegetable and animal oils	Non saline	MRNP	–	–	3.2.5	3 [109]

Source: Own files

Table 18.2 Preconcentration of organic compounds containing oxygen (alcohols, ketones, aldehydes, carboxylic acids, anionic, cationic and non ionic surfactants, phthalates, phenols and dioxane)

Compound type	Type of water	Preconcentration method	Analytical finish	Limit of detection	Section	Chapter [Ref.]
Alcohols	Non saline	MRNP	–	–	3.2.1.1	3 [9]
Alcohols	Non saline	SHA	–	0.008–13	15.1	15 [6,7]
Alcohols	Sea	SHA	–	µg L⁻¹	15.5, 15.2.4	15 [7,25]
Alcohols	Potable	MRNP	–	–	3.2.5	3 [9]
Carboxylic acids	Non saline	MRNP	GC	–	3.2.1.15	3 [65]
Carboxylic acids	Non saline	AER	–	–	5.1.1	5 [1]
Carboxylic acids	Non saline	DHA	GC	0.02M	16.1.1	16 [25]
Carboxylic acids	Non saline	SE	Spec.phot	–	13.1.1	13 [92]
Carboxylic acids	Non saline	SE	TLC	–	13.1.1	13 [91]
Carboxylic acids	Non saline	SE	GC	–	13.1.1, 13.1.7	13 [90,93–96]
Carboxylic acids	Non saline	MRNP	–	–	3.2.1.11	3 [58]
Anionic surfactants	Non saline	OD	HPLC	–	9.1.4.5	9 [5]
Anionic surfactants	Potable	C	–	–	10.1.3.3	10 [4–7,44,45]
Anionic surfactants	Non saline	SE	Spec.photo	–	13.1.1, 13.1.6	13 [193]
Non ionic surfactant	Non saline	SFC	–	–	14.1.3	14 [9]
Non ionic surfactant	Non saline	C	–	ng L⁻¹	10.2.1.2	10 [48]
Non ionic surfactant	Non saline	SE	–	–	13.1.1	13 [182]
Non ionic surfactant	Non saline	MRNP	Spec.photo UV, IR, NMR	<10µg L⁻¹	3.2.1.4	3 [18–20,23,24]
Phthalate esters	Non saline	MRNP	–	–	3.3.1.2	3 [120]
Esters	Non saline	MRNP	–	–	3.2.1.1	3 [7]
Esters	Non saline	MRNP	–	–	3.2.1.6	3 [8]

Table 18.2 continued

Compound type	Type of water	Preconcentration method	Analytical finish	Limit of detection	Section	Chapter [Ref.]
Esters	Non saline	MRNP	–	–	3.2.1.1, 3.2.5	3 [9]
Esters	Non saline	SE	–	–	3.1.1.1	3 [13,14]
Phenols	Non saline	MRNP	–	200µg L^{-1}	3.2.5	3 [11,65,92]
Phenols	Non saline	MRNP	–	–	3.2.5	3 [92]
Phenols	Non saline	AS	HPLC	–	6.31.1	6 [51]
Phenols	Non saline	AS	HPLC	–	6.20.1.1	6 [29]
Phenols	Non saline	P	–	–	7.6.1.1	7 [10]
Phenols	Non saline	P	–	µg L^{-1}	7.2.1.1	7 [3]
Phenols	Non saline	P	–	–	7.5.1.1	7 [8]
Phenols	Non saline	CER	GC	0.5µg L^{-1}	4.1.2	4 [22]
Phenols	Non saline	OD	GC	–	9.1.3	9 [4]
Phenols	Non saline	C	–	–	10.2.1.1	10 [47]
Phenols	Non saline	C	GC	–	10.1.1.1	10 [3]
Phenols	Non saline	SE	HPLC	–	13.1.5	13 [196]
Phenols	Non saline	SE	–	–	13.1.5	13 [89]
Phenols	Non saline	SE	GC	–	13.1.1	13 [83–89]
Phenols	Non saline	COP	–	–	1.1.1.1.1	11 [1–4]
Phenols	Waste	OD	LC	–	9.5.1	9 [32]
Phenols	Waste	CER	–	–	4.3.1	4 [34]
Aldehydes	Non saline	DHA	GC	–	16.1.1, 16.1.8	16 [26]
Aldehydes	Potable	AS	–	–	6.21.2.2	6 [35]
Ketones	Potable	MRNP	–	–	3.2.5	3 [7]
Ketones and aldehydes	Sea	DHA	–	–	15.1.1	15 [90]
					16.2.2	16 [80]

Table 18.2 continued

Compound type	Type of water	Preconcentration method	Analytical finish	Limit of detection	Section	Chapter [Ref.]
Cyclohexane derivs (CO or OH separated by 1 or 2 methylene groups)	Non saline	CER	–	–	4.1.2	4 [9,10]
Cyclohexane deriv. (OH or CO separated by 1 or 2 methylene groups)	Non saline	P	–	–	7.9.1.1	7 [14]
Dioxane	Non saline	DHA	GC–MS	$\mu g\ L^{-1}$	16.1.1, 16.1.7	16 [24]

Source: Own files

Table 18.3 Preconcentration of the halogen containing organic compounds (haloform, aliphatic and aromatic halogens compounds and chlorophenols)

Compound type	Type of water	Preconcentration method	Analytical finish	Limit of detection	Section	Chapter [Ref.]
Haloforms	Non saline	MRNP	GC	—	3.25	3 [86,88–90]
Haloforms	Non saline	MRNP	GC	—	3.2.1.13	3 [62,63]
Haloforms	Non saline	MRNP	GC	—	3.2.1.1	3 [8,9]
Haloforms	Non saline	DHA	GC	—	16.1.10	16 [28,29,35–47]
Haloforms	Non saline	DHA	GC	$1\mu g\ L^{-1}$	16.1.1, 16.1.10	16 [38]
Haloforms	Non saline	DHA	GC	—	16.1.1	16 [28–37,39–47]
Haloforms, aromatic hydrocarbons	Non saline	DHA	GC	$0.5\mu g\ L^{-1}$	16.1.3	16 [72]
Haloforms	Non saline	SHA	GC	$0.1ng\ L^{-1}$	15.1	15 [28]
Haloforms	Non saline	SE	GC	—	13.1.1	13 [110–112]
Haloforms	Non saline	DHA	GC	—	16.1.1.1	16 [61]
Haloforms	Non saline	SE	—	—	13.1.11	13 [201]
Haloforms	Rain	SHA	GC	$0.001\mu g\ L^{-1}$	15.1	15 [46]
Haloforms	Potable	DHA	GC	$<1\mu g\ L^{-1}$	16.3.3	16 [28,32,35–37, 47,95,96, 97,99]
Haloforms	Potable	DHA	GC	$0.1\mu g\ L^{-1}$	16.1.1	16 [35,37]
Haloforms	Potable	DHA	GC	$<0.5\mu g\ L^{-1}$	16.1.1	16 [28]
Haloforms	Potable	DHA	GC	$1\mu g\ L^{-1}$	16.1.1	16 [5,95,96,98]
Haloforms	Potable	SHA	GC	—	15.1	15 [22,29,36–45, 47–52]
Haloforms	Potable	SHA	GC	$1\mu g\ L^{-1}$	15.1	15 [51]
Haloforms	Potable	SHA	GC	$0.1–10\mu g\ L^{-1}$	15.5.1	15 [22,28,29,36–44,49,50,53]

Table 18.3 continued

Compound type	Type of water	Preconcentration method	Analytical finish	Limit of detection	Section	Chapter	[Ref.]
Haloforms	Potable	C	GC	—	10.1.3.4	10	[33,45,46]
Vinyl chloride	Non saline	DHA	GC	4µg L⁻¹	16.1.11	16	[44,46]
Vinyl chloride	Non saline	SHA	Photoionisation detector	—	15.1, 15.1.3	15	[55]
Vinyl chloride	Non saline	DHA	GC	µg L⁻¹	16.1.1	16	[44,46]
Vinyl chloride	Non saline	DHA	GC		16.1.6	16	[64]
Vinyl chloride	Non saline	DHA	GC	µg L⁻¹	6.1.11	16	[45]
Vinyl chloride	Sea	DHA	GC	µg L⁻¹	15.1.1, 16.2.3	15	[46]
Chloroaliphatics	Non saline	MRNP	—	—	3.3.1.2	3	[124]
Chloroaliphatics	Non saline	MRNP	—	—	3.2.4.1	3	[86]
Chloroaliphatics	Non saline	MRNP;TLC, GPC	NAA	—	3.2.1.12	3	[60,61]
Chloroaliphatics	Non saline	MRNP	Schoniger combustion	1–2≤mole L⁻¹	3.2.5	3	[9]
Hexachlorobutadiene	Non saline	MRNP	—	—	3.3.1.2	3	[123]
1,1,1 trichloroethane	Non saline	AS	Kinetics	—	6.31.1.6	6	[52]
Chloroaliphatics	Non saline	C	GC	—	17.2.2.1	17	[4]
Chloroaliphatics	Non saline	SE	GC	—	13.1.1	13	[97–100,102]
Chloroaliphatics	Non saline	SE	NMR	—	13.1.1	13	[62]
Chloroaliphatics	Non saline	DHA	GC	—	16.1.1	16	[53]
Chloroaliphatics	Non saline	DHA	GC	10–50µg L⁻¹	16.1.1	16	[63]
Chloroaliphatics	Non saline	SHA	GC	—	15.1	15	[1,22,28,29, 31–33,37,38, 40,49,50,66, 67]
Chloroaliphatics	Non saline	DHA	GC	—	16.1.1	16	[30,48,50,51, 52,54,60]

Table 18.3 continued

Compound type	Type of water	Preconcentration method	Analytical finish	Limit of detection	Section	Chapter [Ref.]
Chloroaliphatics	Non saline	DHA	GC	1µg L^{-1}	16.1.10	16 [38]
Chloroaliphatics	Non saline	DHA	GC	10µg L^{-1}	16.1.11	16 [63]
Chloroaliphatics	Non saline	DHA	GC	—	16.1.11	16 [30,60,62]
Chloroaliphatics	Non saline	SHA	GC	0.1ppb	15.1.1	15 [29]
Octafluorobutylene	Non saline	SHA	—	10µg L^{-1}	15.1.1, 15.1.3	15 [34,55]
Trichloroethylene, tetrachloroethylene	Non saline	SHA	GC	—	15.1	15 [26]
Bromoalkanes	Non saline	SHA	GC	—	15.1.3, 15.1.1	15 [57]
Chloroalkanes and bromoalkanes	Non saline	DHA	GC	0.5µg L^{-1}	16.1.1	16 [15–17,30, 55–58]
Bromoalkanes	Potable	MRNP	GC	—	3.2.5	3 [9]
Methyl bromide	Non saline	SHA	GC	—	15.1, 15.1.1	15 [57]
Chloroalkanes	Sea	MRNP	—	—	3.2.2.4	3 [78]
Chloroalkanes	Sea	SE	GC	1µg L^{-1}	13.2.3	13 [221,222]
Chloroalkanes	Sea	DHA	HPLC	—	15.1.1	15 [93]
Chloroalkanes	Sea	DHA	GC	µg L^{-1}	15.1.1, 16.2.5	16 [15–17]
Chloroalkanes	Sea	DHA	GC	—	16.2.3	16 [83]
Dichlorodifluoromethane	Sea	SHA	GC	0.05 × 10^{-12}	15.1	15 [35]
Chloroaliphatics	Potable	MRNP	GC	—	3.2.4.1	3 [86]
Chloroaliphatics	Potable	C	—	—	10.1.3.2	10 [34,35]
Chloroaliphatics	Potable	C	—	—	10.1.1.2	10 [33–37]

Table 18.3 continued

Compound type	Type of water	Preconcentration method	Analytical finish	Limit of detection	Section	Chapter [Ref.]
Chloroaliphatics	Potable	DHA	GC	1μg L⁻¹	16.1.1, 16.3.2	16 [59]
Chloroaliphatics	Potable	SHA	GC	–	15.1	15 [23,29]
Chloroaliphatics	Potable	DHA	GC	<0.1–1μg L⁻¹	16.1.1	16 [100]
Chloroaliphatics	Potable	DHA	GC	<1μg L⁻¹	16.1.1	16 [92]
Methyl bromide	Potable	SHA	GC	5ng L⁻¹	15.5.1	15 [56]
Chloroaliphatics	Ground	SHA	–	–	15.1	15 [27]
Chloroaliphatics	Ground	DHA	GC	0.1μg L⁻¹	16.4.1	16 [100]
Chloroaliphatics	Waste	SE	GC	0.01μg L⁻¹	13.4.1,	13 [103]
Chloroaliphatics	Waste	SHA	GC	0.5mg L⁻¹	15.1	15 [30]
Chloroaliphatics	Waste	SHA	GC	0.5mg L⁻¹	15.6.2	15 [45]
Chloroaliphatics	Trade	SHA	GC–MS	–	15.1, 15.7.2	15 [23,29]
Haloaromatics	Non saline	MRNP	GC	1–10μg L⁻¹	3.2.5	3 [95]
Haloaromatics	Non saline	MRNP	GC	–	3.2.1.1	3 [9]
Chloroanisoles	Non saline	SE	GC	–	13.1.1	13 [183]
Chloroamines	Non saline	MRNP	GC	–	3.2.1.14	3 [64]
Pentachloronitrobenzene trichloroacetic acid	Non saline	GC	–	–	6.16.1.1	6 [25]
Trifluoroacetic acid	Rain	–	GC	10μg L⁻¹	15.1	15 [58]
2-Chloronaphthalene	Non saline	–	GC	–	3.3.1.2	3 [123]
Total organic halogen	Non saline	–	–	–	3.2.5	3 [87]
3,4 Dichloropropanide	Non saline	–	GC	–	4.1.2	4 [27]
3-Phenyl-4-hydroxy-6-chloropyridazine	Ground	–	GC	–	4.1.2	4 [26]

Table 18.3 continued

Compound type	Type of water	Preconcentration method	Analytical finish	Limit of detection	Section	Chapter [Ref.]
Chlorolignosulphonic acid	Waste	–	–	–	4.3.2	4 [35]
Haloaromatics	Non saline	DHA	GC	–	16.1.1	16 [49]
Haloaromatics	Non saline	SHA	GC–MS	–	15.1	15 [8–10]
Chlorophenols	Non saline	AS	–	0.001–	6.6.1.1	6 [11]
Chlorophenols	Non saline	AS	GC	5ng L^{-1}	6.4.1.1	6 [8]
Chlorophenols	Non saline	P	–	–	7.1.1.1	7 [1]
Chlorophenols	Non saline	SE	GC	–	13.1.1,	13 [88,104–106]
					13.1.9	13 [105]
Chlorophenols	Non saline	C	GC	sub µg L^{-1}	10.2.1.3	10 [49]
Chlorophenols	Non saline	C	GC	–	10.11.5	10 [21]
Chlorophenols	Ground	AS	–	–	6.18.1.1	6 [27]
Chlorophenols	Potable	AS	GC	0.001–1µg L^{-1}	6.3.1.1	6 [53]
Chlorophenols	Non saline	MRNP	GC	–	3.2.5	3 [93]
Chlorophenols	Non saline	MRNP	GC	–	3.2.1.4	3 [64]

Source: Own files

Table 18.4 Preconcentration of organic nitrogen compounds (EDTA, NTA, aliphatic and aromatic amines, aminophenols, nitrophenols, nitrocompounds, nitrosamines, azarines and amino acids)

Compound type	Type of water	Preconcentration method	Analytical finish	Limit of detection	Section	Chapter [Ref.]
Ethylene diamine tetraacetic acid	Non saline	SE	GC	–	13.1.1	13 [191,192]
Nitriloacetic acid	Non saline	AER	Polarography	0.01 μg L⁻¹	5.1.2	5 [2,3]
Nitriloacetic acid	Non saline	MRNP	GC	0.7 μg L⁻¹	3.2.5	3 [94]
Nitriloacetic acid	Surface water	AER	GC		5.2.1	5 [15]
Aliphatic amines	Non saline	OD	HPLC	0.14–1.8.μg L⁻¹	9.1.7	9 [11]
Aliphatic amines	Non saline	Chelation solvent extraction	AAS	15μg L⁻¹	2.1.1	2 [1]
Aliphatic amines	Non saline	SE	GC–MS	–	13.1.1	13 [185]
Organic bases	Non saline	CER	–	50μg L⁻¹– 1mg L⁻¹	4.1.2	4 [12]
Aromatic bases	Non saline	P	–	–	7.4.1.1	7 [7]
Aminophenols, chloramines	Non saline	AS	HPLC	–	12.1.1.2	12 [8]
Nitrobenzene	Non saline	MRNP	–	–	3.3.1.2	3 [126]
Nitro compounds	Non saline	P	–	–	7.9.1.1	7 [16]
Nitro compounds	Trade	AS	HPLC	1μg L⁻¹	6.11.1.1	6 [2]
Nitro compounds	Non saline	SFC	–	–	14.1.7	14 [13]
Chlorotriazines	Non saline	–	–	–	12.1.1.4	12 [14]
Hydroxytriazines	Non saline	OD	–	–	9.1.10	9 [27]
Nitrocresol	Non saline	MRNP	–	–	3.2.1.8	3 [32]
Urea	Non saline	CER	–	–	4.1.2	4 [20]
Pyrazole	Non saline	UD	–	–	9.1.10	9 [16]
N-methyl carbamates	Non saline	AS	–	–	6.3.1	6 [55]
Nitrophenols	Non saline	SE	–	–	13.1.1	13 [88]

Table 18.4 continued

Compound type	Type of water	Preconcentration method	Analytical finish	Limit of detection	Section	Chapter [Ref.]
N-methylphenol and chlorophenol	Non saline	Off line	–	ng L^{-1}	12.1.1.3	12 [9]
Nucleic acids	Non saline	AS	–	–	6.27.1.1	6 [40]
Nitrosamines	Non saline	C	GC	ng L^{-1}	10.1.1.7	10 [23–25]
Nitrosamines	Non saline	CER	GC	–	4.1.2	4 [23]
Nitrosamines	Non saline	SE	GC	–	13.1.1	13 [180,181]
Azarenes	Sea	MRNP	–	0.5–3µg	3.2.2.5	3 [11,79]
Amino acids	Non saline	CER	GC	–	4.1.1	4 [1,2]
Amino acids	Sea	CER	LC,TLC fluorometry	–	4.2.1	4 [28–33]
Amino acids	Sea	C	–	–	17.2.2.2	17 [16]
Amino acids	Sea	COP	–	–	11.2.1.1	11 [5,6]
		SE	TLC	–	13.2.5	13 [226]

Source: Own files

Table 18.5 Preconcentration of humic and fulvic acids

Compound type	Type of water	Preconcentration method	Analytical finish	Limit of detection	Section	Chapter	[Ref.]
Humic and fulvic acids	Non saline	MRNP	–	–	3.2.5	3	[82]
Humic and fulvic acids	Non saline	P	–	–	7.2.1.3	7	[4]
Humic and fulvic acids	Non saline	C	Fluorescence	–	10.1.1.6	10	[22]
Humic and fulvic acids	Non saline	AER	–	–	5.1.3	5	[4–9]
Humic and fulvic acids	Non saline	CER	–	–	4.1.2	4	[38]
Humic and fulvic acids	Non saline	COP	–	–	11.1.1.2	11	[5]
Humic and fulvic acids	Non saline	Froth flotation	–	–	17.3.1	17	[21,24–28]
Humic and fulvic acids	Sea	MRNP	–	–	3.2.2.6	3	[80–82]

Source: Own files

Table 18.6 Preconcentration of polychlorobiphenyls, polychlorodibenzo-p-dioxins and polychlorodibenzofurans

Compound type	Type of water	Preconcentration method	Analytical finish	Limit of detection	Section	Chapter [Ref.]
Polychlorobiphenyls	Non saline	MRNP	GC	–	3.2.5	3 [20,52,53]
Polychlorobiphenyls	Non saline	MRNP	GC	0.4ng L⁻¹	3.2.1.9	3 [10,19,20, 27, 38,49–51, 55,56]
Polychlorobiphenyls	Non saline	OD	GC	0.01–10µg L⁻¹	9.1.8	9 [12,13]
Polychlorobiphenyls	Non saline	SFC	–	10µg L⁻¹	14.1.4	14 [10]
Polychlorobiphenyls	Non saline	SE	GC	–	13.1.1	13 [107–109]
Polychlorobiphenyls	Non saline	SE	GC	<0.1ng L⁻¹	13.1.10	13 [197–200]
Polychlorobiphenyls	Non saline	MRNP	GC	–	3.2.1.11	3 [58]
Polychlorobiphenyls	Sea	MRNP	GC	<10–<13ng L⁻¹	3.2.2.2	3 [7,74]
Polychlorobiphenyls	Sea	MRNP	GC	0.001 µg absolute	3.2.2.2	3 [19,58]
Polychlorobiphenyls	Sea	MRNP	GC	<1µg L⁻¹	3.3.2.1	3 [199–132]
Polychlorobiphenyls	Sea	PU	–	–	8.2.1	8 [10–12]
Polychlorodibenzo–p–dioxin	Non saline	MRNP	–	–	3.2.1.10	3 [57]
Polychlorodibenzo–p–dioxin	Non saline	AS	–	pg L⁻¹	6.13.1.1	6 [22]
Polychlorodibenzo–p–dioxins	Non saline	SFC	LC	–	14.1.5	14 [11]
polychlorodibenzofurans		Semi permeable membrane	–	–	17.7.1	17 [46]
Polychlorodibenzo–p–dioxins	Sea	AS	–	1ppq	6.31.1	6 [65]
polychlorodibenzofurans	Potable	OD	–	1pg L⁻¹	9.4.2	9 [31]
Chlorinated insecticides	Non saline	AS	GC	µg L⁻¹	6.2.1.1	6 [39]

Source: Own files

Table 18.7 Preconcentration of sulphur containing compounds (linear, alkylbenzene sulphonates, alkylethoxylated sulphates, dimethylsulphide, tetrahydrofuran etc.)

Compound type	Type of water	Preconcentration method	Analytical finish	Limit of detection	Section	Chapter [Ref.]
Linear alkylbenzene sulphonates	Non saline	SE	GC	—	13.1.1	13 [188]
Linear alkylbenzene sulphonates	Non saline	On line SE	LC	—	12.1.1.1,	12 [7]
Linear alkylbenzene sulphonates	Non saline	AS	LC	—	6.8.1.1	6 [13]
Linear alkylbenzene sulphonates	Non saline	C	GC	—	10.1.1.2	10 [4–7]
Linear alkylbenzene sulphonates	Sewage	CER	^{13}CNMR	—	4.1.2	4 [24]
Linear alkylbenzene sulphonates	Waste water	CER	GC	—	4.1.2	4 [21]
Linear alkylbenzene sulphonates	Sewage	OD	GC	—	9.6.1	9 [33]
Alkyl ethoxylated sulphonates	Waste	CER	GC	—	4.3.3	4 [21,24]
Dimethyl sulphide	Non saline	DHA	GC	$0.8\mu g\ L^{-1}$	16.1.1, 16.1.12	16 [66]
Organosulphur compounds	Non saline	DHA	GC microwave induced plasma AAS	$ng\ L^{-1}$	16.1.12	16 [65,67]
Organosulphur compounds	Non saline	DHA	Atomic emission	ppt	16.1.1	16 [67]
Organosulphur compounds	Non saline	DHA	GC	—	16.1.1	16 [65]
Organosulphur compounds	Non saline	MRNP	GC	—	3.2.5	3 [84]
Tetrahydrothiophen	Non saline	SHA	—	—	15.1	15 [59]
Organosulphur compounds	Sea	MRNP	Headspace GC	—	3.2.8	3 [84]

Source: Own files

Table 18.8 Preconcentration of insecticides and herbicides (chloro, phosphorus, carbamate, substituted urea, chlorophenoxy acetic acid, azine types and miscellaneous other types)

Compound type	Type of water	Preconcentration method	Analytical finish	Limit of detection	Section	Chapter [Ref.]
Chloroinsecticides	Non saline	MRNP	–	–	3.2.5	3 [38,50,51]
Chloroinsecticides	Non saline	MRNP	–	–	3.3.1.2	3 [121,122]
Chloroinsecticides	Non saline	MRNP	GC	$0.2\mu g\ L^{-1}$	3.2.1.6	3 [37–41,43]
Chloroinsecticides	Non saline	MRNP	–	–	3.2.1.5	3 [36]
Chloroinsecticides	Non saline	SE (on line)	LC	–	12.1.1.4	12 [11,12]
Chloroinsecticides	Non saline	AS	–	–	6.2.1.1	6 [3–6]
Chloroinsecticides	Non saline	AS	–	–	6.9.1.1	6 [15–17]
Chloroinsecticides	Non saline	AS	GC	–	6.7.1.1	6 [12]
Chloroinsecticides	Non saline	AS	–	–	6.14.1.1	6 [23]
Chloroinsecticides	Non saline	AS	–	$5fgd\ m^{-2}$	6.31.1	6 [54]
Chloroinsecticides	Non saline	AS	GC	$ng\ L^{-1}$	6.26.1	6 [39]
Chloroinsecticides	Non saline	C	GC	$10\mu g\ L^{-1}$	10.1.1.3	10 [1,2,8–17]
Chloroinsecticides	Non saline	OD	TLC	$40\mu g\ L^{-1}$	9.1.9	9 [14,15]
Chloroinsecticides	Non saline	SE	GC	–	13.1.12	13 [109]
Chloroinsecticides	Non saline	SE	–	$ng\ L^{-1}$	13.1.12	13 [138–140,141]
Chloroinsecticides	Non saline	SE	GC	–	13.1.12	13 [139]
Chloroinsecticides	Non saline	SE	NMR	–	13.1.1	13 [62]
Chloroinsecticides	Non saline	SE	TLC	–	13.1.1	13 [137]
Chloroinsecticides	Non saline	SFC	–	–	14.1.6	14 [12]
Chloroinsecticides	Non saline	Dialysis	–	–	17.6.1	17 [43]
Chloroinsecticides	Non saline	Osmosis	–	–	17.5.1	17 [39,40]
Chloroinsecticides and PCBs	Sea	DHA	GC	–	16.2.4	16 [1,2]
Chloroinsecticides and PCBs	Sea	MRNP	GC	–	3.2.2.2	3 [74–77]
Chloroinsecticides and PCBs	Sea	MRNP	GC	$1\mu g\ L^{-1}$	3.3.2.1	3 [129–133]

Table 18.8 continued

Compound type	Type of water	Preconcentration method	Analytical finish	Limit of detection	Section	Chapter [Ref.]
Chloroinsecticides and PCBs	Sea	AS	LC	—	6.20.2.1	6 [30]
Chloroinsecticides and PCBs	Sea	MRNP	—	1ng L⁻¹ (Cl.insect) 10ng L⁻¹ (PCB)	3.2.2.2	3 [71]
Chloroinsecticides and PCBs	Sea	SE	—	<3µg L⁻¹	13.2.4	13 [225]
Chloroinsecticides and PCBs	Sea	SE	GC	—	13.1.1	13 [130]
Chloroinsecticides and PCBs	Groundwater	OD	GC	sub µg L⁻¹	9.2.1	9 [15]
Chloroinsecticides and PCBs	Potable	SE	GC	—	13.1.1	13 [13,14]
Chloroinsecticides	Sea	DHA	—	—	15.1.1	15 [1,2]
Chloroinsecticides	Non saline	SE	—	—	13.1.1	13 [142–148]
Phosphorus insecticides	Non saline	MRNP	Flame photometry	0.1–1ng absolute	3.2.1.5	3 [35]
Phosphorus insecticides	Non saline	MRNP	HPLC	2µg L⁻¹	3.2.1.5	3 [26,36]
Phosphorus insecticides	Saline	AS (on line)	MS	—	12.1.1.4	12 [13]
Phosphorus insecticides	Non saline	AER	GC	—	5.1.5	5 [11]
Phosphorus insecticides	Non saline	MRNP	—	0.001–50µg L⁻¹	3.21.6	3 [37–41,43]
Phosphorus insecticides	Non saline	OD	—	—	9.1.10	9 [18]
Phosphorus insecticides	Saline	SE	GC	—	13.1.1	13 [102,143, 149–153]
Phosphorus insecticides	Sea	SE	GC	0.04–0.09µg L⁻¹	13.1.9	13 [25]
Phosphorus insecticides	Sea	MRNP	—	—	3.2.2.3	3 [29]
Phosphorus insecticides	Sea	AER	—	—	5.2.1	5 [13,14]
Phosphorus insecticides	Sea	Freeze drying	—	—	17.2.2.2	17 [14,15]
Phosphorus insecticides	Sea	MRNP	—	<1µg L⁻¹	3.3.2.1	3 [129–133]
Phosphorus insecticides	Sea	Osmosis	—	—	17.5.2	17 [39,40]
Carbamate insecticides/ herbicides	Non saline	MRNP	N/P–GC	ng L⁻¹	3.2.1.7	3 [45]

Table 18.8 continued

Compound type	Type of water	Preconcentration method	Analytical finish	Limit of detection	Section	Chapter	[Ref.]
Carbamate insecticides/ herbicides	Non saline	SE	Spec.phot	–	13.1.1	13	[102,158, 160–164]
Carbamate insecticides/ herbicides	Non saline	AS (on line)	–	–	12.1.1.4	12	[17]
Aminocarb aldicarb	Non saline	CER	GC	–	4.1.2	4	[19]
Aminocarb aldicarb	Non saline	AER	Spec.phot.	–	5.1.4	5	[10]
Triazine herbicides	Non saline	On line	–	$1ng\ L^{-1}$	12.1.1.4	12	[15,16]
Triazine herbicides	Non saline	MRNP	–	–	3.21.6	3	[37–41]
Carbamate insecticides	Non saline	MRNP	LC	$0.2ng\ L^{-1}$	3.11.6	3	[42]
Azine herbicides	Non saline	OD	GC	$0.01–0.5\mu g\ L^{-1}$	9.1.10	9	[22]
	Non saline	SE	–	–	13.1.1	13	[170–172]
Des/ethylatrazine, simazine atrazine, terbethylazine	Potable	MRNP	–	$15–32\mu g\ L^{-1}$	3.3.3	3	[134]
Urea herbicides	Potable	AS	–	$0.6–2ng\ L^{-1}$	6.23.2.1	6	[37]
Urea herbicides	Non saline	AS	–	–	6.21.1.1	6	[31]
Urea herbicides	Non saline	AS	–	–	6.3.1.1	6	[7]
Urea herbicides	Non saline	SE	GC	–	13.1.1	13	[169,173–175]
2,4 dichloro-phenoxyacetic acid	Non saline	SE	TLC	–	13.1.1	13	[165–169]
Chlorophenoxy acetic acid	Non saline	MRNP	–	–	3.21.6	3	[37–41]
Miscellaneous insecticides and herbicides etc.							
Paraquat, Diquat	Non saline	SE	GC	–	13.1.1	13	[178]
Paraquat, Diquat	Non saline	AER	Spec.photo.	–	5.4.1	5	[16]
Propham, barban, butyrate, picloram	Non saline	MRNP	–	–	9.1.10	9	[20]

Table 18.8 continued

Compound type	Type of water	Preconcentration method	Analytical finish	Limit of detection	Section	Chapter [Ref.]
Propoxur, carbofuran propham, captan, chloro-proton, barbon, butyrate	Non saline	AS	–	–	6.31.1	6 [57]
Fe nitrothion	Non saline	MRNP OD CER	GC GC GC	0.1–0.5ng absolute – 0.5ng absolute	3.2.1.5 9.1.10 4.1.2	3 [27–33] 9 [19] 4 [18]
Fenitrothion	Non saline	MRNP	–	–	3.2.1.8	3 [27–32,46,47]
Aminofenitrothion	Non saline	MRNP	–	–	3.2.1.8	3 [32]
Oxamyl, methomyl, phoxan, 2,4,5 tri-chloro-phenoxy acetic acid, 2,4,DB, MCPB	Non saline	C	–	$0.003\mu g\ L^{-1}$	10.2.1.4	10 [54]
Dicamba	Non saline	SE	GC	–	13.1.1	13 [179]
Acidic herbicides	Non saline	On line solid phase	Electrospray MS	sub $\mu g\ L^{-1}$	12.1.1.4	12 [19]
Polar herbicides	Non saline	On line solid phase	Thermospray LC–MS	$1–100ng\ L^{-1}$	12.1.1.4	12 [20]
Dalapon	Non saline	SE	TLC	–	13.1.1	13 [177]
Fluridone	Non saline	OD	GC	$1–2\mu g\ L^{-1}$	9.1.10	9 [21]
Chlorosulphuron sulphometon	Non saline	OD	–	–	9.1.10	9 [23]
Butachlor, oxiazone chlornitrophen, chlor-methoxynil	Non saline	OD	–	–	9.1.10	9 [24]
Misc. insecticides and pesticides	Non saline	On line solid	GC	–	12.1.1.4	12 [18]

Table 18.8 continued

Compound type	Type of water	Preconcentration method	Analytical finish	Limit of detection	Section	Chapter [Ref.]
Misc. pesticides	Non saline	On line solid phase	-	0.01–0.5µg L⁻¹	12.2.1.4	12 [2]
Misc. insecticides and herbicides	Non saline	On line solid phase	-	1µg L⁻¹	12.2.1.4	12 [10]
Misc. pesticides	Non saline	AS	-	-	6.3.1.1	6 [56,58,59]
Phosphorus–sulphur compounds	Non saline	MRNP	-	S 1ng P 0.1ng	3.2.1.5	3 [34]
Mirex	Sea	MRNP	-	-	3.2.2.2	3 [71]
Misc. pesticides	Non saline	OD	-	10µg L⁻¹	9.1.10	9 [25,26]
Misc. insecticides and pesticides	Non saline	Off line solid phase	LC–MS	sub µg L⁻¹	12.2.1.1	12 [17,24]
Misc. insecticides and pesticides	Non saline	P	-	-	7.8.1.1	7 [13]
Misc. insecticides and pesticides	Non saline	C	HPLC	0.003–0.07µg L⁻¹	12.2.1.4	10 [28,50–53]
Misc. insecticides and pesticides	Non saline	SE	GC	-	13.1.1	13 [154–159]
Misc. insecticides and pesticides	Non saline	SE	-	-	13.1.1	13 [140,141]
Misc. pesticides	Potable	SE	-	-	13.3.1	13 [230]

Source: Own files

Table 18.9 Preconcentration of miscellaneous organic compounds

Compound type	Type of water	Preconcentration method	Analytical finish	Limit of detection	Section	Chapter [Ref.]
Siloxanes	Non saline	SE	ICPAES	–	13.1.1	13 [186]
Geosmin	Non saline	SE	GLC-MS and GLC	–	13.1.1	13 [189,190]
Methyl isoborneol	Non saline	OD	–	10ng L⁻¹	9.1.11	9 [28]
Coprostanol	Non saline	MRNP	GC	–	3.2.1.3	3 [16,17]
Chlorophyll, coprostanol	Sewage	MRNP	Spec.photo. and GLC	–	3.1.1.1	3 [1–3]
Cytokinins	Non saline	MRNP	–	–	3.2.1.7	3 [67]
Carbanilides	Non saline	MRNP	–	–	3.2.1.16	3 [66]
Sterols	Sea	MRNP	LC	–	3.2.27	3 [3,83]
Tween K100, Tween 20	Non saline	AS	Spec.photo	–	6.15.1.1	6 [24]
Alkylphosphates alkylthiophosphates	Non saline	MRNP	GC	–	3.2.5	3 [29]
Squoxin	Non saline	SE	–	–	13.1.1	13 [233]
Mutagenic organics	Non saline	MRNP	–	–	3.2.5	3 [112]
Polar organics	Non saline	AS	–	–	6.23.1.1	6 [36]
Organic priority pollutants	Non saline	SHA	–	–	15.1	15 [10]
Pyrethrins	Non saline	SE	–	–	13.1.1	13 [184]
N-methyl carbamates	Non saline	AS	GC	–	6.31.1	6 [60]
Misc.	Non saline	MRNP	–	–	3.2.5	3 [10,69,75, 99,104,108, 111,113,114, 116–120]

Table 18.9 continued

Compound type	Type of water	Preconcentration method	Analytical finish	Limit of detection	Section	Chapter	[Ref.]
Misc.	Non saline	MRNP	GC	–	3.2.1.1	3	[12,13]
Misc.	Non saline	MRNP	–	ng L⁻¹	3.2.1.1	3	[6,7]
Misc.	Non saline	MRNP	–	20pg	3.2.1.1	3	[4]
Misc.	Non saline	AS	GC	–	6.31.1	6	[63,64]
Misc.	Non saline	AS	LC	–	6.8.1	6	[14]
Misc.	Non saline	AS	–	–	6.17.1.1	6	[26]
Misc.	Non saline	AS	–	–	6.9.1.2	6	[18]
Misc.	Non saline	AS	GC	sub ng L⁻¹	12.1.1.5	12	[21]
Misc.	Non saline	P	–	–	6.2.1.2	6	[6]
Misc.	Non saline	P	–	–	7.2.1.4	7	[5]
Misc.	Non saline	P	–	–	7.3.1.2	7	[6]
Misc.	Non saline	C	–	–	10.1.1.8	10	[26–29,31,32]
Misc.	Non saline	OD	–	–	9.1.12	9	[29]
Misc.	Non saline	CER	GC	–	4.1.2	4	[11,16,17]
Misc.	Non saline	SHA	GC	–	15.1	15	[62]
Misc.	Non saline	SHA	–	1 μg L⁻¹	15.1	15	[11]
Misc.	Non saline	SHA (closed loop)	–	–	15.9.3	15	[80]
Misc.	Non saline	DHA	GC	μg L⁻¹	16.1.1	16	[1–9,15–17, 68–71,73–79]
Misc.	Non saline	DHA	–	–	16.1.1.3	16	[15,17,68]
Misc.	Non saline	DHA	GC	0.5μg L⁻¹	16.1.1	16	[71,74–79]
Misc.	Non saline	DHA	GC	μg L⁻¹	16.1.1	16	[104]

Table 18.9 continued

Compound type	Type of water	Preconcentration method	Analytical finish	Limit of detection	Section	Chapter	[Ref.]
Misc.	Non saline	SE	–	–	13.1.1	13	[1–10,12,13,19–24,194]
Misc.	Non saline	Cryogenic	GC–MS	0.005µg L^{-1}	17.2.2.1	17	[5]
Misc.	Non saline	Cryogenic	GC	–	17.2.2.1	17	[6]
Misc.	Sea	AS	–	–	6.30.1.1	6	[43–46]
Misc.	Sea	C	–	–	10.1.2.1	10	[39–42]
Misc.	Sea	OD	–	–	9.3.1	9	[30]
Misc.	Sea	SHA	GC	–	15.2.5	15	[1,5,6,25,54,69–73,82]
Misc.	Sea	COP	–	–	11.1.2.1	11	[1,2,4]
Misc.	Sea	DHA (closed loop)	GC and GC–MS	–	16.2.5	16	[15–17,91,92]
Misc.	Sea	MRNP	–	–	3.2.2.1	3	[68,69]
Organophosphorus compounds Misc.	Sea	C	Cryogenic	<0.1µg L^{-1}	17.22.1	17	[14]
Misc.	Potable	MRNP	–	–	3.2.5	3	[8,9,12,102,111]
Misc.	Potable	C	GC–MS	–	10.1.3.5	10	[29]
Misc.	Potable	SHA	GC	–	15.1	15	[1,61]
Misc.	Potable	DHA	GC–MS	low µg L^{-1}	16.3.4	16	[5,98,99]
2-methyl isoborneol 2,3,6 trichloroanisole Misc.	Potable	DHA	–	–	16.1.1	16	[27]
Misc.	Surface	MRNP	GC–MS	0.1µg L^{-1}	3.2.5	3	[107]

Table 18.9 continued

Compound type	Type of water	Preconcentration method	Analytical finish	Limit of detection	Section	Chapter [Ref.]
Misc.	Surface	CER	GC–MS	0.1 µg L⁻¹	4.1.2	4 [15]
Misc.	Ground	DHA	GC	<20µg L⁻¹	16.4.2	16 [73]
Misc.	Wellwater	MRNP	—	mg L⁻¹	3.2.5	3 [75]
Misc.	Wellwater	CER	—	—	4.1.2	4 [14]
Misc.	Mineral water	AS	Spec.photo. IR, UV, GLC	—	6.10.1.1	6 [19]
Misc.	Trade	MRNP	—	—	3.2.5	3 [103,110]
Misc.	Waste	DHA	GC	—	16.1.1	16 [94]
Misc.	Waste	SE	GC	—	13.4.2	13 [231]
Volatiles	Waste	SHA	—	—	15.6.3	15 [54,83]
Misc. volatiles	Non saline	MRNP	—	—	3.3.1.1	3 [120]
Misc. volatiles	Non saline	AS	—	—	6.12.1.1	6 [21]
Misc. volatiles	Non saline	OD	—	—	9.1.5	9 [6,7]
Misc. volatiles	Waste	SHA	GC	—	15.1	15 [12,19–21, 57,60,61]
Misc. volatiles	Non saline	SHA	—	—	15.1	15 [2–5,8–9]
Misc.	Resin	MRNP	—	—	3.3.1.2	3 [127]
Misc. volatiles	Non saline	DHA	GC	—	16.1.1.3	16 [70,73]
Misc. volatiles	Non saline	SFC	—	—	14.1.8	14 [14,15]
Misc. volatiles	Non saline	SE	Thermospray	—	13.1.1, 13.1.37	13 [194]
Misc. volatiles	Potable	AS	—	sub µg L⁻¹	6.31.1	6 [61]

Source: Own files

Table 18.10 Preconcentration of mixtures of organic compounds

Compound type	Type of water	Preconcentration method	Analytical finish	Limit of detection	Section	Chapter [Ref.]
Alcohols, esters, aldehydes, alkylbenzenes, polyaromatic hydrocarbons and chloro-compounds	Non saline	MRNP	–	–	3.2.5	3 [99]
Carboxylic acids, alcohols, amines sucrose, aminoacids, quinaldic acid	Non saline	MRNP	–	–	3.2.5	3 [11,75,97,98]
Alkanes, aromatics, phenols, carboxylic acids and chloro-compounds	Non saline	MRNP	–	–	3.2.5	3 [7]
Phenols, carboxylic acids, aldehydes, alcohols	Non saline	MRNP	–	–	3.2.5	3 [100,101]
α, α di chloropropionate, 2,3,6 trichlorobenzoate, 2-methoxy, 3,6 dichlorobenzoate, 2,4.D	Non saline	AER	–	–	5.1.6	5 [12]
Polychlorobiphenyls, polychlorodibenzo-p–dioxins, polychlorodibenzofurans	Non saline	C	–	–	10.1.1.4	10 [19,20]
Phenols, hydrocarbons, phthalates, polyaromatic hydrocarbons	Non saline	MRNP	GC	–	3.2.5	3 [96]
Alkanes, aromatic hydrocarbons, alcohols, ketones, esters, bromocompounds	Non saline	MRNP	–	–	3.2.5	3 [9]
Dioxane, 2-butanone, ketones, alcohols	Non saline	DHA	GC	1 µg L^{-1} (4L sample)	16.1.1	16 [24]
Basic, neutral and phenolic compounds	Non saline	SE	–	–	3.1.1	3 [11]

Table 18.10 continued

Compound type	Type of water	Preconcentration method	Analytical finish	Limit of detection	Section	Chapter [Ref.]
Environmental Protection Agency priority pollutants	Ground water	DHA	MS	<20µg L⁻¹	15.1.1	15 [73]
Alkanes, aromatic hydrocarbons, chlorocompounds, alcohols, ketones, esters, bromocompounds	Potable	MRNP	–	–	3.2.5	3 [9]
Alkanes, aromatic hydrocarbons, ethers, esters, haloforms, polyaromatic hydrocarbons, ketones, alcohols, phenols	Potable	MRNP	GC	–	3.2.5	3 [8]
Alcohols, esters, aldehydes, alkyl benzenes, polyaromatic hydrocarbons	Potable	MRNP	–	–	3.2.5	3 [99]
Phenols, carboxylic acids, aldehydes, alcohols	Waste	MRNP	–	–	3.2.5	3 [100]
Alkanes, aromatic hydrocarbons, chlorocompounds, alcohols, ketones, esters, bromocompounds	Waste	MRNP	–	–	3.2.5	3 [9]
Alcohols, esters, aromatic hydrocarbons, polyaromatic hydrocarbons, chlorocompounds	Waste	MRNP	–	–	3.2.5	3 [99]
Chloroaromatic hydrocarbons	Waste	DHA	GC	1µg L⁻¹	16.1.1	16 [86,100]
Aromatic hydrocarbons, chlorocompounds	Trade	DHA	–	–	16.1.1	16 [8]
					16.6.1	16 [79]
Aliphatic hydrocarbons, alkylbenzene, sulphonates, polyaromatic hydrocarbons, nitrophenol	Sewage	SFC	–	–	14.4.1	14 [36]

Source: Own files

Table 18.11 Preconcentration of organometallic compounds (arsenic, lead, tin and mercury)

Compound type	Type of water	Preconcentration method	Analytical finish	Limit of detection	Section	Chapter [Ref.]
Organolead compounds	Non saline	Chelation–solvent extraction	AAS	0.01µg L⁻¹	2.2.1	2 [2,3]
Organolead compounds	Non saline	COP	ASV	—	11.3.1.2	11 [8]
Organolead compounds	Non saline	SE	GC	0.5µg L⁻¹	13.1.40	13 [213,214]
Organolead compounds	Rain	OD	LC	15µg L⁻¹	9.8.1	9 [36]
Organoarsenic compounds	Sea	CER	GFAAS	0.02µg L⁻¹	4.5.1	4 [37]
Alkyltin compounds	Non saline	OD	GC	—	9.9.1	9 [37,38]
Alkyltin compounds	Non saline	CER	GC	—	4.4.1	4 [36]
Alkyltin compounds	Non saline	SFC	GC	—	14.1.9	14 [16,17]
Alkyltin compounds	Non saline	SE	—	—	13.1.39	13 [208–212]
Alkyltin compounds	Sea	Chelation–solvent extraction	GC-MS	—	2.3.1	2 [5]
Alkyltin compounds	Sea	MRNP	GC	—	3.4.1.1	3 [136–138]
Alkyltin compounds	Sewage	OD	—	—	9.10.1	9 [37]
Organomercury compounds	Non saline	AS	AAS	40–300µg L⁻¹	6.28.1.1	6 [41]
Organomercury compounds	Non saline	AS	—	0.04ng L⁻¹	6.29.1.1	6 [42]
Organomercury compounds	Non saline	AS	GC	0.05ng L⁻¹	6.1.1.1	6 [1,2]
Organomercury compounds	Non saline	On line solid phase	LC	—	12.1.2.1	12 [23]
Organomercury compounds	Non saline	P	AAS	0.2ng L⁻¹	7.7.1.1	7 [12]
Organomercury compounds	Non saline	PU	—	—	8.3.1	8 [13–15]

Table 18.11 continued

Compound type	Type of water	Preconcentration method	Analytical finish	Limit of detection	Section	Chapter [Ref.]
Organomercury compounds	Non saline	AER	AAS	–	5.5.1	5 [18]
Organomercury compounds	Non saline	AER	NAA	–	5.5.1	5 [17]
Organomercury compounds	Non saline	DHA	GLC–FTIR	50–100pg absolute	16.1.14	16 [80]
Organomercury compounds	Non saline	SE	GC	–	13.1.38	13 [203,204, 205,207]
Organomercury compounds	Sea	Chelation–solvent extraction	AAS	5ng L^{-1}	2.3.1	2 [6,7]
Organomercury compounds	Sea	SE	AAS	–	13.2.6	13 [205]

Source: Own files

Chapter 19

Detection limits achievable for organic and organometallic compounds

Detection limits achieved by various methods of preconcentrations are summarised in Tables 19.1–19.3 (based on information given in Tables 18.1–18.11).

19.1 Non saline waters

These include river and stream waters, ground and surface waters, rain, landfill waters, well waters and waste waters.

19.1.1 Methods based on adsorption on non polar macroreticular resins, organic and inorganic solids and solid polymers

These methods are discussed respectively in Chapters 3, 6 and 7.

In each method a large volume of the water sample is contacted with a small cartridge of the adsorbent. When adsorption of the organic or organometallic compounds is complete then to extract the compound the cartridge is either

(a) contacted with a small volume of an organic solvent which is subsequently analysed; or
(b) the cartridge is heated to release the compounds which are then swept directly into the detection instrument; or
(c) the cartridge is subjected to extraction with a supercritical fluid which is then analysed in the detection instrument as discussed in Chapter 14.

There are obvious advantages to using solventless extraction procedures such as those described under (b) and (c) above. Organic solvents frequently contain impurities which interfere in the analysis and in general have an adverse effect on the detection limits achievable.

Reference to Table 19.1 shows that the following compounds can be pre-concentrated to the extent that they can be determined in the 1–5µg L^{-1}

Table 19.1 Review of detection limits achieved by the preconcentration of various types of organic and organometallic compounds in non saline natural waters, also rain (R), surface and ground waters (S,G), potable waters (P), waste waters (W), landfill waters (L), well waters (Well) and soil (S)

Key: (1) Chelation-solvent extraction; (2) adsorption macroreticular non polar resins; (3) absorption organic and inorganic solids; (4) Adsorption on polymers; (5) Adsorption on polyurethane foam; (6) adsorption on octadecyl resin; (7) adsorption on carbon; (8) absorption on cation exchange resins; (9) adsorption on anion exchange resins; (10) solvent extraction; (11) supercritical fluid extraction; (12) static headspace analysis; (13) dynamic headspace analysis; (14) on line adsorption on solid phase; (15) cryogenic methods; (16) electroanalytical method; (17) semipermeable membrane

Type of compound	(1)	(2)	(3)	(4)	(5)	(6)	(7)	(8)	(9)
Aliphatic hydrocarbons		<1µg L⁻¹ min 3µg L⁻¹ max			<50µg L⁻¹		1µg L⁻¹		
Aromatic hydrocarbons		<1µg L⁻¹ min 3µg L⁻¹ max							
Polyaromatic hydrocarbons		<1µg L⁻¹ min 3µg L⁻¹ max <10µg L⁻¹		1µg L⁻¹		1ng L⁻¹ min 4ng L⁻¹ max			
Alcohols		<1µg L⁻¹ min 3µg L⁻¹ max							
Ketones		<1µg L⁻¹ min 3µg L⁻¹ max							
Carboxylic acids									
Aldehydes									
Esters		<1µg L⁻¹ min 3µg L⁻¹ max 3µg L⁻¹							
Ethers									
Dioxans									
Phthalates									
Non ionic surfactants		<10µg L⁻¹							
Trifluoroacetic acid									
Phenols		3µg L⁻¹ min 200µg L⁻¹ max							
Haloforms		<1µg L⁻¹ min 3µg L⁻¹ max							

Table 19.1 continued

Key: (1) Chelation-solvent extraction; (2) adsorption macroreticular non polar resins; (3) absorption organic and inorganic solids; (4) Adsorption on polymers; (5) Adsorption on polyurethane foam; (6) adsorption on octadecyl resin; (7) adsorption on carbon; (8) absorbtion on cation exchange resins; (9) adsorption on anion exchange resins; (10) solvent extraction; (11) supercritical fluid extraction; (12) static headspace analysis; (13) dynamic headspace analysis; (14) on line adsorption on solid phase; (15) cryogenic methods; (16) electroanalytical method; (17) semipermeable membrane

Type of compound	(10)	(11)	(12)	(13)	(14)	(15)	(16)	(17)
Aliphatic hydrocarbons			$1ng\ L^{-1}$ min $<10\mu g\ L^{-1}$ max	$ng\ L^{-1}$ min $10\mu g\ L^{-1}$ max $0.5ng\ L^{-1}$(s) $5\mu g\ L^{-1}$ max(s)				
Aromatic hydrocarbons	sub $\mu g\ L^{-1}$ min $1\mu g\ L^{-1}$ max		sub $\mu g\ L^{-1}$ min $1\mu g\ L^{-1}$ max	$1\mu g\ L^{-1}$ min $5\mu g\ L^{-1}$ max $1\mu g\ L^{-1}$(W)				
Polyaromatic hydrocarbons	$1\mu g\ L^{-1}$ min $100\mu g\ L^{-1}$ max							
Alcohols			$8\mu g\ L^{-1}$	$1\mu g\ L^{-1}$ (4L sample)				
Ketones			$8\mu g\ L^{-1}$	$1\mu g\ L^{-1}$ (4L sample)				
Carboxylic acids				0.02M				
Aldehydes								
Esters								
Ethers								
Dioxans				$\mu g\ L^{-1}$				
Phthalates								
Non ionic surfactants								
Trifluoroacetic acid			$10ng\ L^{-1}$(R)					
Phenols								
Haloforms			$1ng\ L^{-1}$	$0.5\mu g\ L^{-1}$				

Table 19.1 continued

Key: (1) Chelation-solvent extraction; (2) adsorption macroreticular non polar resins; (3) absorbtion organic and inorganic solids; (4) Adsorption on polymers; (5) Adsorption on polyurethane foam; (6) adsorption on octadecyl resin; (7) adsorption on carbon; (8) absorbtion on cation exchange resins; (9) adsorption on anion exchange resins; (10) solvent extraction; (11) supercritical fluid extraction; (12) static headspace analysis; (13) dynamic headspace analysis; (14) on line adsorption on solid phase; (15) cryogenic methods; (16) electroanalytical method; (17) semipermeable membrane

Type of compound	(1)	(2)	(3)	(4)	(5)	(6)	(7)	(8)	(9)
Chloroaliphatics		<1µg L⁻¹ min 1–2µmole L⁻¹ max							
Chloroaromatics		<1µg L⁻¹ min 10µg L⁻¹ max							
Chlorophenols		3µg L⁻¹	0.001ng L⁻¹ min 5ng L⁻¹ max				sub µg L⁻¹		
Nitriloacetic acid									
Aliphatic amines		0.01µg L⁻¹				0.14µg L⁻¹ min 1.8µg L⁻¹ max	0.02µg L⁻¹ min		0.7µg L⁻¹ (S/G)
Organic bases							0.5µg L⁻¹ max	50µg L⁻¹ min 1mg L⁻¹ max	
Aromatic nitrocompounds		1µg L⁻¹	1µg L⁻¹						
Nitrosamines							ng L⁻¹		
Nitroethane									
Azarenes									
Polychlorobiphenyls		0.4ng L⁻¹				0.01µg L⁻¹ min 10µg L⁻¹ max			
Polychlorodibenzo-p-dioxins, Polychlorodibenzofurans									
Dimethyl sulphide									
Organosulphur compounds									
Chloroinsecticides		0.2µg L⁻¹ 0.1ng abs. min 2µg L⁻¹ max	ng L⁻¹						
Organophosphorus insecticides						10µg L⁻¹			

Table 19.1 continued

Key: (1) Chelation-solvent extraction; (2) adsorption macroreticular non polar resins; (3) absorbtion organic and inorganic solids; (4) Adsorption on polymers; (5) Adsorption on polyurethane foam; (6) adsorption on octadecyl resin; (7) adsorption on carbon; (8) absorbtion on cation exchange resins; (9) adsorption on anion exchange resins; (10) solvent extraction; (11) supercritical fluid extraction; (12) static headspace analysis; (13) dynamic headspace analysis; (14) on line adsorption on solid phase; (15) cryogenic methods; (16) electroanalytical method; (17) semipermeable membrane

Type of compound	(10)	(11)	(12)	(13)	(14)	(15)	(16)	(17)
Chloroaliphatics	10ng L⁻¹ min 1µg L⁻¹ max 0.1gL⁻¹		0.1ng L⁻¹min 1mg L⁻¹ max <0.1µg L⁻¹ min µg L⁻¹ max (S/G) 1µg L⁻¹ 500µg L⁻¹(W)	<0.5µg L⁻¹ min 50µg L⁻¹ max		1µg L⁻¹		
Chloroaromatics			1µg L⁻¹(W)	1µg L⁻¹(W)				
Chlorophenols								
Nitriloacetic acid								
Aliphatic amines	15µg L⁻¹							
Organic bases								
Aromatic nitrocompounds								
Nitrosamines								
Nitroethane								
Azarenes								
Polychlorobiphenyls	<0.1ng L⁻¹	10µg L⁻¹						
Polychlorodibenzo-p-dioxins, polychlorodibenzofurans								
Dimethyl sulphide				0.8ng L⁻¹ ng L⁻¹				
Organosulphur compounds								
Chloroinsecticides	ng L⁻¹							
Organophosphorus insecticides	0.04µg L⁻¹ min 0.09µg L⁻¹ max							

Table 19.1 continued

Key: (1) Chelation-solvent extraction; (2) adsorption macroreticular non polar resins; (3) absorbtion organic and inorganic solids; (4) Adsorption on polymers; (5) Adsorption on polyurethane foam; (6) adsorption on octadecyl resin; (7) adsorption on carbon; (8) absorbtion on cation exchange resins; (9) adsorption on anion exchange resins; (10) solvent extraction; (11) supercritical fluid extraction; (12) static headspace analysis; (13) dynamic headspace analysis; (14) on line adsorption on solid phase; (15) cryogenic methods; (16) electroanalytical method; (17) semipermeable membrane

Type of compound	(1)	(2)	(3)	(4)	(5)	(6)	(7)	(8)	(9)
Carbamate insecticides/herbicides									
Azine insecticides/herbicides						$0.01\mu g\ L^{-1}$ min $0.07\mu g$ absolute min $15\mu g$ absolute max			
Substituted urea insecticides/herbicides			$0.6\mu g\ L^{-1}$ min $2ng\ L^{-1}$ max						
Fenitrothion		$0.1ng$ absolute min $0.5ng$ absolute max						$0.5ng$ absolute	
Tamephos									
Acidic herbicides									
Polar pesticides									
Fluridone						$1\mu g\ L^{-1}$ min $2\mu g\ L^{-1}$ max			
Misc. insecticides/herbicides							$0.003\mu g\ L^{-1}$ min $0.07\mu g\ L^{-1}$ max		
Hydroxytriazine						$100\mu g\ L^{-1}$			
Hydroxy or carbonyl cyclohexane derivatives									$50\mu g\ L^{-1}$ min $1mg\ L^{-1}$ max
Methylisoborneol						$10ng\ L^{-1}$			

Table 19.1 continued

Key: (1) Chelation-solvent extraction; (2) adsorption macroreticular non polar resins; (3) absorption organic and inorganic solids; (4) Adsorption on polymers; (5) Adsorption on polyurethane foam; (6) adsorption on octadecyl resin; (7) adsorption on carbon; (8) absorbtion on cation exchange resins; (9) adsorption on anion exchange resins; (10) solvent extraction; (11) supercritical fluid extraction; (12) static headspace analysis; (13) dynamic headspace analysis; (14) on line adsorption on solid phase; (15) cryogenic methods; (16) electroanalytical method; (17) semipermeable membrane

Type of compound	(10)	(11)	(12)	(13)	(14)	(15)	(16)	(17)
Carbamate insecticides/herbicides					$ng\ L^{-1}$ $0.05ng\ L^{-1}$			
Azine insecticides/herbicides								
Substituted urea insecticides/herbicides								
Fenitrothion								
Tamephos					$sub\ \mu g\ L^{-1}$			
Acidic herbicides					$sub\ \mu g\ L^{-1}$			
Polar pesticides					$1ng\ L^{-1}$ min $100ng\ L^{-1}$ max			
Fluridone					$1\mu g\ L^{-1}$			
Misc. insecticides/herbicides								
Hydroxytriazine								
Hydroxy or carbonyl cyclohexane derivatives								$1mg\ L^{-1}$ max
Methylisoborneol								

Table 19.1 continued

Key: (1) Chelation-solvent extraction; (2) adsorption macroreticular non polar resins; (3) absorption organic and inorganic solids; (4) Adsorption on polymers; (5) Adsorption on polyurethane foam; (6) adsorption on octadecyl resin; (7) adsorption on carbon; (8) absorbtion on cation exchange resins; (9) adsorption on anion exchange resins; (10) solvent extraction; (11) supercritical fluid extraction; (12) static headspace analysis; (13) dynamic headspace analysis; (14) on line adsorption on solid phase; (15) cryogenic methods; (16) electroanalytical method; (17) semipermeable membrane

Type of compound	(1)	(2)	(3)	(4)	(5)	(6)	(7)	(8)	(9)
Misc. organics								$mg\ L^{-1}$	
Organolead compounds	$0.1\mu g\ L^{-1}$	$<1pg\ L^{-1}$ $0.1\mu g\ L^{-1}$ max (S/G)	$ng\ L^{-1}$ min $\mu g\ L^{-1}$ max			$15\mu g\ L^{-1}$ (R)			
Organoarsenic compounds									
Organotin compounds									
Organomercury compounds			$0.04ng\ L^{-1}$ min $300\mu g\ L^{-1}$ max	$0.2ng\ L^{-1}$	$0.1ng\ L^{-1}$ min $1\mu g\ L^{-1}$ max				

Type of compound	(10)	(11)	(12)	(13)	(14)	(15)	(16)	(17)
Misc. organics						$0.005\mu g\ L^{-1}$		
Organolead compounds	$0.5\mu g\ L^{-1}$		$1\mu g\ L^{-1}$	$0.5\mu g\ L^{-1}$ $\mu g\ L^{-1}$(L) $<20\mu g\ L^{-1}$ (S/G)				
Organoarsenic compounds								
Organotin compounds								
Organomercury compounds	$0.01\mu g\ L^{-1}$			$50pg$ absolute min $1000pg$ absolute min	$0.05ng\ L^{-1}$			

Source: Own files

Table 19.2 Review of detection limits achieved in the preconcentration of various types of organic and organometallic compounds in sea water

	Adsorption on macroreticular non polar resin	Adsorption on inorganic and organic solids	Adsorption on cation exchange resins	Solvent extraction	Static headspace analysis	Dynamic headspace analysis	Chelation method	Electrolytic preconcentration
Aliphatic hydrocarbons				5μg absolute		ng L^{-1} min μg L^{-1} max		50μg L^{-1}
Alcohols					1μg L^{-1}			
Aldehydes					1μg L^{-1}			
Ketones					1μg L^{-1}			
Vinyl chloride						1μg L^{-1}		
Chloroaliphatics				1μg L^{-1}				
Polychlorobiphenyls	<13ng L^{-1} 0.0001μg absolute							
Azarenes	0.02ng min 3ng max absolute							
Polychlorodibenzo p-dioxin and poly-chlorodibenzofuran		1pg L^{-1}						
Chloroinsecticides	1ng L^{-1}min <0.1μg L^{-1} max			<3μg L^{-1}				
Misc. organics						μg L^{-1}		
Organomercury compounds							5ng L^{-1}	
Organoarsenic compounds			0.02μg L^{-1}					
Organotin compounds			<4ng L^{-1}					

Source: Own files

Table 19.3 Review of detection limits achieved in the preconcentration of various types of organic and organometallic compounds in potable waters

Type of compound	Adsorption macroreticular non polar resins	Adsorption on organic and inorganic solids	Adsorption on octacdecyl resin	Solvent extraction	Static headspace analysis	Dynamic headspace analysis
Aliphatic hydrocarbons				$ng\ L^{-1}$ min $\mu g\ L^{-1}$ max		
Polyaromatic hydrocarbons	$<0.05 ng\ L^{-1}$	$<5\mu g\ L^{-1}$				
Alcohols					low $\mu g\ L^{-1}$	
Ketones					$50\mu g\ L^{-1}$	
Aldehydes						$0.5\mu g\ L^{-1}$
Dioxans					$740\mu g\ L^{-1}$	
Phthalates				$ng\ L^{-1}$ min $\mu g\ L^{-1}$ max		
Haloforms						$0.1\mu g\ L^{-1}$ min $1\mu g\ L^{-1}$ max
Chlorophenols		$1ng\ L^{-1}$ min $1\mu g\ L^{-1}$ max				
Chloroaliphatic compounds					$5ng\ L^{-1}$ min $0.01\mu g\ L^{-1}$ max	$1\mu g\ L^{-1}$
Polychlorodibenzo-p-dioxins, polychlorodibenzofurans			$1pg\ L^{-1}$			
Chloroinsecticides				low $\mu g\ L^{-1}$		
Azine herbicides	$15\mu g\ L^{-1}$ min $32\mu g\ L^{-1}$ max					
Misc. organics Nitroethane				$\mu g\ L^{-1}$	low $\mu g\ L^{-1}$	low $\mu g\ L^{-1}$

Source: Own files

concentration range in the water sample, detection limits which are considerably lower than are achieved by non-preconcentration methods (it should be noted here that detection limits can only be quoted when they are reported in the original publication): aliphatic hydrocarbons; aromatic hydrocarbons; polyaromatic hydrocarbons; alcohols; ketones; carboxylic acids; esters; ethers; non ionic surfactants; phenols; haloforms; haloaliphatic compounds; haloaromatic compounds; chlorophenols; nitriloacetic acid; aromatic nitrocompounds and organolead compounds. Even greater sensitivity has been achieved (ie ng L^{-1} to pg L^{-1} range) in the case of the following compounds:

LD

- chlorophenols   0.001–5ng L^{-1}
- polychlorobiphenyls   0.4ng L^{-1}
- chloroinsecticides   ng L^{-1}
- organophosphorus insecticides   0.1ng absolute
- substituted urea herbicides   0.6ng absolute
- fenitrothion   0.1ng absolute
- misc. organic compounds   <1pg L^{-1}
- organomercury compounds   40pg L^{-1}

On line preconcentration methods as discussed in Chapter 12 in which a large volume of water sample is continuously recirculated through the solid phase adsorbent offers the means of even greater sensitivity as illustrated below.

LD

- carbamate insecticides   ng L^{-1}
- tamephos   sub µg L^{-1}
- fenitrothion   sub µg L^{-1}
- polar pesticides   1ng L^{-1}
- organomercury compounds   0.05ng L^{-1}

19.1.2 Methods based on adsorption on polyurethane foam

This technique (Chapter 8) has been used to determine aliphatic hydrocarbons at the 1–5µg L^{-1}, polyaromatic hydrocarbons at the ng L^{-1} level and organomercury compounds at the 0.1ng L^{-1} level.

19.1.3 Methods based on adsorption on octadecyl silica gel

This technique (Chapter 9) has been used to determine the following compounds at the 1–5µg L^{-1} level: aliphatic amines; chloroinsecticides; fluoridone; hydroxytriazines and organolead compounds.

Sub µg L^{-1} level analysis has been achieved in the case of polychlorobiphenyls (LD 10ng L^{-1}), methyl isoborneol (LD 10ng L^{-1}) and azine herbicides (LD 100ng L^{-1}).

19.1.4 Methods based on adsorption on active carbon

Aliphatic hydrocarbons and chlorophenols have been detected at the 1–5µg L^{-1} level, while nitrosamines and miscellaneous pesticides have been detected, respectively, at the 1ng L^{-1} and 3ng L^{-1} levels.

19.1.5 Adsorption on to cation and anion exchange resins

The following compounds have been detected at the 1–5µg L^{-1} level, using cation exchange resins (Chapter 4 and 5): phenols; organic bases; hydroxy or carbonyl cyclohexyl derivatives and miscellaneous organics, while aliphatic amines have been detected, at the 20ng L^{-1} level. Fenitrothion has been detected in amounts down to 0.5ng absolute. Work on preconcentration on to anion exchange resins is limited to the determination of nitriloacetic acid at the µg L^{-1} level.

19.1.6 Solvent extraction methods

This technique (Chapter 13) has achieved some low detection limits

	LD
• chloroaliphatic compounds	10µg L^{-1}
• polychlorobiphenyls	0.1ng L^{-1}
• chloroinsecticides	ng L^{-1}
• organophosphorus insecticides	40ng L^{-1}
• organomercury compounds	100ng L^{-1}.

19.1.7 Static headspace analysis

The following classes of compounds have been detected at the 1–5µg L^{-1} level by this technique (see Chapter 15): aromatic hydrocarbons; alcohols; ketones; chloroaromatic hydrocarbons and miscellaneous organics.

Aliphatic hydrocarbons, trifluoroacetic acid, haloforms and chloroaliphatic hydrocarbons have been detected, respectively, at the 1, 10, 1 and 0.1ng L^{-1} level.

19.1.8 Dynamic headspace analysis

This technique, discussed in Chapter 16, has been used to detect the following classes of compounds at the 1–5µg L^{-1} level: aromatic hydrocarbons; alcohols; ketones; carboxylic acids; dioxanes; haloforms; chloroaliphatic hydrocarbons; chlorophenols; miscellaneous organics and organolead compounds.

Several classes of compounds have been determined at the ng L $^{-1}$ level:

- aliphatic hydrocarbons   ng L $^{-1}$
- dimethyl sulphide   0.8ng L $^{-1}$
- organosulphur compounds   ng L $^{-1}$
- organomercury compounds   50pg absolute.

19.1.9 Chelation–solvent extraction

This technique, discussed in Chapter 2, has been shown to be capable of determining 10ng L $^{-1}$ of organolead compounds in non saline waters.

19.2 Seawater

19.2.1 Adsorption on non polar macroreticular resins, organic and inorganic solids and solid polymers

This technique is particularly suitable for achieving preconcentration of seawater, as exemplified by the following examples (see Table 19.2):

- polychlorobiphenyls   <13ng L $^{-1}$
- chloroinsecticides   1ng L $^{-1}$
- organoarsenic compounds   20ng L $^{-1}$
- organotin compounds   <4ng L $^{-1}$
- polychlorodibenzo-*p*-dioxins, polychlorodibenzofurans   1pg L $^{-1}$

19.2.2 Solvent extraction

This technique has achieved detection limits of 1µg L $^{-1}$ for aliphatic hydrocarbons, chloroaliphatic hydrocarbons and chloroinsecticides.

19.2.3 Static headspace analysis

Alcohols, aldehydes and ketones have been detected at the 1µg L $^{-1}$ level.

19.2.4 Dynamic headspace analysis

Vinyl chloride has been detected at the 1µg L $^{-1}$ level while aliphatic hydrocarbons have been detected at the 1ng L $^{-1}$ level.

19.2.5 Chelation–solvent extraction

The only report which discussed the preconcentration of organomercury compounds reports a detection limit of 5ng L^{-1}.

19.3 Potable waters

See Table 19.3.

19.3.1 Adsorption on non polar macroreticular resins, organic and inorganic solids and solid polymers

Azine herbicides have been detected at the 1µg L^{-1} level while polyaromatic hydrocarbons and chlorophenols have been detected, respectively, at the 0.05ng L^{-1} and 1ng L^{-1} levels.

19.3.2 Adsorption on octadecyl–silica gel

Polychlorodibenzo-*p*-dioxins and polychlorodibenzofurans have been detected at the 1pg L^{-1} level in seawater.

19.3.3 Solvent extraction

Very low detection limits (ng L^{-1}) have been reported for aliphatic hydro-carbons and phthalate esters.

19.3.4 Static headspace analysis

Alcohols, ketones, dioxanes and nitroethane have been detected at the 1–5µg L^{-1} level while chloroaliphatic compounds have been detected at the 5ng L^{-1} level.

19.3.5 Dynamic headspace analysis

Aldehydes, haloforms, chloroaliphatic hydrocarbons and miscellaneous organics have all been detected in amounts down to 1–5µg L^{-1}.

In conclusion it is pointed out that, frequently, as shown in Table 19.4, the best detection limits reported in the literature for particular types of organic compounds are quite similar regardless of the type of water sample being examined.

Table 19.4 Ranges of detection limits reported for non saline waters, seawater and potable water

	Non saline waters	Seawater	Potable water
Polyaromatic hydrocarbons	$1ng L^{-1}$ to $1\mu g L^{-1}$	$1ng L^{-1}$ to $1\mu g L^{-1}$	$1ng L^{-1}$
Aliphatic hydrocarbons	$1ng L^{-1}$	–	$0.005ng L^{-1}$
Alcohols	$1\mu g L^{-1}$	$1\mu g L^{-1}$	$1\mu g L^{-1}$
Ketones	$1\mu g L^{-1}$	$1\mu g L^{-1}$	$1\mu g L^{-1}$
Dioxanes	$1\mu g L^{-1}$	–	$1\mu g L^{-1}$
Polychlorobiphenyls	$0.1-10ng L^{-1}$	$<13ng L^{-1}$	–
Chloroaliphatic hydrocarbons	$0.1-10ng L^{-1}$	$1\mu g L^{-1}$	$5ng L^{-1}$
Haloforms	$1ng L^{-1}$ to $1\mu g L^{-1}$	–	$1\mu g L^{-1}$
Chloroinsecticides	$1ng L^{-1}$ to $1\mu g L^{-1}$	$1ng L^{-1}$ to $1\mu g L^{-1}$	–
Azine herbicides	$100ng L^{-1}$	–	$1\mu g L^{-1}$
Polychlorodibenzo-p-dioxin, polychloro-dibenzo furan	–	$1pg L^{-1}$	$1pg L^{-1}$
Organomercury	$0.1ng L^{-1}$ to $40pg L^{-1}$	$5ng L^{-1}$	–
Chlorophenols	$1pg L^{-1}$	–	$1ng L^{-1}$

Source: Own files

Chapter 20

Cations: Chelation–solvent extraction techniques

The low concentrations at which metals can occur in certain types of water samples, for example, potable waters, rain water, snow, ice and seawater, preclude their direct determination by even the most recent advanced methods of analysis. To overcome this problem and improve the effective detection limits of these techniques various methods have been devised for the preconcentration of samples prior to analysis.

One such method is complexation–solvent extraction wherein the metals in a large volume of water sample are reacted with an organic complexing agent dissolved in a small volume of an organic solvent. The solvent is then either analysed by direct aspiration into the atomic absorption spectrophotometer, or other suitable instrument or it is back-extracted with a small volume of aqueous acid which is then analysed. Either way, a preconcentration factor is achieved which is approximately equal to the ratio of the original volume of water sample taken to the volume of the final extract analysed. The detection limit of the preconcentration method relative to the original unmodified method will be improved by approximately this ratio.

The considerable difficulty of trace element analysis in a high salt matrix such as seawater, estuarine water or brine is clearly reflected in the literature. Such examples contain extremely high concentrations of the alkali metals, the alkaline earth metals, and the halogens, and contain extremely low levels of the transition metals. Preconcentration procedures based on chelation and solvent extraction assist here by separating the metals to be determined from the sample matrix, and, additionally, serve to lower detection limits to an acceptable level.

An ideal method for the preconcentration of trace metals fromnon saline waters should have the following characteristics: it should simultaneously allow isolation of the analyte from the matrix and yield an appropriate enrichment factor: it should be a simple process, requiring the introduction of few reagents in order to minimise contamination, hence producing a low sample blank and a correspondingly lower detection limit; and it should produce a final solution that is readily matrix matched with solutions used in the calibration method. Various organic chelating agents have been studied, these are discussed below.

20.1 Non saline waters

20.1.1 Dithiocarbamic acid derivatives

Atomic absorption spectrometry

Kinrade and Van Loon [1] as early as 1974 pointed out that although solvent extraction methods for concentrating traces of metal ions in water abound, very little work had been done up to that time on the optimisation of experimental conditions making it impossible to obtain good results on a routine basis. These workers carried out a systematic examination of the factors which have a bearing on the quality of results including pH dependence, variation in the ratio of the two complexing agents used (ammonium pyrrolidinedithiocarbamate and diethylammonium diethyl-dithiocarbamate dissolved in methyl isobutyl ketone), extraction time and reagent stability. They showed that eight metals, cadmium, cobalt, copper, iron, lead nickel, silver and zinc, could be simultaneously extracted by their finally evolved method with good sensitivity and precision (Tables 20.1 and 20.2).

Table 20.1 Precision measurements expressed as per cent coefficient of variation

Solution concentration ($\mu g\ L^{-1}$)	Ag	Cd	Co	Cu	Fe	Ni	Pb	Zn
250 (solutions shaken)	7.16	0.26	0.67	0.52	0.99	0.37	0.41	0.29
25 (solutions inverted)	1.86	0.92	5.23	4.27	6.46	5.51	4.26	2.25

Source: Reproduced with permission from Kinrade, J.D. and Van Loon, J.C.V. [1] American Chemical Society

Table 20.2 Sensitivity and range of linearity

	Ag	Cd	Co	Cu	Fe	Ni	Pb	Zn
Sensitivity* ($\mu g\ L^{-1}$)	0.6	0.8	1.5	0.8	1.3	1.3	2.5	0.6
Range of linearity ($\mu g\ L^{-1}$)	0–200	0–400	0–350	0–400	0–350	0–300	0–5000	0–200

*Sensitivity is defined as the concentration needed to produce a 1% absorption

Source: Reproduced with permission from Kinrade, J.D. and Van Loon, J.C.V. [1] American Chemical Society

Table 20.3 Optimised instrumental parameters for trace metals in the methyl isobutyl ketone phase using graphite furnace atomic absorption spectrometry

Element			Ash		Atom	
	Line (nm)	Slit (nm)	Temp. (°C)*	Time (s)	Temp. (°C)*	Time (s)
Ag	28.1	0.7	300	10	2400	8
Cd	228.8	0.7	by-pass		1600	7
Co	240.7	0.2	500	10	2600	10
Cr	357.9	0.7	500	10	2600	10
Cu	324.8	0.7	600	10	2500	7
Fe	248.3	0.2	600	10	2600	8
Mn	279.5	0.2	500	10	2600	7
Ni	232.0	0.2	500	10	2600	9
Pb	283.3	0.7	400	10	2200	7

*Drying temperatures: Cd, 100°C for 505; other metals, 100°C from 305

Source: Reproduced with permission from Subramanian, K.S. and Meranger, J.C. [2] Gordon AC Breach

Subramanian and Meranger [2] made a critical study of the solution conditions and other factors affecting the reliability of the ammonium pyrrolidinedithiocarbamate–methyl isobutyl ketone extraction system for the determination of silver, cadmium, cobalt chromium, copper, iron manganese, nickel and lead in potable water. Graphite furnace atomic absorption spectrometry was used for the finish. Experimental conditions are reviewed in Table 20.3. The following parameters were investigated in detail: pH of the aqueous phase prior to extraction; amount of ammonium pyrrolidinediethyldithiocarbamate added to the solution following pH adjustment; the length of time needed for complete extraction; and the time stability of the chelate in the organic phase. Except for silver and chromium which were quantitatively extracted only in a very narrow pH range (1.0–2.0 and 1.8–3.0, respectively) and cadmium and lead which were stable in the extracted methyl isobutyl ketone phase only for 2–3h, the solution conditions for quantitative extraction were not critical for the other metals. Simultaneous extraction of all the metals except cadmium and lead was also investigated. Good recoveries (100 ±10%) were obtained for a number of spiked raw treated and distributed potable water samples covering a wide range of hardness (Table 20.4). They concluded that the procedure is reliable and precise under proper solution conditions.

Bone and Hibbert [3] have described a method using ammonium pyrrolidinedithiocarbamate dissolved in 2,6–dimethyl-4-heptanone followed

Table 20.4 Per cent recovery of metals in spiked samples of raw, treated and distributed potable water using APDC–MIBK–GFAA

Concentration of spike (ng M⁻¹)	Recovery (%)								
	Ag	Cd	Co	Cr	Cu	Fe	Mn	Ni	Pb
0.2	–	96 ± 7ᵃ	–	–	–	–	–	–	–
0.4	100 ± 3	101 ± 4	–	–	–	–	94 ± 6	–	–
0.6	104 ± 2	98 ± 3	–	–	–	–	–	–	–
0.8	–	85 ± 4	–	–	–	–	103 ± 5	–	–
2.0	103 ± 3	–	–	92 ± 6	97 ± 4	–	93 ± 4	–	104 ± 5
4.0	100 ± 4	–	94 ± 5	96 ± 3	105 ± 2	98 ± 3	97 ± 3	–	–
6.0	98 ± 3	–	–	–	101 ± 3	–	–	–	95 ± 5
8.0	–	–	96 ± 3	97 ± 2	105 ± 2	–	98 ± 2	–	98 ± 6
10.0	–	–	–	97 ± 4	104 ± 4	96 ± 1	–	–	101 ± 5
20.0	–	–	97 ± 2	91 ± 6	95 ± 2	93 ± 2	–	101 ± 1	–
40.0	–	–	97 ± 4	–	–	102 ± 4	–	104 ± 1	–
50.0	–	–	–	–	–	98 ± 1	–	–	–

ᵃValues given represent the average of the triplicate analyses each of 20 raw, treated and distributed potable water samples ranging in hardness from 1 to 554mg CaCo1 L⁻¹. The values are more or less the same for single as well as simultaneous extractions. The measure of precision is the standard deviation

Source: Reproduced with permission from Subramanian, K.S. and Meranger, J.C. [2] Gordon AC Breach

by atomic absorption spectrometry for the determination of vanadium, chromium, iron, cobalt, nickel, copper, zinc, molybdenum, cadmium and lead in effluents and non saline waters.

This method can be recommended for samples containing less than 3mg of iron. The instability of the manganese chelates made measurement in the organic phase unsatisfactory. However, manganese could be successfully determined in the acid extract of the organic phase. Cerium(IV) sulphate is effective for the oxidation of chromium(III) to chromium(IV) and does not interfere with the extraction. Because of the existence of non-liable species of trace elements, it is advisable to acidify and oxidise samples prior to extraction and cerium(IV) also can be used for this purpose.

Particularly in the analysis of effluents, calibration by the method of standard additions is essential for accurate results. Precisions are between 1.4% (copper) and 6.9% (chromium). Linear ranges are between $0.40\mu g$ L^{-1} (chromium, iron, cobalt, nickel, copper, zinc) and $0-300\mu g$ L^{-1} (manganese). The working pH ranges are 3.4–4.6 (manganese), 3.4–5.1 (iron, vanadium, cobalt, nickel, copper, molybdenum and cadmium), 3.4–5.6 (zinc, lead), and 4.1–5.6 (chromium).

Tessier et al. [4] evaluated an ammonium pyrrolidinedithiocarbamate methyl butyl ketone preconcentration procedure for the determination of traces of cadmium, cobalt, copper, nickel, lead, zinc and molybdenum in river water samples. These workers reported that sample contamination, analyte ions and diverse matrix effects can all adversely affect the reliability of methods. They studied the effect of sample composition (ie its matrix) on trace metal recovery and re-examined the effect of sample pH on metal recovery. The influence of sample pH on the overall efficiency of the chelation–extraction procedure was determined at intervals of 0.5pH units over the pH range 1.5–9.

The results, expressed as relative absorbance show that in the same pH range 3–8 the extraction efficiency is virtually independent of pH for six of the seven metals studied; a slight decrease is observed at high pH values for copper, lead and zinc, whereas a marked reduction in extraction efficiency is noted at pH values less than 3 for all metals but manganese.

The simultaneous extraction of cadmium, cobalt, copper, nickel, lead and zinc is clearly feasible in the pH range 3 to 8. Tessier et al. [4] chose an intermediate sample pH of 4; this corresponds to an extraction pH of approximately 4.9, the shift in pH being due to the addition of ammonium pyrrolidinedithiocarbamate. In the case of molybdenum, however, the marked decrease in relative absorbance in the pH range 2–4 effectively precluded its analysis at a sample pH of 4.

Plots of absorbance as a function of trace metal concentration in the initial aqueous sample showed a linear relationship in the following concentration ranges: copper, nickel and cobalt, $0-100\mu g$ L^{-1}; lead $0-30\mu g$ L^{-1};

Table 20.5 Sensitivity and precision of the chelation–extraction procedure as determined on non saline water samples

Metal	Sensitivity[a] ($\mu g\ L^{-1}$)	Coefficient of variation[b]			
		Filtered sample (%)		Unfiltered sample (%)	
Cadmium	0.6	6.7	(3.4)	2.5	(5.8)
Cobalt	2.9	3.3	(5.6)	2.4	(5.2)
Copper	2.1	2.7	(15.3)	5.1	(39.6)
Nickel	2.2	1.8	(48.1)	2.3	(45.5)
Lead	4.4	10.3	(5.9)	3.2	(49.1)
Zinc	0.7	3.4	(58.3)	1.6	(103.0)

[a]Sensitivity is defined as the concentration needed to produce a 1% absorption.
[b]The coefficient of variation was obtained from 10 replicate analyses; the metal concentration ($\mu g\ L^{-1}$) is given in parentheses

Source: Reproduced with permission from Tessier, A. *et al.* [4] Gordon AC Breach

cadmium and zinc, 0–20$\mu g\ L^{-1}$. The sensitivity of the overall procedure, defined as the metal concentration needed in the original sample to obtain a 1% absorption after chelation and extraction, is given in Table 20.5. Comparison of these values with those obtained on the same instrument by direct aspiration of an aqueous sample shows a 17- to 36-fold increase in sensitivity. The detection limits, defined as the concentrations needed to obtain a signal equal to twice the baseline variation, are cadmium, 0.2$\mu g\ L^{-1}$; copper and zinc, 0.5$\mu g\ L^{-1}$; nickel, 1.5$\mu g\ L^{-1}$; cobalt, 2.0$\mu g\ L^{-1}$; lead 2.5$\mu g\ L^{-1}$.

The conclusions so far reached regarding the use of ammonium pyrrolidinedithiocarbamate must be measured up against the comments made by Smith *et al.* [5] who compared eight different preconcentration techniques for the determination of manganese, cobalt, zinc, europium, caesium and barium in non saline waters. These workers used X-ray fluorescence spectrometry as the means of analysing the concentrates. The preconcentration techniques used were: passage through columns of Dowex A1 chelating resin and silylated silica gel; filtration through laminate membrane filters and chelating diethylenetriamine cellulose filters; precipitation with sodium diethyldithiocarbamate and 1–(2-pyridylazo)-2-naphthol; extraction with ammonium pyrrolidinedithiocarbamate; and chelation by 8–quinolinol (oxine) followed by adsorption on activated carbon. From the results obtained it seems that chelation by oxine and adsorption on activated carbon gives the highest and most reproducible collection yields for all the waters and elements studied, followed by diethyldithiocarbamate and 1–(2-pyridylazonapthol) precipitation; the results of ammonium pyrrolidinedithiocarbamate extractions were least satisfactory. If one considers the average variation of the collection yields from the different water samples studied, then filtration through silylated silica gel and oxine chelation followed by activated carbon adsorption are least influenced by the water characteristics, while ammonium pyrrolidinedithiocarbamate extraction and diethyldithiocarbamate precipitation suffer most. On average, it appeared that zinc is most easily collected, followed by cobalt, europium and manganese.

The sodium diethyldithiocarbamate extraction method has been one of the most widely used techniques of preconcentration for trace metal analysis by atomic absorption spectrometry [1,6–15]. This extraction method can be generally classified into two major categories. The first one involves the extraction of metal dithiocarbamate complexes into oxygenated organic solvents such as methyl isobutyl ketone and then analysing the solvents directly [1,6–9]. The other one is to extract the metal complexes into oxygenated or chlorinated organic solvents such as chloroform, methyl isobutyl ketone, etc., followed by a nitric acid back-extraction, and then analysing the trace elements in an acid solution. The latter category has been the subject of a number of reports [10–14,172]. There are several drawbacks associated with the acid back-extraction of metal dithiocarbamates. The kinetics are generally slow and the efficiency of acid extraction is poor for certain metals such as cobalt, copper and iron [12].

The Analytical Quality Control Committee of the Water Research Centre, UK [15] has organized a comparative study, involving 11 participating laboratories in the UK, of the accuracy of determining cadmium, copper, lead, nickel, and zinc in river waters. Two laboratories used a preconcentration technique involving the use of sodium diethyldithiocarbamate dissolved in chloroform, eight laboratories used a concentration by evaporation, and one laboratory used direct aspiration without preconcentration. Comparison of precisions obtained by chelation–solvent extraction and evaporative methods (Table 20.6) suggests that in no case was the precision of the former method inferior. Also, spiking recoveries tend to be higher by the chelation method. In general, with the possible exception of zinc, bias obtained by the solvent extractions is not unexpectedly lower than that obtained by evaporation methods where negative biases predominate presumably due to mechanical losses of metal salts during the evaporation processes (Table 20.7).

Chakraborti et al. [16] determined traces of cadmium, cobalt, copper, iron, nickel, and lead at the concentrations found in non saline waters by extracting 500ml sample with a carbon tetrachloride solution of sodium diethyldithiocarbamate. The extract was evaporated to dryness and the residue mineralised in one drop of concentrated nitric acid prior to analysis by graphite furnace atomic absorption spectrometry. The

Table 20.6 Precision tests, WRC analytical quality control scheme

Lab No.	River water			Spiked river water			
	Conc. found range ($\mu g\ L^{-1}$)	St range ($\mu g\ L^{-1}$)	RSD range (%)	Conc. found range ($\mu g\ L^{-1}$)	St range ($\mu g\ L^{-1}$)	RSD range (%)	Spike recovery range (%)
Cadmium (target St 5% of concentration or 0.25μg L^{-1} whichever is greater)							
1,2,3,5,7,8,9 (evaporation)	−0.4–1.0	0.04–0.26	17–87	5.2–3000	0.27–3.10	1.0–5.4	93.1–101.0
4,6 (chloroform–diethyldithiocarbamate)	0.2–0.04	0.04–0.18	20–45	11.8–30.0	0.05–0.33	1.6–9.7	98.0–13.8
Copper (target St 5% of concentration or 2.5μg L^{-1} whichever greater)							
1,2,3,5,7,8,9	2.7–10.6	0.21–1.10	4.3–25.9	17.6–317.3	1.13–7.19	1.3–6.4	90.3–104.3
4,6	2.6–9.8	0.29–0.67	6.8–11.1	12.9–23.4	0.42–0.75	3.2–7.1	98.4–101.7
Lead (target as for copper)							
1,2,3,5,7,8,9	0.2–18.5	0.46–2.27	6.2–92.0	25.6–332.6	0.99–12.8	1.2–7.6	92.8–104.9
4,6	10.2–33.1	1.00 1.52	4–6–9.8	20.9–87.5	1.52–2.47	2.8–7.3	98.0–106.9
Nickel (target as for copper)							
1,2,3,5,7,8,9	0.4–12.8	0.33–1.00	3.1–115.0	25.5–306.0	0.07–6.66	1.9–3.7	91.6–101.5
4,6	2.5–9.7	0.52–0.64	6.6–20.8	14.4–36.4	0.58–1.13	3.1–4.0	96.4–119.6
Zinc (target as for copper)							
1,2,3,5,7,8,9	3.3–324.5	0.28–4.25	1.3–28.8	19.6–968.1	0.74–11.7	1.2–5.4	91.3–109.2
4,6	14.3–24.1	1.09–1.63	6.8–7.6	25.1–77.4	2.12–2.48	3.2–8.4	95.9–108.4

Source: Reproduced with permission from Royal Society of Chemistry [15]

Table 20.7 Bias tests WRC analytical quality control scheme

Lab. No.	River water 1		River water 2	
	Determined value ($\mu g\ L^{-1}$)	Maximum possible bias (%)	Determined value ($\mu g\ L^{-1}$)	Maximum possible bias (%)
Cadmium (target 10% of determined or 0.5µg L^{-1} whichever greater)				
1,2,3,5,7,8,9 (evaporation)	2.19	−0.13 to 0.73	5.19	−5.13 to 11.82
4,6 (chloroform–diethyldithiocarbamate)		−0.12 to 0.36	5.19	4.7 to 10.29
Copper (target 10% of determined greater) concentration or 5µg L^{-1} whichever greater)				
1,2,3,5,7,8,9	24.74	−3.00 to 3.45	54.33	−17.88 to 10.54
4,6		−4.42 to 2.97		−12.63 to 5.07
Lead (target as for copper)				
1,2,3,5,7,8,9	24.22	−5.49 to 3.10	51.28	−19.74 to 13.06
4,6		−1.70 to 5.47		3.72 to 5.00
Nickel (target as for copper)				
1,2,3,5,7,8,9	22.76	−3.25 to 5.48	59.21	−20.18 to 20.63
4,6		−2.89 to 3.70		−6.66 to 23.29
Zinc (target as for copper)				
1,2,3,5,7,8,9	20.15	−3.07 to 4.20	50.31	−12.55 to 9.54
4,6		−4.15 to 6.57	50.31	−14.83 to 9.42

Source: Reproduced with permission from Royal Society of Chemistry [15]

detection limits were 10pg of cadmium, 150pg of chromium, 125pg of copper, 100pg of iron and 100pg of lead.

Comparative data for the dithiocarbamate preconcentration atomic absorption spectrometry analyses of metals from various types of non saline waters is presented in Table 20.8.

Inductively coupled plasma atomic emission spectrometry

Tao et al. [39] preconcentrated 100-fold traces of cadmium, cobalt, chromium, iron, manganese, molybdenum, nickel, lead, vanadium and zinc in river and seawater by extraction with ammonium pyrrolidinedithiocarbamate and hexamethyleneammonium hexamethylenedithiocarbamate dissolved in xylene, followed by inductively coupled plasma emission spectrometry of the solvent extract. The detection limits for the elements based on three times the standard deviation of the blank signals and a 100-fold concentration factor range from $0.017\mu g$ L^{-1} (cadmium) to $0.52\mu g$ L^{-1} (lead). The detection limits of this extraction method are sufficient for the determination of metals in non saline waters although higher sensitivity would be desirable for cadmium, cobalt, lead and chromium(VI).

The linearity of the calibration graph after 100-fold concentration was observed from the detection limit up to at least $30ng$ ml^{-1} ($15\mu g$) for most elements. For iron, manganese, and molybdenum the linearity extended to $100ng$ ml^{-1} ($50\mu g$). The ranges proved to be wide enough to encompass the concentrations found in most non saline waters. Deviations from linearity at high concentrations probably arise because the metal carbamates cannot be extracted completely into xylene. However, for only a 20-fold concentration the linearity, as expected, extends to about five times the values for 100-fold concentration. The relative standard deviations were 3–5% for all the elements except for cadmium, cobalt and lead, the lack of precision for which arises from the very low concentrations in non saline waters; but even for these elements the recommended method might be applicable to polluted waters.

Other workers [40–42] have employed dibenzylammonium dibenzylthiocarbamate dissolved in 2–ethyl hexyl acetate to preconcentrate cations prior to inductively coupled plasma atomic emission spectrometry, achieving detection limits down to $20\mu g$ L^{-1}.

Shan et al. [43] preconcentrated the elements cadmium, cobalt, copper, lead, molybdenum and nickel in non saline waters by chelation with ammonium pyrrolidinedithiocarbamate. The elements were determined by inductively coupled plasma atomic emission spectrometry.

Wada et al. [44] also used ammonium pyrrolidinedithiocarbamate to chelate cadmium, copper, nickel and lead from non saline waters prior to extraction with diisobutyl ketone. The metals were determined in the organic extract by inductively coupled plasma atomic emission spectrometry.

Table 20.8 Complexing agent–solvent extraction systems for the dithiocarbamate preconcentration–atomic absorption spectrometry analysis of metals

Organic complexing agent	Sample solvent	Type of water sample	Metal	Analytical finish	Detection limit $\mu g\ L^{-1}$	Ref.
Ammonium pyrrolidinedithiocarbamate	Methyl isobutyl ketone	Non saline	Pb, Cd	AAS	–	[17]
Ammonium pyrrolidinedithiocarbamate	Methyl isobutyl ketone	Brackish	Co, Ni, Cu, Zn, Cd, Pb	AAS	–	[18]
Ammonium pyrrolidinedithiocarbamate	Methyl isobutyl ketone	Ground water	As(III)	AAS	0.17	[19]
Ammonium pyrrolidinedithiocarbamate	Methyl isobutyl ketone	Potable	Cd	AAS	–	[20]
Ammonium pyrrolidinedithiocarbamate	–	Non saline	Cu, Ni, Fe, Co, Cd, Zn, Pb	AAS	–	[21]
Ammonium pyrrolidinedithiocarbamate	Methyl isobutyl ketone	Potable	Pb	AAS	–	[22]
Ammonium pyrrolidinedithiocarbamate	–	Potable	Ni	AAS	–	[23]
Ammonium pyrrolidinedithiocarbamate	2,6–Dimethyl-4-heptanone	Non saline effluents	Misc. metals	AAS	–	[3]
Sodium diethyldithiocarbamate	Methyl isobutyl ketone	Non saline	Pb, Cd	AAS	Pb 10 Cd 0.5	[24]
Sodium diethyldithiocarbamate	Methyl isobutyl ketone	Sea	Cd, Pb	AAS	–	[25]
Sodium diethyldithiocarbamate	Methyl isobutyl ketone	Non saline	Cu, Pb, Cd, Ag, Ni	AAS	–	[26]

Table 20.8 continued

Organic complexing agent	Sample solvent	Type of water sample	Metal	Analytical finish	Detection limit µg L^{-1}	Ref.
Sodium diethyldithiocarbamate	Isoamyl alcohol	River	Cd, Fe, Zn, Cu, Mn, Pb	AAS	–	[27]
Pyrrolidinedithiocarbamate	Methylisobutyl ketone	Non saline	Cu	AAS	1000–100,000	[28]
Ammonium tetramethylene–dithiocarbamate	4-methyl–pent-2-one	Non saline	Pb	AAS	–	[29]
Diethyldithiocarbamate	Methylisobutyl ketone	Non saline	Au	AAS	–	[30]
Ammonium pyrralidinecarbodithioate	Methylisobutyl ketone	Non saline	Cr(VI), Cr(III)	AAS	0.3 0.2	[31,32]
Diethyl dithiocarbamate	Chloroform	Non saline	Pb	AAS	5	[33]
Ammonium pyrrolidine dithiocarbamate	Chloroform	Non saline	Cr(VI) Cr(III)	AAS	–	[34]
Diethyldithiocarbamate	1–chlorotoluene	Non saline	Mn	AAS	–	[35]
Ammonium pyrrolidine dithiocarbamate	MeCl$_3$	Non saline	Cu, Ni, Pb, Cd	AAS	Cu 0.3, Cd 0.02, Pb 0.7, Ni 0.5	[36]
Ammonium pyrrolidine dithiocarbamate	Ethyl isobutyl ketone	Non saline	Cd, Cu, Pb	AAS and ICPAES	–	[37]
Tri-isooctyl phosphorothioate	–	Non saline	Ag	AAS	0.2	[38]

Source: Own files

Mijazaki *et al.* [45] simultaneously determined µg L $^{-1}$ levels of the above elements in river and seawaters by inductively coupled plasma atomic emission spectrometry after extraction as their ammonium pyrrolidine–dithiocarbamate complexes into disobutylketone. This complexing agent formed complexes with all the elements at pH 2.4. Relative standard deviations were less than 4% for all elements except cadmium and lead which had relative standard deviations of about 20% owing to the low concentrations determined. The method blank for all the elements except zinc was reduced to less than 0.005µg by using an acid wash of the equipment and ultra-pure acids and ammonia solution to neutralise the samples. Additional precautions are needed to reduce zinc contamination.

Spectrophotometry

Extraction of chelates by a chloroform solution of sodium diethyldithio-carbamate followed by a spectrophotometric finish has been used to determine total heavy metals [46] and zinc [47].

Neutron activation analysis

Lo *et al.* [48,49] have used chelation with lead diethyldithiocarbamate in chloroform to preconcentrate mercury from samples containing 1–1000µg L $^{-1}$ prior to neutron activation analysis. As well as considerably increasing the sensitivity of the analytical procedure this step eliminates interference from sodium and bromine in the water samples. Irradiation was carried out with a neutron flux of 2×10^{12} n cm $^{-2}$ sec $^{-1}$ for 30h. After cooling for 12h the 77.6kev197 Hg gamma peak was assayed with a 38cm $^{-3}$ Ge(li) detector connected to a 4096 channel pulse height analyser.

Other elements which have been preconcentrated by the solvent extraction of their diethyldithiocarbamates followed by neutron activation analysis include iridium and gallium in amounts down to 1ng L $^{-1}$ [50,51] and lead [52].

Mok *et al.* [53] showed that quantification and speciation of arsenic(III) and arsenic(V) in non saline waters can be achieved by extracting the former species with pyrrolidinecarbodithioate at pH1–1.5 into chloroform followed by nitric acid back-extraction for neutron activation analysis. Besides eliminating interferences from matrix species in non saline waters, the two-step extraction procedure also provides a large preconcentration factor for arsenic. Detection of 10–2µg L $^{-1}$ of arsenic can be achieved by using this extraction method and neutron activation analysis. Applications of this method to arsenic speciation studies in non saline water systems are discussed.

Mok and Wai [54] preconcentrated tri- and pentavalent antimony by a solvent extraction procedure. Samples were saturated with chloroform prior to extraction. Antimony(III) and arsenic(III) were extracted simultaneously from 100ml samples by adding citrate buffer and adjusting the pH between 3.5 and 5.5 with hydrochloric acid or ammonium hydroxide. EDTA, chloroform and ammonium pyrrolidine carbodithioate were then added before vigorously shaking, separating the organic layer and washing this layer several times. Antimony and arsenic ammonium pyrrolidine complexes in the organic phase were back-extracted into nitric acid solution for neutron activation analysis. Antimony(V) and arsenic(V) were extracted by the same procedure after first reducing with thiosulphate and iodide at pH1.0. Detection levels of 1.0ng L^{-1} were achieved.

X-ray fluorescence spectrometry

Preconcentration of selenium by ammonium pyrrolidinedithiocarbamate followed by energy dispersive X-ray fluorescence spectrometry enabled Marcie [55] to determine down to 10μg L^{-1} selenium in non saline water.

Coprecipitation with sodium diethyidithiocarbamate has been used to preconcentrate 0.02–0.1μg levels of iron, manganese, zinc, copper, cadmium, arsenic, lead and zinc prior to determination by X-ray fluorescence spectrometry [56].

X-ray spectrometry

A methyl isobutyl ketone solution of ammonium pyrrolidinedithiocarbamate has been used to preconcentrate various metals in amounts down to 250μg L^{-1} in non saline water prior to determination by X-ray energy spectrometry [57].

Gas chromatography

Rigin and Yurtaev [59] determined bis(trifluoro–ethyl)dithiocarbamate complexes of heavy metals in non saline water samples, after extractive gas chromatographic separation, by atomic fluorescence. To filtered samples (100ml) were added ammonium bis(trifluoroethyl)dithiocarbamate and ammonia solution. The mixture was refluxed at 60°C for 15min before cooling, acidifying (pH2–5), and extracting the complexes with 2ml carbon tetrachloride. The organic extract was dried and the heavy metal complexes separated by capillary gas chromatography with helium as carrier gas. The gas leaving the column was injected into the atomiser of a multi-element atomic fluorescence spectrometer. Metal concentrations were obtained from calibration curves. River water samples containing organic complexing agents (humic acids or surfactants) were pretreated with ozone. Detection limits were slightly inferior to those obtained using atomic absorption but were in good agreement with those obtained from NBS standard water samples.

Supercritical fluid chromatography

Lainz *et al.* [60] and Hseih and Liu [61] respectively used complexation with sodium big(trifluoroethyl) dithiocarbamate and bisthiocarbamate to determine arsenic, bismuth, cobalt, iron, mercury, nickel, antimony and zinc at $\mu g\ L^{-1}$ concentrations in non saline waters.

Anodic stripping voltammetry

Metzger and Braun [62] give details of a procedure for differentiating traces of different species of antimony in non saline waters at the $ng\ L^{-1}$ level. It involved anodic stripping voltammetry after extraction with ammonium pyrrolidinedithiocarbamate into methyl isobutyl ketone (for trivalent antimony) or extraction with *N*-benzoyl-*N*-phenylhydroxyl-amine into chloroform (for pentavalent antimony).

Adeljou and Brown [58] preconcentrated cadmium in non saline waters as its dithiocarbamate by extraction with Freon and subsequently back-extracting into an acidic aqueous medium for determination by anodic stripping voltammetry. The use of a preconcentration procedure considerably reduced the overall time required for the determination. Direct determination of cadmium concentrations of $0.1\mu g\ L^{-1}$ or more was possible using a calibration graph; for lower concentrations the use of the standard addition method was necessary. The minimal amount of cadmium that could be determined reliably was $0.025\mu g\ L^{-1}$.

20.1.2 Dithizone

Atomic absorption spectrometry

Ihnat *et al.* [63] determined copper, zinc, cadmium and lead in natural fresh waters by a preconcentration method based on preconcentration into an *n*-butyl acetate solution of dithizone, 8–hydroxyquinoline, and acetyl-acetone and compared the results obtained with those found by four other methods based on direct electrothermal atomisation atomic absorption spectrometry, heat evaporation followed by flame atomic absorption spectrophotometry, and differential pulse anodic stripping voltammetry.

In the chelation method 100ml aliquot of the sample was preconditioned with 20ml 5% ammonium tartrate and adjusted in the presence of *p*-nitrophenol indicator to pH6 with (1 + 1) ammonium hydr-oxide or tartaric acid crystals. The sample was saturated with 2ml *n*-butyl acetate then extracted for 1min with 8ml of extractant, which contained

Table 20.9 Detection limits of analytical methods

	Detection limit* (µg L^{-1})			
	Cu	Zn	Cd	Pb
Heat evaporation/flame atomic absorption spectrometry	0.8	0.3	0.1–0.4	1.6
Electrothermal atomisation/ atomic absorption spectrometry with Perkin–Elmer heated graphite atomiser	0.5	–	0.01	0.2
Differential pulse anodic stripping voltammetry with hanging mercury electrode and mercury film electrode	0.1 0.05	– –	0.05 0.001	0.05 0.005
Electrothermal atomisation/atomic absorption spectrometry with Varian Technitron carbon rod atomiser	0.1	0.05	0.005	0.05
Solvent extraction/flame atomic	0.5–1.6	0.3–1.1	0.5–1.1	2.1–5.1

*Detection limits usually defined as 2× or 3× standard deviation of replicate analyses of reagent blanks and low level samples; ranges reflect differences in technique (Evap/FAAS) or different runs (Solv. ext/FAAS)

Source: Reproduced with permission from Ihnat, M. *et al.* [63] Gordon AC Breach

0.4g dithizone, 6.0g 8–hydroxyquinoline, and 200ml acetylacetone in 1L *n*-butyl acetate. The aqueous phase was discarded after separation for 20min and the organic phase was brought to 10ml with extractant. Organic solutions were aspirated into an air/acetylene flame of an atomic absorption spectrometer and absorbances were read after 4s integration. Non-atomic absorption was corrected simultaneously by a hollow cathode hydrogen lamp. Concentrations of elements in samples were calculated from linear regression equations of absorbances against concentrations of standards.

Detection limits of the five analytical methods are presented in Table 20.9.

Other applications of dithizone to the preconcentrations of cations are summarised in Table 20.10.

20.1.3 Diantipyrylmethane

Emission spectography

Petrov *et al.* [71] assessed the suitability of 0.05M solution of dianti-pyrylmethane in chloroform or dichloromethane as an extractant for preconcentrating traces of 20 metals in mine waters to which ammonium

Table 20.10 Preconcentration of cations by extraction with solvent solutions of dithizone

Solvent	Type of water	Elements	Finish technique	Detection limit*	Ref.
Ethyl propionate	Non saline	Ag, Be, Cd, Co(II), Ni, Zn, Fe(II), Pb(II), Al	AAS	–	[64]
Carbon tetrachloride	Snow	Ag	GFAAS	0.5ng L $^{-1}$	[65]
Methyl isobutyl ketone	Non saline	Au, Ag	AAS	Ag 5pmol Au 2pmol	[66]
Chloroform	Non saline	Hg	AAS	–	[67]
Methyl isobutyl ketone	Non saline	Hg	AAS	0.4ng absolute	[68]
Methyl isobutyl ketone	Non saline	Ni	AAS	–	[69]
	Non saline	Pd	AAS	0.04ng absolute	[70]

*µg L $^{-1}$ unless otherwise stated

Source: Own files

thiocyanate had been added. This reagent does not complex with nickel, aluminium, iron or manganese but can be used for the preconcentration of copper, zinc, vanadium, tin, molybdenum, niobium, bismuth, tungsten, gadolinium, cobalt, cadmium, and antimony in the presence of 0.2–5.0mg L $^{-1}$ of iron.

For determination of the microcomponents, 200–500ml of mine water, acidified with nitric acid and filtered through a membrane filter with pore diameter of 0.45µm, containing not over 1mg of total iron, was boiled for 20min with 0.5g of ammonium persulphate. Then the sample was evaporated to a volume of 30–40ml, cooled and quantitatively transferred to a separatory funnel. Then 1.4ml of concentrated sulphuric acid, 5ml of 6.5M ammonium thiocyanate solution, and 0.5g of ascorbic acid were added for reduction of the iron and copper, 20ml of 0.05M diantipyryl-methane solution in chloroform was added and the trace elements were extracted for 15min. The extraction was repeated and the extracts were combined and placed in a crucible. The solvent was carefully removed under a lamp, the sample was incinerated with 2ml of sulphuric acid on a plate to remove sulphur trioxide and then in a muffle furnace at 500–600°C for 20min. After the calcination the residue was mixed with 80mg of spectrographic base (composition C:K$_2$SO$_4$: Ni = 2:1:0.002) and transferred to the depression of a carbon electrode. The analytical lines (nm) of the elements were: Co 341.2, Zn 328.2, Cu 327.4, V 318.5, Sn 317.5, Mo 317.0, Nb 316.3, Bi 306.7, Ga 287.4, W 294.7, Cd 228.8, Sb 231.5, the reference element was Ni, 324.3, 313.4, 298.1, 231.1nm.

20.1.4 Thenoyltrifluoroacetone

Atomic absorption spectrometry

Methyl isobutyl ketone solutions of thenoyltrifluoroacetone have been used [72] to preconcentrate, by factors of up to 20, traces of manganese as the Mn(II) thenoyltrifluoroacetone complex in non saline water samples prior to atomic absorption spectrometry.

Chromium, iron, hafnium, niobium, nickel, rhodium, tin, titanium and zirconium interfere strongly in this method even at low concentrations and most other metals (except strontium) give low results when present at high concentrations, but 10mg each of silver, arsenic(V), barium, beryllium, cadmium, caesium, germanium, mercury(II), iridium, lanthanum, molybdenum(VI), lead, palladium(IV), rubidium, selenium(IV), tellurium (VI), thallium(I), and vanadium(V), and 20mg each of potassium and sodium did not interfere. Anionic interferences were restricted to cyanide, fluoride, and thiocyanate; 10mg each of bromide, carbonate, perchlorate, iodide, nitrate, sulphate, and thiosulphate did not interfere, nor did borate (54mg), chloride (20mg), citrate (950mg), phosphate (30mg), and tartrate (730mg). However, addition of tartrate to the aqueous phase before the extraction did not prevent the interferences.

Billah et al. [73] have shown that complexation with theonyltrifluoro-acetone is an easy method for isolating sub-ppb concentrations of manganese in river and lake waters.

Cadmium, manganese, lead, copper and cobalt in non saline waters were preconcentrated by chelation with mono-thiothenoyltrifluoroacetone and trioctylphosphine oxide and extracted into cyclohexane in a method described by Ueda et al. [74]. The metals were determined by electro-thermal atomic absorption spectrometry.

Scintillation counting

Strontium–89 and strontium–90 have been preconcentrated by extraction at pH 10.5 with 2–thenolytrifluoroacetone trioctylphenyl oxide in cyclohexane [75]. Strontium activity in the extract was determined by liquid scintillation counting. Testemale and Leredde [76] preconcentrated strontium–90 using 2–thenoyltrifluoroacetone and tributylphosphate dissolved in carbon tetrachloride. The organic extract was back-extracted with a small volume of aqueous nitric acid prior to counting.

20.1.5 Other chelating agents

A limited amount of work has been carried out on the application of other types of chelating agents in the preconcentrations of cations. This is summarised in Table 20.11.

Table 20.11 Application of miscellaneous chelating agents to the preconcentration of cations in non saline waters

Chelating agent	Extraction solvent	Type of water	Elements	Finish	Detection limit*	Ref.
1-(2-pyridylazo)naphthol	Benzene and isobutyl methyl ketone	Non saline	Zn	AAS	–	[77]
1-(2-pyridylazo)naphthol	Benzene	Non saline	Zn	Spectrometric	–	[78,79]
1-(2-pyridylazo)naphthol	Chloroform	Non saline	Cr(III)	Spectrometric	5.0	[80]
2-mercaptobenzobenzthiazole	Butyl acetate	Non saline	Zn	AAS	0.02	[81]
Heptoxime	Methanol/toluene	Non saline	Ni	Differential pulse polarography	1	[82]
Benzoin–oxine	Chloroform	Non saline	Mo	Spectrophotometric	0.1	[83]
Benzoin–oxine	Methyl isobutyl ketone	Non saline	W	AAS	–	[84]
Tri-iso–octyl phosphorothioate	Methyl isobutyl ketone	Stream water	Ag	AAS	0.0002	[38]
Monoiso–octylmethyl phosphonate	None	Non saline	Yttrium	Spectrometry	–	[85]
Octyl α anilobenzyl phosphonate	Chloroform	Non saline	Zn, Cu	Spectrometry	–	[86]
Tributyl phosphate	Toluene	Non saline	^{233}U	Liquid scintillation counting	–	[87]
Tributyl phosphate	Sodium hydroxide back-extraction	Potable	^{99}Tc	Gas flow proportional counting	–	[88]
N-M-tolyl–z-methyloxybenzo-hydroxamic acid	Chloroform	Non saline	V(V)	Spectrophotometric	–	[89]
N-phenyl-2-naphtholhydroxamic acid	Chloroform	Non saline	U(VI)	Spectrophotometric	–	[90]
Nicotinohydroxamic acid–trioctylmethyl ammonium cation	Methyl isobutyl ketone	Non saline and waste	Mn	AAS	0.1	[91]

Table 20.11 continued

Chelating agent	Extraction solvent	Type of water	Elements	Finish	Detection limit*	Ref.
6-methyl-3-methyl-2-[4-N-methyl-anilophenylazo]benzthiazolium chloride	Benzene/tributyl phosphate	Non saline	Zn	Spectrophotometric	–	[92]
Nerolic (5-amino-2-anilino-benzene sulphonic acid)	Toluene	Potable	V(V)	Spectrophotometric	–	[93]
Sulpharazen (5-nitro-2-3-(4p-sulphophenylazophenyl)-1-trizeno) benzenearsonic acids	Toluene/amyl alcohol	Non saline	Zn, Pb	Spectrophotometric	Zn 5×10^{-3} Pb 0.1	[94]
1,3-diaminothiourea (3-thiocarbohydrazide)	Benzene and chloroform isopropyl ether and ethyl acetate, and amyl acetate and amyl alcohol	Non saline	Hg	–	–	[95]
Mesityl oxide	–	Non saline	V(V)	Spectrophotometric	–	[96]
5,7-dichloro-8-hydroxyquinoline	Butyl acetate	Non saline	V	AAS	10	[97]
5-nitro-o-phenylene diamine	Toluene	Non saline	Se	–	–	[98]
Phenylacetic acid	Chloroform	Non saline	U(VI), Cd	Titration, Anodic stripping voltammetry	–	[99] [58]
Hexahydroze pinium hexahydroazepine-1-carbothioate	Misc.	Non saline	Mn(II), Fe(II), Co, Ni, Zn, Pb(II), Cu(II)	AAS	Mn 6 Fe 40 Co 30 Ni 20 Zn 10 Pb 100 Cu 20	[100]

Table 20.11 continued

Chelating agent	Extraction solvent	Type of water	Elements	Finish	Detection limit*	Ref.
Benzylamine/pelargonic acid	Water/decane	Non saline	Cu, Zn, Fe, Cd, Pb	AAS	Fe 3 Cu 4 Zn 3 Pb 5 Cd 1	[101]
Tropolone	Toluene	Non saline	Sn	AAS	—	[102]
Sb(III) Iodide complex with N, N-diphenylbenzamidine	Chloroform	Industrial waste	Sb(III)	Spectrophoto-metrically	200	[103]
Trilaurylamine-N-oxide	None	Non saline	Hg	AAS	—	[104]
Ammonium thiocyanate	4(-5-noxyl) pyridine	Non saline	Cu	AAS	2	[105]
Napthoquinone thiosemicarbazone	Methyl isobutyl ketone	Non saline	Cu	AAS	—	[106]
Dithiocarbamate-methyllithium (conversion to PbMe₄)	Chloroform	Non saline	Pb	GFAAS	5	[107]
Iodide complex Acetyl acetonate Trifluoroacetyl acetonate	Toluene	Non saline	^{75}Se	AAS	<100	[108]
	Chloroform	Non saline	Be	AAS	—	[109]
	Cyclohexane	Non saline	Be	GLC	0.02	[110]
4-(4-diethylaminophenyl) 930–2,5 dichlorobenzene sulphonate and (1:1) benzo-18-crown-6 reagents	Benzene:2,5-dichloro-benzene	River	K	Spectrophoto-metric by flow injection analysis	—	[111]
Mixed chelates	Hexane	Non saline	Cu, Fe, Co, Cd, Pb, Zn	AAS	—	[112]
Misc. chelates	Misc. solvents	Non saline	Mo	–	—	[113]
Misc. chelates	Misc. solvents	Sea	Mo	AAS	—	[114–118]

Table 20.11 continued

Chelating agent	Extraction solvent	Type of water	Elements	Finish	Detection limit*	Ref.
Misc.	–	Sub-surface waters	Misc. metals	Spectrophotometric, AAS, neutron activation analysis	–	[119]
None	Methyl isobutyl ketone	Nuclear waste processing solutions	K	Spectrophotometric using 2-(5-bromo-2-pyridylazo)-5-diethylamino)phenol by flow injection analysis	–	[120]
8-quinolinol	–	Non saline	Cd, Pb, Zn	AAS	Cd 0.0066 Pb 0.023 Zn 0.019 (400ml sample)	[121]
Diphenyl carbohydrazide	Chloroform	Non saline	Cr(VI)	Spectrophotometric finish	(VI) 1	[122]
Capric acid 1,10 phenanthroline or pyridine	Heptane	Non saline	Zn	AAS	0.03	[123]
N-p-methoxyphenyl-2-furylacryl–hydroxamic acid and 5-(deithylamino)-2-(2-pyridylazo)phenol	–	Natural	Pd	Spectrophotometric finish	0.1	[124]

Table 20.11 continued

Chelating agent	Extraction solvent	Type of water	Elements	Finish	Detection limit*	Ref.
N-p-methoxyphenyl-2-furylacrylolyhydroxamic acid and 5-iodo-5-(dimethylamino)-2-(pyridazo)phenol	Chloroform	Non saline	Bi	Spectrophotometric finish	1	[125]
N-p-methoxyphenyl-2-furylacrylolyhydroxamine acid	Chloroform	Non saline	Nb	Spectrophotometric finish	0.1	[126]
5,5 dimethyl-1,2,3 cyclohexane trione and 1,2 dioxime 3-thiosemicarbazone	Amyl alcohol	Non saline	Fe	Spectrophotometric finish	15	[127]
Erythrosin B	–	Non saline	Cu	–	–	[128]
Polyamidoamine	–	Non saline	Misc. metals	–	–	[129]
1-phenyl-3-methyl-4-benzoyl-5-pyrazolone	–	Non saline	Sr	AAS	7–73	[130]
Heptoxime	Methanol toluene	Non saline	Ni	Differential pulse polarography	1	[82]
Crown ethers	–	Non saline	Y, Sr	–	–	[131]
Crown ethers	–	Non saline	Th, La	–	–	[132]

*μg L^{-1} unless otherwise stated

Source: Own files

20.1.6 Volatile metal chelates suitable for gas chromatography

Various workers [133–139] have preconcentrated selenium(IV) from non saline waters (500ml) by converting it to its 4,6-dibromopiazselanol or 1,2-diamino-3,5-dibromopiazselanol and extracting this with 1ml toluene prior to gas chromatography. Down to $0.002\mu g\ L^{-1}$ selenium(IV) and total selenium can be determined by this procedure.

Cobalt has been determined in amounts down to 4×10^{-11} by a method involving chelation with 6,6,7,7,8,8,8-heptafluoro-2,2-dimethyl-3,5-octanedione (H fod) followed by gas chromatography using an electron capture detector [140].

In this method known aliquots of the benzene solutions are injected into the chromatograph. Peak heights are obtained and compared with the calibration curve produced from the analyses of a standard cobalt 6,6,7,7,8,8,8-heptafluoro-2,2,-dimethyl-3,5-octanedionate solutions. When these particular conditions are used, the *cis* and *trans* isomers of cobalt 6,6,7,7,8,8,8-heptafluoro-2,2-dimethyl-3,5-octanedionate are both eluted as one peak, thereby simplifying the analysis.

Beryllium has been preconcentrated [141] at the $\mu g\ L^{-1}$ level by extraction of 200ml of sample with 1ml of a benzene solution of 1,1,1-trifluoro-2,4-pentadione followed by electron capture gas chromatography of the extract.

The copper(II) and nickel(II) complexes of bis(acetylpivalyl–methane) ethylenediimine (H_2[en (APM)$_2$]) show sufficient volatility and thermal stability to allow for their successful gas chromatography. Although somewhat less volatile than complexes such as 1,1,1-trifluoroacetyl-acetonate, the chelates of H_2–[en(APM)$_2$] possess a remarkable thermal stability and can be eluted undecomposed from a gas chromatograph at temperatures as high as 300°C. The relative ease with which the copper(II) and nickel chelates of H_2[en(APM)$_2$] were formed in aqueous solution and their facile extraction into a wide range of immiscible organic solvents indicated to Belcher *et al.* [142] the possible usefulness of this ligand in quantitative analysis for copper and nickel. Only with this ligand was the separation of the copper(II) and nickel chelates by gas chromatography consistently reproducible and effective enough for quantitative purposes.

To preconcentrate copper and nickel take the aqueous sample (100ml) (buffered to pH7.0) containing copper and nickel (0.02–$0.12mg\ ml^{-1}$) in vials, and add 1M sodium acetate solution (1ml) to each to give pH8.0. A solution of H_2[en(APM)$_2$] in *n*-hexane (1ml) was added and the mixture was heated on a boiling water bath for 15min, add cyclohexane (2ml), ie 50-fold preconcentration, to the cooled mixture and seal the vials. Extract the copper and nickel chelates by shaking the vials for 1h on a mechanical shaker. Subject portions (1–5μL) of the organic phase to flame ionisation

Table 20.12 Complexing agents and concentrations

Metal	Complexing agent	Concentration
Copper	Cupferron	1.0%
	Diethyldithiocarbamates	0.01M
	1(-2-pyridylazo)napthol	0.1%
Cadmium	Dithizone	0.01%
	1(-2-pyridylazo)napthol	0.1%
	8-Hydroxyquinoline (oxine)	0.1M
	8-Isopropyltropolone	0.1%
	1-Nitroso-2-napthol	0.1%
Antimony	Cupferron	1.0%
	Diethyldithiocarbamate	1.0%
	Ammonium pyrrolidinedithiocarbamate	0.2%
	Tri-n-octylamine	5.0%
Arsenic	Tri-n-octylamine	5.0%
	Trioctyl phosphine oxide	0.1M
	Ammonium pyrrolidinedithiocarbamates	0.2%
	Diethyldithiocarbamate	1.0%
Selenium	Diethyldithiocarbamate	1.0%
	Ammonium pyrrolidinedithiocarbamate	0.2%

Source: Reproduced with permission from Chambers, J.C. and McClellan, B.E. [143] American Chemical Society

gas chromatography with the column comprising 5% silicone gum rubber E–350 or Universal B support maintained at 260°C and a carrier gas flow rate of 10ml min $^{-1}$.

20.1.7 Comparison of chelating agents

Chambers and McClellan [143] evaluated several organic complexing agents and solvents for the extraction of copper, cadmium, antimony, arsenic, and selenium from water samples prior to their determination by atomic absorption flame emission spectrometry (Table 20.12). The object of this exercise was to select the best system for each element. Two millilitres of organic extractant per 200ml water were used throughout this study, ie preconcentration factor of 100.

It was necessary to optimise the atomic absorption instrumental variables for each metal solvent pair as these variables will influence the analytical sensitivity. The variables considered were fuel flow rate, oxidant gas flow rate, burner height and lamp current. In general, the optimum oxidant flow rate did not vary a great deal from solvent to solvent and the optimum range was fairly broad. The combustion characteristics of the solvent and the type of flame required for atomisation of the element

Table 20.13 Optimum instrumental conditions

Metal	Solvent	Acetylene flow rate (L min⁻¹)	Air flow rate (L min⁻¹)	Burner height (mm)	Lamp current (mA)
Copper	Isopropyl acetate	1.61	6.94	3.0	8.0
	Butyl acetate	2.32	6.39	1.0	8.0
	Butyraldehyde	1.94	6.39	4.0	8.0
	2-Hepotanone	1.94	5.23	1.0	8.0
	n-Butyl ether	2.68	5.82	1.5	8.0
	Toluene	11.61	6.39	1.0	8.0
Cadmium	Methyl isobutyl ketone	1.94	5.23	1.5	4.0
	n-butyl acetate	1.61	5.82	1.0	4.0
	Cyclohexanone	2.32	6.94	0.0	4.0
	Butyraldehyde	1.94	3.49	2.0	4.0
Antimony	Methyl isobutyl ketone	1.94	6.39	2.5	15.0
	n-butyl acetate	2.32	6.39	1.0	15.0
	2-Octanone	3.04	6.94	1.0	15.0
		Hydrogen flow rate	Nitrogen flow rate		
Arsenic	Water	30.0	10.8	5.5	18.0
Selenium	Water	31.2	10.8	4.0	12.0

Source: Reproduced with permission from Chambers, J.C. and McClellan, B.E. [143] American Chemical Society

determine the optimum fuel flow rate. The optimum range in most cases was narrow. The optimum lamp current for each metal does not vary with the solvent. Therefore, it is necessary to optimise this variable only once for each metal. Table 20.13 lists the optimum instrumental settings for each metal studied.

Several organic solvents were evaluated for enhancement effects in the determination of copper, cadmium, antimony, and selenium. Table 20.14 lists the solvents used and the enhancement values obtained for each of the metal ions. Those solvents giving the greatest sensitivity enhancement, which are insoluble in water, were then used in solvent extraction studies.

Good enhancement of copper sensitivity is obtained using esters, ketones, aromatic hydrocarbons, aldehydes and others. Isopropyl acetate, n-butyl acetate, 2-heptanone, toluene, butyraldehyde and n-butyl ether were considered the most suitable for copper extraction studies. Alcohols depress the absorbance readings for copper because of their high viscosity. Acetate esters, ketones and aldehydes were found to give the best enhancement of cadmium sensitivity. Alcohols depress cadmium sensitivity. Butyraldehyde, n-butyl acetate, methyl isobutyl ketone, and

Table 20.14 Enhancement values with various solvents

Solvent	Enhancement (A_0/A_{aq})			
	Cu	Cd	Sb	Se
n-Butyl acetate	1.3	1.32	1.58	0.21
Isopropyl acetate	2.00	1.44	–	0.29
Butyl butyrate	1.58	0.77	1.33	–
Ethyl acetoacetate	1.51	1.05	1.13	–
Methyl benzoate	1.09	0.44	0.68	–
Butyraldehyde	2.09	1.77	0.62	–
2,4-Pentanedione	1.35	1.16	1.52	–
2-Heptanone	1.77	1.35	1.05	0.26
Methyl isobutyl ketone	1.70	1.35	1.75	0.30
Methyl ethyl ketone	1.80	1.27	1.53	–
Cyclohexanone	1.29	1.05	1.36	–
2-Octanone	1.58	1.07	1.50	–
n-Butanol	1.25	0.81	1.18	–
n-Hexanol	0.45	0.65	0.87	–
n-Octyl alcohol	0.58	–	–	–
Cyclohexanol	0.29	–	–	–
p-Xylene	1.80	–	–	–
Toluene	1.83	–	–	–
Nitrobenzene	0.83	–	–	–
p-Dioxane	1.38	–	–	–
Propylene carbonate	1.12	0.72	1.01	–
n-Butyl ether	2.32	–	–	–
Amyl acetate	–	1.22	–	–
n-Pentanol	–	–	1.00	–

cyclohexanone were chosen for cadmium extraction studies. Enhancements for antimony are highest with acetate esters and ketones. Methyl isobutyl ketone, n-butyl acetate, and 2-octanone give high enhancements of antimony sensitivity and are insoluble in water. These solvents were used in the antimony extraction studies. A suitable flame could not be obtained with aromatic hydrocarbons and n-butyl ether solutions of antimony.

The atomic absorption determination of arsenic and selenium presents a problem due to their low resonance lines. The most sensitive line for selenium is at 1960.3Å while the most sensitive line for arsenic is at 1937.0Å. These wavelengths are in the low ultraviolet region where many gases, including oxygen, absorb over wide bands. Consequently, studies carried out in this region should be made with no air in the light path. Since nitrogen does not absorb at wavelengths above 1850Å,

Table 20.15 Sensitivity and extraction efficiency for copper

Solvent	Chelating or complexing agent	Maximum absorbance	Extraction (%)	Suggested pH range
Butyl acetate	Cupferron	0.478	100	3.50–8.60
	Diethyldithiocarbamate	0.465	100	2.50–5.00
	1-(-2-pyridylazo)napthol	0.450	98	3.20–5.50
n-Butyl ether	Cupferron	0.515	100	2.65–9.60
	Diethyldithiocarbamate	0.555	98	1.00–7.00
	1-(-2-pyridylazo)napthol	0.640	100	6.00–9.90
Butyraldehyde	Cupferron	0.500	98	3.80–5.20
	Diethyldithiocarbamate	0.442	90	1.00–4.10
	1-(-2-pyridylazo)napthol	0.500	98	3.00–4.20
Toluene	Cupferron	0.480	100	4.30–9.55
	Diethyldithiocarbamate	0.520	98	6.50–7.50
	1-(-2-pyridylazo)napthol	0.510	100	4.70–9.20
2-Heptanone	Cupferron	0.500	100	2.60–5.00
	Diethyldithiocarbamate	0.435	80	1.00–2.50
	1-(-2-pyridylazo)napthol	0.520	97	3.20–9.00
Isopropyl acetate	Cupferron	0.620	100	2.70–4.50
	Diethyldithiocarbamate	0.620	98	3.20–5.07
	1-(-2-pyridylazo)napthol	0.630	93	5.55–6.57

Source: Reproduced with permission from Chambers, J.C. and McClellan, B.E. [143] American Chemical Society

measurements may be more accurately made in the vacuum ultraviolet region and by sweeping all the air from the flame region with nitrogen. Acetylene and coal–gas fuels are impractical for arsenic and selenium determinations because the flame species absorb strongly in the low ultraviolet region. Absorbance readings for aqueous arsenic solutions were found to be 2.43 times larger in a nitrogen shielded air entrained hydrogen flame than in an air–acetylene flame. Selenium sensitivity is enhanced by a factor of 2.54 by using a nitrogen shielded hydrogen flame. Organic solvents give poor results with both arsenic and selenium. The absorbance readings are lower than for aqueous solutions and the noise level is considerably larger. Therefore, arsenic and selenium determination are best made by extracting into an organic solvent such as carbon tetrachloride, or chloroform, followed by back-extraction into an aqueous solution for preconcentration purposes.

The best extraction system for atomic absorption determination of copper is the n-butyl ether-1-(2-pyridylazo)naphthol system (Table 20.15). Quantitative extraction is obtained over a wide pH range and the sensitivity is exceptionally high. Fig. 20.1(a) shows the per cent extraction versus pH curve for this system.

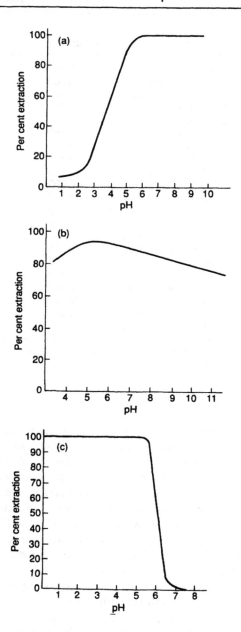

Fig. 20.1 Per cent extraction versus pH curve: (a) for copper using an *n*-butyl ether–PAN system; (b) for cadmium using an MIBK–dithizone system; (c) for selenium using an APDC–chloroform system
Source: Reproduced with permission from Chambers, J.C. and McClellan, B.E. [143] American Chemical Society

In cadmium determinations severe clogging of the aspiration system occurred with the *n*-butyl acetate–oxine extraction system. If the oxine concentration was lowered to the point that a good aspiration rate is maintained, less than 10% extraction of cadmium is obtained. The methyl isobutyl ketone–oxine extraction was carried out using a 1% oxine solution and only 8% extraction resulted. Highest absorbance readings were obtained with a methyl isobutyl ketone–dithizone extraction system. A per cent extraction/pH curve for this extraction system is shown in Fig. 20.1(b).

Ammonium pyrrolidinedithiocarbamate can be used to extract antimony from hydrochloric acid. The extraction is quantitative with a hydrochloric acid concentration of 0.001–1.00M. Methyl isobutyl ketone or *n*-butyl acetate may be used as solvents but methyl isobutyl ketone is more sensitive for atomic absorption determinations.

The atomic absorption sensitivity of arsenic and selenium is best in aqueous solutions. Enhancement of sensitivity is obtained by concentrating the elements by solvent extraction, followed by back-extraction into an aqueous system. Extractions of arsenic with diethyl-dithiocarbamate or ammonium pyrrolidinedithiocarbamate followed by back-extraction with copper gives quantitative recovery of the arsenic. Selenium is also extracted by ammonium pyrrolidinedithio-carbamate in chloroform. Quantitative extraction is obtained over a pH range of 1.5–5.30. Fig. 20.1(c) shows a per cent extraction versus pH curve for this system. Quantitative back-extraction is achieved using a 50mg L^{-1} cyanide ion solution.

The above solvent complexing agent systems were used to prepare a standard curve for each element in the low µg L^{-1} range in order to determine linearity and detection limit. Copper, cadmium and antimony standard curves were prepared by extracting 200ml of aqueous solution with 2ml of organic solvent and aspirating the organic phase. Arsenic and selenium standard curves were prepared by extracting 200ml of aqueous solution with 10ml of organic solvent. The organic phase was then back-extracted with 2ml of the appropriate aqueous solution and aqueous solution aspirated. In all cases, the aqueous and organic phases were pre-saturated with each other. Table 20.16 lists the detectability limit of each metal by this method. The detectability limit is defined as the lowest concentration which will give an absorbance reading twice the noise level. Good linearity and reproducibility was obtained in all cases.

20.2 Seawaters

Generally speaking the sensitivity requirements for the analysis of seawater are greater than those for non saline waters. Consequently, in addition to spectrophotometric and atomic absorption methods, a wide

Table 20.16 Detectability limits by solvent extraction–atomic absorption based on aqueous phase concentration

Metal	Extraction system	Absorbance	Metal ($\mu g\ L^{-1}$)
Copper	1-(-2-pyridylazo)napthol-*n*-butyl ether	0.010	0.01
Cadmium	Dithizone–methyl isobutyl ketone	0.008	0.10
Antimony	Ammonium pyrrolidinedithiocarbamate– methyl isobutyl ketone	0.020	10
Arsenic*	Diethyldithiocarbamate–chloroform	0.011	20
	Ammonium pyrrolidinedithiocarbamate– chloroform	0.010	20
Selenium*	Ammonium pyrrolidinedithiocarbamate– chloroform	0.005	2

*Back-extraction was performed and measurement was made in aqueous medium

Source: Reproduced with permission from Chambers, J.C. and McClellan, B.E. [143] American Chemical Society

variety of other sensitive techniques used in conjunction with preconcentration have been used for the determination of cations in seawater. These include inductively coupled plasma atomic emission spectrometry, X-ray fluorescence spectrometry, neutron activation analysis, anodic stripping voltammetry, and other techniques.

20.2.1 Dithiocarbamaic acid derivatives

Spectrophotometric method

Agrawal *et al.* [144] determined down to 1ppm of uranium in seawater by liquid–liquid extraction with N-phenyl-3-styrylacrylohydroxamic acid followed by a spectrophotometric finish.

Atomic absorption spectrometry

The dithiocarbamate extraction method has been one of the most widely used techniques of preconcentration for trace metal analysis by atomic absorption spectrometry [1,6–10,12,145]. This extraction method can be generally classified into two major categories. The first one comprises conversion of the metals to metal–dithiocarbamate chelates, then the extraction of the metal–dithiocarbamate complexes from a large volume of the aqueous phase into a smaller volume of oxygenated organic solvents such as methyl isobutyl ketone (thereby achieving concentration of metals), and then analysing the solvents directly [1,6–9]. The other one is to extract the metal complexes into oxygenated or chlorinated organic solvents such as chloroform, methyl isobutyl ketone, etc. followed by a nitric acid back-extraction, and then analysing the trace elements in the acid solution. The latter category has been the subject of a number of reports [10,12–14,145]. There are several drawbacks associated with the acid back-extraction of metal dithiocarbamates, the kinetics is generally slow and the efficiency of acid extraction is poor for certain metals such as cobalt, copper and iron [12]. Dithiocarbamate systems can simultaneously extract manganese as well as other trace metals under suitable conditions [2,146,147].

Statham [148] has optimised a procedure based on chelation with ammonium pyrrolidinedithiocarbamate and diethylammonium diethyl-dithiocarbamate for the preconcentration and separation of dissolved manganese from seawater prior to determination by graphite furnace atomic absorption spectrometry. Freon–TF was chosen as solvent because it appears to be much less toxic than other commonly used chlorinated solvents, it is virtually odourless, has a very low solubility in seawater, gives a rapid and complete phase separation and is readily purified. The concentrations of analyte in the back-extracts are determined by graphite furnace atomic absorption spectrometry. This procedure concentrates the trace metals in the seawater 67.3-fold. When a 350ml seawater sample was spiked with ^{54}Mn and taken through the chelation, extraction and back-extraction procedures, the observed recovery of the radio-tracer was 100.6%.

Burton [149] has also described an atomic absorption method for the determination of down to 0.3nmol L^{-1} manganese in seawater. Samples for the analysis of manganese were pressure filtered through 0.4µm nucleopore filters. To 350ml filtrate, 20ml of an aqueous solution of the complexing agents (2% w/v in both ammonium and diethylammonium diethyldithiocarbamate) were added, and the solution extracted first with 35ml and then with 20ml Freon for 6min. The combined extracts were shaken with 100µL of concentrated nitric acid for 30s. After standing for 5min, 5ml distilled water was added and the solution shaken for 30s. The aqueous phase was separated and combined with that from a further back-extraction using the same procedure. The combined aqueous solutions were returned to the shore laboratory and manganese determined by electrothermal atomic absorption spectrophotometry.

Cadmium, copper, and silver have been determined by an ammonium pyrrolidinedithiocarbamate chelation followed by a methyl isobutyl ketone extraction of the metal chelate from the aqueous phase [6,150] followed by graphite furnace atomic absorption spectrometry. The detection limits of this technique for 1% absorption were determined to be: Cu, 0.03µmol L^{-1}; Cd, 2nmol L^{-1}; Ag, 2nmol L^{-1}.

Moore [151] used the solvent extraction procedure of Danielson *et al.* [152] to determine iron in frozen seawater. To a 200ml aliquot of sample was added 1ml of a solution containing sodium diethyldithiocarbamate (1% w/v) and ammonium pyrrolidinedithiocarbamate (1% w/v) in 1% ammonia solution and 0.65ml 1M hydrochloric acid to bring the pH to 4. The solution was extracted three times with 5ml volumes of 1,1,2-trichloro-1,2,2-trifluorethane and the organic phase evaporated to dryness in a silica vial and treated with 0.1ml Ultrex hydrogen peroxide (30%) to initiate the decomposition of organic matter present. After an hour or more, 0.5ml 0.1M hydrochloric acid was added and the solution irradiated with a 1000W Hanovia medium pressure mercury vapour discharge tube at a distance of 4cm for 18min. The iron in the concentrate was then compared with standards in 0.1M hydrochloric acid using a Perkin–Elmer Model 403 Spectrophotometer fitted with a Perkin–Elmer graphite furnace (HGA 2200).

The coefficient of variation of analyses was 21% for seven sub-samples containing 1.6nmol Fe L $^{-1}$ and 30% for eight sub-samples at 0.6nmol Fe L $^{-1}$. The detection limit was estimated to be 0.2nmol Fe L $^{-1}$. The efficiency of the extraction procedure was tested using seawater spiked with iron–59, which indicated a recovery of 97% with stable iron of 86%.

Apte and Gunn [36] have described a method for the preconcentration and determination of copper, nickel, lead and cadmium in small samples of estuarine and coastal waters by liquid–liquid extraction and electro-thermal atomic absorption spectrometry. The metals, in 1.25ml samples, were chelated with ammonium pyrrolidinedithiocarbamate and extracted into 1,1,1–trichloroethane; all the chemical stages were carried out in sample cups of a graphite furnace atomic absorption spectrometer. Detection limits for copper, cadmium, lead, and nickel were 0.3, 0.2 and 0.5µg L $^{-1}$, respectively.

Filippelli [153] determined mercury at the sub-nanogram level in seawater using graphite furnace atomic absorption spectrometry. Mercury(II) was concentrated using the ammonium tetramethylenedithio-carbamate (ammonium pyrrolidinedithiocarbamate, APDC)–chloroform system, and the chloroform extract was introduced into the graphite tube. A linear calibration graph was obtained for 5–1500ng of mercury in 2.5ml chloroform extract. Because of the high stability of the HG(II)–APDC complexes, the extract may be evaporated to obtain a crystalline powder which can be dissolved with a few microlitres of chloroform.

About 84% of mercury was recovered in a single extract (97% in two extractions). The calibration graph was prepared by plotting the peak height against amount of mercury added to 500ml distilled water. The optimised experimental conditions are as follows: lamp current, 6 mA; wavelength, 253.63nm; drying 100°C for 10s; ashing, 200°C for 10s; atomisation, 2000°C for 3s; and purge gas, nitrogen 'stopped flow'. The coefficient of variation of this method was about 2.6% at the $1\mu g$ L^{-1} mercury level. The calibration graph is linear over the range $5-1500\mu g$ mercury.

Lo *et al.* [154] have developed a method of back-extracting metals from their solution as a metal chelate in an organic solvent. This procedure uses dilute mercury(II) solution instead of nitric acid. This back-extraction method is based on the fact that the extraction constant of the mercury(II) ammonium pyrrolidinedithiocarbamate complex is much greater than most of the common trace metals of environmental importance. The substitution of mercury(II) for other metals in the form of dithiocarbamate complex is extremely fast and the efficiency of recovery is nearly 100% for a number of metals including cobalt, copper and iron. In addition, the back-extracted solution contains a low concentration of mercury(II) which is virtually interference free in graphite furnace atomic absorption spectrometry due to its high volatility. This two-step preconcentration method preconcentrates a number of trace metals such as cadmium, cobalt, copper, iron, manganese, nickel, lead and zinc in seawater by graphite furnace spectrometry.

Sturgeon *et al.* [11] determined cadmium, zinc, lead, copper, iron, manganese, cobalt, chromium and nickel in coastal seawater by graphite furnace atomic absorption spectrometry after preconcentration by solvent extraction and use of a chelating ion exchange resin. Following the extraction of the pyrrolidine-N-carbodithioate and oxinate complexes into methyl isobutyl ketone, the trace metals are further preconcentrated by back-extraction into 1.5M nitric acid. Preconcentration on the cheating resin is effected by a combined column and batch technique, allowing greater preconcentration factors to be obtained. Provided samples are appropriately treated to release non-labile metal species prior to preconcentration, both methods yield comparable analytical results with respect to the mean concentrations determined as well as to mean relative standard deviations. Control and treatment of the analytical blank is also described.

Jan and Young [8] present a method for the analysis of trace metals in seawater at concentrations below $1\mu g$ L^{-1}. The method utilises an ammonium pyrrolidinedithiocarbamate methyl isobutyl ketone extraction procedure followed by a nitric acid back-extraction step (to stabilise the metal complexes), and analysis by flameless atomic absorption spectrophotometry. The detection limits for silver, cadmium, chromium, copper, iron, nickel, lead and zinc are 0.02, 0.003, 0.05, 0.05, 0.20, 0.10, 0.03 and $0.03\mu g$ L^{-1} respectively. For those metals occurring below $1\mu g$ L^{-1}, triplicate analyses of three different seawater samples yield mean relative standard deviations ranging from 18 to 25%.

Recoveries obtained in this procedure are listed in Table 20.17.

Some results obtained by this method using two variants of the method are listed in Table 20.18.

Table 20.17 Recovery of metals from spike seawater

Element	µg added[a]	µg found[b] Mean	SD[c]	Average recovery, %
Ag	0.037	0.007	0.0007	19
	0.185	0.024	0.0006	13
Cd	0.020	0.016	0.0025	80
	0.100	0.081	0.0085	81
Cr	0.025	0.020	0.0025	80
	0.125	0.117	0.0279	94
Cu	0.100	0.103	0.0219	103
	0.500	0.373	0.0528	75
Fe	0.100	0.069	0.0095	69
	0.500	0.381	0.0796	76
Ni	0.100	0.080	0.0081	80
	0.500	0.391	0.0099	78
Pb	0.050	0.035	0.0023	70
	0.250	0.163	0.0329	65
Zn	0.050	0.059	0.0064	118
	0.250	0.192	0.0439	77

[a]Filtered island control seawater (200ml) spiked with metal standard solution
[b]After correcting for concentration measured in unspiked sample
[c]SD = standard deviation, $n = 3$

Source: Reproduced with permission from Jan, T.K. and Young, D.R. [8] American Chemical Society

Sturgeon *et al.* [155] compared five different methods for the preconcentration of metals in coastal seawater utilising the instrumental technique. Results obtained using isotope dilution spark source mass spectrometry, graphite furnace atomic absorption spectrometry, and inductively coupled plasma atomic emission spectrometry following trace metal separation–preconcentration (using ion exchange and chelation–solvent extraction), or direct analysis by graphite furnace atomic absorption spectrometry were collated. Comparison of data between suitably different analytical methods is a practical way of testing the validity of those methods, gives increased confidence in the results obtained, and is especially valuable when standard reference materials are not available.

Table 20.19 shows the results obtained for a seawater sample. The mean concentrations and standard deviations of replicates (after rejection of outliers on the basis of a simple c test function) are given for each method of analysis. Each mean reflects the result of four or more separate determinations by the indicated method.

Table 20.18 Trace metals analysis of seawater using MIBK single extraction and MIBK–HNO$_3$ successive extraction methods

Method	Ag, µg L^{-1}	Cr, µg L^{-1}	Cu, µg L^{-1}	Fe, µg L^{-1}	Ni, µg L^{-1}	Pb, µg L^{-1}
MIBK–HNO$_3$	<0.02	0.14	0.78	3.28	0.59	0.29
extraction (a)	<0.02	0.17	0.73	3.57	0.59	0.25
	<0.02	0.16	0.82	3.51	0.59	0.28
	<0.02	0.12	0.82	3.10	0.59	0.31
mean	<0.02	0.15	0.79	3.37	0.59	0.28
SD^a	–	0.022	0.043	0.22	0.00	0.025
% RSD^b	–	7.3	5.4	6.5	0.0	8.9
MIBK single	<0.01	0.11	0.95	3.40	0.51	0.32
extraction (b)	<0.01	0.15	0.75	3.47	0.54	0.40
	<0.01	0.15	0.67	3.15	0.54	0.34
	<0.01	0.19	0.84	4.05	0.56	0.43
mean	<0.01	0.15	0.80	3.52	0.54	0.37
SD	–	0.033	0.120	0.380	0.021	0.051
%RSD	–	22	15	11	3.9	14

^{a}SD = standard deviation; ^{b}RSD = relative standard deviation

Source: Reproduced with permission from Jan, T.K. and Young, D.R. [8] American Chemical Society

Table 20.19 Analysis of seawater sample

	Concentration, µg L^{-1}				
	GFAAS			ICPES	IDSSMS
Element	Direct	Chelation–extraction^e	Ion exchange^f	Ion exchange^f	Ion exchange^f
Fe	3.7 ± 0.3^a	3.2 ± 0.2	3.4 ± 0.4	3.2 ± 0.2	3.3 ± 0.3
Mn	2.5 ± 0.2	1.9 ± 0.2	2.2 ± 0.3	2.3 ± 0.1	ND^b
Cd	0.05 ± 0.01	0.06 ± 0.01	0.053 ± 0.007	ND	0.07 ± 0.01
Zn	1.8 ± 0.3	1.8 ± 0.1	2.0 ± 0.1	1.6 ± 0.2	1.9 ± 0.1
Cu	ND	0.05 ± 0.1	0.51 ± 0.03	0.73 ± 0.06	0.61 ± 0.04
Ni	ND	0.46 ± 0.03	0.45 ± 0.05	0.38 ± 0.02	0.43 ± 0.03
Pb	ND	0.06 ± 0.02	0.10 ± 0.01	ND	0.11 ± 0.02
Cr	ND	0.29 ± 0.03	0.25 ± 0.02	ND	ND
Co	ND	0.015 ± 0.003	0.018^c ± 0.008	ND	0.028^d ± 0.001

^{a}Precision expressed as standard deviation; ^{b}Not determined; ^{c}Preconcentrated 100-fold; ^{d}Spark source mass spectrometry–internal standard method; ^{e}Using ammonium pyrrolidinedithiocarbamate methyl isobutyl ketone extraction in combination with back-extraction with nitric acid; ^{f}Using Chelex 100 cation exchange resin

Source: Reproduced with permission from Sturgeon, R.E. *et al.* [155] American Chemical Society

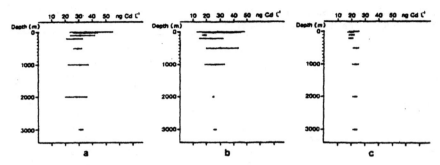

Fig. 20.2 Depth profiles for cadmium at four stations in the Norwegian Sea. The concentration range and mean value of four data are plotted for each depth. Below 1000m there are only two data at each depth. (a) ASV method, (G) MIBK extraction method, (c) Freon extraction method
Source: Reproduced with permission from Brugmann, L. *et al.* [156] Elsevier Science

Copper, nickel, lead, chromium and cobalt could not be measured by direct graphite furnace atomic absorption spectrometry because of their inherently low concentrations (below graphite furnace atomic absorption spectrometry detection limits) and/or pronounced physicochemical matrix interference effects.

Manganese could not be determined by isotope dilution spark source mass spectrometry because it is mono-isotopic. Furthermore, an internal standard method of calibration was not attempted for this element because the yield of manganese by the Chelex 100 ion exchange preconcentration technique was variable (strongly dependent on the pH of the buffer solution).

Overall there is good agreement between the elemental values in relation to the method of analysis. The precision of replicate determination between methods for all elements is comparable. Complete analysis of each sample by all of the methods indicated usually required about two months. The spread in the results may therefore reflect both the real spread inherent in the analytical methods as well as any minor changes that may have occurred in the sample composition during this time period.

Brugmann *et al.* [156] compared preconcentration methods using ammonium pyrrolidinedithiocarbamate/methyl isobutyl ketone and ammonium pyrrolidine thiocarbamate/Freon extraction followed by atomic absorption spectrometry or anodic stripping voltammetry in the analysis of North Sea and North East Atlantic waters.

The results illustrated in Figs. 20.2 and 20.3 give depth profiles obtained for two elements (cadmium and copper) by these various procedures.

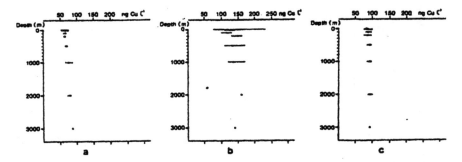

Fig. 20.3 Depth profiles for copper at four stations in the Norwegian Sea. The concentration range and mean value of four data are plotted for each depth. At 200m there are two data and at 300m one datum. (a) ASV method, (b) MIBK extraction method, (c) Freon extraction method
Source: Reproduced with permission from Brugmann, L. *et al.* [156] Elsevier Science

Other applications of chelation–solvent extraction followed by atomic absorption spectrometry to the determination of low levels of metals in seawater are reviewed in Table 20.20.

Detection limits (defined as 2 or 2.5 times the standard deviation of the blank) are in the ranges shown in Table 20.21 and as such are often suitable for the analysis of background levels in seawater.

Inductively coupled plasma atomic emission spectrometry

Sugimae [147] developed a method for lead, zinc, cadmium, nickel, manganese, iron, vanadium and copper in which they chelated with diethyldithiocarbamic acid and the chelates extracted with chloroform and the chelate decomposed prior to determination by inductively coupled plasma atomic absorption spectrometry. When 1L water samples are used, the lowest determinable concentrations are: Mn, 0.063g L $^{-1}$; Zn, 0.13µg L $^{-1}$; Cd, 0.25µg L $^{-1}$; Fe, 0.25µg L $^{-1}$; V, 0.38µg L $^{-1}$; Ni, 0.5µg L $^{-1}$; Cu, 0.5µg L $^{-1}$; Pb, 2.5µg L $^{-1}$. Above these levels, the relative standard deviations are better than 12% for the complete procedure.

Muyazaki *et al.* [173] found that di-isobutyl ketone is an excellent solvent for the extraction of the 2,4-pyrrolidinedithiocarbamate chelates of cadmium, lead, zinc, iron, copper, nickel, molybdenum and vanadium from seawater. Unlike halogenated solvents, it does not produce noxious substances in the inductively coupled plasma, has a very low aqueous solubility and gives 100-fold concentration in one step. Detection limits are 0.02µg L $^{-1}$ (cadmium) to 0.60µg L $^{-1}$ (lead). The results indicate that the proposed procedure should be useful for the precise determination of

Table 20.20 Preconcentration of metals in sea water chelation–solvent extraction techniques followed by direct atomic absorption spectrometry and graphite furnace atomic absorption spectrometry

Metals	Chelating agent	Solvent	Detection limit*	Ref.
Direct atomic absorption spectrometry				
Mn, Fe, Co, Ni, Zn, Pb, Cu	Hexahydroazepine-l-carbodithioate	Butylacetate	Mn 0.2 Fe 1.5 Co 0.6 Ni 0.6 Zn 0.4 Pb 2.6 Cu 0.5	[157]
Fe, Pb, Cd, Co, Ni, Cr, Mn, Zn, Cu	Diethyldithiocarbamate	MIBK or xylene		[158]
Fe, Cu	Ammonium pyrrolidinedithiocarbamate	MIBK	Cu <1 Fe <1	[159]
Cd, Cu, Pb, Ni, Zn	(a) Ammonium dipyrrolidinedithiocarbamate (b) Ammonium dipyrrolidinedithiocarbamate plus diethyldithiocarbamate	MIBK	Cu 10 Cd 2 Pb 4 Ni 16 Zn 30	[156]
Graphite furnace atomic absorption spectrometry				
Cu, Ni, Cd	Ammonium pyrrolidinedithiocarbamate	MIBK	Ag 0.02 Cd 0.03 Cr 0.05 Cu 0.05 Fe 0.20 Ni 0.10	[160]
Ag, Cd, Cr, Cu, Fe, Ni, Pb, Zn	Ammonium dipyrrolidinedithiocarbamate			[8]

Table 20.20 continued

Metals	Chelating agent	Solvent	Detection limit*	Ref.
Cu, Cd, Zn, Ni	Diethyldithiocarbamate plus ammonium pyrrolidinedithiocarbamate	Chloroform	Pb 0.03, Zn 0.05, Cu 1.0, Cd 0.2, Zn 2, Ni 10	[14]
Cd, Pb, Ni, Cu, Zn	Ammonium pyrrolidinedithiocarbamate plus diethyldithiocarbamate	Freon	Not stated	[161]
Cu	Ammonium pyrrolidinedithiocarbamate	MIBK	<0.5	[162]
Cd, Cu, Fe	Ammonium pyrrohidinedithiocarbamate plus diethyldithiocarbamate	Freon	Not stated	[152]
Cd, Zn, Pb, Cu, Fe, Mn, Co, Cr, Ni	Ammonium pyrrolidine-N-carbodithioate plus 8-hydroxyquinoline	MIBK	Fe 0.08, Cu 0.10, Pb 0.06, Cd 0.02, Zn 0.34	[11]
Cd	Ammonium pyrrolidinedithiocarbamate	Carbon tetrachloride	Cd 0.006	[163, 164]
Cd Zn, Pb, Fe, Mn, Cu, Ni, Co, Cr	Dithiocarbamate	MIBK	Not stated	[165]
Cd, Co, Cu, Fe, Mn, Ni, Pb, Zn	Ammonium pyrrolidinedithiocarbamate	Chloroform	Cd <0.0001, Cu <0.012, Fe <0.02, Mn <0.004, Ni <0.012, Pb <0.016, Zn <0.08	[154]

Table 20.20 continued

Metals	Chelating agent	Solvent	Detection limit*		Ref.
Cd, Co, Cu, Fe, Mn, Ni, Pb, Zn	Ammonium pyrrolidinedithiocarbamate	Chloroform	Cd Cu Fe Mn Ni Pb Zn	0.02 0.24 0.24 0.02 0.08 0.04 1.0	[154]
Mn, Cd	Ammonium pyrrolidinedithiocarbamate and diethylammonium diethyldithiocarbamate	Freon	Mn Cd	0.07 0.027	[148]
Cd, Cu, Fe, Pb, Ni, Zn	Ammonium pyrrolidinedithiocarbamate and diethylammonium diethyldithiocarbamate	Freon	Not quoted		[166]
Bi	Ammonium pyrrolidinedithiocarbamate	Xylene	0.003		[167]
Pb, Cd, Co, Cu Sn, As, Mo	Ammonium pyrrolidinedithiocarbamate	–	–		[168]
Cd, Co, Cu, Ni, Pb, Zn	Sodium bis (2-hydroxyethyl) dithiocarbomate—	–	–		[169]
Cu, Bi, Cd, Zn, Pb	Diethyl and dibutyl dithiophosphate	Carbon tetrachloride	Cu Bi Cd Zn Pb	0.6 0.5 0.8 0.8 0.5	[170]
Cd	Ammonium pyrrolidine dithiocarbamate	Organohalides	–		[171]

*µg L^{-1} unless otherwise stated

Source: Own files

Table 20.21 Detection limits for metals in seawater

	Lowest detection limit reported ($\mu g\ L^{-1}$)	Ref.	Highest detection limit reported ($\mu g\ L^{-1}$)	Ref.
Manganese	0.004	[154]	0.2	[157]
Iron	<0.02	[154]	1.5	[157]
Cobalt	0.4	[172]	0.6	[157]
Nickel	0.012	[154]	16	[156]
Lead	0.016	[154]	4	[156]
Copper	<0.012	[154]	10	[156]
Silver	0.02	[8]	0.05	[172]
Cadmium	0.0001	[154]	2	[156]
Zinc	0.03	[8]	30	[156]
Chromium	0.05	[8]		
Bismuth	0.003	[167]	0.5	[170]

Source: Own files

metals in oceanic water, although a higher sensitivity would be necessary for lead and cadmium. The relative standard deviations were 4% for all elements except cadmium and lead, which had relative standard deviations of about 20% owing to the low concentrations determined.

Bloekaert et al. [174] applied inductively coupled plasma atomic emission spectrometry with ammonium pyrrolidinedithiocarbamate preconcentration to the determination of cadmium, copper, iron, manganese and zinc in highly saline waste waters. The application of inductively coupled plasma atomic emission spectrometry to the analysis of brines containing up to 37.5mmol L^{-1} sodium chloride has been discussed [175,176]. Detection limits as low as 4μg L^{-1} have been claimed.

Cathodic and anodic scanning voltammetry

Van der Berg [177] determined zinc complexing capacity in seawater by cathodic stripping voltammetry of zinc–ammonium pyrrolidinedithiocarbamate complex by cathodic ions. The successful application of cathodic stripping voltammetry, preceded by adsorptive collection of complexes with ammonium pyrrolidinedithiocarbamate for the determination of zinc complexing capability in seawater is described. The reduction peak of zinc was depressed as a result of ligand competition by natural organic material in the sample. Sufficient time was allowed for equilibrium to occur between the natural organic matter and added ammonium, pyrrolidinedithiocarbamate. Investigations of electro-chemically reversible and irreversible complexes in seawater of several salinities are detailed, together with experimental measurements of ligand concentrations and conditional stability constants for complexing ligands. Results obtained were comparable with those obtained by other equilibrium techniques but the above method had a greater sensitivity.

Brugmann *et al.* [156] compared results obtained by anodic stripping voltammetry and atomic absorption spectrometry in the determination of cadmium, copper, lead, nickel, and zinc in seawater. The methods consisted of atomic absorption spectrometry but with preconcentration using either Freon or methyl isobutyl ketone. Anodic stripping voltammetry was used for cadmium copper and lead only. Inexplicable discrepancies were found in almost all cases. The exceptions were the cadmium results by the two methods and the lead results from the Freon with atomic absorption spectrometric methods and the anodic stripping voltametric methods.

Clem and Hodgson [178] discuss the temporal release of traces of cadmium and lead in bay water from EDTA, ammonium pyrrolidine-thyldithiocarbamate, humic acid and tannic acid after treatment of the sample with ozone. Anodic scanning voltammetry was used to determine these elements.

Neutron activation analysis

Yusov *et al.* [179] separated arsenic(III) and arsenic(VI) in seawater using a chloroform solution of ammonium pyrrolidinediethyldithiocarbamate. The separated fractions were then analysed by neutron activation analysis.

α-Activity

Shannon and Orden [180] determined polonium–210 and lead–210 in seawater. These two elements are extracted from seawater (at pH2) with a solution of ammonium pyrrolidinedithiocarbamate in isobutyl methyl ketone (20ml organic phase to 1.5L of sample). The two elements are back-extracted into hydrochloric acid and plated out of solution by the technique of Flynn [181], but with use of a PTFE holder in place of the Perspex one, and the α-activity deposited is measured. The solution from the plating-out process is stored for 2–4 months, then the plating–out and counting are repeated to measure the build-up of polonium–210 from lead–210 decay and hence to estimate the original lead–210 activity.

X-ray spectrometry

Tseng *et al.* [182] determined cobalt–60 in seawater by successive extractions with tris (pyrrolidinedithiocarbamate) bismuth(III) and ammonium pyrrolidinedithiocarbamate and back-extraction with bismuth(II). Filtered seawater adjusted to pH1.0–1.5 was extracted with chloroform and 0.01M tris(pyrrolidinedithiocarbamate) bismuth(III) to remove certain metallic contaminants. The aqueous residue was adjusted to pH4.5 and re-extracted with chloroform and 2% ammonium pyrrolidinedithiocarbamate, to remove cobalt. Back-extraction with bismuth(III) solution removed further trace elements. The organic phase was dried under infrared and counted in a germanium/lithium detector coupled to a 4096 channel pulse height analyser. Indicated recovery was 96%, and the analysis time excluding counting was 50min per sample.

High performance liquid chromatography

Boyle et al. [183] preconcentrated cobalt from 100ml samples of surface seawater using an ammonium pyrrolidinedithiocarbamate–carbon tetrachloride extraction system. Cobalt was determined in the extract by high performance liquid chromatography using a luminal post column chemiluminescence detection system. The detection limit of this method was 5pmol cobalt L $^{-1}$.

X-ray fluorescence spectrometry

Murata et al. [184] give details of equipment and a procedure for determination of traces of heavy metals by solvent extraction using di-isobutyl ketone and isobutyl methyl ketone, combined with microdroplet analysis by X-ray fluorescence spectrometry using a specially designed filter paper, sodium diethyldithiocarbamate is used as chelating agent. The limits of detection for manganese, iron, cobalt, nickel, copper, zinc and lead were 15, 16, 8, 8, 13, 13 and 40µg L $^{-1}$ respectively for a 100µL sample volume. Table 20.22 shows that the results are in fair agreement with the reference values determined by atomic absorption spectrometry.

Spectrophotometric method

Yang et al. [185] have described a spectrophotometric method for the determination of dissolved titanium in seawater after preconcentration using sodium diethyldithiocarbamate.

20.2.2 Dithizone

Atomic absorption spectrometry

Hirao et al. [186] concentrated lead in seawater using a chloroform solution of dithizone and determined it in amounts down to 40µg L $^{-1}$ by

Table 20.22 Results of analyses (µg L⁻¹) of liquid samples by X-ray fluorescence spectrometry, with reference values obtained by atomic absorption spectrometry

Ion analysed	Waste water						Seawater	
	Sample A		Sample B		Sample C			
	XRFᵃ	AAS	XRFᵇ	AAS	XRFᵇ	AAS	XRFᵇ	AAS
Mn	120	130	–	–	240	240		
Fe	170	200	130	150	100	90	60	60
Co	220	240	–	–	–	–	–	–
Ni	130	140	20	24	70	80	–	–
Cu	–	–	40	40	30	30	20	20
Zn	–	–	140	140	60	50	–	–
Pb	–	–	70	70	50	40	–	–

Sample: concentrated 10-fold times. ᵃDDTC–IBMK extraction; ᵇDDTC–DIBK extraction

Source: Reproduced with permission from Murata, M. et al. [184] John Wiley & Sons Ltd, New York

graphite furnace atomic absorption spectrometry. Lead in 1kg acidified seawater was equilibrated with lead–212 of a known radioactivity, extracted with dithizone in chloroform, back-extracted with 0.1M hydrochloric acid, and subjected to graphite furnace atomic absorption spectrometry by a two-channel spectrometer. Recovery yield of lead was found to be 60–90% from the radioactivity of lead–212 in the back-extract. Lead concentrations were thus determined with about 10% precision.

This technique has also been applied by other workers (Table 20.23).

20.2.3 8-Hydroxyquinoline

Atomic absorption spectrometry

Klinkhammer [188] and Landing and Bruland [189] have described methods for determining manganese in a seawater matrix for concentrations ranging from about 30 to 5500ng L⁻¹. The samples are extracted with 4nmol L⁻¹ 8-hydroxyquinoline in chloroform and the manganese in the organic phase is then back-extracted into 3M nitric acid [188]. The manganese concentrations are determined by graphite furnace atomic absorption spectrophotometry. The blank of the method is about 3.0ng L⁻¹ and the precision from duplicate analyses is ±9%. The theoretical yield of the method is less than 100% since only 80–90% of the aqueous phase is removed after the back-extraction. The actual yield obtained by ⁵⁴Mn counting was 69.5 ± 7.8% and this can be allowed for in

Table 20.23 Applications of dithizone extraction–atomic absorption spectrometry to the preconcentration of cations

Element	Complexing agent	Extraction solvent	Detection limit*		Ref.
Cd, Cu, Ni, Zn	Dithizone	Chloroform	Cu	0.006	[187
			Cd	0.0004	
			Ni	0.032	
			Zn	0.016	
Cd, Zn, Pb, Ca, Ni, Cu, Ag	Dithizone	Chloroform	Ag	0.05	[14]
			Cd	0.05	
			Zn	0.6	
			Pb	0.05	
			Cu	0.06	
			Ni	0.3	
			Ca	0.04	

*$\mu g\ L^{-1}$ unless otherwise stated

Source: Own files

the calculation of results. Environmental Protection Agency standard seawater samples of known manganese content ($4370 ng\ L^{-1}$) gave good manganese recoveries ($4260 ng\ L^{-1}$).

Atomic absorption spectrometry coupled with solvent extraction of iron complexes has been used to determine down to $0.5 \mu g\ L^{-1}$ iron in seawater [151,190]. Hiiro et al. [190] extracted iron as its 8-hydroxy-quinoline complex. The sample is buffered to pH3–6 and extracted with a 0.1% methyl isobutyl ketone solution of 8-hydroxyquinoline. The extract is aspirated into an air–acetylene flame and evaluated at 248.3nm.

Chau and Lum-Shue-Chan [115] investigated the use of atomic absorption in conjunction with solvent extraction using 1% 8-hydroxy-quinoline in methyl isobutyl ketone for preconcentration. The detection limit is $3 \mu g\ L^{-1}$, in which a preconcentration factor of 20 is employed. The disadvantages of the system are that there are interferences, although some of these can eliminated.

X-ray fluorescence spectrometry

Armitage and Zeitlin [191] converted uranium, copper, nickel, cobalt, iron and manganese to the 8-hydroxyquinolates and extracted these with chloroform. The extract was applied to a filter paper disc in a ring oven at 160°C and the metals separated prior to final determination by X-ray fluorescence spectrometry. Morris [192] separated microgram amounts of vanadium, chromium, manganese, iron, cobalt, nickel, copper and zinc from 800ml seawater by precipitation with ammonium tetramethylene-dithiocarbamate and extraction of the chelates at pH2.5 with methyl isobutyl ketone. Solvent was removed from the extract and the residue dissolved in 25% nitric acid and the inorganic residue dispersed in powdered cellulose. The mixture was pressed into a pellet for X-ray fluorescence measurements. The detection limit was 0.14µg or better, when a 10min counting period is used.

Electron spin resonance spectroscopy

Background copper levels in seawater have been measured by electron spin resonance techniques [193]. The copper was extracted from the seawater into a solution of 8-hydroxyquinoline in ethyl propionate (3ml extractant per 100ml seawater) and the organic phase (1ml) was introduced into the electron spin resonance tube for analysis. Signal-to-noise ratio was very good for the four-line spectrum of the sample and of the sample spiked with 4 and 8ng Cu^{2+}, the graph of signal intensity versus concentration of copper was rectilinear over the range 2–10µg L $^{-1}$ of seawater, and the coefficient of variation was 3%. Traces of copper and lead have been separated [194] from macro amounts of calcium, magnesium, sodium and potassium by adsorption from the sample on to active carbon modified with hydroxyquinoline, dithizone, or diethyldithiocarbamate.

20.2.4 Dimethylglyoxime

Spectrophotometric methods

The concentration of nickel in non saline waters is so low that one or two enrichment steps are necessary before instrumental analysis. The most common method is graphite furnace atomic absorption after preconcentration by solvent extraction [14] or co-precipitation [160]. Even though this technique has been used successfully for the nickel analyses of seawater [195,196], it is vulnerable to contamination as a consequence of the several manipulation steps and of the many reagents used during preconcentration.

This element has been determined spectrophotometrically in seawater in amounts down to 0.5µg L $^{-1}$ as the dimethylglyoxime complex [197, 198]. In one procedure [197] dimethylglyoxime is added to a 750ml sample and the pH adjusted to 9–10. The nickel complex is extracted into chloroform. After extraction into 1M hydrochloric acid, it is oxidised with aqueous bromine, adjusted to pH10.4 and dimethylglyoxime reagent added. It is made up to 50ml and the extinction of the nickel complex measured at 442nm. There is no serious interference from iron, cobalt, copper, or zinc but manganese may cause low results. In another procedure [198] the sample of seawater (0.5–3L) is filtered through a membrane filter (pore size 0.7µm) which is then wet ashed. The nickel is separated from the resulting solution by extraction as the dimethylglyoxime complex and is then determined by its catalysis of the reaction of tiron and diphenylcarbazone with hydrogen peroxide with spectrophotometric measurement at 413nm. Cobalt is first separated as the 2-nitroso-1-naphthol complex and is determined by its catalysis of the oxidation of alizarin by hydrogen peroxide at pH12.4. Sensitivities are 0.8µm L^{-1} (nickel) and 0.04µg L^{-1} (cobalt).

Atomic absorption spectrometry

Rampon and Cavalier [199] used atomic absorption spectrometry to determine down to 5µg L^{-1} nickel in seawater. Nickel is extracted into chloroform from seawater (500ml) at pH9–10, as its dimethylglyoxime complex. Several extractions and a final washing of the aqueous phase with carbon tetrachloride are required for 100% recovery. The combined organic phases are evaporated to dryness and the residue is dissolved in 5ml of acid for atomic absorption analysis.

20.2.5 4-(2-pyridylazo)resorcinol

Spectrophotometric method

Nishimura et al. [200] described a spectrophotometric method using 2-pyridylazoresorcinol for the determination of down to 0.025µg L^{-1} vanadium in seawater. The vanadium was determined as its complex with 4-(2-pyroiylazo)resorcinol formed in the presence of 1,2-diamino-cyclohexane–N, N, N', N'-tetraacetic acid. The complex was extracted into chloroform by coupling with zephiramine. Difficulties due to turbidity in the chloroform layer and incomplete masking of some cations by 2-pyridylazoresorcinol were overcome by addition of potassium cyanide and washing the chloroform layer with sodium chloride solution. The extinction of the chloroform layer was measured at 560nm against water as was that of a blank prepared with vanadium-free artificial seawater. Sixteen foreign ions were investigated and no interferences were found at 5–100 times their usual concentration in seawater.

Atomic absorption spectrometry

Monien and Stangel [201] studied the performance of a number of alternative chelating agents for vanadium and their effect on vanadium analysis by atomic absorption spectrometry with volatilisation in a graphite furnace. Two promising compounds were evaluated in detail, namely 4-(2-pyridylazo)resorcinol in conjunction with tetraphenyl-arsonium chloride and tetramethylenedithiocarbamate. These substances, dissolved in chloroform, were used for extraction of vanadium from seawater, and after concentrating the organic layer 5µL were injected into a pyrolytic graphite furnace coated with lanthanum carbide. For both reagents a linear concentration dependence was obtained between 0.5 and 7µg L^{-1} after extraction of a 100ml sample. Using the 2-pyridylazo-resorcinol–tetraphenylarsonium chloride system a concentration of 1µg L^{-1} could be determined with a relative standard deviation of 7%.

20.2.6 Nitrosophenols

Spectrophotometric methods

Various methods have been proposed for the determination of traces of cobalt in seawater and brines, most necessitating preconcentration. Solvent extraction followed by spectrophotometric measurements [191,202–208] is the most popular method. In many cases, excess of reagent and various metal complexes are co-extracted with cobalt and cause errors in determining the absorbance of the cobalt complex.

In one spectrophotometric procedure Motomizu [203] added to the sample (2L) 40% (w/v) sodium citrate dihydrate solution (10ml) and a 0.2% solution of 2-ethylamino-5-nitrosophenol in 0.01M hydrochloric acid (20ml). After 30min, add 10% aqueous EDTA (10ml) and 1,2-dichloroethane (20ml), mechanically shake the mixture for 10min, separate the organic phase and wash it successively with hydrochloric acid (1:2) (3 × 5ml), potassium hydroxide (5ml), and hydrochloric acid (1:2) (5ml); filter and measure the extinction at 462nm in a 50mm cell. Determine the reagent blank by adding EDTA solution before the citrate solution. The sample is either set aside for about 1 day before analysis (the organic extract should then be centrifuged) or preferably, it is passed through a 0.45µm membrane filter. The optimum pH range for samples is 5.5–7.5. From 0.07 to 0.16µg L^{-1} of cobalt was determined; there is no interference from species commonly present in seawater.

20.2.7 Pyrocatechol violet

Spectrophotometric method

Korenaga et al. [209] described an extraction procedure for the spectrophotometric determination of trace amounts of aluminium in seawater with pyrocatechol violet. The extraction of the ion-associate formed between the aluminium/pyrocatechol violet complex and the quaternary ammonium salt, zephiramine (tetradecyldimethylbenzyl-ammonium chloride), is carried out with 100ml seawater and 10ml chloroform. The excess of reagent extracted is removed by back-washing with 0.25M sodium bromide solution at pH9.5. The calibration graph at 590nm obeyed Beer's law over the range 0.13–1.34µg aluminium. The apparent molar absorptivity in chloroform was 9.8×10^4 L $^{-1}$ mol $^{-1}$ cm $^{-1}$.

Several ions – such as manganese, iron(II), iron(III), cobalt, nickel, copper, zinc, cadmium, lead, and uranyl – react with pyrocatechol violet and to some extent are extracted together with aluminium. The interferences from these ions and other metal ions generally present in seawater could be eliminated by extraction with diethyldithiocarbamate as masking agent. With this agent most of the metal ions except aluminium were extracted into chloroform and other metal ions did not react in the amounts commonly found in seawater.

20.2.8 Volatile metal chelates suitable for gas chromatography

Shimoishi [138] determined selenium by gas chromatography with electron capture detection. To 50–100ml seawater add 5ml concentrated hydrochloric acid and 2ml 1% 4-nitro-o-phenylenediamine and, after 2h, extract the product formed into 1ml of toluene. Wash the extract with 2ml 7.5M hydrochloric acid, then inject 5µL into a glass gas–liquid chromatography column (1 × 4mm) packed with 15% of SE-30 on Chromosorb W (60–80mesh) and operated at 200°C with nitrogen (53ml min $^{-1}$) as carrier gas. There is no interference from other substances present in seawater.

An example of a gas chromatographic method is that of Lee and Burrell [210]. In this method the aluminium is extracted by shaking a 30ml sample (previously subjected to ultraviolet radiation to destroy organic matter) with 0.1M trifluoroacetylacetone in toluene for 1h. Free reagent is removed from the separated toluene phase by washing it with 0.01M aqueous ammonia. The toluene phase is injected directly on to a glass column (15cm × 6mm) packed with 4.6% of DC710 and 0.2% of Carbowax 20M on Gas–Chrom Z. The column is operated at 118°C with nitrogen as carrier gas (285ml min $^{-1}$) and electron capture detection. Excellent results were obtained on 2µL of extract containing 6pg of aluminium.

Isotope dilution gas chromatography–mass spectrometry has also been used for the determination of µg L $^{-1}$ levels of total chromium in seawater [211–213]. The samples were reduced to ensure Cr(III) and then extracted and concentrated as tris(1, 1, 1-trifluoro-2,4-pentanediono)chromium(III) (Cr(tfa)$_3$) into hexane. The isotopic distribution of mass fragments were monitored into a selected ion monitoring (SIM) mode (Fig. 20.4).

Isotope dilution techniques are attractive because they do not require quantitative recovery of the analyte. One must, however, be able to monitor specific isotopes which is possible by using mass spectrometry. Table 20.24 shows results of two seawater sample analyses. Agreement with data obtained by isotope dilution spark source mass spectrometry

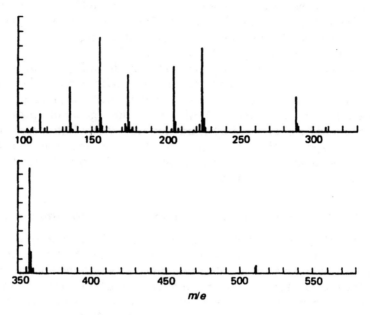

Fig. 20.4 Mass spectrum of Cr(tfa)₃
Source: Reproduced with permission from Siu, W.M. *et al.* [211] American Chemical Society

Table 20.24 Mean (±SD) chromium concentration in seawater (µg L⁻¹) (n≥3)

ID–GC/MS	ID–SSMS	GFAAS
0.177 ± 0.09	0.17 ± 0.03	0.19 ± 0.03
0.19 ± 0.01ᵃ	0.18 ± 0.01	ND

ᵃSeawater reference material NAAS–I. ND = not determined

Source: Reproduced with permission from Siu, W.M. *et al.* [211] American Chemical Society

[214] and graphite furnace [215] was excellent. Lee and Burrell [216] have used a toluene solution of trifluoroacetylacetone to extract cobalt, iron, indium and zinc from seawater.

Other complexing agents that have been used in solvent extraction–solvent extraction preconcentration of cations in seawater are reviewed in Table 20.25.

Table 20.25 Chelating agents used in solvent extraction and preconcentration methods

Element	Complexing agent	Solvent	Finish	Ref.
Cs	4-tertbutyl-2(α-methyl benzyl) phenol	–	AAS	[217]
Mn	bis(2-ethylhexyl) phosphate	Heptane	Spectrophoto-metric	[218]
U	Trioctylphosphine oxide	Ethyl ether	Fluorometric or spectrophotometric	[219, 220]

Source: Own files

20.3 Estuary waters

20.3.1 Dithiocarbamic acid derivatives

Apte and Gunn [36] used liquid–liquid extraction of the ammonium pyrrolidinedithiocarbamate to preconcentrate copper, nickel, lead and cadmium for estuary waters prior to atomic absorption spectrometry. Detection limits achieved were respectively, 0.3, 0.02, 0.7 and 0.5µg L^{-1} (see section 20.2.1).

Sturgeon *et al.* [155] used solvent extraction with ammonium pyrrolidine in methyl isobutyl ketone and preconcentration on Chelex– 100 resin to determine low levels of iron, manganese, cadmium, zinc, copper, nickel, lead and cobalt in coastal waters. They determined these metals at levels in the 0.01 to 0.4µg L^{-1} range, depending on the particular metal.

Kingston *et al.* [221] preconcentrated eight transition elements (cadmium, cobalt, copper, iron, manganese, nickel, lead and zinc from estuarine and seawater using solvent extraction/chelation and determined at sub ng L^{-1} levels by graphite furnace atomic absorption spectrometry.

Yamamoto *et al.* [222] have studied the differential determination methods of heavy metals according to their oxidation states by flameless atomic absorption spectrometry combined with solvent extraction with ammonium pyrrolidinedithiocarbamate or sodium diethyldithio-carbamate.

Danielsson *et al.* [166] preconcentrated cadmium, copper, iron, lead, nickel and zinc in estuary waters of salinity 0–35% by extraction of the dithiocarbamate complexes into Freon.

Results from a set of measurements of total trace metal concentrations in the Göta River estuary obtained by this method are shown in Table 20.26. Blank levels found by running Milli–Q water through the full procedure are also given in Table 20.26. The levels shown hold for the determination of total concentrations as well as for dissolved concentrations.

Table 20.26 Total trace metal concentration in the Göta River estuary* and blank levels with Milli–Q water

Salinity (‰)	Cd (ng L⁻¹)	Cu (μg L⁻¹)	Fe (μg L⁻¹)	Ni (μg L⁻¹)	Pb (μg L⁻¹)	Zn (μg L⁻¹)
0.5	20, 20	1.2, 1.2	170, –	1.2, 1.3	0.30, 0.36	7.6, 8.4
5	25, 25	1.7, 1.6	195, 191	1.2, 1.0	0.55, 0.44	8.1, 7.6
13.5	45, 43	1.1, 1.1	50, 52	1.1, 1.1	0.18, 0.20	2.8, 3.1
18.5	23, 24	0.7, 0.7	15, 15	0.6, 0.7	0.07, 0.07	0.7, 0.8
32	19, 20	0.3, 0.3	16, 16	0.4, 0.4	0.07, 0.06	0.5
Blank.	1	0.01, 0.02	0.2	0.04, 0.1	0.002, 0.01	0.01, 0.06

*Duplicate values are the results of separate determinations on aliquots from the same sample bottle

Source: Reproduced with permission from Danielsson, L.G. *et al.* [166] Elsevier Science, UK

In estuaries, where large ranges in concentrations are often found, carry-over can cause problems. Extracting samples in the order of decreasing salinity limits this problem. Between series, funnels should be rinsed with acid and thoroughly washed with Milli–Q water. Detection limits are adequate for estuarine studies. If still better detection limits are required, larger concentration factors can be used. Increased sample volumes should however be used with caution if the iron concentrations are high (>100μg L⁻¹).

20.3.2 Carboxylated polyethylenemine–polymethylene phenylene isocyanate

Carboxylated polyethylenemine–polymethylene phenylene isocyanate has been used for preconcentrating metals from estuary and seawaters [223] prior to analysis by inductively coupled plasma spectrometry. The uptake of copper, cadmium, lead and zinc by the resin was quantitative in the presence of high concentrations of ammonia, calcium, magnesium, potassium and sodium and in the presence of acetate and citrate buffers. The collection of other heavy metals and rare earths was also investigated.

20.4 Potable waters

20.4.1 Dithiocarbamic acid derivatives

An application of hexamethyleneammonium hexamethylenedithio-carbamate as a complexing agent involves the extraction of nanogram amounts of cadmium, silver, bismuth, cobalt, copper, nickel, lead,

Table 20.27 Extraction of cadmium

Aqueous solution Concentration (µg L^{-1})	Volume (ml)	Extraction solution volume (ml)	Radio of phases	Recovery %	Measurement technique[a]
0.1	20	1	20:1	>95	Graphite furnace
0.2	20	1	20:1	>95	Graphite furnace
0.5	800	16	50:1	>95	Flame
1	800	16	50:1	>95	Flame
2.5	400	20	20:1	>95	Flame
3	800	16	50:1	>95	Flame
6	800	16	50:1	>95	Flame
10	400	20	20:1	>95	Flame

[a]Measurement: Perkin–Elmer, Model 420, with either graphite furnace HGA500 accessory, or an air–acetylene flame

Source: Reproduced with permission from Dohreman, A. and Kleist, H. [224] Royal Society of Chemistry

Table 20.28 Inter-laboratory comparative test, DIN, October 1978

	Element				
	Cd	Co	Cu	Ni	Pb
Number of laboratories	10	10	10	10	10
Outliers	1	–	–	1	–
Arithmetic mean of concentrations found (µg L^{-1})	2.8	16	126	20	28
Standard deviation					
Absolute (µg L^{-1})	0.19	1.6	10	1.7	3.5
Relative (variance) (%)	6.7	9.9	7.9	8.6	12.4

Source: Reproduced with permission from Dohreman, A. and Kleist, H. [224] Royal Society of Chemistry

thallium and zinc in potable water samples [224]. Metals are determined in the solvent extract (xylene–di-isopropyl ketone) by atomic absorption spectrometry or graphite furnace atomic absorption spectrometry.

Table 20.27 presents results obtained by this procedure in the determination of cadmium. In all cases recoveries are better than 95%.

An inter-laboratory comparative test organised by DIN (German Institute for Standardisation) involved the determination of five metals in potable water, spiked with five heavy metals; ten laboratories participated in this test. The results are given in Table 20.28. Application of the Dixon

Table 20.29 Applications of solvent–extraction to the preconcentration of cations in potable water

Metal	Complexing agent	Solvent	Analytical	LD*	Ref.
Cd, Cu, Pb, Zn	Ammonium pyrrolidinedithiocarbamate	Methyl isobutyl ketone	AAS	Cd 0.05 Cu 0.01 Pb 0.9 Zn 6	[225]
Cd	Ammonium pyrrolidinedithiocarbamate	Methyl isobutyl ketone	AAS	–	[20]
Ni	Ammonium pyrrolidinedithiocarbamate	Methyl isobutyl ketone	AAS	–	[23]
Pb	Ammonium pyrrolidinedithiocarbamate	Methyl isobutyl ketone	AAS	–	[225]
V	Nerolic acid (S–amine-2-aniline–benzene sulphane acid)	Toluene	Spectro-photometric	–	[228]

*μg L^{-1} unless otherwise stated

Source: Own files

test for 5% significance level led to the rejection as outliers of one cadmium value and one nickel value. After elimination of these outliers the variance for lead was 12.4% and for the other metals below 10%.

Further limited applications of chelation–solvent extraction to the preconcentration of cations in potable waters are reviewed in Table 20.29.

20.5 Sewage effluents

A mixture of ammonium pyrrolidinedithiocarbamate and diammonium diethyldithiocarbamate in 4-methyl–pentan-2-one has been used as a means of preconcentrating cadmium, zinc, copper, iron, lead and nickel from sewage effluents [227]. Down to 1µg L^{-1} of these elements can be determined satisfactorily in sewage samples.

20.6 Trade effluents

Petrov *et al.* [71] assessed the suitability of 0.05M solution of diantipyryl methane in chloroform or dichloromethane as an extractant for preconcentrating traces of 20 metals in mine waters to which ammonium thiocyanate has been added. This reagent does not complex with nickel, aluminium, iron or manganese but can be used for the preconcentration of copper, zinc, vanadium, tin, molybdenum, niobium, bismuth, tungsten, gadolinium, cobalt, cadmium and antimony in the presence of 0.2–50mg L^{-1} of iron.

References

1 Kinrade, J.D. and Van Loon, J.C.V. *Analytical Chemistry*, **46**, 1894 (1974).
2 Subramanian, K.S. and Meranger, J.C. *International Journal of Environmental Analytical Chemistry*, **7**, 25 (1979).
3 Bone, K.M. and Hibbert, W.D. *Analytica Chimica Acta*, **107**, 219 (1979).
4 Tessier, A., Campbell, P.G.C. and Bisson, M. *International Journal of Environmental Analytical Chemistry*, **7**, 41 (1979).
5 Smith, J., Nelissen, J. and Van Grieken, R. *Analytica Chimica Acta*, **111**, 215 (1979).
6 Brooks, R.R., Presley, B.J. and Kaplan, I.R. *Talanta*, **14**, 809 (1967).
7 Kremligg, K. and Peterson, H. *Analytica Chimica Acta*, **70**, 35 (1974).
8 Jan, T.K. and Young, D.R. *Analytical Chemistry*, **50**, 1250 (1978).
9 Stolzberg, R.J. In *Analytical Methods in Oceanography*, ed. T.R.P. Gibb, Jr., American Chemical Society, Washington, DC, Advanced Chem. No. 147, p.30 (1975).
10 Danielson, L., Magnusson, B. and Westerlund, S. *Analytica Chimica Acta*, **98**, 45 (1978).
11 Sturgeon, R.E., Berman, S.S., Desauiniers, A. and Russel, D.S. *Talanta*, **27**, 85 (1980).
12 Magnusson, B. and Westerlund, S. *Analytica Chimica Acta*, **131**, 63 (1981).
13 Armansson, H. *Analytica Chimica Acta*, **88**, 89 (1977).

14 Bruland, K.W., Franks, R.P., Knauer, G.A. and Martin, J.H. *Analytica Chimica Acta*, **105**, 233 (1979).

15 Analytical Quality Control (Harmonized Monitoring) Committee, Water Research Centre, Harlow, Bucks, UK. *Analyst (London)*, **110**, 109 (1985).

16 Chakraborti, D., Adams, F., Van Mol, W. and Irgolic, K.J. *Analytica Chimica Acta*, **196**, 23 (1987).

17 Cockroft, H.R., Nield, D. and Ramson, L. Technical Report TR 59. Atomic Absorption Spectrometric Method for the Determination of Lead and Cadmium in Water. Water Research Centre, Medmenham, UK (1977).

18 British Standards Institution UK. BS 6068 Section 2.29. Determination of cobalt, nickel, copper, zinc, cadmium in flame atomic absorption spectrometric methods. (1987).

19 Subramanian, K.S., Meranger, J.C. and McCurdy, R.F. *Atomic Spectrometry*, **5**, 192 (1984).

20 Department of the Environment. Methods for the examination of waters and associated materials. Cadmium in potable water by atomic absorption spectrophotometry. Tentative method. HMSO, London (1976).

21 Pakalns, P. *Water Research*, **15**, 7 (1981).

22 Department of the Environment. Methods for the examination of waters and associated materials. Lead in potable waters by atomic absorption spectrophotometry. HMSO, London (1976).

23 Department of the Environment/National Water Council Standing Committee of Analysts. Methods for the examination of waters and associated materials. Nickel in potable waters. Tentative methods. HMSO, London (1981).

24 Childs, E.A. and Gaffke, J.N. *Journal of Association of Official Analytical Chemists*, **57**, 360 (1974).

25 Shiraishi, N., Hasegawa, T., Hisayuki, T. and Takahashi, H. Japan *Analyst*, **21**, 705 (1972).

26 Sourova, J. and Capkova, A. *Vodni. Hospodarstvi, Series B*, **30**, 133 (1980).

27 Tweeten, T.N. *Analytical Chemistry*, **48**, 64 (1976).

28 Bradshaw, S., Gascoigne, A.J., Headbridge, J.B. and Moffett, J.H. *Analytica Chimica Acta*, **197**, 323 (1987).

29 Regan, J.G.T. and Warren, J. *Analyst (London)*, **103**, 447 (1978).

30 Byrko, V.M., Vizhenskii, V.A. and Molchanova, T.P. *Zh. Anal. Khim.*, **42**, 15767 (1987).

31 Subramanian, K.S. *Analytical Chemistry*, **60**, 11 (1988).

32 Subramanian, K.S. *Journal of Research of the National Bureau of Standards (US)*, **93**, 305 (1988).

33 Breuggemeyer, T.W. and Caruso, J. *Analytical Chemistry*, **54**, 872 (1982).

34 Fujinaga, T. and Takamatsu, T. *Journal of Chemical Society of Japan. Pure Chemistry Section*, **91**, 1165 (1970).

35 Shijo, Y., Watenabe, J., Aklyama, S., Shimizu, T. and Sakai, K. *Bunseki Kagaku*, **36**, 59 (1987).

36 Apte, S.C. and Gunn, A.M. *Analytica Chimica Acta*, **193**, 147 (1987).

37 Rubio, R., Haguet, J. and Rauret, G. *Water Research*, **18**, 423 (1984).

38 Chao, T.T. and Ball, J.W. *Analytica Chimica Acta*, **54**, 166 (1971).

39 Tao, H., Miyazaki, A., Bansho, K. and Umezaki, Y. *Analytica Chimica Acta*, **156**, 159 (1984).

40 Moore, R.V. *Analytical Chemistry*, **54**, 895 (1982).

41 Sugiyama, M., Fujino, O., Kihara, S. and Matsui, M. *Analytica Chimica Acta*, **181**, 159 (1986).

42 Smith, C.L., Matoaka, J.M. and Willson, W.R. *Analytical Letters*, **17**, 1715 (1984).
43 Shan, X., Tie, J. and Xie, G. *Journal of Analytical Atomic Spectroscopy*, **3**, 259 (1988).
44 Wada, K., Matsuchita, T., Hizumi, S. and Kojima, K. *Bunseki Kagaku*, **37**, 405 (1988).
45 Mijazaki, A., Kimura, A., Bansho, K. and Umezaki, Y. *Analytica Chimica Acta*, **144**, 213 (1982).
46 Sokolovich, V.B., Lel'chuk Yu, L. and Detkova, G.A. Izv. Tonsk. Politekh. Inst, 163, 130. Ref: Zhur. Khim (1971), 199D (16). Abstract No. 16G188 (1970).
47 Sadilikova, M. *Mikrochimica Acta*, **5**, 934 (1968).
48 Lo, J.M., Wei, J.C., Yang, M.H. and Yeh, S.S. *Journal of Radioanalytical Chemistry*, **72**, 571 (1982).
49 Lo, J.M., Wei, J.C. and Yeh, S.J. *Analytical Chemistry*, **49**, 1146 (1977).
50 Shatipov, E.B. and Khudaibenganov, A.I. Izv., Akad. Nauk. Uzbek. S.S.R. Ser. Fiz water Nauk, 6, 55 (1970).
51 Ya, J.C. and Wai, C.M. *Analytical Chemistry*, **56**, 1689 (1984).
52 Lo, J.G. and Yang, J.Y. *Journal of Radioanalytical and Nuclear Chemistry Letters*, **94**, 311 (1985).
53 Mok, W.M., Shah, N.K. and Wai, C.M. *Analytical Chemistry*, **58**, 110 (1986).
54 Mok, W.M. and Wai, C.M. *Analytical Chemistry*, **59**, 233 (1987).
55 Marcie, F.J. *Environmental Science and Technology*, **1**, 164 (1967).
56 Wanatabe, H., Berman, S. and Russel, D.S. *Talanta*, **19**, 1363 (1972).
57 Tisue, T., Suls, C. and Keel, R.T. *Analytical Chemistry*, **57**, 82 (1985).
58 Adeljou, S.B. and Brown, K.A. *Analyst (London)*, **112**, 221 (1987).
59 Rigin, V.I. and Yurtaev, P.V. *Soviet Journal of Water Chemistry and Technology*, **8**, 77 (1986).
60 Laintz, K.E., Yu, J.J. and Wai, C.M. *Analytical Chemistry*, **64**, 311 (1992).
61 Hsieh, T. and Liu, L.K. *Analytica Chimica Acta*, **282**, 221 (1993).
62 Metzger, M. and Braun, H. *Analytica Chimica Acta*, **189**, 263 (1986).
63 Ihnat, M., Gordon, A.D., Gaynor, L.D. *et al. International Journal of Environmental Analytical Chemistry*, **8**, 259 (1980).
64 Sachdev, S.L. and West, P.W. *Environmental Science Technology*, **4**, 749 (1970).
65 Chormann, F.H., Spencer, M.J., Lyons, W.B. and Mayewski, P.A. *Chemical Geology*, **53**, 25 (1985).
66 Woodriff, R., Culner, B.R., Shrader, D. and Super, A.B. *Analytical Chemistry*, **45**, 230 (1973).
67 Chau, Y.K. and Saitoh, H. *Environmental Science and Technology*, **4**, 839 (1970).
68 Shevchuk, I.A. and Metel, N.I. *Soviet Journal of Water Chemistry and Technology*, **9**, 247 (1987).
69 Nakamura, T. and Sato, J. *Onsen Kogakkaishi*, **20**, 37 (1986).
70 Guo, R., Chem, N. Silundka, C. and Lai, E.P.C. *Analyst (London)*, **113**, 1105 (1988).
71 Petrov, B.I., Oshchepkova, A.P., Zhipovistev, U.P. and Nemkovskii, B.B. *Soviet Journal of Water Chemistry and Technology*, **3**, 51 (1981).
72 Kato, K. *Talanta*, **24**, 503 (1977).
73 Billah, M., Honjo, T. and Terade, K. *Analytical Science*, **9**, 251 (1993).
74 Ueda, K., Kitahara, S., Kubo, K. and Yamamoto, Y. *Bunseki Kagaku*, **36**, 728 (1987).
75 Lapid, J., Munster, M.T., Forhi, S., Erni, M. and Kaloucher, L. *Journal of Radioanalytical and Nuclear Letters*, **86**, 321 (1984).
76 Testemale, G. and Leredde, S.L. In Report CEA–R–3908. Centre of Nuclear Studies, Forenay-aux Roses, France (1970).

77 Komarek, J., Horak, J. and Sommer, L. *Collection Czechoslovakian Chemical Communications*, **39**, 92 (1974).
78 Nikolaeva, E.M. Trudy Perm med. Inst, 108, 17. Ref: Zhur Khim (1973) 199D(4) Abstract 4G84 (1972).
79 Yotsuyanagi, T., Takeda, Y., Yamashita, R. and Aomura, K. *Analytica Chimica Acta*, **67**, 297 (1973).
80 Fujinaga, T. and Takamatsu, T. *Journal of Chemical Society of Japan, Pure Chemistry Section*, **91**, 1165 (1970).
81 Doolan, K.J. and Smythe, L.E. *Talanta*, **20**, 241 (1973).
82 Gemmer Colos, V., Tuss, H., Saur, D. and Neeb, R. *Fresenius Zeitschrift für Analytische Chemie*, **307**, 347 (1981).
83 Wenger, R. and Hogel, O. *Mitt. Geb. Lebensmittelunters u. Hygiene*, **62**, 1 (1971).
84 Karrey, J.S. and Goulden, P.D. *Atomic Absorption Newsletter*, **14**, 33 (1975).
85 Andukinova, M.M., Mordberg, G.L. and Nakhorossheva, M.P. In *The Isolation of Strontium–90. Collection of Radiometric and Gamma Spectrometric Methods of Analysing Materials in the Environment*, Leningrad, p.26 (1970).
86 Tamhina, B., Herak, M.J. and Jayodic, V. *Craot. Chem. Acta*, **45**, 593 (1973).
87 Gorbushina, L.V., Zhil'tsova, L.Y., Matveeva, E.N. *et al. Journal of Radioanalytical Chemistry*, **10**, 165 (1972).
88 Garcia-Leon, M., Piazza, C. and Madunga, G. *International Journal of Applied Radiation and Isotopes*, **35**, 957 (1984).
89 Agrawal, Y.K., Chrattopadhyaya, M.C., Abbasi, S.A. and Bodas, M.G. *Separation Science*, **8**, 613 (1973).
90 Agrawal, Y.K. *Separation Science*, **8**, 709 (1973).
91 Abbasi, S.A. *International Journal Environmental Analytical Chemistry*, **33**, 113 (1988).
92 Kish, P.P. and Zimomrya, I.I. *USSR Zavod Lab.*, **35**, 541 (1969).
93 Lazazev, A.I. and Lazareva, V.I. *Zhur. Analit. Khim*, **24**, 395 (1969).
94 Yagodnitsyr, M.A. Gig Savit (11), 62. Ref: Zhur Analit Khim, 19GD (10), Abstract No. 10G190 (1971).
95 Joshi, S.R., Srivanstava, P.K. and Tandon, S.N. *Journal of Radioanalytical Chemistry*, **13**, 343 (1973).
96 Shinde, V.M. and Khopar, S.M. *Chemia. Analit.*, **14**, 749 (1969).
97 Chau, Y.K. and Lum Shue Chan, L. *Analytica Chimica Acta*, **50**, 201 (1970).
98 Talmi, Y. and Audrin, A.W. *Analytical Chemistry*, **46**, 2122 (1974).
99 Adam, J. and Pribil, R. *Talanta*, **20**, 1344 (1973).
100 Alimarin, I.P., Tarasevich, N.I. and Isalev, D.L. *Zhur Analit. Khim*, **27**, 647 (1972).
101 Savitskii, V.N., Paloshenko, V.I. and Osadchii, V.I. *Journal of Analytical Chemistry of USSR*, **42**, 677 (1987).
102 Weber, G. *Analytica Chimica Acta*, **186**, 49 (1986).
103 Golwelker, A., Patel, K.S. and Mishra, R.K. *International Journal of Environmental Analytical Chemistry*, **33**, 185 (1988).
104 Rashid, M. and Ejaz, M. *Mikrochemica Acta*, **No. 3/4**, 191 (1986).
105 Ejaz, M., Zuha, S., Dit, W., Akhtar, A. and Chandri, S.A. *Talanta*, **28**, 441 (1981).
106 Silva, M. and Valcarcel, M. *Analyst (London)*, **107**, 511 (1982).
107 Brueggemeyer, T.W. and Caruso, J.A. *Analytical Chemistry*, **54**, 872 (1982).
108 Ejaz, M. and Qureshi, M.A. *Talanta*, **34**, 337 (1987).
109 Korkisch, J., Sorio, E. and Stelstan, F. *Talanta*, **23**, 289 (1976).
110 Christianson, T.F., Busch, J.E. and Krogh, S.C. *Analytical Chemistry*, **48**, 1051 (1976).
111 Motomizu, S., Onada, M., Oshima, M. and Iwachido, T. *Analyst (London)*, **113**, 743 (1988).

112 Savistky, V.N., Peleshenko, V.I. and Osadchiy, C. *Hydrobiological Journal*, 1, 60 (1986).
113 Tervero, M. and Gracia, I. *Analyst (London)*, 108, 310 (1983).
114 Butler, L.R.P. and Matthews, P.M. *Analytica Chimica Acta*, 36, 319 (1966).
115 Chau, Y.K. and Lum-Shue-Chan, K. *Analytica Chimica Acta*, 48, 205 (1969).
116 Akama, Y., Nakai, T. and Kawamura, F. *Nippon Kaisui Gakkai-shi*, 33, 180 (1979).
117 Fujinaga, T., Kusaka, Y., Koyama, M. *et al. Journal of Radioanalytical Chemistry*, 13, 301 (1973).
118 Kulathilake, A.I. and Chatt, A. *Analytical Chemistry*, 52, 828 (1980).
119 Moore, P.J. *Transactions of the Institute of Minerals and Metallurgy Section B*, 79, 107 (1970).
120 Atallah, R.H., Christian, G.D. and Hartenstein, S.D. *Analyst (London)*, 113, 463 (1988).
121 Akatsuka, K., Nobuyama, N. and Atsuya, K. *Analytical Science*, 4, 281 (1988).
122 Ai, Y., Zing, D. *Fenxl Huaxue*, 16, 478 (1988).
123 Onishchenko, T.A., Onishchenko, Y.K., Sukhan, V.V. and Knyazeva, E.U. *Ukr. Khim. Zn (Russian edition)*, 53, 855 (1987).
124 Abbasi, S.A. *Analytical Letters (London)*, 20, 1013 (1987).
125 Abbasi, S.A. *Analytical Letters (London)*, 21, 461 (1988).
126 Abbasi, S.A. *International Journal of Environmental Analytical Chemistry*, 33, 43 (1988).
127 Salinas, F., Galeano, Diaz, T. and Jiminez Sanchez, J.C. *Talanta*, 34, 655 (1987).
128 Goudhi, M.N. and Khopkar, S.M. *Mikrochimica Acta*, 111, 93 (1993).
129 Pesavento, M., Soldi, T., Riolo, C., Profumo, A. and Barbucci, R. *Environmental Protection Engineering*, 16, 49 (1991).
130 Honjo, T. and Nakata, T. *Bulletin of the Chemical Society of Japan*, 60, 2271 (1987).
131 Du, H.S., Wood, D.J., Elshani, S. and Wal, C.M. *Talanta*, 40, 173 (1993).
132 Wood, D.J., Eishani, S., Du, H.S., Natale, N.R. and Wal, C.M. *Analytical Chemistry*, 65, 1350 (1993).
133 Shimoishi, Y. and Toei, K. *Analytica Chimica Acta*, 100, 65 (1978).
134 Uchida, H., Shimoishi, Y. and Toei, K. *Environmental Science and Technology*, 14, 541 (1980).
135 Saitoh, K., Kabayashi, M. and Suzuki, N. *Analytical Chemistry*, 53, 2309 (1981).
136 Nakashima, S. and Toei, K. *Talanta*, 15, 1476 (1968).
137 Gosink, T.A. and Reynolds, P.J. *Journal of Marine Science Communications*, 1, 10 (1975).
138 Shimoishi, Y. *Analytica Chimica Acta*, 64, 465 (1973).
139 Young, J.W. and Christian, G.D. *Analytica Chimica Acta*, 65, 127 (1973).
140 Ross, W.D., Scribner, W.G. and Sievers, R.E. In Reprint of 8th Int. Symposium on Gas Chromatography, Ballsbridge, Dublin, Ireland, September (1970).
141 Measures, C.I. and Edmond, J.M. *Analytical Chemistry*, 58, 2065 (1986).
142 Belcher, R., Khalique, A. and Stephen, W.L. *Analytica Chimica Acta*, 100, 503 (1978).
143 Chambers, J.C. and McClellan, B.E. *Analytical Chemistry*, 48, 2061 (1976).
144 Agrawal, Y.K., Upadhyaya, D.B. and Chudasama, S.P. *Journal of Radio-analytical and Nuclear Chemistry*, 170, 79 (1993).
145 Sturgeon, R.E., Berman, S.S., Desaulniers, A. and Russell, D.S. *Talanta*, 27, 85 (1980).
146 Moore, R.M., Burton, J.D., Williams, P.J. le B. and Young, M.L. *Cosmochimica Acta*, 43, 919 (1979).

147 Sugimae, A. *Analytica Chimica Acta*, **121**, 331 (1980).
148 Statham, P.J. *Analytica Chimica Acta*, **169**, 149 (1985).
149 Burton, J.C. In *Trace Metals in Seawater*. Procedures of a NATO Advanced Research Institute on Trace Metals in Seawater. 30/3–3/4/81. Sicily, Italy, eds. C.S. Wong *et al.* Plenum Press, New York, p. 419 (1981).
150 Brewer, P.G., Spencer, D.W. and Smith, C.L. *American Society of Testing Materials*, **443**, 70 (1969).
151 Moore, R.M. In *Trace Metals in Seawater*. Proceedings of a NATO Advanced Research Institute on Trace Metals in Seawater. 30/3–3/4/81, Sicily, Italy, eds. C.S. Wong *et al.* Plenum Press, New York (1981).
152 Danielson, L., Magnusson, G.B. and Westerlund, S. *Analytica Chimica Acta*, **98**, 47 (1978).
153 Filippelli, M. *Analyst (London)*, **109**, 515 (1987).
154 Lo, J.M., Yu, J.C., Hutchinson, F.I. and Wal, C.M. *Analytical Chemistry*, **54**, 2536 (1982).
155 Sturgeon, R.E., Berman, S.S., Desauliniers, A.P. *et al. Analytical Chemistry*, **52**, 1585 (1980).
156 Brugmann, L., Danielsson, L.G., Magnusson, B. and Westerlund, S. *Marine Chemistry*, **13**, 327 (1983).
157 Tsalev, D.L., Alimarin, T.P. and Neiman, S.I. *Znur Analit. Khim.*, **27**, 1223 (1972).
158 El-Enamy, F.F., Mahmond, K.F. and Varma, M.M. *Journal of the Water Pollution Control Federation*, **51**, 2545 (1979).
159 Pellenberg, R.E. and Church, T.M. *Analytica Chimica Acta*, **97**, 81 (1978).
160 Boyle, E.A. and Edmond, J.M. *Analytica Chimica Acta*, **91**, 189 (1977).
161 Rasmussen, L. *Analytica Chimica Acta*, **125**, 117 (1981).
162 Ediger, R.D., Peterson, G.E. and Kerber, J.D. *Atomic Absorption Newsletter*, **13**, 61 (1974).
163 Sperling, K.R. *Fresenius Zeitschrift für Analytische Chemie*, **301**, 294 (1980).
164 Sperling, K.R. *Fresenius Zeitschrift für Analytische Chemie*, **54**, 2536 (1982).
165 Sturgeon, R.E., Berman, S.S., Desauliniers, J.A.H. *et al. Analytical Chemistry*, **52**, 1585 (1980).
166 Danielsson, L.G., Magnusson, B. and Westerlund, S. *Analytica Chimica Acta*, **144**, 183 (1982).
167 Shijo, Y., Mitsuhashi, M., Shimizu, T. and Sakurai, S. *Analyst (London)*, **117**, 1929 (1992).
168 Jin, L., Wu, D. and Ni, Z. *Huaxue Xuebao*, **45**, 808 (1987).
169 Van Geen, A. and Boyle, E. *Analytical Chemistry*, **62**, 1705 (1990).
170 Radionova, T.V. and Ivanov, V.M. *Zh. Anal. Khim*, **41**, 2181 (1986).
171 Sperling, K.R. *Fresenius Zeitschrift für Analytische Chemie*, **310**, No. 3/4, 254 (1982).
172 Armansson, H. *Analytica Chimica Acta*, **110**, 21 (1979).
173 Muyazaki, A., Kimuka, A., Bansho, K. and Amezaki, Y. *Analytica Chimica Acta*, **144**, 213 (1981).
174 Bloekaert, J.A.C., Leis, F. and Laguna, K. *Talanta*, **28**, 745 (1981).
175 Jones, J.S., Harrington, D.E., Leone, B.A. and Branstedt, W.R. *Atomic Spectroscopy*, **4**, 49 (1983).
176 Buchanan, A.S. and Hannaker, P. *Analytical Chemistry*, **56**, 1379 (1984).
177 Van der Berg, C.M.G. *Marine Chemistry*, **16**, 121 (1985).
178 Clem, R.G. and Hodgson, A.T. *Analytical Chemistry*, **50**, 102 (1978).
179 Yusov, A.M., Ishsan, Z.B. and Wood, A.K.H. *Journal of Radioanalytical and Nuclear Chemistry*, **179**, 277 (1994).

180 Shannon, L.L. and Orden, M.J. *Analytica Chimica Acta*, **52**, 166 (1970).
181 Flynn, A. *Analytical Abstracts*, **18**, 1624 (1970).
182 Tseng, C.L., Hsieh, Y.S. and Yong, M.H. *Journal of Radioanalytical and Nuclear Chemistry Letters*, **95**, 359 (1985).
183 Boyle, E.A., Handy, B. and Van Green, A. *Analytical Chemistry*, **59**, 1499 (1987).
184 Murata, M., Omatsu, M. and Muskimoto, S. *X-ray Spectrometry*, **13**, 83 (1984).
185 Yang, C.Y., Shih, J.S. and Yeh, Y.C. *Analyst (London)*, **106**, 385 (1981).
186 Hirao, Y., Fukumoto, K., Sugisaki, H. and Kimura, K. *Analytical Chemistry*, **51**, 651 (1979).
187 Smith, R.G. and Windom, H.L. *Analytica Chimica Acta*, **113**, 39 (1980).
188 Klinkhammer, G.P. *Analytical Chemistry*, **42**, 117 (1980).
189 Landing, W.M. and Bruland, K.W. *Earth Planet Science Letters*, **49**, 45 (1980).
190 Hiiro, K., Tanaka, T. and Sawada, T. *Japan Analyst*, **21**, 635 (1972).
191 Armitage, B. and Zeitlin, H. *Analytica Chimica Acta*, **53**, 47 (1971).
192 Morris, A.W. *Analytica Chimica Acta*, **42**, 397 (1968).
193 Virmani, Y.P. and Zeller, E.J. *Analytical Chemistry*, **46**, 324 (1974).
194 Zharikov, V.F. and Senyavin, M.K. Trudy gos okeanogr. Inst. (101). Ref: Zhur Khim 19GD, (7) Abstract No. 7G189 (1970).
195 Bruland, K.W. *Science Letters*, **47**, 176 (1980).
196 Boyle, E.A., Huested, S.S. and Jones, S.P. *Journal of Geographical Research*, **86**, 8048 (1981).
197 Kentner, E., Armitage, D.B. and Zeitlin, H. *Analytica Chimica Acta*, **45**, 343 (1969).
198 Yatsimirskii, K.B., Ewel'Yakov, E.M., Pavlova, V.K. and Savichenko, Ya. S. Okeanologiya, III, 10. Ref: Zhur Khim. 19GD Abstract No. 11G, 203 (11) (1979).
199 Rampon, H. and Cavalier, R. *Analytica Chimica Acta*, **60**, 226 (1972).
200 Nishimura, M., Matsunaga, K., Kudo, T. and Obara, F. *Analytica Chimica Acta*, **65**, 446 (1973).
201 Monien, H. and Stangel, R. *Fresenius Zeitschrift für Analytische Chemie*, **311**, 209 (1982).
202 Kentner, E. and Zeitlin, H. *Analytica Chimica Acta*, **49**, 587 (1970).
203 Motomizu, S. *Analytica Chimica Acta*, **64**, 217 (1973).
204 Forster, W. and Zeitlin, H. *Analytica Chimica Acta*, **34**, 211 (1966).
205 Riley, J. and Topping, G. *Analytica Chimica Acta*, **44**, 234 (1969).
206 Going, J., Wesenberg, G. and Andrejat, G. *Analytica Chimica Acta*, **81**, 349 (1976).
207 Korkisch, J. and Sorio, A. *Analytica Chimica Acta*, **79**, 207 (1975).
208 Gurtler, O. *Fresenius Zeitschrift für Analytische Chemie*, **284**, 206 (1977).
209 Korenaga, T., Motomizu, S. and Toei, K. *Analyst (London)*, **105**, 328 (1980).
210 Lee, M.L. and Burrell, D.C. *Analytica Chimica Acta*, **66**, 245 (1973).
211 Siu, W.M., Bednas, H.E. and Berman, S.S. *Analytical Chemistry*, **55**, 473 (1983).
212 Heumann, K.G. *Toxicological Environmental Chemical Review*, **3**, 111 (1980).
213 Colby, B.N., Rosecrance, A.E. and Colby, M.E. *Analytical Chemistry*, **53**, 1907 (1981).
214 Mykytiuk, A.P., Russell, D.S. and Sturgeon, R.E. *Analytical Chemistry*, **52**, 1281 (1980).
215 Sturgeon, R.E., Berman, S.S., Willie, S.N. and Desauliniers, J.A.H. *Analytical Chemistry*, **53**, 2337 (1981).
216 Lee, M.G. and Burrell, D.C. *Analytica Chimica Acta*, **62**, 153 (1972).
217 Shen, Z. and Li, P. *Fenxi Huaxue*, **14**, 55 (1986).
218 Flynn, W.W. *Analytica Chimica Acta*, **67**, 129 (1973).

219 Korkisch, J. and Koch, W. *Mikrochimica Acta*, **1**, 157 (1973).
220 Korkisch, J. *Mikrochimica Acta*, **6**, 87 (1972).
221 Kingston, H.M., Barnes, I.L., Brady, T.J. and Rains, T.C. *Analytical Chemistry*, **50**, 2064 (1978).
222 Yamamoto, M., Urata, K., Murashige, K. and Yamamoto, Y. *Spectrochimica Acta*, **36B**, 671 (1981).
223 Horvath, Z. and Barnes, R.M. *Analytical Chemistry*, **58**, 1352 (1986).
224 Dohreman, A. and Kleist, H. *Analyst (London)*, **104**, 1030 (1979).
225 Yhanez, N., Moutoro, R., Catela, R. and Cervera, M.L. *Rev. Agroquim. Tenol. Allment.*, **27**, 270 (1987).
226 Krishnamurty, K.V. and Reddy, M.M. *Analytical Chemistry*, **49**, 222 (1977).
227 Webster, T.B. *Water Pollution Control*, **79**, 511 (1980).
228 Lazarev, P.I. and Lazarev, V.I. *Zhur. Analitic Khim.*, **24**, 395 (1969).

Chapter 21

Cations: Adsorption on immobilised chelators

21.1 Non saline waters

21.1.1 Silica and glass beads

Metal chelating resins and immobilised (adsorbed or chemically bonded) chelates have found widespread application for the concentration and/or separation of trace metals from a variety of matrices. Dimethylglyoxime, alkylamines, diamines, xanthates, and dithiocarbamates, propylene-diaminetetraacetic acids, *n*-butylamides, *N*-substituted hydroxylamine, hexylthioglycolate, ferroin-type chelating agents, iminodiacetate, amidoxime, dithizone, and 8-hydroxyquinoline have all been immobilised on various substrates. Tailoring chemically bonded chelating agents to specific needs allows the use of selective or general concentration schemes and also permits the 'recycling' of the chelating agent. Their major use lies in the preconcentration of trace metal ions from aqueous and saline media.

Bonded silicas, widely used in liquid chromatographic separations, have recently shown recognised potential for sample preconcentration or matrix isolation. The so-called 'extraction columns', employing bonded silicas, have become popular for sample preparation prior to chromatographic separations, virtually replacing the more laborious liquid–liquid extraction procedures. In fact, bonded-phase sample preparation has been identified as a growing technological trend and has received prevalent attention for the selective preconcentration of trace metal ions. The immobilisation of reagents on silica supports offers some distinct advantages over immobilisation on organic polymer supports. First, the silica is readily modified by a variety of silylating agents allowing for a myriad of functional groups to be immobilised. Second, since the bound group is at the surface of the support, high exchange rates are generally observed whereas some highly cross-linked organic polymer matrices may require hours for equilibration. Third, silica offers excellent swelling resistance with changes in solvent composition having little effect on the support at pH <9. Therefore, silica surfaces are excellent choices for the immobilisation of analytical reagents.

Silica-immobilised 8-quinolinol has proven to be a particularly useful material for sample preparation, matrix isolation, and preconcentration of trace metal ions. The selectivity of 8-hydroxyquinoline for transition–metal ions over the alkali and alkaline–earth metal ions makes it useful for samples containing large quantities of the latter (eg seawater, non saline waters, etc.).

21.1.1.1 Silica immobilised 8-hydroxyquinoline

Marshall and Mottola [1] evaluated the use of silica-immobilised 8-quinolinol as a preconcentration material for trace metal ions in flow systems. Breakthrough capacities were evaluated under different flow, temperature and geometric characteristics of the preconcentrating column. Mass transfer limitations under flow conditions explain the dependence of breakthrough capacities on these variables. The capabilities of this material for on-line preconcentration of copper(II) using flow injection analysis for sample processing and atomic absorption spectrometry for detection was also evaluated. The relatively high capacities of these simply and reproducibly prepared materials as well as the absence of swelling complications afforded by the inorganic silica framework allow for their effective use in flow injection analysis–atomic absorption spectrometry by implementation of simple manifolds. Results obtained for the determination of ng/ml levels of copper(II) in some EPA water samples agreed very well with reported values.

More recently, Esser et al. [2] used 8-hydroxyquinoline immobilised on silica gel and RE–Spec, a supported organophosophorus extractant, to preconcentrate and purify rare earth elements from non saline waters prior to their determination by isotope-dilution inductively coupled plasma mass spectrometry. Preconcentration onto silica-8-hydroxy-quinoline is applicable to a wide range of trace metals, making it suitable for multielement isotope dilution inductively coupled plasma mass spectrometry studies. The silica-8-hydroxyquinoline, RE–Spec technique concentrates rare earths from 1L or less of water into 1mL of salt-free 0.1% nitric acid. The technique is rapid and has high rare earth yields (>80%) and low rare earths blanks (<2–6pg). In addition, barium separation is high, allowing determination of lanthanum and europium by isotope dilution – ≤300pg of barium is present in the final concentrates of sample solutions initially containing >4μg of barium.

Further applications involving 8-hydroxyquinoline are reviewed in Table 21.1.

21.1.1.2 Silica immobilised miscellaneous chelating agents

These applications are reviewed in Table 21.2.

Table 21.1 Preconcentration of cations using silica immobilised 8-hydroxyquinoline

Element	Analytical finish	Detection limit μg L^{-1}	Ref.
Yttrium–90	β counting	–	[3]
Co	Catalytic action of cobalt on oxidation of protocatechoic acid by hydrogen peroxide	5	[4]

Source: Own files

Table 21.2 Various chelating agents immobilised on silica gel used in preconcentration techniques

Element	Chelating agent	Analytical finish	Detection limit*	Ref.
Cu, Pb, Ni, Zn, Cd, Co	2,2'dipyridyl 4-amino 3-hydrazino-5-mercapto-1,2,4 triazole	AAS	–	[5,6]
Sb	2-mercapto-N-2-naphthyl–acetamide	Hydride generation AAS	0.4	[7]
Cd	Dithizone	AAS	10μg absolute	[8]
Hg	Dithiocarbamate	AAS	0.1	[9]
Fe(II)	o-phenanthroline	Spectrophotometric	–	[10]
Misc.	Diphenylcarbazone	–	–	[11]

*μg L^{-1} unless otherwise stated

Source: Own files

21.1.2 Immobilised C$_{18}$ bonded silica

Sturgeon et al. [12] preconcentrated selenium(IV) by adsorption of their ammonium pyrrolidinedithiocarbamate chelates on to C$_{18}$ bonded silica prior to desorption and determination by graphite furnace atomic adsorption spectrometry. The detection limit was 7ng L^{-1} selenium(VI), based on a 300ml water sample, respectively buffered to pH6.8 (with phosphate buffer, 10mM), separated and detected by an ultraviolet photodiode array detector. This procedure allowed determination of sub-nanogram quantities of metal ions, including copper(II), and mercury(II) ions in potable water and of cadmium(II), lead(II), cobalt(II), nickel(II) and bismuth(II).

Comber [13] used a C_2 column coated with an ammonium pyrrolidine-1-ylddithioformate–cetyltrimethyl ammonium bromide ion pair to preconcentrate copper, nickel and cadmium from seawater. The metal dithiocarbamate complexes were then separated on a C_{18} column by high performance liquid chromatography using an ultraviolet detector. The detection limit with a 10ml sample was 0.5ppb.

21.1.3 Immobilised glass beads

Allen *et al.* [14] compared methods involving solvent extraction using ammonium pyrrolidinedithiocarbamate–isobutyl methyl ketone with column chelation procedures using immobilised 8-hydroxyquinoline on a controlled glass pore support for the determination of lead and copper in sea and river water. The final determination of the preconcentrated element was accomplished by atomic absorption spectrophotometry using a flame source. Results at the µg L^{-1} level for standard solutions of copper gave recoveries of better than 98% from both procedures. The determination of copper in non saline waters showed higher results by the column procedure, suggesting that column extraction was more efficient than solvent extraction. The column procedure was less time consuming and less costly than solvent extraction.

The determination of lead at the µg L^{-1} level with copper present gave a recovery of better than 99% when employing 8-hydroxyquinoline column separation. Copper, however, was only separated to the extent of 70% from the same solution of mixed elements.

The results obtained with synthetic solutions of copper(II) following ammonium pyrrolidinedithiocarbamate solvent extraction are given in Table 21.3. In the absence of a 'pure' and stable organocopper standard the calibration solutions were treated in the same manner as was the test samples. Any inaccuracy in the extraction procedure was therefore cancelled out and empirical rather than absolute values were obtained.

It was found that the methyl isobutyl ketone solutions of the copper ammonium pyrrolidinedithiocarbamate extracts were stable for up to 2h and after that time a noticeable drop in the copper values was observed. The blank value obtained was less than 1µg L^{-1} of copper(II) and hardly measurable on the chart recorder plot. Any such blank was taken into account in calculating the percentage recovery.

The results for a river water sample following the two extraction procedures are summarised in Table 21.4. Each set of analyses were carried out on different days. The blank values were minimal and taken into account. Ammonium pyrrolidinedithiocarbamate extraction yielded a mean value of 1.98µg L^{-1} of copper(II) and a standard deviation of 0.433µg L^{-1} compared with a mean value of 2.93µg L^{-1} and a standard deviation of 0.564µg L^{-1} for column extraction procedure with respect to the river water sample.

Table 21.3 APDC extraction of synthetic samples

Day	Sample concentration (µg L^{-1})	Concentration found (µg L^{-1})	Recovery (%)[a]
1st	30	28.5	95.00
	30	27.0	90.00
	30	30.0	100.00
	30	30.5	101.66
	30	30.0	100.00
	30	31.5	105.00
2nd	30	30.0	100.00
	30	30.0	100.00
	30	30.0	100.00
	30	28.0	93.33
	30	30.0	100.00

[a] x = 98.63%; s = 2.56%

Source: Reproduced with permission from Allen, E.A. et al. [14] Royal Society of Chemistry

Table 21.4 Cu–ammonium and Cu–CPG-8-hydroxyquinoline column extraction of river water sample

Day	Ammonium pyrrolidine-dithiocarbamate Conc. found (µg L^{-1})[a]	Day	Column CPG-8-HOQ Concn. found (µg L^{-1})[b]
1st	2.25	1st	2.25
	1.75		2.25
	1.75		2.5
	2.25		2.75
	2.25		2.75
	2.75		3.25
2nd	1.5		
	2.05	2nd	3.0
	1.5		2.88
	3.12		
3rd	1.75		
	2.0	3rd	3.0
	2.25		3.5
	2.25		3.5
	2.25		3.5

[a] x = 1.98ppb; s = 0.43ppb
[b] x = 2.93ppb; s = 0.56ppb

Source: Reproduced with permission from Allen, E.A. et al. [14] Royal Society of Chemistry

Correct pH conditions are essential for a satisfactory extraction of metal ions in both separation procedures. In the solvent extraction ammonium pyrrolidinedithiocarbamate metal complexes are extractable and stable between pH1.0 and 5.0 with some selectivity according to the metal ion in question. A controlling buffer or an acid may be employed to adjust the pH of the sample. Hydrochloric acid was found to be satisfactory for pH adjustment. The chelation of metal ions with 8-hydroxyquinoline is pH dependent. The extraction efficiency is also dependent on the flow rate of the sample in contact with it.

Copper ion and lead ion in their pure solutions were satisfactorily extracted at pH4.6 at flow rates of 5ml min⁻¹, but in admixture the extraction pattern changed. At pH2.0 the uptake of lead was very small, but lead was completely extracted in the presence of copper at pH2.0, at which value only about 70% of the copper was extracted in the presence of lead. At pH4.6, while lead in the mixture was still 99% retained, the copper uptake had decreased to about 15%, presumably at the expense of lead uptake. A possible explanation of this discrepancy is that there is competition between lead and copper ions, and possibly hydrogen ions, for the 8-hydroxyquinoline sites of the column material. Reducing the flow rate might have increased the copper uptake but one remedy would be to pass the extracted solution (now lead free) through the column a second time to extract the copper ion remaining in the sample.

Allen *et al.* [15] also compared the efficiencies of 8-hydroxyquinoline and EDTA immobilised on controlled pore size glass for the preliminary concentration of traces of aluminium in non saline waters, prior to determination by flame atomic absorption spectrometry. Both chemicals were satisfactory at sample pH values greater than 4.6.

21.1.4 Other immobilised glass bead chelates

Various other applications of immobilised glass beads are reviewed in Table 21.5.

21.1.5 Other immobilised chelators

Modified cellulose, microporous polymers and polythiocarbamate resins have also been used to preconcentrate cations in non saline waters (Table 21.6).

21.2 Seawater

21.2.1 Silica immobilised chelator

Marshall and Mottola [1] have used silica immobilised 8-quinolol as a means of preconcentrating metals for analysis by flow injection atomic

Table 21.5 Application of glass bead immobilised chelations to the preconcentration of cations

Element	Chelating agent	Analytical finish	Detection limit µg L^{-1}	Ref.
Sb	Fructose-6-phosphate	AAS	–	[16]
Cu, Ni, Zn, Co	Polyamine–polyurea resin on glass beads	AAS	Preconcentration factor up to 1000	[17]
Cu, Pb, Cd, Zn	2-mercaptobenzthiazole	AAS	–	[18]
Ag, As, Co, Cr, Cu, Fe, Hg, Mn, Pb, Zn	Sodium diethyl–dithio-carbamate	AAS	–	[19]
Al, Ga, In	Quinolin-8	–	–	[20]

Source: Own files

absorption spectrometry. This has proved to be a particularly useful material for sample preparation, matrix isolation, and preconcentration of silica of trace metal ions [42–46]. The selectivity of silica immobilised 8-quinolol for transition metal ions over the alkali and alkaline earth metal ions makes it useful for samples containing large quantities of the latter such as seawater. These workers evaluated breakthrough capacities under different flow, temperature and geometric characteristics of the preconcentration column. The columns have relatively high capacities for metals and do not suffer from complications due to swelling. Excellent agreement was obtained in determinations of copper on standard environmental samples (Table 21.7).

Sturgeon *et al.* [47] preconcentrated cadmium, copper, zinc, lead, iron, manganese, nickel and cobalt from seawater on to silica immobilised 8-hydroxyquinoline prior to determination by graphite furnace atomic absorption spectrometry. Results for the analyses of a near-shore seawater sample are given in Table 21.8. Near-shore samples were concentrated 50-fold, the open-ocean samples 90-fold. Calibration was achieved by spiking an aliquot of the concentrate with the element of interest, thereby obtaining an exact matrix match. Results obtained by using the immobilised 8-hydroxyquinoline concentration procedure were compared to 'accepted' values for these samples. Good agreement with accepted values is evident for all three samples.

Diphenylcarbazone and diphenylcarbazide have been widely used for the spectrophotometric determination of chromium [48]. The nature of these complexation reactions has been elucidated. Cr(III) reacts with diphenylcarbazone where Cr(VI) reacts (probably via a redox reaction combined with complexation) with diphenylcarbazide [49]. Although

Table 21.6 Chelating agents supported on various inert supports used for the preconcentration of cations

Element	Solid absorbent	Analytical finish	Detection limit*	Ref.
Pb	Modified cellulose	AAS	sub µg L $^{-1}$	[21]
Mo, V	Indon or Induron loaded cellulose	–	<1	[22]
Be	Hyphan cellulose or pyrogallol cellulose	AAS	<0.001	[23]
Hg	P-phenylene diamine cellulose	–	–	[24]
Hg(I), Hg(II)	o-aminophenol cellulose ether	–	–	[25]
Hg(I), Hg(II), Cd	Glyoxal-bis (2 mercapto amine)	–	–	[25]
Se(IV), Se(VI)	2,2'diamino-diethylamine cellulose	X-ray spectrometry	0.05	[26,27]
Misc.	Cellulose modified with triethylene tetra-amine pentaacetic acid	–	–	[28]
Cu, Fe, Zn	Polythiocarbamate resins	–	–	[29]
Hg	Polythiocarbamate resins	AAS	0.0002	[30]
Hg	Polythiocarbamate resins	AAS	<1	[31]
Fe	Microporous PVC	–	–	[32]
Pu	Microporus polyethylene-triphenyl phosphine oxide	–	–	[33]
Fe	2,4,6 tri-2-pyridinyl-1,3,5 triazine tetraphenyl borate	Spectro-photometric	<40	[34]
Ti, V	Chelating ion exchanger	ICPAES	<1000	[35]
Ni	Modified XAD resin	AAS	100	[36]
Misc.	Butane 2,3, dionebis (N-pyridinoacetyl hydrazone)	ICPAES	–	[37]
Cs	Prussian blue impregnated resin	–	–	[38]
Cd, Pb	Resins	–	Cd <0.1 Pb <2	[39]
Rare earths	Iminodiacetate resin	ICPAES	0.2 $^{-1}$ng L $^{-1}$	[40]
Sb	Glass immobilised fructose-6-phosphate	AAS	–	[85]
Cd, Cu, Mg, Ca, Mn(II), Ni(II), Pb(II), Zn(II)	Immobilised 8-oxine units	–	1–5	[42]

*µg L $^{-1}$ unless otherwise stated

Source: Own files

Table 21.7 Results of copper determination samples (Environmental Protection Agency)

Sample	Cu concentration (ng ml^{-1})	
	Determined	Reported
EPA–1	51 ± 4.6	50
EPA–2	252 ± 2.6	250
EPA–3	38 ± 3.0	40
EPA–4	8.9 ± 1.6	8.3
Tap water	26 ± 2.0	NA

Other elements known to be present (concentration in ng ml^{-1} in parentheses); EPA–1, Al (450), As (60), Be (250), Cd (13), Cr (80), Co (80), Fe (80), Pb (120), Mn (75), Hg (3.5), Ni (80), Se (30), V (250), Zn (200); EPA–2, Al (700), As (200), Be (750), Cd (50), Cr (150), Co (500), Fe (600), Pb (250), Mn (350), Hg (7.5), Ni (250), Se (40), V (750), Zn (200); EPA–3, Al (350), As (40), Be (150), Cd (10), Cr (60), Co (70), Fe (50), Pb (80), Mn (55), Hg (3.0), Ni (70), Se (20), V (200), Zn (60); EPA–4, Al (106); As (27), Be (29), Cd (9.1), Cr (7.1), Co (43), Fe (22), Pb (43), Mn (13), Hg (0.7), Ni (17), Se (11); tap water, unknown. Stillwater, OK
NA = not available

Source: Reproduced with permission from Marshall, M.A. and Mottola, A. [1] American Chemical Society

Table 21.8 Analyses of near-shore seawater (salinity 29.5‰) (means ± SD; n = 3)

Element	Concentration (µg ml^{-1})			
	Sample 1		Sample 2	
	1-8-HOQ	Accepted value	1-8-HOQ	Accepted value
Cd	0.020 ± 0.001	0.024 ± 0.004	0.025 ± 0.001	0.023 ± 0.001
Pb	0.22 ± 0.01	0.22 ± 0.06	0.014 ± 0.003	0.018 ± 0.001
Zn	0.44 ± 0.01	0.41 ± 0.05	0.29 ± 0.03	0.28 ± 0.01
Cu	1.03 ± 0.06	0.96 ± 0.04	0.17 ± 0.01	0.22 ± 0.02
Fe	1.0 ± 0.1	1.03 ± 0.04	7.2 ± 0.9	6.9 ± 0.02
Mn	0.71 ± 0.02	0.68 ± 0.05	1.06 ± 0.01	1.13 ± 0.02
Ni	0.33 ± 0.01	0.31 ± 0.04	0.39 ± 0.01	0.34 ± 0.01
Co	0.018 ± 0.002	0.015 ± 0.007	0.017 ± 0.001	ND

ND Not determined

Source: Reproduced with permission from Sturgeon, R.E. et al. [47] American Chemical Society

speciation would seem a likely prospect with such reactions, commercial diphenylcarbazone is a complex mixture of several components, including diphenylcarbazide, diphenylcarbazone, phenylsemicarbazide, and diphenylcarbadiazone with no stoichiometric relationship between the diphenylcarbazone and diphenylcarbazide [50]. As a consequence, use of diphenylcarbazone to chelate Cr(III) selectively also results in the sequestration of some Cr(VI). Total chromium can be determined with diphenylcarbazone following reduction of all chromium to Cr(III).

Use of immobilised chelating agents for sequestering trace metals from aqueous and saline media presents several significant advantages over chelation–solvent extraction approaches to this problem [51,52]. With little sample manipulation, large preconcentration factors can generally be realised in relatively short times with low analytical blanks. As a consequence of these considerations Willie et al. [11] developed an approach to the determination of total chromium. This involves preliminary concentration of dissolved chromium from seawater by means of an immobilised diphenylcarbazone chelating agent, prior to determination by atomic absorption spectrometry. A Perkin–Elmer Model 500 atomic absorption spectrometer fitted with a HGA–500 furnace with Zeeman background correction capability was used for chromium determinations. Chromium was first reduced to Cr(III) by addition of 0.5ml aqueous sulphur dioxide and allowing the sample to stand for several minutes. Aliquots of seawater were then adjusted to pH9.0 ± 0.2 by using high-purity ammonium hydroxide and gravity fed through a column of silica at a nominal flow rate of 10ml min^{-1}.

The sequestered chromium was then eluted from the column with 10.0ml 0.2M nitric acid. More than 93% of chromium was recovered in the first 5ml of eluate by this method. Extraction of 80ng spikes of Cr(III) from 200ml aliquots of seawater was quantitative. Neither Cr(III) nor Cr(V) could be quantitatively extracted. Results for the analysis of a near-shore sample of seawater and open ocean trace metal reference seawater, NASS–1 are given in Table 21.9.

21.2.2 Immobilised C₁₈ bonded silica

Ke and Chuen-Lin [53] carried out flame laser enhanced detection of lead in seawater by flow injection analysis with on-line preconcentration and separation. The flow injection manifold is incorporated with a microcolumn packed with a C_{18} bonded silica. The chelating agent DDPA is used to form the Pb–DDPA complex, which may be sorbed in the microcolumn and then eluted with methanol. The preconcentrated lead is then detected by the flame laser enhanced ionisation technique with either single-step or two-step excitation. At 5 and 15mL volume–fixed sample loading, the detection limits of 11 and 3.3ppt and enrichment factors of 16 and 48 are achieved, respectively, using a two-step flow injection–flame laser enhanced ionisation. The sensitivity of the current system proves to be better by at least 1 order of magnitude than that of conventional flame laser enhanced ionisation method. The flow

Table 21.9 Concentration analysis (µg L^{-1}) of seawater for total chromium

Trial	Coastal water (salinity = 29.5%)	Open-ocean. NASS–1 (salinity = 35.0%)
1	0.100	0.19
2	0.096	0.15
3	0.095	0.18
4		0.19
Mean (± SD)	0.097 ± 0.003	0.18 ± 0.02
Accepted value	0.10 ± 0.01	0.184 ± 0.016

Source: Reproduced with permission from Willie, S.N. et al. [11] American Chemical Society

injection–laser enhanced ionisation system also increases the tolerance of matrix interference. The flame laser enhanced ionisation signal is slightly reduced to 80% intensity as 10,000µg ml (ppm) sodium and potassium matrixes are mixed in the lead solution. The resistance to the alkali matrixes is enhanced ~4 times that reported previously using a similar water-immersed probe as a laser enhanced ionisation collector. Finally, the flow injection laser enhanced ionisation system is for the first time applied to detect the lead content in seawater, achieving a result of 0.0112 ± 0.0006µg L^{-1} (ppb) consistent with the certified value of 0.013 ± 0.005µg L^{-1} (ppb).

Sturgeon *et al.* [12] preconcentrated antimony(III) and antimony(V) from coastal and seawaters by adsorption of their ammonium pyrrolidinedithiocarbamate chelates into C_{18} bonded silica prior to determination by graphite furnace atomic absorption spectrometry. A detection limit of 0.05µg L^{-1} was achieved.

21.2.3 XAD resin immobilised chelator

Wan *et al.* [54] have described a two-column method for the preconcentration of trace metals in seawater on acrylate resin.

Conditions for the direct preconcentration on XAD–7 columns were determined for several trace metal ions in aqueous solutions of low and high salinity. Low breakthrough volumes in the presence of humic materials necessitated their prior removal at low pH by means of a pre-column of XAD–7. The two-column procedure was applied to the determination of trace metals in seawater. The final effluent for measurement by graphite furnace atomic absorption spectrometry is readily matrix matched and permits use of the standard calibration curve

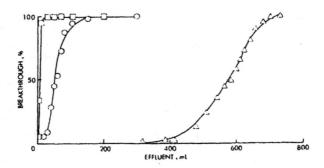

Fig. 21.1 Breakthrough curves for solutions of 10mg L^{-1} each of Cu(II) (Δ), Na$^+$ ($\square$), and Ca(II) (O) on a 1 × 6cm XAD–7 column (80–100mesh); flow rate, 1ml min^{-1}; pH7.0
Source: Reproduced with permission from Wan, C-C. *et al.* [54] American Chemical Society

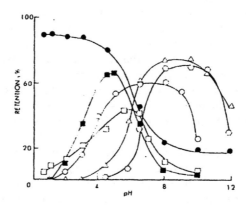

Fig. 21.2 Retention of humic substances (•, 10mg L^{-1}) and of metal ions (■, Fe(III); $\square$, Cr(III), O, (Cu(II); Δ, Pb(II); and O, Cd(II) in the presence of humic substances as a function of pH; 1 × 6cm XAD–7 column (20–50mesh); metal–ion concentration, 20µg L^{-1}; flow rate, 1ml min^{-1}; sample volume, 100ml
Source: Reproduced with permission from Wan, C-C. *et al.* [54] American Chemical Society

procedure for the majority of trace metals determined. This results in a reduction of sample consumption and analysis time.

The results in Fig. 21.1 show that the breakthrough volumes for sodium and calcium(II) are very low compared to that of copper(II). Ten millilitres of 0.01M ammonia/ammonium nitrate (pH8) was found adequate to rinse the sodium and calcium chlorides from the XAD–7 column without

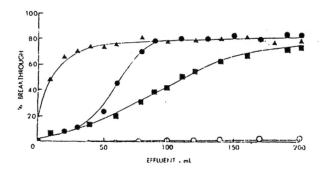

Fig. 21.3 Breakthrough curves for Fe(III) (O), Cu(II) (O), and Mn(II) (O) in the absence of humic materials and for Fe(III) (▲), Cu(II) (•), and Mn(II) (■) in the presence of 10mg L⁻¹ of humic materials: 1 × 6cm XAD–7 column (20–50mesh); flow rate, 1ml min⁻¹; metal–ion concentration, 5µg L⁻¹ each; pH6.3 for Fe(III); pH8.0 for Cu(II) and Mn(II)
Source: Reproduced with permission from Wan, C-C. et al. [54] American Chemical Society

removing trace heavy metals after passage of seawater samples. This rinsing procedure was subsequently used after preconcentration of seawater samples and prior to elution of the trace metals with the 1% nitric acid.

The preponderance of dissolved organic ligand in non saline waters is humic substances (humic and fulvic acids). The binding of metal ions with humic material is well-known and its presence has a strong effect on the retention of metal ions on XAD–7 resin, as illustrated in Fig. 21.2. With increasing concentration of humic material, the percent retention is decreased further than shown. In correspondence with Fig. 21.2, Fig. 21.3 illustrates that the percent breakthrough for metal ions is negligible in the absence of humics but in their presence, it is high for all three test ions, ie iron(III), copper(II) and manganese(II). There is no question that the major effect of the humics on breakthrough volume is at least in part related to thermodynamic factors since in separate batch equilibrium experiments the distribution ratio for copper(II) was found to be 21 in the presence of humic material vs 319 in its absence (for 20–50mesh resin). Similar observations were made for cadmium(II), cobalt(II) and manganese(II). Therefore, the stability of the ion–resin complex is not high enough to cause complete displacement of the metal ions from their humate complexes and the breakthrough volume is thus substantially lowered. It is also clear that the lability of the metal humate complexes relative to residence time in the column (3min/mL) is too low to cause complete displacement of the metal from the humate complex to the resin. Therefore, for total metal determination, the metal humates must first be dissociated and the humic ligands removed from the solution, otherwise a considerable fraction of metal ions will not be retained by the column. The metal humates are readily dissociated in acid solution (pH <2) and as shown in Fig. 21.2, the liberated humic material is adsorbed on XAD–7 at low pH. Since the metal ions pass through at low pH, the effluent can be raised to pH8 and passed through a second XAD–7 column to preconcentrate the trace metal ions. The effectiveness of this two-column procedure was tested with synthetic solutions containing humic material (10mg L^{-1}) and several trace elements, each present at 20µg L^{-1}. When humic materials were removed by the pre-column, quantitative recovery was obtained except for chromium(III). When the pre-column was not used, recovery for all elements was well below 100%.

21.2.4 Chromosorb immobilised chelator

Knockel and Prange [55] converted metals in seawater into their diethyldithiocarbamates prior to X-ray fluorescence analysis of the preconcentrated solids. Membrane filtration of the precipitates resulted in carbamate-loaded filters, which could be directly measured by using radioisotope-excited X-ray fluorescence analysis. Furthermore, elution of Chromosorb columns loaded with the dithiocarbamate complex, by the passage of chloroform gave chloroform solutions in which the trace metals could be determined by X-ray fluorescence analysis using totally reflecting sample supports. Similarly the precipitate on the membrane filter could be dissolved in chloroform and determined in solution. The sensitivity of this method and the pH dependence of the reaction was also investigated.

21.2.5 Polyethylene immobilised chelator

Delle Site *et al.* [56] have used extraction chromatography to determine plutonium in seawater sediments, and marine organisms. These workers used double extraction chromatography with Microthenes–210 (microporous polyethylene) supporting tri-*n*-octylphosphine oxide; a technique that has been used previously to isolate plutonium from other biological and environmental samples [57]. Plutonium–236 and plutonium–242 were tested as the internal standards to determine the overall plutonium recovery, but plutonium–242 was generally preferred because plutonium–236 has a shorter half-life and an α-emission (5.77MeV) which interferes strongly with the 5.68 MeV (95%) α-line of radium–224, the daughter of thorium–228. However, the 5.42–MeV-lines of thorium–228 interfere with those of plutonium–238 (5.50MeV) and so a complete purification from thorium isotopes is required.

Plutonium sources were counted by an α-spectrometer with good resolution, background, and counting yield. The counting apparatus used had a resolution of 40KeV. The mean (±so) background value was 0.0004 ± 0.0003cpm in the plutonium–239 and plutonium–240 energy range and 0.0001 ± 0.0001cpm in the plutonium–238 energy range. The mean (±so) counting yield, obtained with plutonium–239, plutonium–240 reference sources counted in the same geometry, was found to be 25.08 ± 0.72%.

To determine the overall recovery obtained by this procedure (chemical recovery and electrodeposition yield) a known activity of plutonium–242 was added to the different samples; the plutonium sources were counted by α-spectrometry for 3000min and the percentage overall recovery was calculated from the area of plutonium–242 peak. The percentage overall recovery for 3L samples of seawater was 63 ± 10%. Owing to the very low activity of the samples, the determination of plutonium–239, plutonium–240 and plutonium–238 in the reagents is very important in calculating the net activity of the radionuclides.

The method proposed was checked by analysing some seawater reference samples prepared by the IAEA Marine Radioactivity Laboratory (Monaco) for intercomparison programmes. The values reported by IAEA and the experimental values obtained by Delle Site et al. [56] were in good agreement.

21.2.6 Cellulose immobilised chelator

Shreedhara et al. [58] introduced ethylediamine, 2,2'diaminodiethylamine and triethylene–tetramine onto microcrystalline cellulose after tosylation. Dithiocarbamate groups were introduced by reaction with carbon disulphide. The metal uptake behaviour of these amine and dithiocarbamate derivatives were compared for copper(II), cadmium(II), lead(II), mercury(II), cobalt(II) and uranium(VI). The outcome of this work was that triethylene–tetramine–dithiocarbamate–cellulose was used to preconcentrate copper(II), cadmium(II), mercury(II) and lead(II) from seawater. Recoveries for the four metals were in the range 89 to 99%. Recoveries for uranium and cobalt were respectively nil and 55 to 63%.

A limited amount of work has been carried out on the determination of molybdenum in seawater by atomic absorption spectrometry [59,60] and graphite furnace atomic absorption spectrometry [61]. In a recommended procedure [61] a 50ml sample of seawater at pH2.5 is preconcentrated on a column of 0.5g p-aminobenzylcellulose, then the column is left in contact with 1M ammonium carbonate for 3h, after which three 5ml fractions are collected. Finally, molybdenum is determined by atomic absorption at 313.2nm with use of the hot graphite rod technique. At the 10mg L^{-1} level the standard deviation was 0.13µg.

Murthy and Ryan [86] preconcentrated copper, mercury and lead from seawater on a column of a dithiocarbamate cellulose derivative. Metal

Table 21.10 Miscellaneous supported chelators used for preconcentrations in seawater

Element	Chelating agent	Solid adsorbent support	Analytic finish	Detection limit*	Ref.
Cd, Pb, Zn, Cu, Fe, Mn, Ni	8-hydroxyquinoline	Silica	AAS	0.2 Co 40 Fe (50ml samples)	[62]
Cd, Cu, Pb, Zn	(3–mercapto propyl) trimethyl oxysilane	Silica	–	–	[63]
Ag(I),Au(III), Pd(II)	p-dimethyl–amino benzyldine–rhodanine	Silica	–	–	[64]
Cs	Copper–hexacyano–ferrate(II)	Silica	AAS	–	[65]
Cu, Ni, Cd	8-quinolinol	Silica	ICPAES	0.016–0.07	[66]
Co(II)	8-quinolinol	Silica	Chemical luminescence detection	–	[67]
Fe, Co, Zn, Cs, Zr	Zirconium hexacyanoferrate(II)	Cation exchange resin	γ spectrometry	–	[68]
Fe	Ferrozine	C_{18} silica	Spectrophotometric	0.1mM	[69]
Zn	Methyl tri-capryl ammonium chloride	C_{18} silica	AAS	2.4ng dm^3	[70]
Rare earths	Bis (2-ethyl–hexyl) hydrogen phosphate and 2-ethyl dihydrogen phosphate	C_{18} silica	ICPMS	–	[71]
Cd, Cu	Sodium diethyl dithiocarbamate	C_{18} silica	AAS	0.004–0.024	[72]
Pd	Liquid membrane	Tri-N—ctyl–amine	–	–	[73]
Hg, Pb	Tubular membrane	Dithiocarbamate	–	–	[74]
Mo	Pyrrolidinedithiocarbamate	Charcoal	–	–	[75]

Table 21.10 continued

Element	Chelating agent	Solid adsorbent support	Analytic finish	Detection limit*	Ref.
Cd, Cu, Mn, Ni, Pb, Zn	Chelamine resin	–	–	–	[76]
Ni	Polytriaminophenyl glyoxal resin (Ni extracted from resin with methyl isobutyl ketone solution of pyrrolidinedithio-carbamate)	–	AAS	–	[77]
V	Chelating resin	–	ICPAES	–	[78]
Ga, Ti, In	8-hydroxyquinoline immobilised resin	–	ICPAES	0.001–0.004	[79]
Fe(III)	8-hydroxyquinoline immobilised resin	–	Chemiluminescence detection	$0.05nM\ L^{-1}$	[80]
Sr	Chelation with 1-phenyl-3 methyl-4-benzoyl-5-pyrazolone	–	AAS	0.007–0.073	[81]
Fe, Ni, Cu, Cd, Mo, Cr, V	Chelating	Silica	ICPAES	–	[82]

*$\mu g\ L^{-1}$ unless otherwise stated

Source: Own files

concentrations on the adsorbent material were determined by neutron activation analysis. The recovery of added spikes to sea and tap water suggest that copper(II), mercury(II) and lead(II) can be quantitatively collected.

21.2.7 Miscellaneous immobilised chelators

Various other immobilised chelators that have been studied are reviewed in Table 21.10.

21.3 Groundwaters

Kerr *et al.* [83] have developed a technique using high performance liquid chromatography and trace enrichment techniques to measure trace levels of uranium in solutions containing high concentrations of dissolved salts. This procedure is required to support research into the feasibility of deep geological disposal of used nuclear fuel, which includes studying the leaching of uranium from fuel by non saline groundwaters. After conditioning, several millilitres of sample are passed through a small reversed-phase enrichment cartridge, where the uranium is concentrated and separated from the bulk of other constituents. The uranium is then back flushed from the column onto a reversed-phase analytical column where further separation is achieved. The separated species are monitored spectrophotometrically after post-column reaction with the chromogenic reagent Arsenazo(III). Analysis of simulated groundwaters has shown the procedure to be free from major interferences. Automation of the system using automatic switching valves and an automated sample injector allows approximately 40 samples per day to be analysed with a measurement precision of about 4%. Detection limits are in the 1–2µg L $^{-1}$ range.

21.4 Waste waters

Samchuk *et al.* [84] preconcentrated copper, lead, cadmium, nickel, cobalt and zinc from waste water on fibrous thioamide substituted poly-acrylonitrile, prior to desorption and determination by atomic absorption spectrometry.

References

1 Marshall, M.A. and Mottola, H.A. *Analytical Chemistry*, **57**, 729 (1985).
2 Esser, B.K., Volpe, A., Kenneally, J.M. and Smith, D. *Analytical Chemistry*, **66**, 1736 (1994).
3 Szabo, G., Guczi, J. and Stur, D. *Journal of Radioanalytical and Nuclear Chemistry*, **111**, 441 (1987).

4 Yamare, T., Wanatabe, K. and Mottola, H.A. *Analytica Chimica Acta*, **207**, 331 (1988).
5 Samara, C. and Kouimtzis, T.A. *Chemosphere*, **16**, 405 (1987).
6 Samara, C. and Kouimtzis, T.A. *Analytica Chimica Acta*, **174**, 305 (1985).
7 Fukada, H., Tsunoda, J., Matsumoto, K. and Tecada, K. *Bunseki Kagaku*, **36**, 683 (1987).
8 Genova, N., Crespi, V., Maggi, L. and Brandove, A. *International Journal of Applied Radiation and Isotopes*, **34**, 757 (1983).
9 Jin, L., Jin, S., He, P. and Xu, T. *Fenxi Huaxue*, **16**, 233 (1988).
10 Kuselman, I. and Lev, O. *Talanta*, **40**, 749 (1993).
11 Willie, S.N., Sturgeon, R.L. and Berman, S.S. *Analytical Chemistry*, **55**, 981 (1983).
12 Sturgeon, R.E., Willie, S.N. and Berman, S.S. *Analytical Chemistry*, **57**, 6 (1985).
13 Comber, S. *Analyst (London)*, **118**, 505 (1993).
14 Allen, E.A., Bartlett, P.K.N. and Ingram, G. *Analyst (London)*, **109**, 1075 (1984).
15 Allen, E.A., Boardman, H.C. and Punkett, B.A. *Analytica Chimica Acta*, **196**, 323 (1987).
16 de la Calle-Guntinas, M.B., Madrid, Y. and Camera, C. *Journal of Analytical Atomic Spectrometry*, **8**, 745 (1993).
17 Dingman, J., Siggia, S. and Barton, C. *Analytical Chemistry*, **44**, 1351 (1972).
18 Terada, K., Matsumoto, K. and Inaba, T. *Analytica Chimica Acta*, **170**, 225 (1985).
19 Laydon, D.E., Wegscheider, W. and Bodnar, W. *International Journal of Environmental Analytical Chemistry*, **7**, 85 (1979).
20 Mohammed, B., Une, A.M. and Littlejohn, D. *Journal of Analytical Atomic Spectroscopy*, **8**, 325 (1993).
21 Naghmugh, A.M., Pyrzynska, K. and Trojanowicz, M. *Talanta*, **42**, 851 (1995).
22 Burba, P., Willmer, P.G. *Fresenius Zeitschrift für Analytische Chemie*, **324**, 298 (1986).
23 Burber, P., Willmer, P.G., Betz, M. and Fuchs, S. *International Journal of Environmental Analytical Chemistry*, **13**, 177 (1983).
24 Masri, M.S. and Friedman, M. *Environmental Science and Technology*, **6**, 745 (1972).
25 Goldbach, K. and Lieser, K.H. *Fresenius Zeitschrift für Analytische Chemie*, **311**, 183 (1982).
26 Smits, J. and Van Grieken, R. *Analytica Chimica Acta*, **123**, 9 (1981).
27 Smits, J. and Van Grieken, R. *International Journal of Environmental Analytical Chemistry*, **57**, 6 (1985).
28 Burba, P., Rocha, J.C. and Schulte, A. *Fresenius Journal of Analytical Chemistry*, **346**, 414 (1993).
29 Ylbra-Biurrun, M.C., Bermejo-Borrera, A. and Bermejo-Barrara, P. *Microchimica Acta*, **109**, 243 (1992).
30 Minagawa, K., Takazawa, Y. and Kufune, I. *Analytica Chimica Acta*, **115**, 1103 (1980).
31 Yamagami, E., Tateishi, S. and Hashimoto, A. *Analyst (London)*, **105**, 491 (1980).
32 Saito, T. *Journal of AOAC International*, **77**, 1031 (1994).
33 Testa, C. and Staccoli, L. *Analyst (London)*, **97**, 527 (1972).
34 Puri, B., Sakata, M., Kano, G. and Usami, S. *Analytical Chemistry*, **59**, 1850 (1987).
35 Yang, K.L., Jiang, S.J. and Hwang, T.J. *Journal of Analytical Atomic Spectroscopy*, **11**, 139 (1996).

36 Olbrych-Slesynska, E., Braster, K., Matuszewski, W., Trojanowicz, M. and Frenzel, W. *Talanta*, **39**, 779 (1992).
37 Yang, H., Huang, K., Jiang, S., Wu, C. and Chou, C. *Analytica Chimica Acta*, **282**, 437 (1993).
38 Godoy, J.H., Guimaraes, J.R.P. and Carvalho, Z.L. *Journal of Environmental Radioactivity*, **20**, 213 (1993).
39 Isozaki, A., Ueki, K., Sazaki, H. and Utsumi, R. *Bunseki Kagaku*, **36**, 672 (1987).
40 Hall, G.E.M., Vaive, J.E. and McConnell, J.W. *Chemical Geology*, **120**, 91 (1995).
41 Abboli, A.O., Mentasti, E., Porta, V. and Sarzanin, C. *Analytical Chemistry*, **62**, 21 (1990).
42 Leydon, D.E. and Underwood, A.L. *Journal of Physical Chemistry*, **68**, 2093 (1964).
43 Florence, J.M. and Batley, G.E. *Talanta*, **23**, 179 (1976).
44 Riley, J.P. and Taylor, D. *Analytica Chimica Acta*, **41**, 175 (1968).
45 Abdullah, M.J. and Royle, L.G. *Analytica Chimica Acta*, **58**, 283 (1972).
46 Figura, P. and McDuffle, B. *Analytical Chemistry*, **49**, 1950 (1977).
47 Sturgeon, R.E., Berman, S.S., Willie, S.N. and Desauliniers, J.A.H. *Analytical Chemistry*, **53**, 2337 (1981).
48 Sandell, E.B. In *Colorimetric Determination of Traces of Metals*, Interscience, New York (1959).
49 Marchart, H. *Analytica Chimica Acta*, **30**, 11 (1984).
50 Willems, G.J. and de Ranter, C.J. *Analytica Chimica Acta*, **68**, 111 (1974).
51 Myasoedova, G.V. and Savvin, S.G. *Zhur. Analytica Khim.*, **37**, 499 (1982).
52 Leyden, D.E. and Wegscheider, W. *Analytical Chemistry*, **63**, 1059A (1981).
53 Ching-Bin Ke and King Chuen-Lin. *Analytical Chemistry*, **71**, 1561 (1999).
54 Chi-Chung Wan, Chiang, S. and Carsini, A. *Analytical Chemistry*, **57**, 719 (1985).
55 Knockel, A. and Prange, A. *Fresenius Zeitschrift für Analytische Chemie*, **306**, 252 (1981).
56 Delle Site, A., Marchionni, V. and Testa, C. *Analytica Chimica Acta*, **117**, 217 (1980).
57 Testa, C. and Delle Site, A. *Journal of Radioanalytical Chemistry*, **34**, 121 (1976).
58 Shreedhara Murthy, R.S. and Ryan, D.E. *Analytica Chimica Acta*, **140**, 163 (1982).
59 Nakahara, T. and Chakrabarti, C.L. *Analytica Chimica Acta*, **104**, 99 (1979).
60 Van den Sloot, H.A., Wals, G.D. and Das, H.A. *Progress in Water Technology*, **8**, 193 (1977).
61 Muzzarelli, R.A.A. and Rochetti, R. *Analytica Chimica Acta*, **64**, 371 (1973).
62 Nakashima, S., Sturgeon, R.E., Willie, S.N. and Berman, S.S. *Fresenius Zeitschrift für Analytische Chemie*, **330**, 592 (1988).
63 Volkan, M., Ataman, O.Y. and Howard, A.G. *Analyst (London)*, **112**, 1409 (1987).
64 Terada, K., Morimoto, K. and Kiba, T. *Analytica Chimica Acta*, **116**, 127 (1980).
65 Ganzerli, V.M.T., Stella, R., Maggi, L. and Ciceri, G. *Journal of Radioanalytical and Nuclear Chemistry*, **114**, 105 (1987).
66 Lau, C.R. and Yang, M. *Analytica Chimica Acta*, **287**, 111 (1994).
67 Hirata, S., Hashimoto, Y., Aihara, M. and Mallika, G.V. *Fresenius Zeitschrift für Analytische Chemie*, **355**, 676 (1996).
68 Kawamura, S., Sadao, S. and Katsumi, K. *Analytica Chimica Acta*, **56**, 405 (1971).
69 Blain, S. and Treguer, D. *Analytica Chimica Acta*, **308**, 425 (1995).
70 Akatsuka, K., Katoh, T., Nobuyama, N. *et al. Analytical Science*, **12**, 109 (1996).

71 Shabani, M.B., Akagi, T. and Masuda, A. *Analytical Chemistry*, **64**, 737 (1992).
72 Wang, M., Yuzetovsky, A.I. and Michel, R.G. *Microchemical Journal*, **48**, 326 (1993).
73 Wang, Z., Li, Y. and Guo, Y. *Analytical Letters (London)*, **27**, 957 (1994).
74 Johansson, M., Emteborg, H., Glad, B., Reinholdsson, F. and Baxter, D.C. *Fresenius Journal of Analytical Chemistry*, **351**, 461 (1995).
75 Van der Sloot, H.A., Wals, G.D. and Das, H.A. *Analytica Chimica Acta*, **90**, 193 (1977).
76 Blain, S., Appriou, P. and Handel, H. *Analytica Chimica Acta*, **272**, 91 (1993).
77 Zlatkus, A., Bruening, W. and Bayer, E. *Analytica Chimica Acta*, **56**, 399 (1971).
78 Dupont, V., Auger, Y., Jeandel, C. and Wartel, M. *Analytical Chemistry*, **63**, 520 (1991).
79 Orians, K.J. and Boyle, E.A. *Analytica Chimica Acta*, **282**, 63 (1993).
80 Obata, H., Karatani, H. and Nakayama, E. *Analytical Chemistry*, **65**, 1524 (1993).
81 Honjo, T. and Nakata, T. *Bulletin of the Chemical Society of Japan*, **60**, 2271 (1987).
82 Hirayama, K. and Unchara, N. Nihon. *Dalgaku Kogakubu Kiyo Bunrul, K.*, **28**, 149 (1987).
83 Kerr, A., Kupferchmidt, W. and Atlas, M. *Analytical Chemistry*, **60**, 2729 (1988).
84 Samchuk, A.I., Kazakevich, Y.E., Romanov, N.N. *et al.* *Soviet Journal of Water Chemistry and Technology*, **10**, 63 (1988).
85 de la Calle-Guntinas, M.B., Madrid, Y. and Camera, A. *Journal of Analytical Atomic Spectroscopy*, **8**, 745 (1993).
86 Murthy, R.S.S. and Ryan, D.E. *Analytica Chimica Acta*, **140**, 163 (1982).

Chapter 22

Cations: Adsorption on solid non chelating adsorbents

22.1 Non saline waters

22.1.1 XAD–2 and XAD–4 non polar resins

Mackey [1] studied the adsorption of copper, zinc, iron and magnesium ions on both these resins. Resolution was enhanced and time saved by using a multi-channel non-dispersive purpose-built atomic fluorescence detector in batch experiments. Both resins can adsorb appreciable amounts of iron, copper, and zinc ions and XAD–2 adsorbs large amounts of copper and zinc relative to its surface area. Adsorption of copper and zinc is independent of pH and, on XAD–1, independent of the amount of the other ion present. EDTA complexes are not adsorbed and a number of electrically charged metal–organic species can rapidly and completely remove iron, copper and zinc from the resins.

Sakai and Mori [2] preconcentrated cobalt with N-(dithiocarboxy) sarcosine and Amberlite XAD–4 resin. Cobalt reacted with N-(dithiocarboxy)sarcosine to form a 1:3 cobalt N-(dithiocarboxy) sarcosine complex which was stable in 4M hydrochloric acid. The complex so formed was adsorbed on a column of Amberlite XAD–4 copolymer from acid solution and eluted with 10ml of a 1:1:3 v/v, mixture of 1.0M ammonia solution (pH9), 0.1M EDTA, and methanol. The absorbance of the eluted chelate was determined at 320nm. The extinction coefficient was 21500L mol $^{-1}$ cm $^{-1}$. Interferences were eliminated by the addition of EDTA after chelation of cobalt. The copper complex with N-(dithiocarboxy)sarcosine was partly adsorbed because of the slow rate of decomposition by EDTA. Most chelated copper could be eluted with hydrochloric acid and any co-eluted with the cobalt chelate could be decomposed by heating. Cobalt enrichment factors of at least 100 were obtained and the method could be applied to determination of cobalt at the ng ml $^{-1}$ level.

Metal ions have been preconcentrated by complexation with sodium *bis* (2-hydroxyethyl)dithiocarbamate and sorption on XAD–4 resin [3]. Hydroxy groups present in sodium *bis*(2-hydroxyethyl)dithiocarbamate caused its metal complexes to be soluble in water at low concentrations. A procedure is described for the concentration of metal ions from solution on to XAD–4 resin, using sodium *bis*(2-hydroxyethyl)dithiocarbamate. Recovery of metals from an artificial seawater spiked with metal salts is reported for solution pH varying from 3.0 to 10.0. Recovery rates approximated 100% with lower recovery rates at extreme pH values. Selectivity could be achieved by pH adjustment and the use of masking agents such as cyanide or ethylenediaminetetraacetic acid. Advantages of the method included high concentration factors, high sample throughput, formation of soluble metal complexes, and easy elution of adsorbed metals.

Brajter *et al.* [4] modified XAD–2 resin by adsorbing pyrocatechol violet on to its hydrophobic surface. They demonstrated that of iron(III), cobalt(II), nickel(II), copper(II), indium(III), lead(II) and bismuth(III), only indium(III) and lead(II) were firmly retained at the appropriate pH and they used this fact to preconcentrate these two elements from potable water prior to their determination by atomic absorption spectrometry. Hiraide *et al.* [5] demonstrated that indium treated XAD–2 resin retained heavy metals complexed with humic acid but not inorganic cations and anions, EDTA complexes or colloidal, hydrated ferric oxides. The sorbed heavy metal–humic acid complexes could be desorbed from the indium treated resin with aqueous nitric acid and they utilised this observation as the basis of a method for the preconcentration and determination of the humic acid complexed cations and anions.

22.1.2 Polyurethane resins

Omar and Bowen [6] have investigated a procedure for the isolation and determination of tin in lake and other non saline waters which involves preliminary preconcentration by adsorption of tin onto polyurethane foam soaked in toluene 3,4,dithiol. A column of polymethane foam extracts tin(II) from non saline waters over the pH range 2–8. The tin is converted to the tetravalent state to remove interferences and determined spectrophotometrically at 650nm as its complex with catechol violet and cetyltrimethyl ammonium bromide.

The extraction of metals from water samples by polyurethane foam has been discussed by several other workers [7–9]. Cobalt was precon-centrated by percolating the water sample through an open cell polyurethane support loaded with 1–(2–pyridylazo)naphthol prior to spectrophotometric determination at 510nm with the 4-(2-pyridylazo) resorcinol complex. Various heavy metals have been precipitated [9] on to polyurethane foam discs loaded with ammonium diethylthiocarbamate to increase the sensitivity of XRF spectrometry and the foam subsequently analysed by X-ray fluorescence spectrometry. Gawargious

Table 22.1 Distribution coefficients (w/v) on different celluloses for metal ions at the 20µg L^{-1} level (pH 11.0)

Element	log K_d					
	Cellulose S&S* 123/3	Cellulose S&S* 123ag	Linters S&S* 124a	Powder S&S* 180a	Powder Avicel* Merck 2330	Native cellulose Merck 2351
Cd^{2+}	3.73	3.58	3.51	3.80	3.85	3.74
Co^{2+}	4.99	4.30	3.85	4.60	4.57	4.98
Cu^{2+}	4.25	4.12	4.10	4.50	3.75	4.75
Fe^{3+}	≥5.0	≥5.0	≥5.0	≥5.0	≥5.0	≥5.0
Mn^{2+}	4.0	3.95	3.72	4.05	3.94	4.40
Pb^{2+}	3.67	3.90	3.70	3.80	3.85	3.85

*Schleicher & Schüll

Source: Reproduced with permission from Burba, P. and Willmer, P.G. [11] Springer Verlag, Heidelberg

et al. [10] adsorbed the coloured bismuth iodide complex from potable water on to thin layers of polyurethane foam prior to measurement of its absorbance at 495 and 600nm against a blank foam in benzene. Recoveries were between 103 and 107%. Copper and silver interfered.

22.1.3 Cellulose and modified cellulose

Natural cellulose

While this material is not suitable for the preconcentration of metals from water it does have some sorptive properties for metals occurring in the environment and these have been studied [11]. Distribution coefficients of various metal ions on different natural cellulose are tabulated in Table 22.1. The distribution coefficients of about 10^5 that can be reached on cellulose in alkaline solutions (pH10–12) are very interesting for analytical trace enrichment. Another remarkable result is the very different sorption of chromium(III) and chromium(V) species on cellulose. For example, it makes possible their analytical separation and speciation in non saline waters.

The K_d values found for individual species, eg copper and zinc, at neutral pH (eg pH8) in 0.5M sodium chloride prove to be nearly constant (eg 10^3 for the two just mentioned) over large concentration ranges (eg from 200ng to 1mg L^{-1}). However, in the presence of other trace metals (eg iron(III) at concentrations of some 100µg L^{-1}) a considerable synergistic enhancement of the distribution coefficients occurs.

2,2'-Diaminodiethylaminecellulose

This material $\left(\text{cellulose—N}\begin{smallmatrix}\diagup\ CH_2CH_2NH_2\\ \diagdown\ CH_2CH_2NH_2\end{smallmatrix}\right)$

has been examined by several workers [12,13] for the preconcentration of transition elements in non saline and potable waters prior to analysis by X-ray fluorescence spectrometry [12,13] and neutron activation analysis [14].

Smits and Van Grieken [12] showed that a 2,2'-diaminodiethylamine–cellulose filter is a very simple and effective method for preconcentrating metals prior to X-ray fluorescence spectrometry and it is an ideal target material in this technique.

Table 22.2 lists some typical detection limits of the system (three times the square root of the background count rate depending on scatter, on the characteristic peaks from blank filter impurities, and proportional to the filter thickness). These hypothetical detection limits assume no mutual interference of the elements; in practical cases the detection limits can be somewhat less favourable. Most elements can be measured down to $0.5\mu g\ L^{-1}$.

Table 22.2 Interference-free detection limits for some elements (collected as cations or anions) for 3000s XRF–measurement after preconcentration on a DEN filter

| Element | Detection limits ($\mu g\ L^{-1}$) for a 1L water sample collected on a 10cm² DEN filter | | | |
| | Secondary fluorescer system | | | |
	Ge	Mo	Ag	Sn
V	0.07	0.25	0.8	0.9
Cr	0.07	0.2	0.6	0.8
Mn	0.05	0.1	1.4	0.8
Fe	2.2	2.5	2.5	3.3
Cu		0.2	0.8	1.0
Zn		0.4	0.4	0.4
As		0.1	0.3	0.2
Se		0.05	0.1	0.1
La*		1.4	3.5	4.9
W*		0.35	0.3	0.5
Hg*		0.3	0.25	0.4
Pb*		0.2	0.4	0.2
U*		0.2	0.2	0.3

*Detection through L lines, K lines with other cases

Source: Reproduced with permission from Smits, J. and Van Grieken, R. [12] Gordon AC Breach

Hyphan cellulose

Burba *et al.* [15] used these celluloses to preconcentrate by factors of 100–200, nanogram quantities of beryllium in non saline water and seawater prior to determination by graphite furnace atomic absorption spectrometry.

Natural levels of beryllium found by this method ranged from less than 1µg L^{-1} in seawater with a 90% recovery to $-$14µg L^{-1} in river water, with a 95% recovery. Less than 1µg L^{-1} beryllium was found in potable water samples. Both hyphan cellulose and pyrogallol cellulose were equally effective in preconcentrating beryllium.

Cellulose hyphan has been used for the preconcentration of lead, copper, nickel, cadmium, and zinc from water [16]. The pH of the samples was shown not to be critical over the range 5–8. The metal ions could be eluted from the column with 1M nitric acid. No interference was caused by salts commonly present in non saline potable waters. Burba *et al.* [17] evaluated this resin for the determination of trace uranium in non saline and potable waters using atomic absorption spectrophotometry, inductively coupled plasma emission spectrometry, and X-ray fluorescence spectrometry for the analytical finish. Down to µg quantities of uranium can be determined by this procedure, without interference from inorganic matter or organic matter including humic acid and lignin. Following the enrichment, traces of uranium are directly determined on the ion exchanger by wavelength dispersive X-ray fluorescence analysis (detection limit (3α) 1.5µg L^{-1} U). On the other hand, after elution by 2M hydrochloric acid uranium can be determined by means of spectrophotometry (U-complex of Arsenazo(III) or Chlorphosphonazo(III)), detection limit 0.1µg L^{-1} U, at 10µg L^{-1} U, or ICPAES (detection limit 0.5µg L^{-1} U, at 10µg L^{-1} U).

p-Phenylenediaminecellulose

p-Phenylenediaminecellulose has been used for the preconcentration of mercury in water [18]. *o*-Aminophenocellulose ether and glyoxal–*bis*(2-mercaptoaniline)cellulose both chelate mercury(I), mercury(II), methylmercury and cadmium(II) [19] which can be subsequently desorbed.

Cellulose derivatives

Cellulose piperazine dithiocarboxylate Imai *et al.* [20] preconcentrated 18 trace elements by passing the non saline water sample through a column packed with this reagent. The solid was then ashed and elements determined by neutron activation analysis. This procedure was applied to fresh water samples.

Nitrocellulose

Taguchi *et al.* [21] preconcentrated cadmium on a nitrocellulose membrane filter as the 1-(2-pyridylazo)-2-naphthol complex. This chelate was then dissolved in 0.5ml of warm concentrated sulphuric acid together with the membrane filter. No interferences were observed in the presence of high levels of sodium chloride, calcium, magnesium, sulphate, and bromide or in the presence of sub-mg L $^{-1}$ levels of zinc(II), iron(III), manganese(II) and copper(II). The lower limit of detection for cadmium was 1.5ng L $^{-1}$.

Ohzeki *et al.* [22] preconcentrated trace amounts of aluminium from potable water by collection on a nitrocellulose membrane filter, prior to spectrophotometric determination using Chrome Azurol S and Zephiramine. Zephiramine, Chrome Azurol S and sodium perchlorate were added to an acidified solution of aluminium(III). The aluminium (III)–Chrome Azurol S–Zephiramine complex and the Zephiramine–perchlorate ion pair precipitate were collected on a membrane filter by filtration under suction. The selectivity of the method was increased by adding a ligand buffer solution of *trans*-cyclohexane-1,2-diaminetetra-acetic acid. The solid-state absorbance of the complex was measured at 640 and 700nm against the blank thin layer and the difference calculated. The method could be used to determine levels of filterable aluminium in tap water (about 2.3ng ml $^{-1}$) with reproducible results.

22.1.4 C₁₈ bonded silica

Heavy metals determinations in potable waters have been carried out by a procedure involving the simultaneous formation of metal dithiocar-bamates and on-line preconcentration using a C_{18} bonded silica precolumn followed by reversed phase high performance liquid chromatographic separation [41]. The C_{18} precolumn was previously loaded with cetyltrimethylammonium bromide (cetrimide)–dithiocarbamate ion pair. Metal dithiocarbamates formed and retained on the precolumn were eluted directly to the analytical column with a gradient of acetonitrile and water containing cetrimide (10mM and 2mM respectively) buffered to pH6.8 (with phosphate buffer, 10mM), separated, and detected by an ultraviolet photodiode array detector. This procedure allowed deter-mination of sub-nanogram quantities of metal ions, including copper(II) and mercury(II) ions in potable water and of cadmium(II), lead(II), cobalt(II), nickel(II) and bismuth(II).

22.1.5 Other non chelating solid adsorbents

Information available on other non chelating solid adsorbents is reviewed in Table 22.3.

Table 22.3 Non chelating solid adsorbents used to preconcentrate cations in non saline waters

Cation	Adsorbent	Analytical finish	Detection limit*	Ref.
Mo	Chitosan	–	–	[23–25]
Fe	PVC containing bathophenathroline	–	–	[26]
Hg	Aniline sulphur resin	–	I	[27]
Sr90 Y90	Bentonite matrix	–	0.02p Ci L^{-1}	[28,29]
Misc.	Misc.	–	–	[30]
Heavy metals	Derivativise on 4-(2-pyridyl azo) resorcinol, then collect on C$_{18}$ SPE cartridge	–	–	[31]
U	TiO$_2$	Luminescence method	0.02	[32]
U	SiO$_2$		–	[33]
Misc.	Immobilised algae, cells and silica immobilised litchen and seaweed biomass	–	–	[34,35]
Pb	SiO$_2$	AAS	0.01	[36]
Trace metals	SiO$_2$	ICPAES	–	[37]
Cs	Prussian blue impregnated ion exchange resin	–	–	[38]
Misc.	Dibenzoyl methane loaded polyurethane foam	–	–	[39]

*µg L^{-1} unless otherwise stated

Source: Own files

22.2 Seawater

22.2.1 C$_{18}$ bonded silica

This preconcentration resin can be used in one of two ways. The method most commonly used is to convert the cations to a chelating complex such as ammonium pyrrolidinedithiocarbamates and then pass the solution down a C$_{18}$ bonded silica column which retains the complex. The concentrated cation complexes are then desorbed with a small volume of aqueous or aqueous organic solvent. In the second method the C$_{18}$ bonded silica (carbon–18 Sep–pak) is loaded with a chelating agent and the solution of cations passed through. Metal chelates are formed and retained on the column.

Table 22.4 Applications of C$_{18}$ bonded silica in preconcentration of cations in seawater

Cation		Analytical finish	Detection limit*	Ref.
Free Cu^{2+}	Speciation study	–	–	[43]
Co	–	Laser excited atomic fluorescence spectroscopy with graphite electrothermal atomiser	–	[44]
Hg	–	AAS	0.016	[45]
Zn	–	AAS	2.4ng dm^3	[46]
Pb	Complexing with ammonium pyrroli-dinedithiocarbamate followed by collection on C$_{18}$ silica column	AAS	–	[47]

*µg L^{-1} unless otherwise stated

Source: Own files

Sturgeon et al. [40] preconcentrated antimony(III) and antimony(V) from coastal and seawaters by adsorption of their ammonium pyrrolidine-dithiocarbamate chelates on to C$_{18}$ bonded silica prior to determination by graphite furnace atomic absorption spectrometry. A detection limit of 0.05µg L^{-1} was achieved.

Sturgeon et al. [40,42] preconcentrated selenium and antimony from seawater on C$_{18}$ bonded silica gel prior to determination by graphite furnace atomic absorption spectrometry. The method was based on the complexation of selenium(IV), antimony(III) and antimony(V) with ammonium pyrrolidinedithiocarbamate. These complexes were adsorbed on to a column of C$_{18}$ bonded silica gel, then eluted with methanol, followed by evaporation to near dryness. The residue was taken up in 1% nitric acid. Concentration factors of 200 could be obtained. Detection limits for selenium(IV), antimony(III) and antimony(V) were 7, 50 and 50ng L^{-1} respectively, based on a 300ml sample volume.

Other applications of C$_{18}$ bonded silica are reviewed in Table 22.4.

22.2.2 XAD resins

Fujita and Iwashima [48] preconcentrated mercury compounds in seawater by first forming the diethyldithiocarbamate and then concentrating this on XAD–2 resin. The resin was eluted with methanol/3M hydrochloric acid; the organic mercury was extracted with benzene and then back-extracted with cysteine solution. The organic mercury in the cysteine solution and the total mercury adsorbed on the resin were determined by flameless atomic absorption spectrometry. The method was applied to determinations of mercury levels in seawater in and around the Japanese archipelago. The lower limit of detection in seawater is 0.1ng L $^{-1}$ for organic mercury, using 80L samples.

Hirose and Sugimura [49] investigated the speciation of plutonium in seawater using adsorption of plutonium(IV)–xylenol orange and plutonium–arsenazo(III) complexes on the macroreticular synthetic resin XAD–2. Xylenol orange was selective for plutonium(IV) and arsenazo(III) for total plutonium. Plutonium levels were determined by α-ray spectrometry.

Isshiki and Nakayama [50] preconcentrated cobalt in seawater by complexing it with 4-(2-thiazolylazo)resorcinol to form a water soluble inert complex and passing this solution down a column of XAD–4 resin. The complex was then eluted with methanol/chloroform, the eluate digested with nitric and perchloric acids, and cobalt determined by graphite furnace atomic absorption spectrometry.

Mackey [51] investigated the suitability of Amberlite XAD–1 resin for extracting organic complexes of copper, zinc and iron from seawater. At low flow rates and at loading capacities far below theoretical values, the adsorption of these metals is not reproducible and the results are reminiscent of the behaviour observed when the adsorption capacity is being exceeded or flow rates are too high. It is suggested that the resin also adsorbs small but significant amounts of inorganic ions from seawater and that this effect makes the resin unsuitable for quantitative measurements of trace metal speciation.

22.2.3 Polyacrylamidoxime resin

Collela *et al.* [52] studied the use of this resin for the preconcentration of low levels of iron, manganese, copper, zinc, cobalt, nickel, lead, cadmium and silver in seawaters.

The metal ions were sequestered on a poly(acrylamidoxime) resin from laboratory prepared solutions. Recovery of the sequestered metals was achieved by equilibration of the resin matrix with either a 1:1 nitric acid/water mixture, a 1:1 hydrochloric acid/water mixture, or a 1M thiosulphate solution. Sequestered metals were at least 90% recovered with a standard deviation of ≤5%. Regeneration of the resin was achieved by equilibration of the resin with a 3M ammonia solution after the acid equilibration. The resin was applied for the separation and simultaneous concentration of iron(III), copper(II), cadmium(II), lead(II) and zinc(II) from seawater. Metals were removed from the resin by equilibrating with a 1:1 hydrochloric acid/water mixture and their concentrations

Table 22.5 Determination of trace metals in seawater sample[a]

Metal	Determination by resin preconcentration,[b] $\mu g\ L^{-1}$	Direct determination,[c] $\mu g\ L^{-1}$
Fe(III)	92 (5.0)	62 (4.0)
Cu(II)	63 (6.0)	62 (4.0)
Cd(II)	2.8 (0.30)	3 (0.5)
Pb(II)	3.2 (0.60)	3 (0.4)
Zn(II)	7.0 (0.70)	10 (0.60)

[a]Standard deviation of three determinations in parentheses
[b]Cd(II) and Pb(II) determined by differential pulse polarography, others by atomic absorption spectrometry
[c]Determined by differential pulse anodic stripping voltammetry

Source: Reproduced with permission from Collela, M.B. et al. [52] American Chemical Society

determined by atomic absorption spectrometry. Metal concentrations as determined by the resin method are in good agreement with the values determined directly on samples by either differential pulse polarography or differential pulse anodic stripping voltammetry.

Recoveries of metals ranged from 17% (cadmium) to 98% (iron). Good agreement was obtained between the preconcentration method and direct analysis (Table 22.5).

22.3 Miscellaneous solid non chelating adsorbents

Other non chelating solids that have been used in the preconcentration of solid cations in seawater are reviewed in Table 22.6.

22.4 Estuary waters

22.4.1 Chelex–100

Paulson [64] has examined the effects of flow rate and pretreatment on the extraction of iron, manganese, copper, nickel, cadmium, lead and zinc from estuarine and coastal seawater by Chelex–100. During the extraction of previously acidified estuarine samples, organic material still retains some capacity to inhibit the extraction of trace metals by Chelex–100. Previous studies have indicated that heating or ultraviolet oxidation of samples reduces the capacity of this organic matter to inhibit the extraction of trace metals by Chelex–100. The results of this study, using fresh samples indicate that decreasing the flow rate to 0.2mL min $^{-1}$ is also

Table 22.6 Miscellaneous solids used in the preconcentration of cations in seawater

Cation	Solid adsorbent	Analytical finish	Detection limit*	Ref.
Mo,V	Cellulose phosphate	ICPAES	–	[53]
Se(IV), Se(VI)	Thiol cotton (subsequent desorption with nitric and magnesium nitrate)	Cathodic stripping voltammetry	Se(IV) 0.009 Se 0.018	[54]
Cd, Co, Cu, Fe, Mn, Ni, Pb, Zn	Maleic acid/ammonium hydroxide buffer system	–	–	[55]
Cu, Cd, Pb, Zn	Carboxylated polyethylene–imine–polymethylene–phenylene isocyanate	–	–	[56]
Ni	Poly(triamino phenyl) glyoxal (nickel desorbed from above resin and extracted into methyl isobutyl ketone solution of ammonium pyrrolidinedithiocarbamate)	AAS	–	[57]
Cu	Chitosan, chelated copper extracted with 1,10 phenrinthroline	AAS	<1	[58,59]
Zn	–	Anodic stripping voltammetry	–	[60]
Heavy metals	Silica gel	X-ray fluorescence spectroscopy	–	[61]
Cr(III)	Silica gel	Voltammetry	1pmole L^{-1}	[62]
Pb	Chelex-100	ICPAES	0.0006	[63]

*µg L^{-1} unless otherwise stated

Source: Own files

an effective means of increasing the retention of trace metals by Chelex–100. Additional benefits of the slow–flow column extraction method include improvements in precision and the elimination of pretreatment procedures that could cause contamination or reduce the extractability of iron. Aged acidified sample require heating of the sample prior to extraction. Controlled contamination can be minimised for most metals by the pre-extraction of the buffer solution.

22.5 Potable waters

22.5.1 Polystyrene supported poly(maleic anhydride) resin

This resin has been used to preconcentrate lead in potable water prior to determination by flame atomic absorption spectrometry [65]. Chelex–100 in the calcium form proved unsuitable for lead preconcentration. This method met the European Community directive limiting the content of lead in potable water to $50\mu g$ L $^{-1}$ [66].

A plumbo solvent potable water having a low alkalinity (18mg L $^{-1}$ as $CaCO_3$) and low pH (6.58) had most of the lead in the smaller size fractions but a significant proportion is particulate (ie >$0.4\mu m$). The Ca–Chelex–100 extraction was not 100% efficient for lead removal from any of the sub-samples. Incomplete recoveries may be due to the presence of colloidal species or to non-Chelex labile organic complexes of lead or both. Slow dissociation of complexes relative to the solution/resin contact time may also have an effect. The polystyrene supported poly (maleic anhydride) resin behaved in a significantly different manner. The column extraction (50ml of sample) gave 100% lead removal from the 0.4 and $0.08\mu m$ filtered waters but particulate lead was not recovered from the unfiltered and $12\mu m$ filtered samples. A batch extraction on the unfiltered sample after passage through the column chelated all the remaining lead from solution thereby indicating its potential for total lead determinations.

Some potable water samples of relatively high alkalinity were examined because such waters can attack lead. Total lead in unfiltered samples was determined by atomic absorption spectrometry directly and following a 48h batch extraction of 50ml of water. Lead removal by the poly(maleic anhydride) resin was complete for both high and low alkalinity samples.

22.5.2 8-quinolinol supported on poly(chlorotrifluoroethylene) resin

Yamaguchi et al. [67] preconcentrated copper, cadmium, iron, manganese, nickel and zinc as 8–quinolinol complexes, silver, cadmium, copper and zinc as bismuthiol(II) complexes and cadmium, copper and zinc as 8-

Table 22.7 Applications of non chelating solid adsorption to the preconcentration of cations in effluents

Cation	Type of sample	Adsorption medium	Analytical finish	Ref.
Mo	Nuclear fuel solutions	Chitosan	–	[23,24,25, 59,68,69]
Zr, Nb, Ce, Ru	Nuclear fuel solutions	Chitosan	–	[69]
Hg	Waste waters	Liquid chelating exchange resin (Kebucleric or Versatic–10)	AAS	[70]

Source: Own files

quinolinol-5-sulphonic complexes on poly(chlorotrifluoroethylene) resin prior to the determination of these elements in potable waters.

22.5.3 Chitosan

Jha *et al.* [28] investigated the efficiency of removal of cadmium from potable water using chitosan. Studies with potable water showed that cadmium removal capacity was not affected by the presence of calcium, bicarbonate or chloride ion.

22.6 Effluents

Some applications of non chelating solid adsorption techniques to the preconcentration of cations are reviewed in Table 22.7.

22.7 Ground and surface waters

200–450µg L^{-1} bismuth have been determined [71] by sorbing the coloured bismuth iodide complex on to a polyether polyurethane foam and measuring its net absorbance at 495 and 600nm against a blank foam in benzene. Copper and silver ions interfere in this procedure.

Fitzgerald *et al.* [72] reported a cold trap preconcentration technique for the determination of trace amounts of mercury in water. Kramer and Neidhart [73] determined µg L^{-1} levels of mercury by using an aniline–S resin for the selective enrichment of mercury from surface waters.

References

1 Mackey, D.J. *Journal of Chromatography*, **236**, 81 (1982).

2 Sakai, Y. and Mori, N. *Talanta*, **33**, 161 (1986).
3 King, J.N. and Fritz, J.S. *Analytical Chemistry*, **57**, 1016 (1985).
4 Brajter, K., Olbrych-Sleszynska, E. and Stastiewicz, M. *Talanta*, **35**, 65 (1988).
5 Hiraide, M., Arima, Y. and Mizurka, A. *Analytica Chimica Acta*, **200**, 171 (1987).
6 Omar, M. and Bowen, H.J.M. *Analyst (London)*, **107**, 654 (1982).
7 Maloney, M.P., Moody, G.J. and Thomas, J.D.R. *Analyst (London)*, **105**, 1087 (1980).
8 Braun, T. and Abbas, M.N. *Analytica Chimica Acta*, **119**, 113 (1980).
9 Torak, S., Braun, P., Van Dyck, P. and Van Grieken, R. *X-ray Spectrometry*, **15**, 7 (1986).
10 Gawargious, Y.A., Abbas, M.N. and Hassan, H.N.A. *Analytical Letters (London)*, **21**, 1477 (1988).
11 Burba, P. and Willmer, P.G. *Fresenius Zeitschrift für Analytische Chemie*, **324**, 298 (1986).
12 Smits, J. and Van Grieken, R. *International Journal of Environmental Analytical Chemistry*, **9**, 81 (1981).
13 Reggeks, G. and Van Grieken, R. *Fresenius Zeitschrift für Analytische Chemie*, **317**, 520 (1984).
14 Voutsa, D., Samara, K. and Fyticaros, K. *Fresenius Zeitschrift für Analytische Chemie*, **330**, 596 (1988).
15 Burba, P.L., Willmer, P.G., Betz, M. and Fuchs, S. *International Journal of Environmental Analytical Chemistry*, **13**, 177 (1983).
16 Brajster, K. and Slonawka, B. *Analytica Chimica Acta*, **185**, 271 (1986).
17 Burba, P., Cebulo, M. and Broekaert, J.A.C. *Fresenius Zeitschrift für Analytische Chemie*, **318**, 1 (1984).
18 Masri, M.S. and Friedman, M. *Environmental Science and Technology*, **6**, 745 (1972).
19 Goldbach, K. and Lieser, K.H. *Fresenius Zeitschrift für Analytische Chemie*, **311**, 183 (1982).
20 Imai, S., Muroi, M., Hamaguchi, A. and Kuyama, H. *Analytical Chemistry*, **55**, 1215 (1983).
21 Taguchi, S., Yamazaki, S. Yamamoto, A. *et al. Analyst (London)*, **113**, 1695 (1988).
22 Ohzeki, K., Uno, T., Nukatsuku, I. and Ishida, R. *Analyst (London)*, **113**, 1545 (1988).
23 Muzzarelli, R.A.A. and Rochetti, R. *Analytica Chimica Acta*, **64**, 371 (1973).
24 Muzzarelli, R.A.A. *Analytica Chimica Acta*, **54**, 133 (1971).
25 Vanderborght, B.M. and Van Grieken, R.E. *Analytical Chemistry*, **49**, 311 (1977).
26 Saito, T. *Journal of AOAC International*, **77**, 1031 (1994).
27 Kramer, H.J. and Neidhart, J. *Journal of Radioanalytical Chemistry*, **37**, 835 (1977).
28 Jha, I.N., Iyengor, L. and Rao, A.V.S.P. *Journal of Environmental Engineering*, **114**, 962 (1988).
29 Mundschenk, H. *Gewasserkundliche Mittalungen*, **23**, 64 (1979).
30 Nickson, R.A., Hill, S.J. and Worsfold, P.J. *Analytical Proceedings (London)*, **32**, 387 (1995).
31 Leepipatpikoon, V. *Journal of Chromatography*, **697**, 137 (1995).
32 Nikitina, S.A. and Stephanov, A.V. *Radiokhimilya*, **25**, 606 (1986).
33 Havel, J., Vrchlabsky, M. and Kohu, Z. *Talanta*, **39**, 795 (1992).
34 Mahan, C.A. and Holcombe, J.A. *Spectrochimica Acta, Part B*, **47B**, 1483 (1992).
35 Ramelow, G.J., Lui, L., Himel, C. *et al. International Journal of Environmental Analytical Chemistry*, **53**, 219 (1993).
36 Rodriguez, D., Fernandez, P., Perez-Conde, C., Gutierrez, A. and Camera, C. *Fresenius Journal of Analytical Chemistry*, **349**, 442 (1994).
37 Najiri, Y., Kawai, T., Otsuki, A. and Fuwa, K. *Water Research*, **19**, 503 (1985).

38 Godoy, J.M., Guimaraes, J.R.D. and Carvalho, Z.L. *Journal of Environmental Radioactivity*, **20**, 213 (1993).
39 Aziz, M. Beheir, G. and Shakir, J. *Journal of Radioanalytical and Nuclear Chemistry*, **172**, 319 (1993).
40 Sturgeon, R.E., Wilie, S.N. and Berman, S.S. *Analytical Chemistry*, **57**, 6 (1985).
41 Irth, H., De Jong, G.J., Brinkman, U.A.T. and Frei, R.W. *Analytical Chemistry*, **59**, 98 (1987).
42 Sturgeon, R.E., Willie, S.N. and Berman, S.S. *Analytical Chemistry*, **57**, 2311 (1985).
43 Sunda, W.G. and Hanson, A.K. *Limnology and Oceanography*, **325**, 37 (1987).
44 Yuzetovsky, A., Lonardo, R.F., Wang, M. and Michel, R.G. *Journal of Analytical Atomic Spectroscopy*, **9**, 1195 (1994).
45 Fernandez Garcia, M., Pereiro Garcia, R., Bondel Garcia, N. and Sanz-Medel, A. *Talanta*, **41**, 1833 (1994).
46 Akatsuka, K., Katoh, T., Nobuyama, N. *et al. Analyticl Science*, **12**, 109 (1996).
47 Liu, Z.S. and Huang, S.D. *Spectrochimica Acta, Part B*, **50B**, 197 (1995).
48 Fujita, M. and Iwashima, K. *Environmental Science and Technology*, **15**, 929 (1981).
49 Hirose, K. and Sugimura, Y.J. *Radioanalytical and Nuclear Chemistry Articles*, **92**, 363 (1985).
50 Isshiki, K. and Nakayama, E. *Analytical Chemistry*, **59**, 291 (1978).
51 Mackey, D.J. *Marine Chemistry*, **11**, 169 (1982).
52 Collela, M.B., Siggia, S. and Barnes, R.M. *Analytical Chemistry*, **52**, 2347 (1980).
53 Ogura, H. and Oguma, K. *Microchemical Journal*, **49**, 220 (1994).
54 Yang, Y., Huang, W. and Liu, D. *Xiamen Daxue, Xuebao, Ziran Kexueban*, **26**, 96 (1987).
55 Dai, S.E., Chen, T.C. and Wong, G.T.F. *Analytical Chemistry*, **62**, 774 (1990).
56 Horvath, Z. and Barnes, R.M. *Analytical Chemistry*, **58**, 1352 (1986).
57 Zlatkis, A., Bruening, W. and Bayer, E. *Analytica Chimica Acta*, **56**, 399 (1971).
58 Muzzarelli, R.A.A. and Rochetti, R. *Analytica Chimica Acta*, **69**, 35 (1974).
59 Ricardo, A.A., Muzzarelli, R.A.A. and Tubertini, O. *Journal of Chromatography*, **47**, 414 (1970).
60 Muzzarelli, R.A.A. and Sipos, L. *Talanta*, **18**, 853 (1971).
61 Prange, H., Omatsu, M. and Muskimoto, S. *X-ray Spectrometry*, **13**, 83 (1984).
62 Boussewart, M. and Van den Berg, C.M.G. *Analyst (London)*, **119**, 1349 (1994).
63 Reimer, P.A. and Miyazaki, A. *Journal of Analytical Atomic Spectroscopy*, **7**, 1238 (1992).
64 Paulson, A.J. *Analytical Chemistry*, **58**, 183 (1986).
65 De Mora, S.J. and Harrison, R.M. *Analytica Chimica Acta*, **153**, 307 (1983).
66 Official Journal of the European Community. No. L229/11(80/778/EEC), 23 (1980).
67 Yamaguchi, T., Zhang, L., Matsumoto, K. and Terada, K. *Analytical Science*, **8**, 85 (1992).
68 Yoshimura, K., Hiraoka, S. and Tarutani, T. *Analytica Chimica Acta*, **142**, 101 (1982).
69 Muzzarelli, R., Rochetti, R. and Marangio, G. *Journal of Radioanalytical Chemistry*, **10**, 17 (1972).
70 Yamada, E., Yamada, T. and Sato, M. *Analytical Science*, **8**, 863 (1992).
71 Gawagious, Y.A., Abbas, M.N. and Hassan, H.N.A. *Analytical Letters (London)*, **21**, 1477 (1988).
72 Fitzgerald, W.F., Lyons, W.B. and Hunt, C.D. *Analytical Chemistry*, **46**, 1882 (1974).
73 Kramer, L.J. and Neidhart, J. *Journal of Radioanalytical and Nuclear Chemistry*, **37**, 836 (1977).

Chapter 23

Cations: Adsorption on active carbon

23.1 Non saline waters

Two approaches have been used for the preconcentration of cations on active carbon. Either the metals are chelated with an organic complexing agent and passed down a column of active carbon which adsorbs the metal complexes, or active carbon is modified by reaction with an organic chelating agent and then the solution of metal ions is passed through the column and thereby adsorbed. Examples of the former approach are the preconcentration of molybdenum as its ammonium pyrrolidinedithio-carbamate complex on charcoal [1] and the preconcentration of manganese, iron, cobalt, nickel, copper and zinc as their 8-quinolinates on activated carbon [2]. An example of the latter approach is the separation and preconcentration of traces of copper and lead from macro amounts of calcium magnesium, sodium and potassium by adsorption from the sample on to active carbon modified with 8-hydroxyquinoline, dithizone or diethyidithiocarbamate [3].

23.1.1 Atomic adsorption spectrometry

Samchuk [4] developed an atomic absorption method for the determination of gold, silver, gallium, indium, thallium, cadmium, molybdenum, nickel and copper in non saline waters, with pre-concentration on carbon modified by complex-forming organic reagents and on chelated sorbents. The methods for preparing the modified carbons and chelated sorbents are described. The work was done on a two-beam atomic absorption spectrometer with a graphite atomiser and deuterium background corrector. Hall *et al.* [5] preconcentrated gold from non saline water on activated carbon discs prior to analysis by graphite furnace atomic absorption spectrometry. This enabled 10–15ng L $^{-1}$ of gold to be determined.

23.1.2 Inductively coupled plasma atomic emission spectrometry

Hall *et al.* [6] preconcentrated tungsten and molybdenum from non saline spring water on to activated charcoal prior to analysis by inductively coupled plasma atomic emission spectrometry and inductively coupled plasma mass spectrometry. The detection limit for both elements by inductively coupled plasma mass spectrometry was 0.06µg L^{-1}, and by ICPAES the detection limits were 1.2 and 0.4µg L^{-1}, respectively, for tungsten and molybdenum.

23.1.3 Neutron activation analysis

Lieser *et al.* [7] applied neutron activation analysis to the determination of 15 trace elements with particular reference to the limits of detection obtained for different elements when employing a preconcentration technique based on adsorption of the metals as their dithizonates or diethyldithiocarbamates on charcoal. Metal analysis was carried out directly on the solid charcoal by neutron activation analysis. The method was applied to trace element determinations in river waters and North Sea waters. The 15 elements determined were cobalt, chromium, europium, iron, mercury, lanthanum, uranium, molybdenum, scandium, selenium, uranium, zinc, silver, gold, cadmium and cerium. Recoveries were between 75 and 100%. For arsenic and antimony, recoveries were less satisfactory (55–70%).

The following procedure was found to be most efficient for preconcentrating elements in estuary and seawater. 1L of water was first shaken at pH5.5 with a suspension of 30mg charcoal (Carbopurvor, 4R, Degussa) and 1ml of an acetone solution containing 1mg sodium diethyldithiocarbamate. After filtration the solution is brought up to pH8.5 by addition of 0.1M ammonia and shaken again with a suspension of 30ml charcoal and 1ml of an acetone solution containing 1mg dithizone. Both fractions of charcoal are combined, dried and filled into a quartz ampoule for irradiation.

23.1.4 X-ray fluorescence spectrometry

Van der Bought and Van Grieken [2] optimised the conditions of the procedure for the preconcentration on carbon of some 20 elements including manganese iron, cobalt, nickel, copper, zinc and cadmium as their 8-quinolinates (8-hydroxyquinolates). For a mixture of about 20 elements, they obtained an enrichment factor of 10000 a precision of 5 to 10%, and an element recovery of 85–100% depending on the element. Both X-ray fluorescence spectrometry and neutron activation analysis were used for the determination of preconcentrated metal 8-quinolinates on the activated carbon.

Van der Bought and Van Grieken [8] preconcentrated transition metals by conversion to oxinates and subsequent adsorption on to activated carbon which was then examined by X-ray fluorescence spectrometry or neutron activation analysis.

Robberecht and Van Grieken [9] reported that selenate and selenite in various environmental waters can be determined by X-ray energy spectrometry after preconcentration of elemental selenium on activated carbon. Selenite is reduced to elemental selenium with ascorbic acid. Selenate plus selenite is determined after refluxing the samples with thiourea in sulphuric acid and then adsorbing the elemental form on activated carbon. Selenate is determined by difference. The limit of detection is 50ng L^{-1} for selenite and 60ng L^{-1} for total selenium. The coefficient of variation is approximately 10% for both species at the 0.5–1µg L^{-1} selenium level. Humic material, common abundant ions and oxidising substances do not interfere.

Orvini et al. [10] presented an extended speciation scheme for the determination of selenium in polluted river waters. They used filtration, charcoal adsorption, selective reduction by ascorbic acid followed by charcoal adsorption and collection on anion exchange resin for determination of suspended and colloidal compounds, selenite and selenate, respectively. They found an important fraction of the selenium to be present in the colloidal fraction. This was probably due to the fact that at pH6.5 organoselenium compounds and some selenite adsorb on the charcoal and are included in this fraction.

23.2 Seawater

23.2.1 Neutron activation analysis

Lieser et al. [7] studied the application of neutron activation analysis to the determination of trace elements in seawater with particular reference to the limits of detection and reproducibility obtained for different elements when comparing different preliminary concentration techniques, such as adsorption on charcoal (also cellulose and quartz), and complexing agents such as dithizone and sodium diethyldithiocarbamate. In these procedures 1L of seawater was shaken with 60mg charcoal for 15min. Complexing agents were added in amounts of 1mg, dissolved in 1ml acetone. The pH was 5.5 or it was adjusted to 8.5 by addition of 0.1M ammonia. The charcoal was filtered off and irradiated. Results of three sets of experiments with charcoal alone, charcoal in the presence of dithizone, and charcoal in the presence of sodium diethyldithiocarbamate are presented in Table 23.1. The following elements are adsorbed to an extent of 75 to 100%: silver, gold, cerium, cadmium, cobalt, chromium, europium, iron, mercury, lanthanum, scandium, uranium and zinc. The amount of

Table 23.1 Mean values and standard deviations found by adsorption on charcoal (KF = correction factor)

Element	Conc. in the standardised water samples (g L⁻¹)	Found by adsorption on activated charcoal, pH8.5; three determinations			Found by adsorption on activated charcoal in presence of dithizone pH8.5; seven determinations			Found by adsorption on activated charcoal in presence of NaDDTC, pH5.5; seven determinations		
		g L⁻¹	%	KF	g L⁻¹	%	KF	g L⁻¹	%	KF
Ag	1×10^{-7}	$(8.5 \pm 1.4) \times 10^{-8}$	85	1.18	$(8.5 \pm 0.3) \times 10^{-8}$	85	1.18	$(5.5 \pm 1.5) \times 10^{-8}$	55	1.82
As	1×10^{-6}	$(6.6 \pm 1.2) \times 10^{-7}$	66	1.52	$(6.7 \pm 1.5) \times 10^{-7}$	67	1.49	$(3.2 \pm 1.9) \times 10^{-7}$	32	3.13
Au	1×10^{-9}	$(1.1 \pm 0.3) \times 10^{-9}$	100	1.00	$(1.1 \pm 0.1) \times 10^{-9}$	100	1.00	$(1.0 \pm 1.7) \times 10^{-9}$	100	1.00
Br	5×10^{-7}	$(5.3 \pm 1.0) \times 10^{-7}$	10^{-3}	–	$(2.1 \pm 0.5) \times 10^{-7}$	10^{-3}	–	$(3.6 \pm 1.7) \times 10^{-7}$	10^{-3}	–
Ca	1×10^{-1}	$(1.2 \pm 2.3) \times 10^{-3}$	–	–	$(1.7 \pm 0.2) \times 10^{-3}$	2	–	$(1.6 \pm 0.8) \times 10^{-3}$	2	–
Cd	1×10^{-6}	$(4.9 \pm 1.3) \times 10^{-7}$	49	2.04	$(9.5 \pm 2.7) \times 10^{-7}$	95	1.05	$(7.7 \pm 2.7) \times 10^{-7}$	77	1.30
Ce	1×10^{-6}	$(8.7 \pm 2.1) \times 10^{-7}$	87	1.15	$(8.2 \pm 1.1) \times 10^{-7}$	82	1.22	$(4.3 \pm 1.6) \times 10^{-7}$	43	2.33
Co	1×10^{-6}	$(4.0 \pm 0.6) \times 10^{-7}$	40	2.50	$(8.1 \pm 0.8) \times 10^{-7}$	81	1.23	$(7.6 \pm 1.0) \times 10^{-7}$	76	1.32
Cr	1×10^{-6}	$(9.1 \pm 0.3) \times 10^{-7}$	91	1.10	$(9.6 \pm 0.3) \times 10^{-7}$	96	1.04	$(3.6 \pm 0.2) \times 10^{-7}$	36	2.78
Eu	5×10^{-7}	$(5.2 \pm 0.4) \times 10^{-7}$	100	1.00	$(4.7 \pm 0.3) \times 10^{-7}$	95	1.05	$(3.8 \pm 1.7) \times 10^{-7}$	76	1.32
Fe	1×10^{-4}	$(7.4 \pm 1.5) \times 10^{-5}$	74	1.35	$(7.7 \pm 0.8) \times 10^{-5}$	77	1.30	$(7.0 \pm 0.6) \times 10^{-5}$	70	1.43
Hg	1×10^{-7}	$(9.7 \pm 0.2) \times 10^{-8}$	97	1.03	$(1.0 \pm 0.3) \times 10^{-7}$	100	1.00	$(1.0 \pm 0.1) \times 10^{-7}$	100	1.00
K	4×10^{-1}	$(3.2 \pm 0.4) \times 10^{-5}$	10^{-2}	–	$(2.1 \pm 0.3) \times 10^{-1}$	10^{-2}	–	$(4.3 \pm 0.9) \times 10^{-5}$	10^{-2}	–
La	1×10^{-6}	$(1.0 \pm 0.1) \times 10^{-6}$	100	1.00	$(1.0 \pm 0.1) \times 10^{-4}$	100	1.00	$(9.1 \pm 0.3) \times 10^{-7}$	91	1.10
Mo	1×10^{-6}	$(5.0 \pm 3.3) \times 10^{-7}$	50	2.00	$(2.1 \pm 0.9) \times 10^{-7}$	21	4.76	$(1.0 \pm 0.1) \times 10^{-6}$	100	1.00
Na	5	$(1.4 \pm 0.3) \times 10^{-5}$	10^{-1}	–	$(1.6 \pm 0.2) \times 10^{-5}$	10^{-1}	–	$(3.2 \pm 1.5) \times 10^{-5}$	10^{-1}	–
Sb	2×10^{-7}	$(1.8 \pm 0.5) \times 10^{-7}$	18	5.56	$(4.0 \pm 0.9) \times 10^{-7}$	40	2.50	$(5.6 \pm 2.0) \times 10^{-7}$	56	1.79
Sc	2×10^{-7}	$(1.9 \pm 0.2) \times 10^{-7}$	95	1.05	$(2.0 \pm 0.1) \times 10^{-7}$	100	1.00	$(1.4 \pm 0.1) \times 10^{-7}$	70	1.43
Se	1×10^{-6}	$(7.7 \pm 1.8) \times 10^{-7}$	77	1.30	$(6.1 \pm 1.7) \times 10^{-7}$	61	1.64	$(4.0 \pm 0.9) \times 10^{-7}$	40	2.50
U	1×10^{-7}	$(1.1 \pm 0.1) \times 10^{-7}$	100	1.00	$(1.2 \pm 0.2) \times 10^{-7}$	100	1.00	$(7.8 \pm 0.2) \times 10^{-8}$	78	1.28
Zn	1×10^{-6}	$(9.6 \pm 0.4) \times 10^{-7}$	96	1.04	$(1.0 \pm 0.1) \times 10^{-6}$	100	1.00	$(1.0 \pm 0.1) \times 10^{-6}$	100	1.00

Source: Reproduced with permission from Lieser, K.H. et al. [7] Elsevier Sequoia, Lausanne

Table 23.2 Determination of trace elements in seawater (mean ± SD). Samples taken at 54°3' North, latitude, 6°30' east, longitude in the North Sea

Element	Water without suspended material	Suspended material	Water with suspended material
Ag	$(8.8 \pm 0.4) \times 10^{-9}$	$(3.6 \pm 0.3) \times 10^{-9}$	$(8.7 \pm 0.4) \times 10^{-9}$
As	$(3.5 \pm 0.3) \times 10^{-7}$	$(3.4 \pm 1.2) \times 10^{-8}$	$(4.7 \pm 0.7) \times 10^{-7}$
Au	$(3.4 \pm 0.3) \times 10^{-10}$	$(4.5 \pm 2.0) \times 10^{-10}$	$(3.9 \pm 0.2) \times 10^{-10}$
Ba	$(5.7 \pm 0.1) \times 10^{-7}$	$(4.1 \pm 0.6) \times 10^{-7}$	$(1.3 \pm 0.5) \times 10^{-6}$
Br	$(5.5 \pm 0.1) \times 10^{-7}$	$(4.0 \pm 3.8) \times 10^{-6}$	$(3.7 \pm 0.2) \times 10^{-6}$
Ca	$(3.6 \pm 0.2) \times 10^{-5}$	$(1.5 \pm 0.3) \times 10^{-4}$	$(2.0 \pm 0.5) \times 10^{-4}$
Cd	$< 10^{-6}$	$< 10^{-6}$	$< 10^{-6}$
Ce	$(3.4 \pm 0.3) \times 10^{-4}$	$(2.4 \pm 1.5) \times 10^{-6}$	$(1.0 \pm 0.2) \times 10^{-6}$
Co	$(4.5 \pm 0.3) \times 10^{-4}$	$(2.3 \pm 0.8) \times 10^{-4}$	$(6.7 \pm 0.2) \times 10^{-8}$
Cr	$(1.4 \pm 0.1) \times 10^{-7}$	$(1.3 \pm 0.3) \times 10^{-7}$	$(2.9 \pm 0.2) \times 10^{-7}$
Eu	$(8.2 \pm 1.4) \times 10^{-10}$	$(4.4 \pm 1.9) \times 10^{-9}$	$(5.6 \pm 1.7) \times 10^{-9}$
Fe	$(1.5 \pm 1.2) \times 10^{-5}$	$(3.5 \pm 1.2) \times 10^{-5}$	$(6.8 \pm 1.5) \times 10^{-5}$
Hg	$(2.2 \pm 0.2) \times 10^{-6}$	$< 5 \times 10^{-9}$	$< 5 \times 10^{-9}$
K	$(3.6 \pm 1.9) \times 10^{-5}$	$(2.9 \pm 0.2) \times 10^{-5}$	$(4.7 \pm 1.4) \times 10^{-5}$
La	$(3.2 \pm 0.1) \times 10^{-8}$	$(2.8 \pm 2.1) \times 10^{-8}$	$(7.4 \pm 1.4) \times 10^{-8}$
Mo	$(4.4 \pm 1.4) \times 10^{-8}$	$(2.1 \pm 0.1) \times 10^{-8}$	$(6.1 \pm 0.5) \times 10^{-8}$
Na	$(3.6 \pm 0.3) \times 10^{-4}$	$(4.7 \pm 6.9) \times 10^{-4}$	$(1.7 \pm 0.7) \times 10^{-4}$
Sb	$(5.7 \pm 0.4) \times 10^{-9}$	$(2.3 \pm 0.8) \times 10^{-9}$	$(1.3 \pm 0.2) \times 10^{-8}$
Sc	$(4.5 \pm 0.3) \times 10^{-8}$	$(1.9 \pm 2.0) \times 10^{-8}$	$(2.3 \pm 0.4) \times 10^{-8}$
Se	$(4.5 \pm 0.3) \times 10^{-8}$	$(3.5 \pm 0.9) \times 10^{-8}$	$(6.3 \pm 0.7) \times 10^{-8}$
U	$(3.3 \pm 1.0) \times 10^{-8}$	$(1.5 \pm 0.8) \times 10^{-8}$	$(5.6 \pm 0.9) \times 10^{-8}$
Zn	$(2.3 \pm 0.1) \times 10^{-6}$	$(3.3 \pm 1.5) \times 10^{-7}$	$(3.9 \pm 0.2) \times 10^{-6}$

Source: Reproduced with permission from Lieser, K.H. *et al.* [7] Elsevier Sequoia, Lausanne

sodium is reduced to about 10^{-6}, bromine to about 10^{-5}, and calcium to about 10^{-2}. Analyses of North Sea water and suspended solids obtained by this procedure are tabulated in Table 23.2.

Van den Shoot *et al.* [1] preconcentrated molybdenum from seawater as its pyrrolidinedithiocarbamate chelate on charcoal.

Okutani *et al.* [11] adsorbed beryllium in seawater onto activated carbon as its acetylacetone complex prior to determination by graphite furnace atomic absorption spectrometry.

In this method the beryllium–acetylacetonate complex is adsorbed easily onto activated carbon at pH8–10. The activated carbon which adsorbed the beryllium–acetylacetonate complex was separated and dispersed in pure water. The resulting suspension was introduced directly into the graphite furnace atomiser. The determination limit was 0.6ng L^{-1} (S/N = 3), and the relative standard deviation at a concentration of 0.25µg L^{-1} was 3.0–4.0% (n = 6). Not only was there no interference from the major ions such as sodium, potassium, magnesium, calcium, chloride and sulphate in seawater but there was also no interference from other minor ions. The method was applied to the determination of nanogram per millilitre levels of beryllium in seawater and rainwater.

References

1 Van den Sloot, H.A., Wals, G.D. and Das, H.A. *Analytica Chimica Acta*, **90**, 193 (1977).
2 Van der Bought, B.M. and Van Grieken, R.E. *Analytical Chemistry*, **49**, 311 (1977).
3 Zharikov, C.F. and Senyavin, M.M. Trudy gas Okeanography Institute (1070) (101). Ref. Zhur. Khim. 19GD (7) Abstract No 7G189 (1970).
4 Samchuk, A.I. *Soviet Journal of Water Chemistry and Technology*, **9**, 57 (1987).
5 Hall, G.E.M., Vaive, J.E. and Ballantye, S. *Journal of Geochemical Exploration*, **26**, 191 (1986).
6 Hall, G.E.M., Jefferson, C.W. and Michel, F.A. *Journal of Geochemical Exploration*, **30**, 63 (1988).
7 Lieser, K.H., Calmano, W., Heuss, E. and Neitzert, V. *Journal of Radioanalytical Chemistry*, **37**, 717 (1977).
8 Van der Bought, B.H. and Van Grieken, R.E. *Talanta*, **27**, 417 (1980).
9 Robberecht, H. and Van Grieken, R. *Talanta*, **29**, 823 (1982).
10 Orvini, E., Ladola, L., Gallorini, M. and Zerlia, T. In *Heavy Metals in the Environment*, Amsterdam, p. 657 (1981).
11 Okutani, T., Tsurta, Y. and Sakuragwa, A. *Analytical Chemistry*, **65**, 1273 (1993).

Chapter 24

Cations: Adsorption on anion exchange resins

24.1 Introduction

Strong base anion exchange resins are manufactured by chloromethylation of sulphonated polystyrene followed by reaction with a tertiary amine:

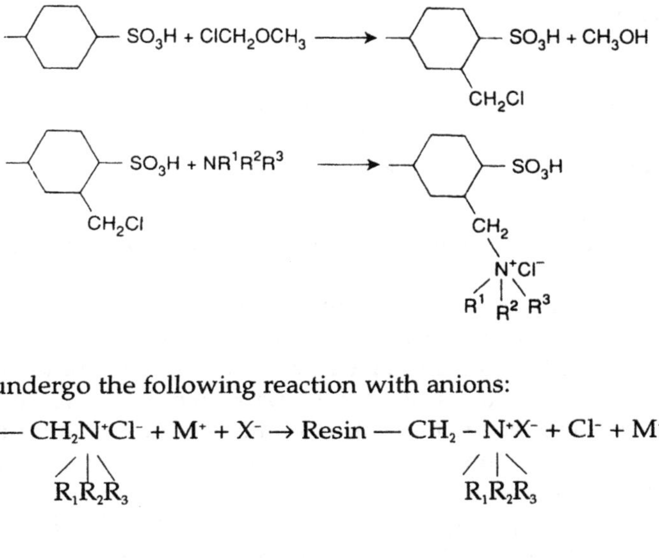

They undergo the following reaction with anions:

$$\text{Resin} - \text{CH}_2\text{N}^+\text{Cl}^- + \text{M}^+ + \text{X}^- \rightarrow \text{Resin} - \text{CH}_2 - \text{N}^+\text{X}^- + \text{Cl}^- + \text{M}^+$$

$$R_1R_2R_3 \qquad\qquad R_1R_2R_3$$

or

$$\text{Resin} - \text{CH}_2 - \text{N}^+\text{Cl}^- + \text{MX}^- \rightarrow \text{Resin} - \text{CH}_2 - \text{N}^+\text{MX}^- + \text{Cl}^-$$

$$R_1R_2R_3 \qquad\qquad R_1R_2R_3$$

where MX^- is a metal containing anion.

Weak base anion exchange resins are manufactured by chloromethylation of sulphonated polystyrene followed by reaction with a primary or secondary amine:

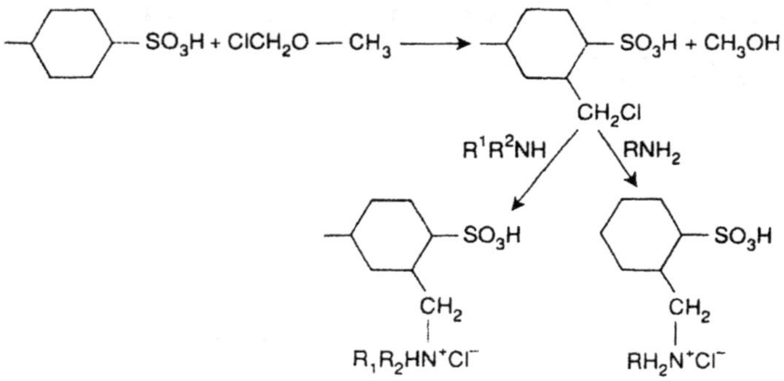

$$\text{—}\bigcirc\text{—SO}_3\text{H} + \text{ClCH}_2\text{O}\text{—CH}_3 \longrightarrow \text{—}\bigcirc\text{—SO}_3\text{H} + \text{CH}_3\text{OH}$$

with CH$_2$Cl substituent, reacting via R^{1}R^{2}NH and RNH$_2$:

$$\bigcirc\text{—SO}_3\text{H} \quad \text{(CH}_2\text{—R}_1\text{R}_2\text{HN}^+\text{Cl}^-) \qquad \bigcirc\text{—SO}_3\text{H} \quad \text{(CH}_2\text{—RH}_2\text{N}^+\text{Cl}^-)$$

They undergo the following reaction with anions:
eg

$$\text{Resin} \text{—} \overset{\underset{R^1R^2H}{|}}{N^+Cl^-} + X^- + M^+ \rightarrow \text{Resin} - \overset{\underset{R^1R^2H}{|}}{N^+MX^-} + Cl^-$$

or

$$\text{Resin} \text{—} \overset{\underset{R^1R^2H}{|}}{N^+Cl^-} + MX^- \rightarrow \text{Resin} \text{—} \overset{\underset{R^1R^2H}{|}}{N^+MX^-} + Cl^-$$

where MX$^-$ is a metal containing anion.

Some basic properties of the various types of anion exchange resins and suppliers are tabulated in Table 24.1.

The ligand 8-hydroxyquinoline-5-sulphonic acid forms anionic complexes with cobalt(II), zinc(II), cadmium(II) and lead(II). A technique has been described [1] for the separation and preconcentration of these metals in non saline waters prior to measurement by graphite furnace atomic absorption spectrometry. At optimum conditions, all four metals are quantitatively retained as their negatively charged 8-hydroxy-quinoline-5-sulphonic acid complexes by the column. Zinc, cadmium and lead(II) ions are completely eluted with 11ml or less of 2M nitric acid; cobalt(II) is totally removed by 12M hydrochloric acid. All four anionic complexes can be left on the column for 7d and still be quantitatively (99%) recovered. The technique is well suited for on-site work.

This procedure gives 99% recovery for zinc(II), lead(II) and cadmium (II) and a 94–99% recovery for cobalt(II) at the μg L^{-1} concentration range (0.2–40μg L^{-1}).

Mandel and Das [2] applied an anion exchange resin to the determination of traces of mercury as an anionic complex in spring waters by cold vapour atomic absorption spectrometry using a reduction

Table 24.1 Properties of anion exchange resins

Resin type	Functional group	Water content (approx)[a] (g g⁻¹ dry resin)	Exchange capacity (approx)[a] (mol equiv. g⁻¹ dry resin)	Packing density (approx)[a] (g ml⁻¹)	Regeneration	Washing of salt forms	Trade names of same commercial examples
Strong base types	Quaternary ammonium $-CH_2NR_3 + Cl^-$	1	4 at all pH values	0.7	Excess strong base	Stable	(c) Dowex 1 Dowex 2 Dowex AG1–X2 Dowex AG1–X8 Dowex AG1–X4 (b) Amberlyst P1–27 Amberlite IRA–400 Amberlite GC–400 Amberlite IRA–410 (a) Deacidite FF Lewatite M5080 Biorad AG1 Biorad 140–AG1–X2 Biorad X8
Weak base types	Secondary or primary amine $-CH_2N^+HR_2Cl^-$ or $-CH_2-N^+H_2R \ Cl^-$	0.3	4 at low pH values	0.7	Readily regenerated with sodium carbonate	Anion slowly hydrolyses	(b) Amberlite IR–45 (a) Deacidite G

[a]Depends on grade and does not necessary include resins available from (a) Permitit Co., London W4; [b]Rohm and Haas Co., Philadelphia, USA; [c]Dow Chemical Co., Midland, Michigan, USA

Source: Own files

Table 24.2 Determination of mercury in spring water samples

Sample	Sample vol (L)	Mercury added (ng)	Mercury found (ng)*
Spring water A	5		47
	5	100	143
Spring water B	5		55
	5	100	152
Spring water C	5		392
	5	100	488
Stream water A	5		41
	5	100	139

*Mean of three replicate determinations

Source: Reproduced with permission from Mandel, S. and Das, A.K. [2] Perkin–Elmer Ltd

aeration method. In this method water samples are collected in polythene bottles previously cleaned by soaking overnight in 10% nitric acid. The samples are preserved with 0.1N hydrochloric acid containing 0.01% potassium dichromate. The samples are irradiated with ultraviolet light for 15min before analysis.

5ml of sample are transferred to a beaker and treated with hydroxyl-ammonium chloride solution (10% w/v) added drop by drop until the yellow colour due to potassium dichromate disappears. 50g of sodium chloride are added and the sample stirred to dissolve the salt. The solution is passed through the ion exchange column at a rate of 5–6ml min $^{-1}$. The column is washed with redistilled water until the eluates are acid free. 80ml of 4N nitric acid are passed through the column at a rate of 3ml min $^{-1}$ and the eluate collected, in a 100ml volumetric flask containing 1ml of 1% potassium dichromate solution. This solution is made up to volume with 4N nitric acid and used for cold vapour atomic absorption measurements.

The method was applied to the determination of trace amounts of mercury in spring water and stream water samples. 5L water samples were used in each determination. Recovery of mercury from these samples was studied by addition of known amounts of mercury as mercury(II) chloride. The results are shown in Table 24.2.

Dowex A–1 has a serious disadvantage in that the resin also retains the alkaline earth metal ions but it has been shown experimentally that concentrations of sodium up to 5×10^{-1}M and calcium up to 4×10^{-3}M do not affect the recovery of other trace metals.

Anionic metal species

Simple cationic metal ions do not react with, ie are not retained by, anion exchange resins. However, if the metal cation M^+ is first reacted with a reagent with which it forms a negatively charged anionic complex then the resulting negatively charged metal containing ions are retained on an anionic exchange column. Thus cadmium(II) forms a soluble anionic complex upon reaction with potassium cyanide:

$$CdCl_2 + 4KCN \rightarrow K_2[Cd(CN)_4]^{2-}$$

This complex is retained on a column of strong base anionic exchange resin which, for convenience, is represented as follows:

$$2\text{ Resin } CH_2R_3N^+Cl^- + [Cd(CN)_4]^{2-}$$

$$\rightarrow \text{Resin } CH_2R_3N^+]_2[Cd(CN)_4]^{2-} + 2Cl^-$$

The cadmium complex can then be dissolved off the column with a small volume of aqueous acetic acid to procedure the acetate form of the resin and cadmium acetate:

$$[\text{Resin} - CH_2R_3N^+]_2(Cd(CN)_4)^{2-} + 4CH_3COOH \rightarrow 2\text{ Resin}$$

$$- CH_2R_3N^+[OOCCH_3]^- + Cd(OOCCH_3)_2 + 4HCN$$

High preconcentration factors can be achieved by such techniques, some further examples of which are now discussed.

24.2 Non saline waters

24.2.1 Atomic absorption spectrometry

Silver has been preconcentrated from 1L sample at pH1 on an AG–1–X8 anion exchange column [3]. Silver was eluted from this column with acetone: nitric acid: water (20: 1: 1) and acetone removed from the elute by evaporation. After pH adjustment to 2–3 the solution was treated with ammonium pyrrolidinedithiocarbamate and the silver chelate extracted into a small volume of methyl ethyl ketone. Evaluation at 328.1µm by atomic absorption spectrometry enabled extremely low levels of silver to be determined.

Riley and Siddiqui [4] preconcentrated thallium by adsorption of the tetrachlorothallate ion on to a strongly basic anion exchange resin, followed by elution with sulphur dioxide and determination by graphite furnace atomic absorption spectroscopy or differential pulse anodic stripping voltammetry. De Ruck et al. [5] oxidised thallium(I) to thallium(III) with cerium(IV) sulphate and passed the solution through an anion exchange column which retained thallium as the tetrachloro-thallate(III) ion. Thallium was then eluted with ammonium sulphate solution and determined by electrothermal atomic absorption spectrometry. A concentration factor of 400 was achieved and the detection limit was $3.3ng\ L^{-1}$.

Vazquez-Gonzalez et al. [6] have described a method for preconcentrating and determining molybdenum by electrothermal atomisation atomic absorption spectrometry after preconcentration by means of anion exchange using Amberlite IRA–400 in resin citrate form. The optimal analytical parameters were established by drying, carbonisation, charring, atomisation and cleaning in a graphite furnace. The precision and accuracy of the method were investigated. Less than $0.2\mu g\ L^{-1}$ molybdenum could be determined by this procedure. The separation eliminates significant interference from cadmium, copper, iron, vanadium and tungsten.

Neutron activation analysis

Bergerioux et al. [7] compared various methods including anion exchange resin techniques for the preconcentration of 22 elements. In this technique the metals were adsorbed on to the resin which was then examined directly by neutron activation analysis and gamma spectrometry (when radiotracers were used). Asamov [8] used preconcentration on anion exchange resin followed by neutron activation analysis to determine total gold in non saline water. Becknell et al. [9] adsorbed mercury from acidified non saline water on to anion exchange resin loaded paper prior to analysis by neutron activation analysis. Down to $0.005\mu g$ mercury in the original sample could be determined.

24.2.2 Other analytical finishes

Further applications of anion exchange resins to the preconcentration of metals in non saline waters are reviewed in Table 24.3.

24.3 Seawater

24.3.1 Spectrophotometric methods

Shriadah and Ohzeki [31] determined iron in seawater by spectrophotometry after enrichment as a bathophenanthroline disulphonate complex on a thin layer of anion exchange resin. Seawater samples (50ml) containing iron(II) and iron(III) were diluted to 150ml with water followed by sequential addition of 20% hydrochloric acid (100μL), 10% hydroxylammonium chloride (2ml), 5M ammonium solution (to pH3.0 for iron(III) reduction), bathophenanthroline disulphonate solution (1.0ml), and 10% sodium acetate solution (2.0ml) to give a mixture with a final pH of 4.5. A macroreticular anion exchange resin, Amberlyst A27, in the chloride form was added, the resultant coloured thin layer was scanned by a densitometer and the adsorbance measured at 550nm.

Kodama and Tsubota [32] determined tin in seawater by anion exchange chromatography and spectrophotometry with catechol violet. After adjusting to 2mol L $^{-1}$ in hydrochloric acid 500ml of the sample is adsorbed on a column of Dowex 1–X8 resin (Cl form) and elution is then effected with 2M nitric acid. The solution is evaporated to dryness after adding 1M hydrochloric acid and the tin is again adsorbed on the same column. Tin is eluted with 2M nitric acid. Tin is determined in the eluate by the spectrophotometric catechol violet method. There is no interference from 0.1mg of each of aluminium, manganese, nickel, copper, zinc, arsenic, cadmium, bismuth and uranium, any titanium, zirconium and antimony are removed by the ion exchange. Filtration of the sample through a Millipore filter does not affect the results, which are in agreement with those obtained by neutron activation analysis.

Kiriyama and Kuroda [33] combined ion exchange preconcentration with spectrophotometry using 2-pyridylazoresorcinol in the determination of vanadium in seawater. The sample (2L) made 0.1M in hydrochloric acid, filtered, and made 0.1M in ammonium thiocyanate, is passed through a column of Dowex 1–X8 resin anion exchange (thiocyanate form). The vanadium is retained and is eluted with concentrated hydrochloric acid. Thiocyanate in the elute is decomposed by heating with nitric acid and the solution is evaporated to fuming with sulphuric acid. A solution of the residue is neutralised with aqueous ammonia and evaporated nearly to dryness. The residue is treated with water and aqueous sodium hypobromite and after 30min with phenol, phosphate buffer solution of pH6.5, and aqueous 1,2-diaminocyclohexane–N, N, N′ N′–tetraacetic acid, and the vanadium is determined spectrophotometrically at 545nm with 4-(2-pyridylazo) resorcinol. Vanadium was determined in seawater at levels of 1.65µg L $^{-1}$. After boiling such samples under reflux with potassium permanganate and sulphuric acid (to establish the concentration of organically bound vanadium), values for vanadium were 30–60% higher than corresponding values obtained without oxidation.

Anion exchange resins have been used to preconcentrate molybdenum in seawater prior to its spectrophotometric determination as the Tiron complex [34–36]. Kawabuchi and Kuroda [36] have concentrated molybdenum by anion exchange from seawater containing acid and thiocyanate or hydrogen peroxide [36,37] and determined it spectrophotometrically. Korkische et al. [38] have concentrated molybdenum from sea waters on Dowex 1–X8 anion exchange resin in the presence of thiocyanate and ascorbic acid. A sodium citrate and ascorbic acid system has also been used for the concentration of molybdenum on Dowex 1–X8 (citrate form) as a citrate complex from tap and mineral waters.

Table 24.3 Preconcentration of anionic metal species from non saline water

Metal	Type of water	Resin	Medium	Eluting agent	Detection limit*	Analytical finish	Ref.
Al	Potable	Anion exchange	Pyrocatechol violet	—	—	—	[10]
Fe	Non saline	Anion	—	Acetic acid	1–2ng	Spectrophotometric	[11]
Cd	Non saline	Amberlite IRA–400 EDE–10P	Cyanide	Hydrochloric acid	—	Spectrophotometric	[12]
Zn	Sewage		Hydrochloric acid	Hydrochloric acid	—	Spectrophotometric	[13]
Co	Non saline	Dowex AGI–X8	Hydrochloric acid–thiocyanate	Hydrochloric acid	—	Spectrophotometric	[14]
Cd	Non saline	Dowex AGI–X8 and Amberlite IRA 400	Hydrochloric or hydrobromic acid Acetic and nitric acid	Nitric acid	—	Spectrophotometric	[15]
Th	Non saline	Dowex AGI–X8	Nitric acid	Hydrochloric acid	—	Spectrophotometric	[16]
Ni	Mineral water	Dowex A–I	Sodium acetate	Hydrochloric acid	0.5µg	Spectrophotometric	[17]
U	Non saline	Dowex AGI–X8	Sulphuric acid	Sulphuric acid	$0.3\mu g\ L^{-1}$	Fluorometric	[18]
U	Non saline	Amberlite LA–I	—	—	$0.4\mu g\ L^{-1}$	Fluorometric	[19–21]
^{239}Pu ^{240}Pu	—	Anion exchange	—	Hydrofluoric acid–hydrochloric	—	Deposited on Pt–γ-ray spectrometry	[22]
Various radio nuclides	—	Single bead of anion exchange resin	—	—	—	Point source	[23]

Table 24.3 continued

Metal	Type of water	Resin	Medium	Eluting agent	Detection limit*	Analytical finish	Ref.
Mo	Non saline	Biorad AGI	–	Ammonia–ammonium chloride	10μg L^{-1}	AAS	[24]
Co, Mo, V	Mineral	Ion exchange	–	–	–	Inductively coupled plasma atomic emission spectrometry	[25]
Se	Non saline	Ion pair reversed phase or anion exchange chromatography	–	–	sub μg L^{-1}	Inductively coupled plasma atomic emission spectrometry	[26]
Cr, Ni	Non saline	In-line anion exchange resin	–	–	sub mg L^{-1}	Inductively coupled plasma atomic emission spectrometry	[27]
Au	Non saline	Anion exchange resin	–	–	Femti molar	Flow injection analysis	[28]
Se(IV)	Non saline	Anion exchange resin (interferences removed on Chelex–100)	–	–	μg L^{-1}	Differential pulse polarography	29
Hg	–	ARA–8P–T–40	–	–	0.02μg L^{-1}	Neutron activation analysis	[30]

*μg L^{-1} unless otherwise stated

Source: Own files

Table 24.4 Determination of molybdenum in seawater

Sample[a]	Sample vol. (L)	Mo added (µg)	Mo found (µg)	Original content (µg L^{-1})
Seawater[a]	0.5	0	4.27, 4.51	8.54, 9.02
			4.39	8.78
	1.0	0	8.90	8.90
	0.5	4.24	8.73	8.98
	0.5	8.48	12.80	8.64
				av. 8.81 ± 0.19

[a]Collected at Kamioke Harbour, Kagoshima Bay, Japan, on 23 June 1983, salinity 33.48%

Source: Reproduced with permission from Kiriyama, R. and Kuroda, R. [39] Elsevier Science, UK

In a method described by Kiriyama and Kuroda [39] molybdenum is sorbed strongly on Amberlite CG 400 anion exchange resin (Cl form) at pH3 from seawater containing ascorbic acid and is easily eluted with 6M nitric acid. Molybdenum in the effluent can be determined spectrophotometrically with potassium thiocyanate and stannous chloride. The combined method allows selective and sensitive determination of traces of molybdenum in seawater. The precision of the method is 2% at a molybdenum level of 10µg L^{-1}. To evaluate the feasibility of this method Kiriyama and Kuroda [39] spiked a known amount of molybdenum into seawater and analysed it by the procedure, the results are given in Table 24.4. As can be seen, the recoveries for the molybdenum added to 500 or 1000ml samples are satisfactory.

Shriadah *et al.* [34] determined molybdenum(V1) in seawater spectrophotometrically after enrichment as the Tiron complex on a thin layer of anion exchange resin. There were no interferences from trace elements or major constituents of seawater, except for chromium and vanadium. These were reduced by the addition of ascorbic acid. The concentration of dissolved molybdenum(VI) determined in Japanese seawater was 11.5µg L^{-1} with a relative standard deviation of 1.1%.

Kuroda *et al.* [40] observed that uranium is strongly adsorbed from acidified saline solutions by a strongly basic ion exchange resin in the presence of azide ions. The distribution coefficient of uranium with 0.5M sodium chloride increased rapidly with an increase in azide concentration, and was much higher than the coefficient obtained with hydrazoic acid alone. The sorbed uranium was easily eluted with 1M hydrochloric acid, and was determined spectrophotometrically. Recoveries of 98.3–99.6% were obtained from artificial seawater containing 3.4µg L^{-1} uranium. Selenium(IV) has been preconcentrated on a bismuthiol(II) modified anion exchange resin (Amberlite IRA–400) [41] followed by fluorometric estimation using diamino naphthalene. Selenium(IV) adsorbed on the column as selentotrisulphate was desorbed with a small volume of 0.1M penicillamine or 0.1M cysteine prior to fluorometric determination of selenium.

24.3.2 Atomic absorption spectrometry

A limited amount of work has been carried out using polyacrylamidoxine resin. Collela et al. [42] agitated seawater samples with polyacrylamidoxime resin preparatory to the determination of iron(III), copper(II), cadmium(II), lead(II) and zinc(II).

Metals were removed from the filtered OH form resin by equilibrating with 1:1 hydrochloric acid/water mixture and their concentrations determined by atomic absorption spectrometry. Metal concentrations as determined by the resin method were in good agreement with the values determined directly on samples by either differential pulse polarography or differential pulse anodic stripping voltammetry.

Kiriyama and Kuroda [43] determined vanadium, cobalt, copper, zinc and cadmium in seawater by adsorption in an anion exchange resin. Preconcentration from seawater was achieved in thiocyanate medium. A strongly basic anion exchange resin in the thiocyanate form preconcentrated five metals from seawater adjusted to 1M thiocyanate and 0.1M hydrochloric acid. Sorbed metals were recovered simultaneously by elution with 2M nitric acid prior to determination by graphite furnace atomic absorption spectrometry.

Adsorption of metals on a single bead of ion exchange resin has been used as a means of effecting preconcentration [44,45]. Koide et al. [44] showed that cadmium, palladium, iridium, gold, plutonium and technetium can be concentrated from seawater on to a single bead of anion exchange resin. This process eliminated salt interference. The beads acted as point sources during subsequent analytical determination. Optimal conditions for the adsorption of cadmium–109, palladium–103, iridium–192, gold–105, plutonium–237 and technetium–95m on to a single bead were determined. Two types of applications of the techniques were investigated, with no prior concentration and with preconcentration; increasing the yield of plutonium and technetium on to a single bead for improved sensitivity in mass spectrometric analysis. Two types of anion exchange resin (gel type and macroporous type) were tested. Hodge et al. [45] determined platinum and iridium in marine waters by preconcentration by anion exchange, purification by uptake on a single anion exchange bead, and determination by graphite furnace atomic absorption spectrometry. All steps were followed by radiotracers (platinum–191 and iridium–192). Yields varied between 35 and 90% for determination of platinum and iridium in sediments, manganese nodules, seawater and micro-algae.

Kuroda *et al.* [46] preconcentrated trace amounts of molybdenum from acidified seawater on a strongly basic anion exchange resin (Bio-Rad AGI–X8 in the chloride form) by treating the water with sodium azide. Molybdenum(VI) complexes with azide were stripped from the resin by elution with ammonium chloride/ammonium hydroxide solution (2M/2M). Relative standard deviations of better than 8% at levels of 10µg L $^{-1}$ were attained for seawater using graphite furnace atomic absorption spectrometry.

Chelation solvent extraction or resin–adsorption preconcentration procedures coupled with graphite furnace atomic absorption spectrometry or neutron activation analysis are capable of determining many elements at the baseline concentrations occurring in deep seawater samples. All workers in this field emphasise the need for extreme precautions during sampling of seawater for analysis at these very low concentrations.

Two methods for the determination of vanadium in seawater have been studied which use neutron activation analysis and atomic absorption spectrometry [47]. In the atomic absorption spectrometry procedure [47], potassium thiocyanate (10g) and ascorbic acid (5g to reduce to V$_{(vi)}$) were dissolved in 1L of seawater and the solutions were left to stand for 2–3h. These samples (1–3L) were passed through a Dionex 1–X8 anion exchange column at a flow rate of 1.7ml min $^{-1}$. The resin was then washed with 20ml distilled water and vanadium eluted with 150ml eluent solution. The vanadium eluate was slowly evaporated under an infrared lamp, the residue dissolved in 10ml 6M hydrochloric acid containing 1ml of the aluminium chloride solution [48] and vanadium determined by atomic absorption spectrometry. For calibration, suitable standard solutions were aspirated before and after each batch of samples.

The analysis of the seawater samples by both methods is shown in Table 24.5. The average concentration and standard deviation of the Pacific Ocean waters (µg L $^{-1}$) were 2.00 ± 0.09 by neutron activation analysis and 1.86 ± 0.12 by atomic absorption spectrometry. For the Adriatic water the corresponding values were about 1.7µg L $^{-1}$. The difference between the values for the same seawater is within the range to be expected from the standard deviations observed. Though the neutron activation analysis is inherently more sensitive than the atomic absorption spectrometry, both procedures give a reliable measurement of vanadium in seawater at the natural levels of concentration.

24.3.3 Neutron activation analysis

Matthews and Riley [49] have described the following procedure for determining down to 0.06µg L $^{-1}$ rhenium in seawater. From 6 to 8µg L $^{-1}$

Table 24.5 Results of vanadium determinations in seawater samples

Sample	Vanadium concentration (µg L^{-1})		
	Vol. (L)	NAA	AAS
Pacific Ocean (Scripps Pier)	1		1.80
	1		2.00[a](2.0)
	2		1.90[a]
	3		1.73
	0.098	1.99[b]	
	0.098	2.00[c](0.20)	
Adriatic Sea			
(Shore near Lignano	3		1.71
Sabbiadoro, Italy)	0.041	1.69	
(Shore near Ancona, Italy)	3		1.73
	0.043	1.64	

[a]Results after subtraction of the quantity of vanadium added to the sample before the ion exchange or coprecipitation step. The amount added (in µg) is shown in parentheses
[b]Average of 12 determinations, standard deviation 0.10µg L^{-1}
[c]Average of two samples, average deviation 0.01µg L^{-1}

Source: Reproduced with permission from Weiss, H.V. *et al.* [47] Elsevier Science, UK

rhenium was found in Atlantic seawater. The rhenium in a 15L sample of seawater acidified with hydrochloric acid, is concentrated by adsorption on a column of Deacidite FF anion exchange resin (Cl-form), followed by elution with 4M nitric acid and evaporation of the eluate. The residue (0.2ml) together with standards and blanks, is irradiated in a thermal neutron flux of at least 3×10^{12} neutrons cm^{-2} s^{-1} for at least 50h. After a decay period of 2d, the sample solution and blank are treated with potassium perrhenate as carrier and evaporated to dryness with a slight excess of sodium hydroxide. Each residue is dissolved in 5M sodium hydroxide. Hydroxylammonium chloride is added (to reduce Tc(VII)) which arises as ^{99m}Tc from activation of molybdenum present in the samples, and the Re(VII) is extracted selectively with ethyl methyl ketone. The extracts are evaporated, the residue is dissolved in formic acid–hydrochloric acid (19:1), the rhenium is adsorbed on a column of Dowex 1, and the column is washed with the same acid mixture followed by water and 0.5M hydrochloric acid; the rhenium is eluted at 0°C with acetone–hydrochloric acid (19:1) and is finally isolated by precipitation as tetraphenylarsonium perrhenate. The precipitate is weighed to determine the chemical yield and the ^{186}Re activity is counted with an end-window Geiger–Müller tube. The irradiated standards are dissolved in water together with potassium perrhenate. At a level of 0.057µg L^{-1} rhenium the coefficient of variation was ±7%.

Kawabuchi and Riley [50] used neutron activation analysis to determine silver in seawater. Silver in a 4L sample of seawater was concentrated by ion exchange on a column (6cm × 0.8cm) containing 2g of Deacidite FF–IP anion exchange resin, previously treated with 50ml 0.1M hydrochloric acid. The silver was eluted with 20ml 0.4M aqueous thiourea and the eluate was evaporated to dryness, transferred to a silica irradiation capsule, heated at 200°C, and ashed at 500°C. After sealing, the capsule was irradiated for 24h in a thermal–neutron flux of 3.5×10^{12} neutrons cm^{-2} s^{-1}, and after a decay period of 2–3d, the ^{100m}Ag arising from the reaction ^{199m}Ag(n, γ) ^{100m}Ag was separated by a conventional radiochemical procedure. The activity of the ^{110m}Ag was counted with an end-window Geiger–Müller tube, and the purity of the final precipitate was checked with a Ge(Li) detector coupled to a 400-channel analyser. The method gave a coefficient of variation of ±10% at a level of 40ng silver per litre.

24.4 Potable water

Sarzanini et al. [10] compared two ion exchange methods for the preconcentration of aluminium from potable water. In the first method, the aluminium pyrocatechol blue complex was formed then eluted through an anion exchange column. In the other the pyrocatechol was loaded onto the resin then the sample passed through the column. The latter method gave the better recovery of aluminium.

24.4.1 Atomic absorption spectrometry

Korkische and Sario [51] have described a procedure for the determination of cadmium(II), copper(II) and lead(II) in potable waters. The method involved acidification with hydrobromic acid; filtration of the water sample; addition of ascorbic acid; ion exchange using the strongly basic anion exchange resin, Dowex 1–X8, which adsorbs cadmium, copper and lead as anionic bromide complexes, of the type H_2M Br$_4^-$ (where M is metal); elution with nitric acid; and determination by atomic absorption spectrometry. When the methanolic hydrobromic acid solution was used, the following sensitivities for 1% absorption were obtained; Cd = 0.033mg L^{-1}; Cu = 0.04mg L^{-1}; Pb = 0.28mg L^{-1}.

In the determination of extremely small quantities of cadmium, copper and lead, it is necessary to run a reagent blank through the whole procedure (starting with the addition of concentrated hydrobromic acid and finally to deduct its concentration of cadmium, copper and lead from those contents measured in the water samples.

To investigate the effect of the volume of water sample on the recovery and accuracy of the determinations of cadmium, copper and lead,

Korkische and Sario [51] analysed varying volumes of potable water for these three elements. The volume of water has no effect provided that it does not exceed 500ml. If larger volumes of the water sample are passed through the anion exchange column, the adsorption of copper and lead decreases with increasing volume. This is because of the salt effect, ie the displacement of the anionic bromo complexes of these metals by sulphate and bromide ion contained in the water (bromides of calcium and magnesium are formed from the carbonates on acidification of the water sample with hydrobromic acid). With respect to the accuracy of the determinations of copper and lead, the sample volume has practically no influence (in the range of 0.1–0.5L). It is impossible, however, to determine cadmium when only 100–500ml of the sample is used, in this case, at least 1L of the sample has to be passed through the anion exchange column to obtain an eluate which contains sufficient cadmium to be measurable with some accuracy. Therefore, relatively large volumes of water have to be processed in order to obtain reliable cadmium values.

The main constituents of non saline waters ie calcium, magnesium, the alkali metals and iron, are not retained on Dowex 1 from 0.15M hydrobromic acid. The effect of various metal ions on the anion exchange separation of cadmium, copper and lead was studied with respect to some trace elements. It was found that 1mg amounts of most common elements did not affect the recoveries of cadmium, copper and lead. Among these foreign metal ions, only zinc may be present in non saline waters to any larger extent but even then it would not interfere because zinc is not retained by the resin from 0.15M hydrobromic acid.

24.5 Sewage effluents

Zinc in sewage has been preconcentrated on anion exchange resin prior to desorption with hydrochloric acid and spectrometric determination [13].

References

1 Borge, D.G., Going, J.E. *Analytica Chimica Acta*, **123**, 19 (1981).
2 Mandel, S. and Das, A.K. *Atomic Spectroscopy*, **3**, 56 (1982).
3 Chau, I.T., Fishmann, M.J. and Ball, T.W. *Analytica Chimica Acta*, **43**, 189 (1969).
4 Riley, J.P. and Siddiqui, S.A. *Analytica Chimica Acta*, **181**, 177 (1986).
5 De Ruck, A., Vandecasteele, C. and Dams, R. *Mikrochimica Acta*, **416**, 187 (1987).
6 Vazquez-Gonzalez, J.F., Bermejo-Barrera, P. and Bermejo-Martinez, F. *Atomic Spectroscopy*, **8**, 159 (1987).
7 Bergerioux, C., Blan, P.C. and Haerdi, W. *Journal of Radioanalytical Chemistry*, **39**, 823 (1977).
8 Asamov, K.A., Abdullah, A.A. and Zakhidov, A.S. Doklady Acad. Nauk. Uzbek. SSR, (3), 26 (1969).
9 Becknell, D.E., Marsh, R.H. and Allie, W. *Analytical Chemistry*, **43**, 1230 (1971).

10 Sarzanini, C., Mentasi, E., Porta, V. and Gennaro, M.C. *Analytical Chemistry,* **59**, 484 (1987).
11 Shi Yu, L. and Wei Ping, G. *Talanta,* **31**, 844 (1984).
12 Ashizawa, T. and Hosoya, K. *Japan Analyst,* **20**, 1416 (1971).
13 Kurochkina, M.I., Lyakh, V.I. and Perelyaeva, G.L. Nauch. Trudy. irkutsh. gos. Nauchno issled. Inst. Tedk. Esvet. Metall, (24), 149 (1972). Ref: Zhur. Khim. 19 GD (13) Abstract No. 13 G 148 (1972).
14 Korkische, J. and Dimitriades, D. *Talanta,* **20**, 1287 (1973).
15 Korkische, J. and Dimitriades, D. *Talanta,* **20**, 1295 (1973).
16 Korkische, J. and Dimitriades, D. *Talanta,* **20**, 1303 (1973).
17 Nevoral, V. and Okae, A. *Czlka Form,* **17**, 478 (1988).
18 Danielsson, A., Roennholm, B., Kiellstoem, L.E. and Ingman, F. *Talanta,* **20**, 185 (1973).
19 Brits, R.J.S. and Smit, M.C.B. *Analytical Chemistry,* **49**, 67 (1977).
20 Gladney, E.S., Owens, J.W. and Starner, J.W. *Analytical Chemistry,* **48**, 973 (1976).
21 Zielinski, R.A. and McKoun, M. *Journal of Radioanalytical and Nuclear Chemistry Articles,* **84**, 207 (1984).
22 Golchert, N.V. and Sedlet, J. *Radiochemical and Radioanalytical Letters,* **12**, 215 (1972).
23 Carter, J.A., Walker, R.L., Smith, D.H. and Christie, W.H. *International Journal of Environmental Analytical Chemistry,* **8**, 241 (1980).
24 Kuroda, R., Matsumoto, N. and Oguma, K. *Fresenius Zeitschrift für Analytische Chemie,* **330**, 111 (1988).
25 Steffan, I. and Vujicic, G. *Mikrochimica Acta,* **110**, 89 (1993).
26 Cai, Y., Cabanas, M., Fernandez, M. *et al. Analytica Chimica Acta,* **314**, 183 (1995).
27 Petrucci, F., Alimanti, A., Lasztity, A., Horvath, Z. and Caroli, S. *Canadian Journal of Applied Spectroscopy,* **39**, 113 (1994).
28 Falkner, K.K. and Edmond, J.M. *Analytical Chemistry,* **62**, 1477 (1990).
29 Batley, G.E. *Analytica Chimica Acta,* **187**, 108 (1986).
30 Belova, N.I. and Vetrov, V.A. *Gidrokhim. Mater.,* **97**, 127 (1987).
31 Shriadah, M.M.A. and Ohzeki, K. *Analyst (London),* **111**, 555 (1986).
32 Kodama, Y. and Tsubota, H. *Japan Analyst,* **20**, 1554 (1971).
33 Kiriyama, T. and Kuroda, R. *Analytica Chimica Acta,* **62**, 464 (1972).
34 Shriadah, H.M.A., Katoeka, M. and Ohzeri, K. *Analyst (London),* **110**, 125 (1985).
35 Riley, J.P. and Taylor, D. *Analytica Chimica Acta,* **40**, 479 (1968).
36 Kawabuchi, K. and Kuroda, K. *Analytica Chimica Acta,* **46**, 23 (1969).
37 Kuroda, R. and Tarui, T. *Fresenius Zeitschrift für Analytische Chemie,* **269**, 22 (1974).
38 Korkische, J., Godl, L. and Gross, H. *Talanta,* **22**, 669 (1975).
39 Kiryama, R. and Kuroda, R. *Talanta,* **31**, 472 (1984).
40 Kuroda, R., Oguma, K., Mukai, N. and Imamoto, M. *Talanta,* **34**, 433 (1987).
41 Wu, T.L., Lambert, L., Hastings, D. and Banning, D. *Bulletin Environmental Contamination and Toxicology,* **24**, 411 (1980).
42 Colella, M.B., Siggia, S. and Barnes, R.M. *Analytical Chemistry,* **52**, 2347 (1980).
43 Kiriyama, T. and Kuroda, R. *Mikrochimica Acta,* **1**, 405 (1985).
44 Koide, M., Lee, D.S. and Stallard, M.O. *Analytical Chemistry,* **56**, 1956 (1984).
45 Hodge, V., Stallard, M., Koide, M. and Goldberg, E.D. *Analytical Chemistry,* **58**, 616 (1986).

46 Kuroda, R., Matsumoto, N. and Ogmura, K. *Fresenius Zeitschrift für Analytische Chemie*, **330**, 111 (1988).
47 Weiss, H.V., Guttman, H.A., Korkische, J. and Steffan, I. *Talanta*, **24**, 509 (1977).
48 Korkische, J. and Gross, H. *Talanta*, **20**, 1153 (1973).
49 Matthews, A.D. and Riley, J.P. *Analytica Chimica Acta*, **51**, 455 (1970).
50 Kawabuchi, K. and Riley, J.P. *Analytica Chimica Acta*, **65**, 271 (1973).
51 Korkische, J. and Sario, A. *Analytica Chimica Acta*, **76**, 393 (1977).

Chapter 25

Cations: Adsorption on cation exchange resins

25.1 Introduction

In addition to the non-polar macroreticular Rohm and Hass and Amberlite XAD resins (XAD–2 and XAD–4) a wide range of polar resins exist which are very useful for the preconcentration of anionic and cationic species and have some applications in the preconcentration of ionic organic substances. Intermediate and highly polar types of resins are commonly referred to as ion exchange resins. These may be subdivided into cationic types (cation exchange resins) that are discussed in this chapter and anionic types (anion exchange resins) that have been discussed in Chapter 24. Cationic exchange resins carry a negative charge and this reacts with positively charged metallic ions (cations) or cationic organic species. Anionic ion exchange resins carry a positive charge and this reacts with negatively charged anions or anionic organic species.

Strong acid cation exchange resins manufactured by the sulphonation of polystyrene or polydivinyl benzenes undergo the following reaction with cations:

$$\text{Resin } SO_3^-H^+ + M^+ X^- \rightarrow \text{Resin } SO_3^-M^+ + H^+ + X^-$$

Weak acid cation exchange resins manufactured, eg by the polymerisation of methacrylic acid undergo the following reaction with cations:

$$\text{Resin } COO^-H^+ + M^+ + X^1 \rightarrow \text{Resin } COO^-M^+ + H^+ + X^-$$

Some basic properties of the various types of cation exchange resins available and their suppliers are tabulated in Table 25.1.

Preconcentration is achieved by passing a large volume of water sample, suitably adjusted in pH and reagent composition down a small column of the resin. The adsorbed ions are then desorbed with a small volume of a suitable reagent in which the metals or metal complexes or anionic species dissolve. This preconcentrated extract can then be analysed by any suitable means.

Table 25.1 Properties of cation exchange resins

Resin type	Functional group	Water content (approx)[a] ($g\ g^{-1}$ dry resin)	Exchange capacity (approx)[a] ($mol\ equiv\ g^{-1}$ dry resin)	Packing density (approx)[a] ($g\ ml^{-1}$)	Regeneration	Washing of salt forms	Trade volumes of some commercial examples
Strong acid types	$-SO_3H$	0.7	4 at all pH values	0.8	Excess strong acid	Stable	(c) Dowex 50 Dowex 50W–X8 Dowex 50W–X4 Dowex AI (b) Amberlite IR–120 Amberlite GC–120 Amberlite XAD–12 (a) Zeocarb 225 Cationite KB–4P–2
Weak acid types	$-COOH$	1	9–10 at high pH	0.7	Readily regenerated	Cation slowly hydrolyses off	(b) Amberlite IRC–50 (a) Zeocarb 226 (c) Dowex XAD–7 Dowex XAD–8

[a]Depends on grade and does not necessarily include resins available from (a) Permutit Co., London W4; (b) Rohm and Haas Co., Philadelphia, USA; (c) Dow Chemical Co., Midland, Michigan, USA

Source: Own files

25.2 Non saline waters

25.2.1 Miscellaneous cation exchange resins

25.2.1.1 Atomic absorption spectrometry

Treit *et al.* [1] linked a cation exchange resin preconcentration column (Dowex 50W–X8) directly to the nebuliser tube of an atomic absorption spectrometer in their method for the determination of free metal ions including copper. They used a miniaturised ion exchange column. The metal ion is eluted from the resin as a narrow peak, the area of which is proportional to the free metal ion concentration in the initial sample solution. The spectrophotometer is thus used as an ion selective probe. Precision and accuracy were better than 1%. The method allows free metal determination in the sample and both sample volumes and measurement times are reduced.

The apparatus used by these workers is shown in Fig. 25.1. To determine free copper ion prepare all copper containing samples and standards in 0.100M sodium nitrate. Use a flow rate of 5.5ml min⁻¹. Pump sample solution through the 2mm resin column via V_2 and V_3 (solid lines in Fig. 25.1). Conduct this resin equilibration step for 7.5min. Divert the effluent load solution to the flow meter by valve V_4 during this step to prevent fouling of the nebuliser and burner by the high salt concentration.

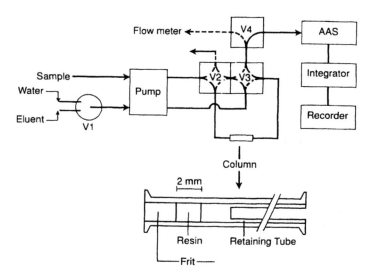

Fig. 25.1 Diagram of instrument system and ion exchange column; see text for details
Source: Reproduced with permission from Treit, J. *et al.* [1] American Chemical Society

After the effluent is loaded, switch V_2, V_3 and V_4 (dashed lines in Fig. 25.1) and backwash the resin with water for 2.5min to remove interstitial solution and to establish a base line AA signal. Next elute copper on the resin by switching V_1 to 0.02M EDTA, pH3.5 eluent for 1.5min. Detect the eluted copper as an AA peak and measure the peak area. After elution, again wash the column with water for 2.5min to re-establish a base line. Then return valves V_2, V_3 and V_4 to their initial positions for the next sample.

A significant advantage of this technique is that the ion exchange resin is brought to equilibrium with the sample in such a way as to allow free metal determination in the unperturbed sample. Furthermore, the use of very small columns permits faster measurements with smaller sample volumes. Still smaller sample volumes may be realised by reducing resin particle size, lowering flow rates during the loading step, and decreasing the amount of resin. Use of a graphite furnace in place of a flame would also permit reduced sample volumes but would require indirect coupling to the column. The semi-automated system is especially useful for measuring free ion concentrations and complexation capacities of biological fluids, non saline waters, waste water and other aquatic samples.

Other workers have devised equipment based on ion exchange enrichment of metals followed by direct desorption of the concentrated metal into an atomic absorption spectrometer [2,3]. Fang et al. [2] used a flow injection system with ion exchange preconcentration for the determination of trace amounts of heavy metals in non saline waters. A multi-functional rotary valve, incorporating two parallel miniature ion exchange columns packed with a chelating resin, containing salicylic acid functional groups, was used for sequential sampling, injection, ion exchange and elution. The sensitivity was increased 20 to 28-fold compared to direct aspiration of samples, for nickel, copper, lead and cadmium at a sampling rate of 40 per hour.

Fang et al. [3] found that the sensitivity of the method was comparable to that of the graphite furnace technique and was cheaper and simpler. Exact timing was used for sample metering and injection, in combination with constant pumping rates. The detection limits ranged from 0.05 to 0.5µg L $^{-1}$ with relative standard deviations 0.2–3.2%. Samara and Kouimtzia [4] preconcentrated cobalt, chromium, copper, iron, nickel and zinc from reactor cooling water with an Acropane resin prior to desorption with acid and determination by atomic absorption spectrometry.

Pilipenko et al. [5] determined the adsorption properties of five cation exchangers for copper(II), zinc(II), lead(II), manganese(II), cobalt(II), nickel(II) and cadmium(II) in the same solution at different pH values. After sorption of the cations, the exchangers were transferred to a column and eluted with hydrochloric acid for subsequent analysis by atomic absorption spectrometry. Type KU–2 and KU–23 cation exchangers were the most convenient exchangers to use, as type KU–1 and KB–4 required a long contact time and type KB–4 was also strongly affected by the pH. The effects of other cations and anions present in non saline waters was negligible.

Werner [6] used Dowex 50WX–4 resin in the calcium form to adsorb cadmium and zinc from a nitriloacetic acid buffer and samples rich in humic acid prior to desorption and analysis by atomic absorption spectrometry. pH had a profound effect on the binding capacity of the resin for cadmium in humic acid rich samples. Between 1 and 3µg L $^{-1}$ cadmium and zinc could be determined by this method.

Sweilek et al. [7] used an ion exchange column to preconcentrate copper(II) species from solutions containing citrate, glycinate, phthalate, salicylate, chloride and fulvate. The preconcentrated metal was desorbed and determined by atomic absorption spectrometry. The method was subject to interference from cationic and neutral copper complexes as well as from filterable colloid copper–hydroxo species at higher pH values. The method would be particularly useful for determination of divalent copper in non saline waters where the copper was often present in low concentrations, where cationic and neutral copper complexes were likely to be absent, and where humates and fulvates were the principal complexing agents.

25.2.1.2 Neutron activation analysis

Preconcentration on a cation exchange membrane followed by neutron activation analysis has been used to determine extremely low levels of cadmium in non saline waters [8]. Linstedt and Kruger [9] preconcentrated vanadium(V) from non saline waters on to Dowex 50–X8 cation exchange resin. The vanadium was then desorbed with nitric acid prior to determination in amounts down to 0.1µg L $^{-1}$ in the sample by neutron activation analysis. Duffy et al. [10] preconcentrated aluminium, calcium, potassium, magnesium, manganese and sodium from soil pore water on to a cation exchange column, having first removed humic materials on an anion exchange column. Cations were then desorbed and determined by neutron activation analysis.

25.2.1.3 X-ray fluorescence spectrometry

Levesque and Mallet [11] preconcenrated zinc, cadmium, nickel, lead, cobalt, manganese, iron and chromium on a column of Amberlite IR–120 resin prior to desorption with sodium chloride solution and determination by X-ray fluorescence spectrometry.

25.2.1.4 Other analytical finishes

Various other methods have also been employed as analytical finishes in cation preconcentration techniques (see Table 25.2).

25.2.2 Aminodiacetate resins, Chelex–100

The cheating resin Chelex–100, a purified form of the strongly acidic Dowex A1 cation exchange resin, has been increasingly used for the separation and preconcentration of trace metals from non saline waters. It is particularly suited to this application because optimum metal removal occurs in the pH range of non saline waters, pH6–8, where distribution coefficients of the order of 10^5 have been measured [33].

Several papers have reported distribution data for selected metal ions on Chelex–100 as a function of pH [37,38]. Batch equilibration studies reported by Dow Chemical Co [39] show complete removal of cobalt, nickel and copper at pH4, while similar results have been obtained for zinc, cadmium and manganese and for iron(III) and lead [40]. Exchange kinetics are slow at low pH values where the imino nitrogen of the resin functional group is protonated, however, at the pH of non saline waters, exchange is rapid and efficient column operation is observed in a 1.5ml column with flow rates of 1.5–2.5ml min^{-1}.

Voutsa et al. [41] compared three different chelating resins, Chelex–100, Hyphan cellulose and Amberlite IRC 718 with respect to their preconconcentration efficiency for traces of copper from fresh waters. Interferences by some complexing agents such as EDTA, NTA, sodium tripolyphosphate and humic acids were also investigated. The retention of copper on the three chelating resins was quantitative above pH3. The complete elution of copper from Chelex–100 and Hyphan resins was only achieved by a mixture of nitric and hydrochloric acids. None of the eluting solutions tested completely eluted copper from Amberlite.

25.2.2.1 Atomic absorption spectrometry

Pakalns et al. [42] have described the use of Chelex–100 for the preconcentration of metals from unpolluted and polluted non saline waters. In particular, they looked at the effect of some organic surface active pollutants on the efficiency of metal removal. These pollutants include cationic, anionic and non ionic detergents, formulated detergents, detergent additives and a soap, which are all likely domestic and industrial discharges into river systems. These workers showed that trace amounts of zinc, cadmium, copper, nickel, manganese, cobalt and lead could be separated from non saline waters on Chelex–100 resin (50–100mesh) in the presence of all the aforementioned pollutants.

Table 25.2 Preconcentration on cation exchange resins, non saline waters

Metals	Type of water	Resin	Medium	Eluting agent	Detection limit*	Analytical finish	Ref.
Mn	River	Amberlite CG–120	Acetic acid	M–ammonium chloride	3μg	Spectrophotometric	[13]
Zn	Non saline	Amberlite CG–120	Acetic acid	M–ammonium thiocyanate Ammonium chloride	2	Spectrophotometric	[14]
Cr(III)	Non saline seawater	Cationite KB–4P–2	Sodium chloride	–	–	Spectrophotometric	[15]
Cu(II)	Non saline	Cationic resin	–	–	–	Spectrophotometric	[16]
Co, Ni, Cu, Zn, Pb, Sn	Non saline	KU–2	None	Hydrochloric acid	–	Spectrographic	[17]
Cu	Non saline	–	–	–	–	Ion selective electrode atomic absorption spectrometry	[7]
²²Na	Non saline	Cationic resin (H form)	–	Hydrochloric acid	–	β–γ counting	[18,19]
¹³⁷Cs	Non saline	Amberlite IR 120 CP (H form)	–	–	–	Radiochemical	[20–22]
⁹⁰Sr	Non saline	KP–P4–2	–	–	–	Radiochemical	[21–23]
¹³⁷Cs	Non saline	KU–2	–	–	–	Radiochemical	[21–23]
⁸⁹Sr	Non saline	Cationic resin	–	–	5 femto Ci L⁻¹	–	[21,22]
²²⁶Ra	Non saline	Cationic	–	–	0.03Ci L⁻¹	Liquid scintillation counting	[24]

Table 25.2 continued

Metals	Type of water	Resin	Medium	Eluting agent	Detection limit*	Analytical finish	Ref.
Cr(III)	Non saline	Cationic	–	–	–	Spectrophotometric	[25]
Ni	Non saline	Dowex Al	–	Hydrochloric acid	–	Spectrophotometric	[26]
U	Non saline	Cellex P	–	–	µg amounts	AAS, inductively coupled plasma atomic emission spectrometry, X-ray fluorescence spectroscopy	[27]
U	Non saline	Dowex 1–X8	–	Sulphuric acid	0.3	Spectrofluorometric	[28]
Ag	Non saline	Amberlite IRA-40 I5CD	Potassium permanganate	Hydrochloric acid	–	Spectrophotometric	[29]
Fe(III)	Non saline	Cation exchange resin	–	0.7M perchloric acid	0–20	UV spectrophotometry	[30]
Ni(II)	Non saline	Modified XAD resin	–	–	100	AAS	[31]
Mo	Non saline	Zeokarb 225	–	–	–	–	[32]

*µg L⁻¹ unless otherwise stated

Source: Own files

Metal recoveries are better than 92% but are poor in the presence of soap or the detergent additive, nitrilotriacetic acid. Although strong adsorption of cationic and to a lesser extent anionic and non ionic detergents, occurs on the resin surface, low recoveries can be attributed to incomplete metal elution rather than to blockage of adsorption sites. Total metal present in non saline waters is not adsorbed by Chelex–100 unless metal ions are first released from colloids or strong complexes. Destruction of the complexes by ultraviolet light or an acid digestion before the sample is applied to the Chelex–100 column results in a complete recovery of metals.

The efficiency of Chelex–100 in concentrating metals in the presence of surfactants and the subsequent effectiveness of the eluant were studied with radiotracers. The tracer solutions ^{54}Mn, ^{65}Zn, ^{60}Co and ^{115}Cd were equilibrated with the corresponding standard metal solutions, surfactant was added and the recommended procedure followed. Effluent and eluate fractions were collected and the distribution of the metals determined by gross γ-counting in a well crystal γ-counter. Blocking of adsorption sites by surfactants did not occur and metal adsorption was very good in the presence of the surfactants studied. Low metal recoveries could in general be attributed not to initial blockage of adsorption sites but to the incomplete removal of metal from the column with the acid eluant. This could result from metal being trapped within with the resin structure as the volume of the resin decreased (normally twofold) on elution with acid. During this step, detergents, particularly the anionic and non ionic species, could hinder elution by partially blocking the resin pores. In the presence of nitriloacetic acid, incomplete metal tracer adsorption was observed. Similar results were obtained in the presence of humic acids.

To gain insight into the fraction of total metal determined in non saline waters after concentration by Chelex–100, small quantities of manganese–54, zinc–65, cobalt–60 and cadmium–115 traces were mixed with four different filtered non saline water samples. These were allowed to stand for up to 7d before the mixtures were put through the resin column. The proportion of metal tracer adsorbed by the resin decreased with standing time. Because up to 28% of tracer was not recoverable by the resin after storage for 7d, the metal concentrations could not be regarded as total metal concentrations. Moreover, any metal present either as strong complexes which do not (or only slowly) exchange with the iminodiacetate resin groups, or as metal associated with colloidal particles, would not be retained by the chelating resin. Colloidal particles cannot penetrate the pore network of Chelex–100 which has an average pore diameter of only 6nm.

Corsini *et al.* [43] separated soluble manganese into two fractions. A sequential preconcentration on XAD–7 resin, followed by Chelex–100 left manganese(II) on the XAD–7 column. The second fraction, consisting of at least one unidentified form of manganese, was retained on the Chelex–100 after elution. The manganese was determined by electrothermal atomic absorption spectrometry.

25.2.2.2 Neutron activation analysis

Gladney et al. [44] preconcentrated uranium–235 and uranium–238 in non saline waters on Chelex–100 resin and subsequently measured the concentration of these elements in the resin by neutron activation analysis. Hinse et al. [45] and Greenberg and Kingston [46] have also used neutron activation analysis to determine metals preconcentrated from non saline waters on to cation exchange resins.

25.2.2.3 X-ray fluorescence spectrometry

This technique has also been used in the analysis of preconcentrated cations [47,48]. Thus, Van Espen et al. [48] preconcentrated toxic metals in non saline water on Chelex–100 filters and determined the elements directly on the filters in amounts down to 1µg L $^{-1}$ by energy dispersive X-ray fluorescence spectrometry.

25.2.2.4 Anodic scanning voltammetry

Figura and McDuffie [49] preconcentrated copper, cadmium, lead and zinc by adsorption on Chelex–100 resin, enabling sub µg L $^{-1}$ quantities of these elements to be determined. The Chelex column effluents were analysed by anodic scanning voltammetry.

Chelex–100 resins have also been used for the preconcentration of molybdenum [35], zinc, cadmium, mercury and lead [50]; iron, copper, nickel, cadmium, cobalt, zinc, lead and manganese [51]; and uranium [52].

25.2.3 Macroreticular polyacrylic acid and SM–7 ester resin

Corsini et al. [53] have used a column of this resin to preconcentrate unchelated trace metals in water samples prior to determination by atomic absorption spectrometry. It is claimed to be simple, rapid and economical. A study of the effect of pH on the uptake of cadmium(II), cobalt(II), copper(II), manganese(II), nickel(II), lead(II), chromium(III) and iron(II) showed that with 100ml of sample containing 1ml 1M ammonium acetate buffer for the divalent ions at levels ranging from 1–20µg L $^{-1}$ in water a maximum metal uptake (85–100%) occurred in the pH range 7–9, and for the trivalent ions in the pH range 4–5. Nitric acid 1% was the most satisfactory eluting agent for removing the metals from

Table 25.3 Comparison of SM–7 and Chelex–100 methods in analysis of lake water (results in µg L⁻¹)

	SM–7ᵃ	Chelex–100	Direct
Fe(III)	2.4 ± 0.4	2.3	2.4
Cr(III)	0.9 ± 0.1	0.79	1.0
Pb(II)	0.12 ± 0.01	0.14	<0.2
Cu(II)	1.4 ± 0.1	1.1	1.5
Ni(II)	2.0 ± 0.2	2.3	2.6
Co(II)	0.09 ± 0.01	0.05	<0.2
Mn(II)	0.60 ± 0.01	0.39	0.4
Cd(II)	0.072 ± 0.001	0.093	<0.1

ᵃSM–7 data are the average of two results; data for Chelex–100 and direct analysis represent a single determination. Flow rates for uptake and elution for both columns were 1.0–1.2ml min⁻¹

Source: Reproduced with permission from Corsini, A. et al. [53] American Chemical Society

the column. More concentrated solutions of nitric acid (eg 10%) increase the blank value considerably due to contamination. For a 7cm bed height, the chromatographic elution profiles of all ions are essentially the same.

Analytical results for the lake water sample are given in Table 25.3. Data obtained for preconcentration by SM–7 and Chelex–100 are compared. Included are data obtained by direct analysis. The concentration of some elements present is high enough for measurement by graphite furnace atomic absorption without preconcentration. The method of standard additions was used for all data in Table 25.3. The results represent total soluble metal. The agreement among the methods is acceptable in consideration of the low concentrations involved.

On the whole, the blank values for the SM–7 results were 2–4 times lower than with Chelex–100, the blank for iron(III) in particular being most improved. The lower blanks arise because the SM–7 resin is cleaner and because less nitric acid is required for elution. The lesser amount of nitric acid may be at least in part a consequence of a lower affinity of the SM–7 resin than that of the Chelex–100 for metal ions, in view of the fact that metal ion uptake by a strong chelating mechanism cannot occur with SM–7. The disadvantage is that two sample aliquots are used, one for uptake of trivalent ions at pH4–5 and another for divalent ions at pH8–9.

Whereas percentage recovery of metal ion spikes from distilled waters were essentially quantitative by the above method, spike recoveries from lake water were considerably lower than quantitative, for example, 80% recovery of copper to as low as 48% recovery of lead. Subsequently these workers [54] devised a two column procedure for preconcentrating these metals from lake water and seawater in which organic ligands, particularly humic materials, were removed by passing the water through a pre-column at pH1–2 before preconcentrating the trace metals at pH8. The final effluent for measurement by graphite furnace atomic absorption spectrometry was readily matrix matched and allowed the use of the standard calibration curve procedure for the majority of the trace metals.

25.3 Seawater

25.3.1 Miscellaneous

Wan *et al.* [54] determined conditions for the direct preconcentration of cadmium, manganese, chromium, copper, nickel, iron, cobalt and lead from seawater samples using a two column Amberlite XAD–7 resin system. Low breakthrough volumes in the presence of humic materials necessitated their prior removal at a pH of 1–2 prior to preconcentration of the trace metals on a second column of XAD–7 at pH8. Metals were subsequently desorbed from the second column with 1% nitric acid by means of a pre-column of XAD–7. The final effluent for measurement by graphite furnace atomic absorption spectrometry is readily matrix matched and permits use of the standard calibration curve procedure. Preconcentration factors of 40 were obtained by this procedure permitting the analysis of coastal seawaters for the eight elements mentioned earlier.

Dehairs *et al.* [55] give details of a procedure for the determination of barium in seawater, involving separation of barium from manganese by collection on a cation exchange resin in the ammonium form and extraction of the barium from the resin into nitric acid for determination by graphite furnace atomic absorption spectrometry.

Atomic absorption spectrometry has been used to determine caesium in seawater [56]. The method uses preliminary chromatographic separation on a strong cation exchange resin, ammonium hexacyanocobalt ferrate, followed by electrothermal atomic absorption spectrometry. The procedure is convenient, versatile and reliable, although decomposition products from the exchanger, namely iron and cobalt can cause interference. Caesium is fully retained by a chromatographic column of ammonium hexacyanocobalt ferrate and can then be recovered by dissolution of the ammonium hexacyanocobalt ferrate in hot 12M sulphuric acid. The resin is stable in strong acid solutions.

A further method for the determination of caesium isotopes in saline waters [57] is based on the high selectivity of ammonium hexacyano-cobalt ferrate for caesium. The sample (100–500ml) is made 1M in hydrochloric acid and 0.5M in hydrofluoric acid, then stirred for 5–10min with 100mg of the ferrocyanide. When the material has settled, it is collected on a filter (pore size 0.45μm), washed with water, drained, dried under an infrared lamp, covered with plastic film and β-counted for ^{137}Cs. If ^{131}Cs is also present, the γ-spectrometric method of Yamamoto [58] must be used. Caesium can be determined at levels down to 10pCi L $^{-1}$.

Nowicki et al. [59] collected zinc in seawater on a cation exchange column, complexed it with p-tosyl-8-aminoquinoline then determined zinc in amounts down to 0.1nM (4.4ml sample) by fluorescence spectroscopy.

Boyle et al. [60] developed a method for the analysis of cobalt in seawaters by cation exchange liquid chromatography using luminol chemiluminescence detection. Cobalt can be determined directly in freshwater samples using 500μL samples with a detection limit of 20pmol kg $^{-1}$; larger samples provide proportionately lower detection limits. Seawater samples can be analysed on 100ml samples following ammonium pyrrolidinedithiocarbamate solvent extraction; the detection limit of this method is 5pmol kg $^{-1}$. The precision of the method is ±5%. The method should also be applicable to the analysis of vanadium, copper and iron in non saline and saline waters. Equipment is low in cost and transportable and can be used in the field.

In most instances, Chelex–100 has been used in the H$^+$ form, although the effluent pH under these conditions may be as low as 2.8. Florence and Batley [34] showed that the H$^+$ form did not completely remove labile zinc, cadmium, lead and copper from seawater until the passage of more than 500ml of sample had increased the pH of the effluent to 6.5. Treatment of the column with sodium acetate before use raised the effluent pH to 7.1. The ammonium or calcium forms also have been used to circumvent this problem [34–37]. Chelex–100 is also used in the sodium form.

25.3.2 Imido diacetate resins, Chelex–100

Pai et al. [61] examined the preconcentration efficiency of Chelex–100 for heavy metals in seawater. They showed that the metal-chelating efficiency of this resin was lower in seawater than in fresh water owing to the complicated speciation of metals in seawater and competition from high levels of magnesium and calcium. The optimal pH for seawater samples prior to loading was about 6.5, but optimal pH values for column operation were strongly affected by the salt matrix. Care should be exercised in the choice of conditions to minimise losses of cadmium and manganese from seawater. Pai [62] also examined the distribution of heavy metals in seawater on a Chelex–100 column. Columns were divided into small sections and spiked heavy metals recovered from each section. A simplified model was derived to describe the effect of column size and flow rate on the recovery of heavy metals (chelating efficiency) at various pH values. Columns containing Chelex–100 resin (2g) in the magnesium form with a flow rate of 4ml min $^{-1}$ were suitable for the preconcentration

Table 25.4 Application of Chelex–100 resin to the preconcentration of metals in seawater prior to analysis by graphite furnace atomic spectrometry

Elements	Concn. factor	Eluent	Detection limit		Ref.
Cd, Co, Cu, Fe, Mn, Ni, Pb, Zn	100:1	2.5M nitric acid	subnanogram µg L^{-1}		[66]
Cu, Cd, Zn, Ni	120:1	2M nitric acid	Cu	0.006	[67]
			Cd	0.006	
			Zn	0.015	
			Ni	0.015	
Cd, Zn, Pb, Cu, Fe, Mn, Co, Cr, Ni	20:1	2.5M nitric acid	Not stated µg L^{-1}		[63]
Cd, Pb, Ni, Cu, Zn	100:1	2.5M nitric acid	Cd	0.01	[65]
			Pb	0.16–0.28	
			Ni	0.24–0.68	
			Cu	0.6	
			Zn	1.8	
Zn	–	Nitric acid	0.5		[68]
Cr, Cu, Mn	–	–	–		[69]
Cd, Cu, Ni, Pb	–	Nitric acid (pH5) plus 7.5M ammonium acetate	–		[70]

Source: Own files

of cadmium, copper, cobalt, manganese, nickel, lead and zinc in seawater after adjustment of samples to pH6.5.

25.3.2.1 Atomic absorption spectrometry

Studies of the use of ion exchange resins for the preconcentrations of metals from seawater have been mainly concerned with the use of Chelex–100 cation exchange resin [63–65]. Chelex–100 is the most commonly employed chelating resin for the removal and preconcentration of trace heavy metals from seawater. Work on the use of Chelex–100 cation exchange resin for the preconcentration of metals from seawater is reviewed in Table 25.4. In each case metals are desorbed from the resin with nitric acid (2–2.5M) and then determined in the extract by graphite furnace atomic absorption spectrometry. Preconcentration factors of up to 100–120 [63] have been reported by this technique enabling metals to be determined at the µg L^{-1} of ng L^{-1} level.

Early Chelex–100 procedures only partially separated the alkali and alkaline earth metal components of seawater prior to the analysis of the eluted elements of interest. A further separation procedure utilising the

Table 25.5 Applicability of preconcentration procedures to seawater analysis. Detection limits based on 2SD of blank

Element	Concn. in seawater consensus values ($\mu g\ L^{-1}$)	Best achievable values by preconcentration		
		Chelation–solvent extraction–graphite furnace AAS	Ion exchange separation Chelex–100 resins	
			Graphite furnace AAS	Neutron activation analysis
Cr	0.03	0.05	–	0.14
Mn	0.02	0.004	–	0.16
Fe	0.2	0.02	–	1.2
Co	0.005	0.04	–	0.006
Ni	0.17	0.012	0.015	–
Cu	0.05	0.006	0.006	0.08
Zn	0.49	0.016	0.015	0.20
Cd	–	0.0001	0.0006	–
Pb	–	0.016	–	–
Th	0.01			0.0004
V	2.5			0.06

Source: Reproduced with permission from Paulson, A.J. [71] American Chemical Society

Chelex resin produced a sample devoid of alkali, alkaline earth and halogen elements, and left a dilute nitric acid/ammonium nitrate matrix containing only the trace elements of the seawater samples [66]. While this procedure produces a highly desirable and appropriate matrix for most spectroscopic methods of analysis, a solid sample would be more appropriate for other instrumental techniques such as X-ray fluorescence or neutron activation analysis. In addition, the above separation procedure also makes it difficult or impossible to analyse several elements which are held strongly by the resin but cannot be quantitatively eluted. Chromium and vanadium exhibit this type of behaviour and attempts to reproducibly elute these elements from Chelex–100 have not been successful.

Paulson [71] studied the effects of flow rate and pre-treatment on the extraction of manganese, cadmium and copper from estuarine and coastal seawater by Chelex–100 resin. Decreasing the flow rate for column extraction of estuarine samples by Chelex–100 to 0.2ml min $^{-1}$ increases the yield of trace metals and improves the precision for determination of these elements. Detection limits achievable by chelation–solvent extraction and ion exchange separation on Chelex–100 resin are summarised in Table 25.5.

Kingston et al. [66] and Bender and co-workers [72,73] used Chelex–100 followed by stripping with nitric acid to preconcentrate total and soluble manganese from seawater. Samples were collected into 500ml polyethylene bottles. All samples were brought to pH2 with nitric acid free of trace metals and stored in individual zip-lock plastic bags to minimise contamination. When the samples were returned to the laboratory the pH was adjusted to approximately pH8 using concentrated ammonia (Ultrapure, G. Frederick Smith). Chelex–100 cation exchange resin in the ammonia form (20ml) was added to the samples and they were batch extracted on a shaker table for 36h. The resin was decanted into columns and the manganese eluted using 2N nitric acid [66]. The eluant was then analysed by graphite furnace atomic absorption spectrophotometry. Replicate analyses of samples indicate a precision of about 5%.

Batley and Farrah [74] and Gardner and Yates [75] used ozone to decompose organic matter in samples and thus break down metal complexes prior to atomic absorption spectrometry. By this treatment metal complexes of humic acid and EDTA were broken down in less than 2min. These observations led Gardner and Yates [75] to propose the following method for the determination of cadmium in seawater: The sample is filtered immediately after collection, acidified to about pH2, and transferred to a 1L Pyrex storage bottle. Prior to extraction the sample is ozonised in the sample bottle for 30min. Nitrogen is passed through the sample for 5min to remove excess ozone, then the pH is carefully raised to about 5 by addition of ammonia solution and about 5ml Chelex–100 resin in the ammonia form is added. After stirring for at least 1h, the resin is collected in a Pyrex chromatography column and washed with the calculated quantity of an appropriate buffer to elute calcium and magnesium. After further washing with 50ml deionised water, the resin is eluted with 2M nitric acid, to a volume of 25ml. The eluate is analysed by graphite furnace atomic absorption spectrometry.

Boniforti et al. [76] compared several preconcentration methods in the determination of metals in seawater. A comparison was made of ammonium pyrrolidinedithiocarbamate-8-quinolinol complexation followed by extraction with methyl isobutyl ketone or Freon–113, co-precipitation with magnesium hydroxide or iron(II) hydroxide or chelating by batch treatment with Chelex–100 for the determination of manganese, iron, cobalt, zinc, nickel, copper and chromium. Atomic absorption spectrometry and inductively coupled plasma atomic emission spectrometry were used for analysis. Interferences, recovery, precision, accuracy and detection limits were compared. The Chelex–100 resin method was most suitable for the preconcentration of all determinands except chromium, whereas preconcentration of $Cr(III)$ and $Cr(V)$ was achieved only by coprecipitation with iron(II) hydroxide.

Olsen et al. [77] gave details of equipment and procedure for preconcentrating and determining traces of cadmium, lead, copper and zinc in seawater by atomic absorption spectrophotometry combined with flow injection analysis. Preconcentration was achieved by passing 2ml of sample through a microcolumn packed with Chelex–100 resin. Metals were desorbed with 180µL 2M aqueous nitric acid which was passed to an atomic absorption spectrometer, thereby achieving a concentration factor of about 10. Seen from a practical viewpoint this combination results in timesaving because it allows an unprecedented sample throughput at the µg L^{-1} level. As the analytical readout is available within 5s for the direct assay and at the latest within 110s for the system including preconcentration, smaller sample series can be treated expediently by manual injection.

Since the original work of Olsen et al. [77] many improvements have been proposed to flow injection analysis equipment. Valve designs have been improved, in addition to Chelex–100 other ion exchange resins have been tested, and different flow systems have been proposed to increase the efficiency of the process [2,78–80]. These developments indicate that the new approach not only increases the speed of the preconcentration process, but could also ultimately rival the sensitivity and speed of graphite furnace atomic absorption spectrometry.

Fang et al. [3] described a flow injection system with on-line ion exchange preconcentration on dual columns for the determination of trace amounts of heavy metals at µg L^{-1} and sub µg L^{-1} levels by flame atomic absorption spectrometry. The degree of preconcentration ranges from 50- to 105-fold for different elements at a sampling frequency of 60 samples hourly. The detection limits for copper, zinc, lead and cadmium are 0.07, 0.03, 0.5 and 0.05 µg L^{-1} respectively. Relative standard deviations were 1.2–3.2% at µg L^{-1} levels. These workers studied the behaviour of the different chelating exchangers used with respect to their preconcentration characteristics, with special emphasis on interferences encountered in the analysis of seawater.

Fang et al. [3] studied closely the swelling properties of Chelex–100 resin, and showed that depending on the pH of the element, the resin swelled by a factor up to two. To limit the maximum change in volume of a resin in a column to 25% during a single cycle of operation (which is imperative to avoid excessive pressure and void volume variations), the sampling period (during which chelation takes place) at a flow rate of 6ml min^{-1} should not exceed 50s for a 0.5M ammonium acetate buffer, 75s for a 0.1M buffer or 100s for a 0.05M buffer. Secondly, the column should be packed about three-quarters full with resin in the H$^+$ form (washed with water) after conversion from the NH$_4^+$ form (in which state it should be equilibrated with buffer of the same concentration as that used in the ensuing procedure). Packing in the NH$_4^+$ form would otherwise result in an excessively loose column packing giving rise to large dispersion and degrading the sensitivity. Finally, resin columns in the NH$_4^+$ form should never be washed with water, lest excessive pressure develop in the column, causing blockages, leakages or dislodgement of the nylon retaining gauzes. When not in use, the columns should be converted to the H^+ form by normal elution with acid and washed thoroughly with water.

Fang *et al.* [3] concluded that the flow-injection atomic absorption spectrometry system with on-line preconcentration will challenge the position of the graphite furnace technique, because it yields comparable sensitivity for much lower cost by using simpler apparatus and separation mode. The method offers unusual advantages when matrices with high salt contents such as seawater are analysed because the matrix components do not reach the nebuliser.

25.3.2.2 Inductively coupled plasma atomic emission spectrometry

Berman *et al.* [81] attempted to determine nine elements in seawater by a combination of ion exchange preconcentration on Chelex–100 [66,82] and inductively coupled plasma atomic emission spectrometry using ultrasonic nebulisation. Preconcentration factors of between 25 and 100 were obtained by this technique. Table 25.6 compares the ICPAES results with data generated for the same sample by two other independent methods – isotope dilution spark source mass spectrometry (IDSSMS) and graphite furnace atomic absorption spectrometry (GFAAS). The IDSSMS method also uses a preconcentration of the metals and matrix separation using the ion exchange procedure, following isotope addition. The atomic absorption determinations were preceded by an MIBK extraction [82]. In general the agreement is very good, but one discrepancy merits comment. The spark source mass spectrometry result for manganese is not as reliable as the other data by this method. Since manganese is mono-isotopic, a less accurate internal standardisation method of calibration has to be used. The ICPAES result for manganese is in close agreement with the GFAAS result.

Manganese, iron, zinc, copper and nickel can be determined by preconcentration ICPAES giving good agreement with the other methods. No consistently reliable results were obtained for cadmium, chromium, cobalt and lead when samples up to 1L were processed (ie no preconcentration). Chromium is only weakly retained by the Chelex–100 resin under the conditions used so that the seawater concentration (about $0.3\mu g\ L^{-1}$ is not sufficiently enhanced.

Reimer and Myazaki [83] preconcentrated lead onto Chelex–100 resin prior to desorption and determination by inductively coupled plasma atomic emission spectrometry in amounts down to $0.006\mu g\ L^{-1}$ in seawater.

25.3.2.3 Stable isotope dilution spark source mass spectrometry

Mykytiuk *et al.* [84] have described a stable isotope dilution spark source mass spectrometric method for the determination of cadmium, zinc,

Table 25.6 Analysis of Sandy Cove seawater

Element	ICPAES	GFAAS	IDSSMS
Mn	1.5 ± 0.1	1.1 ± 0.2	1.8 ± 0.2*
Fe	1.5 ± 0.6	1.5 ± 0.1	1.4 ± 0.1
Zn	1.5 ± 0.4	1.9 ± 0.2	1.6 ± 0.1
Cu	0.7 ± 0.2	0.6 ± 0.2	0.7 ± 0.1
Ni	0.4 ± 0.1	0.33 ± 0.08	0.37 ± 0.02
Pb	–	0.22 ± 0.04	0.35 ± 0.03
Cd	–	0.24 ± 0.04	0.28 ± 0.02
Cr	–	–	0.34 ± 0.06
Co	–	–	0.02*

Precision expressed as 95% confidence intervals
*Spark source spectrometry, internal standard method

Source: Reproduced with permission from Berman, S.S. *et al.* [81] American Chemical Society

copper, nickel, lead, uranium and iron in seawater and compared results with those obtained by graphite furnace atomic absorption spectrometry and inductively coupled plasma emission spectrometry. These workers found that to achieve the required sensitivity it was necessary to preconcentrate elements in the seawater using Chelex–100 [63] followed by evaporation of the desorbed metal concentrate on to a graphite or silver electrode for isotope dilution mass spectrometry.

Results obtained on a seawater sample by three procedures are given in Table 25.7. Isotope dilution results agree well with those obtained by graphite furnace atomic absorption spectrometry and inductively coupled plasma emission spectrometry. The preconcentration for all three techniques was achieved by Chelex–100 ion exchange. However, since solvent extraction with ammonium pyrrolidinedithiocarbamate is the most commonly accepted method, the values obtained using it with graphite furnace atomic absorption spectrometry are also included for comparison.

One of the advantages of the isotope dilution technique is that the quantitative recovery of the analytes is not required. Since it is only their isotope ratios that are being measured, it is necessary only to recover sufficient analyte to make an adequate measurement. Therefore, when this technique is used in conjunction with graphite furnace atomic absorption spectrometry, it is possible to determine the efficiency of the preconcentration step. This is particularly important in the analysis of seawater where the recovery is very difficult to determine by other techniques since the concentration of the unrecovered analyte is so low. In

Table 25.7 Analysis of seawater sample B (concentrations (ng ml^{-1}) expressed as means ± SD)

| | IDSSMS | GFAAS | | ICPAES |
	ion exchange	Solvent extraction	Ion exchange	ion exchange
Fe	3.4 ± 0.3	3.2 ± 0.2	3.4 ± 0.4	3.2 ± 0.2
Cd	0.07 ± 0.01	0.06 ± 0.01	0.053 ± 0.007	ND
Zn	1.9 ± 0.1	1.8 ± 0.1	2.0 ± 0.1	1.6 ± 0.2
Cu	0.61 ± 0.04	0.5 ± 0.1	0.51 ± 0.03	0.73 ± 0.06
Ni	0.43 ± 0.03	0.46 ± 0.03	0.45 ± 0.05	0.38 ± 0.02
Pb	0.11 ± 0.02	0.06 ± 0.02	0.10 ± 0.01	ND
Co	0.028 ± 0.001*	0.015 ± 0.003	0.018 ± 0.008	ND
U	2.6 ± 0.2	ND	ND	ND

*By internal standard
ND = not determined

Source: Reproduced with permission from Mykytiuk, A.P. *et al.* [84] American Chemical Society

using this technique, one must assume that isotopic equilibrium has been achieved with the analyte regardless of the species in which it may exist.

25.3.2.4 Flame photometry

Blake *et al.* [85] have described a flame photometric method for the preconcentration of calcium in solutions of high sodium content. The method was applied to simulated seawater. In the method Chelex–100 chelating resin (Na$^+$ form) was stirred with 2N hydrochloric acid for 5min, the acid was decanted and the resin washed with water, stirred with 2N sodium hydroxide for 5min, and again washed with water. The procedure was repeated five times then the resin was dried at 100°C. A neutral solution containing up to 50µg L^{-1} of calcium and up to 4% of sodium was passed through a column of the resin, a specified amount of hydrochloric acid (pH2.4) was passed through and the percolate containing the sodium was discarded. Elution was then effected with 2N hydrochloric acid (5ml) and the column washed with water (25ml), the combined eluate and washings were diluted to 100ml and calcium determined by flame photometry at 622nm. There was no interference from magnesium, zinc, nickel, barium, mercury, manganese, copper or iron present separately in concentrations of 25µg L^{-1} or collectively in concentrations of 5µg L^{-1} each. Aluminium depresses the amount of calcium found.

Table 25.8 Trace elements in water–SRM 1643[a] (µg L $^{-1}$)

Element	Neutron activation analysis[a]	Certified[b]
Co	19.1 ± 1	19 ± 2
Cr	16 ± 2	17 ± 2
Cu	19.1 ± 0.6	18 ± 2
Fe	88 ± 16	88 ± 4
Mn	30.9 ± 0.6	31 ± 2
Mo	97 ± 6	95 ± 6
Ni	56 ± 8	55 ± 3
V	52 ± 1	53 ± 3
Zn	68 ± 5	72 ± 4

[a]Uncertainties are 2SD; [b]Uncertainties are 95% confidence limit

Source: Reproduced with permission from Greenberg, R.R. and Kingston, H.M. [65] Elsevier Sequoia, Lausanne

25.3.2.5 Neutron activation analysis

This technique is intrinsically sensitive and can be made more so by preconcentrating the metals on to a resin such as Chelex–100 which can be directly analysed by the neutron activation technique. With these facilities analysis for many metals at the ultra-low background levels at which they occur in seawater becomes a possibility.

Greenberg and Kingston [65] described a method to prepare solid samples from 100ml estuarine or seawater using Chelex–100 resin, followed by the determination of trace elements in the solid resin by neutron activation analysis. Using this procedure, typical decontamination factors of $\geq 10^7$ for sodium, $\geq 10^5$ for chlorine and $\geq 10^3$ for bromine are observed. They used this procedure to determine cobalt, chromium, copper, iron, manganese, molybdenum, nickel, scandium, thorium, uranium, vanadium and zinc in NBS. Standard Reference Material 1643a (Trace Elements in Water) (Table 25.8). They also analysed a sample of seawater taken in Chesapeake Bay by neutron activation analysis and compared these results with those obtained by other techniques. The good agreement is evident in Table 25.9. Some further applications of neutron activation analysis to seawater analysis are summarised in Table 25.11.

25.3.2.6 γ ray spectrometry

In a procedure [86] employing preconcentration of the metals on a column containing a mixture of Chelex–100 and Pyrex glass powder the problems associated with swelling of the Chelex–100 were overcome and

Table 25.9 Trace elements in seawater sample (µg L⁻¹)*

Element	Neutron activation analysis	GFAAS	XRF
Co	0.444 ± 0.003	<0.1	
Cr	3.3 ± 0.2		
Cu	2.01 ± 0.05	2.0 ± 0.1	2.0 ± 0.2
Fe	2.1 ± 0.3	2.1 ± 0.5	
Mn	1.89 ± 0.03	2.0 ± 0.1	2.0 ± 0.1
Mo	5.3 ± 0.1		
Ni	1.3 ± 0.2	1.2 ± 0.1	1.3 ± 0.2
Se	<0.00095 ± 0.0005		
Th	<0.0002		
U	1.90 ± 0.04		
V	0.45 ± 0.01		
Zn	4.9 ± 0.2	4.8 ± 0.3	4.5 ± 0.4

*All uncertainties are at the ISD level; GFAAS Graphite furnace atomic absorption spectrometry; XRF X-ray fluorescence spectrography

Source: Reproduced with permission from Greenberg, R.R. and Kingston, H.M. [65] Elsevier Sequoia, Lausanne

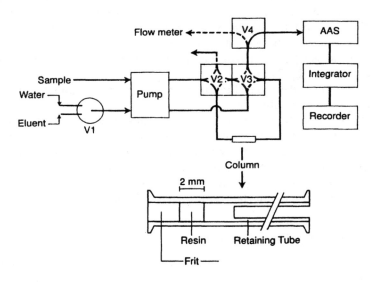

Fig. 25.2 Gamma ray spectrum of preconcentrated river water after short irradiation. Irradiation time 3min; thermal–neutron flux 1 × 10¹³n cm⁻², s⁻¹ decay time 3min; counting time 100s
Source: Reproduced with permission from Lee, C. et al. [86] Elsevier Science UK

Table 25.10 Recovery of trace elements from spiked samples of stripped waters

Element	Seawater Added (µg)	Found (µg)	Yield (%)
Ba	1.44	1.48	103
Ca	–	–	–
Cd	30.0	27.6	90.0
Ce	1.48	1.45	98.0
Co	1.50	1.49	99.3
Cr	230	230	100
Cu	189	187	98.9
Fe	615	615	100
La	0.830	0.832	100
Mg	–	–	–
Sb	–	–	–
Sc	0.140	0.142	101
U	0.160	0.157	98.1
V	27.0	27.0	100
Zn	401	397	99.0

Source: Reproduced with permission from Lee, C. et al. [86] Elsevier Science, UK

constant flow rates of sample down the column achieved. The water sample were passed through the resin column and eluted with 100ml 0.01M nitric acid and the eluate was discarded. Trace elements were collected from the column by eluting with 50ml 4M nitric acid.

Portions of the sample were irradiated for 5min and 3d and subject to γ ray spectrometry (Fig. 25.2).

To determine the half-lives of the nuclides produced, the counting was repeated at appropriate intervals.

The recovery ratios shown in Table 25.10 indicate that the added traces of cadmium, cerium, copper, lanthanum, manganese, scandium and zinc are quantitatively recovered. The recoveries of barium, cobalt, bromium, iron, uranium and vanadium were also satisfactory. Table 25.10 shows the recoveries of calcium and magnesium were very poor for seawater. The reasons for this may be connected with matrix effects.

25.3.2.7 Anodic stripping voltammetry

A significant fraction of the copper, lead, cadmium and zinc in seawater exists in a form which is not retained by a chelating resin (Chelex–100) or extracted by ammonium pyrrolidinedithiocarbamate [34]. Anodic stripping voltammetry suggests that the major part of the unavailable trace metal is adsorbed on, or occluded in, organic or inorganic colloidal particles. An ionic equilibria computer program was used to predict the

Table 25.11 Application of neutron activation analysis to the determination of metals in seawater

Elements	Sample preconcentration	Ref.
Co, Cr, Ca, Fe, Pb, Se, Sr	Sea salts Freeze dried	[87]
As, Cu, Sb	Thionalide precipitation	[88]
Misc elements	Frozen sample	[89]
Al, V, Cu, Mo, Zn, U	Organic coprecipitation (8-hydroxyquinoline)	[65]
Ag, Au, Cd, Ce, Co, Cr, Eu, Fe, Hg, La, Mo, Sc, Se, U, Zn, As, Sb	Adsorption and charcoal	[46]
Ba, Ca, Cd, Ce, Co, Cr, Cu, Fe, La, Mg, Mn, Se, U, V, Zn	Adsorption of Chelex–100 glass powder	[90]
Hg, Au, Cu	Coprecipitation with lead diethyidithio-carbamate	[91]
12 elements	Chelating resin	[92]
Transition metals	Chelating resin	[93]
As, Mo, U, V	Colloid flotation	[94]
Th	Ion exchange	[84]
Co, Cu, Hg	Coprecipitation with lead pyrrolidine dithiocarbarmate	[63]
In	Ion exchange chromatography	[85]
Co	Ion exchange	[95]

Source: Own files

effect of various complexing agents on trace metal species in seawater. Citric acid and amino-acids, with the exception of *l*–cysteine, were shown to be unimportant as complexing agents, and the ability of an EDTA-like ligand to complex copper, lead, cadmium and zinc is controlled entirely by the concentrations of the ligand and of 'labile' iron and chromium in seawater, since these two ions together will react quantitatively with EDTA.

25.4 Potable water

Subramanian *et al.* [96] have discussed on-site sampling with preconcentration for the determination of some Chelex–100 levels of labile metals (cadmium, copper, lead and zinc) in potable water. The on-site pump integrated water sampler coupled with a Chelex–100 preconcentration unit is described in detail. Metals in the concentrate were determined by graphite furnace atomic absorption spectrophotometry.

Subramanian *et al.* [96] confirm that Chelex–100 will remove only that fraction of the trace metal level that is 'Chelex labile' and found for example, that only 44–63% of lead and copper in potable water is labile, presumably due to the presence of humic acid and possibly other chelators in the water. They do not discuss decomplexing using ultraviolet light or acid digestion methods.

Kempster and Van Vliet [97] have described a semi–automated resin concentration method for the preconcentration of trace metals (chromium, manganese, iron, cobalt, nickel, copper, zinc, cadmium and lead) in potable water, prior to atomic absorption analysis. A peristaltic pump was used to control the flow of water samples through columns of a cation exchange resin (Amberlite IR–120/H form), the samples being stabilised with ascorbic acid (0.5g L^{-1}) at a pH of 2.5 during the sorption stage. A 26 channel peristaltic pump was used to pump the contents of 25 samples and one blank simultaneously through the resin columns, using 1.6mm id polyethylene for transmission lines and 0.86mm id tygon tubing in the peristaltic pump, which gave a constant flow rate of 0.42ml min^{-1} and a flow-through time in the resin column of just under 5min. Before use the resin columns were freed, if necessary, of any entrapped air by detaching the columns from the flow lines and aspirating deionised water rapidly through each column with a syringe. The resin columns were then cleaned by pumping 12ml 5M acid (redistilled AR) followed by 12ml deionised water through the columns.

Of the nine elements mentioned above, eight gave a recovery through the whole procedure of between 88 and 99% while iron had a recovery of 75%. Detection limits (µg L^{-1}) were as follows: Cr, 3; Mn, 0.5; Co, 1; Cu, 0.5; Zn, 2; Cd, 0.1; Pb, 6. This method is useful for the preconcentration of a large number of samples, does away with the tedium characteristic of manual enrichment techniques, and gives good recovery for the nine metals tested. Difficulty was experienced in obtaining consistent results for iron but it was found that with the addition of ascorbic acid to the samples, prior to sorption on to the resin, more consistent results were obtained. Ascorbic acid serves as a agent to counteract hydrolysis of iron.

Ho and Lin [12] preconcentrated calcium, iron, cobalt, nickel, copper, zinc, lead, mercury, chromium and selenium by passing the potable water sample repeatedly through a cation exchange resin. The resin was subsequently analysed by energy dispersive X-ray fluorescence spectrometry. These workers used Amberlite IR–120 paper for calcium, iron, cobalt, nickel, copper, lead and zinc, and Amberlite LRS–400 paper for mercury, selenium, and chromium. Detection limits ranged from 0.5µg L^{-1} (iron) to 9µg L^{-1} (selenium).

25.5 Sewage effluents

Morrison [98] estimated bioavailable metal uptake from sewage and storm water using dialysis with Chelex–100 resin in the calcium form. The resin was incorporated within the dialysis bag sealed at both ends.

The bag was immersed in the sample for 1–4d before releasing the chelated resin to a separation column. Metals were eluted from the resin with 1M nitric acid and metal concentrations determined by graphite furnace and flame atomic absorption spectrometry. Zinc, cadmium, lead and copper were studied.

References

1 Treit, J., Nielson, J.S., Kratochvil, B. and Cantwell, F.F. *Analytical Chemistry*, **55**, 1650 (1983).
2 Fang, Z., Xu, S. and Zhang, S. *Analytica Chimica Acta*, **164**, 41 (1984).
3 Fang, Z., Ruzicka, J. and Hansen, E.H. *Analytica Chimica Acta*, **164**, 23 (1984).
4 Samara, C. and Kouimtzia, T.A. *Chemosphere*, **16**, 405 (1987).
5 Pilipenko, A.T., Safronova, V.G. and Zakrevskaya, L.V. *Soviet Journal of Water Chemistry and Technology*, **9**, 74 (1987).
6 Werner, J. *Science of the Total Environment*, **62**, 281 (1987).
7 Sweilek, J.A., Lucky, D., Kratochvil, B. and Cantwell, F.F. *Analytical Chemistry*, **59**, 586 (1987).
8 Mark, H.B., Eisner, U. and Rothschafler, J.M. *Environmental Science and Technology*, **3**, 165 (1969).
9 Linstedt, K.D. and Kruger, D. *Analytical Chemistry*, **42**, 113 (1970).
10 Duffy, S.J., Hay, G.W., Micklethwaite, R.K. and Van Loon, G.W. *Science of the Total Environment*, **76**, 203 (1988).
11 Levesque, D. and Mallet, V.N. *International Journal of Environmental Analyses*, **16**, 139 (1983).
12 Ho, J.S.Y. and Lin, P.C.L. *International Laboratory*, **12**, 44 (1982).
13 Matsui, H. *Analytica Chimica Acta*, **62**, 216 (1974).
14 Matsui, H. *Analytica Chimica Acta*, **66**, 143 (1973).
15 Suranova, Z.P., Oleinck, G.M. and Morozov, A.A. Izvestia Vyssh. ucheb Zavob Khim. Tekhnology, 12, 149. Ref: Zhur Khim. 19GD (14) Abstract No. 14G82 (1969).
16 Yoshimura, K., Nigo, S. and Tarutani, T. *Talanta*, **29**, 173 (1982).
17 Marshall, M.A. and Mottola, A. *Analytical Chemistry*, **57**, 729 (1985).
18 Yasulenis, R.Y., Luyanas, V.Y. and Kekite, V.P. *Soviet Radiochemistry*, **14**, 673 (1972).
19 Burden, B.A. *Analyst (London)*, **93**, 715 (1968).
20 Stewart, M.L., Pendleton, R.E. and Lords, J.L. *International Journal of Applied Radiation and Isotopes*, **23**, 345 (1972).
21 Sevagacnik, M. and Paljk, S. *Zeitung Analytical Chemistry*, **244**, 306 (1969).
22 Sevagacnik, M. and Paljk, S. *Zeitung Analytical Chemistry*, **244**, 375 (1969).
23 Kapustin, V.K., Egorov, A.I. and Leonov, V.V. *Soviet Journal of Water Chemistry and Technology*, **3**, 119 (1981).
24 Higachi, H., Uesugi, M., Satoh, K. Ohashie, N. and Norguchi, N. *Analytical Chemistry*, **56**, 761 (1984).
25 Subramanian, K.S. *Analytical Chemistry*, **60**, 1413 (1988).
26 Nevoral, V. and Okae, A. *Czlka. Form.*, **17**, 478 (1968).
27 Burba, P., Cebulo, M. and Broekaert, J.A.C. *Fresenius Zeitschrift für Analytische Chemie*, **318**, 1 (1984).
28 Danielson, A., Roennholm, B., Kielstrom, L.E. and Pughman, F. *Talanta*, **20**, 185 (1973).
29 Sandhu, S.S. and Nelson, P. *Environmental Science and Technology*, **43**, 476 (1979).

30 Tanaka, T., Higashi, K., Kawabara, A. *et al. Osaka Kogyo Gijitsu Shikensho Kiho,* **38**, 212 (1987).
31 Olbrych-Slesynska, E., Brajter, K., Matuszewski, W., Trojanowicz, M. and Frenzel, W. *Talanta,* **39**, 779 (1992).
32 Yoshimura, K., Hiraoka, S. and Tarutani, T. *Analytica Chimica Acta,* **142**, 101 (1982).
33 Leydon, D.E. and Underwood, A.L. *Journal of Physical Chemistry,* **68**, 2093 (1964).
34 Florence, J.M. and Batley, G.E. *Talanta,* **23**, 179 (1976).
35 Riley, J.P. and Taylor, D. *Analytica Chimica Acta,* **41**, 175 (1968).
36 Abdullah, M.J. and Royle, L.G. *Analytica Chimica Acta,* **58**, 283 (1972).
37 Figura, P. and McDuffle, B. *Analytical Chemistry,* **49**, 1950 (1977).
38 Loewenschuss, H. and Schmuckler, G. *Talanta,* **11**, 2093 (1964).
39 Dowex A–1 Chelating resin. Dow Chemical Co, Midland, Michigan, USA (1959).
40 LeMeur, J.F. and Courtot-Coupez, J. *Bulletin of the Chemical Society of France,* **929** (1973).
41 Voutsa, D., Samara, K., Fyticaros, K. and Kouimtzis, T. *Fresenius Zeitschrift für Analytische Chemie,* **330**, 596 (1988).
42 Pakalns, P., Batley, G.E. and Cameron, A.J. *Analytica Chimica Acta,* **99**, 233 (1978).
43 Corsini, A., Wade, G., Wan, C.C. and Prasad, S. *Canadian Journal of Chemistry,* **65**, 915 (1987).
44 Gladney, E.S., Peters, R.J. and Perrni, D.R. *Analytical Chemistry,* **55**, 976 (1983).
45 Hinse, A., Kabori, K. and Ishii, D. *Analytica Chimica Acta,* **97**, 303 (1978).
46 Greenberg, R.R. and Kingston, H.M. *Analytical Chemistry,* **55**, 1160 (1983).
47 Van Grieken, R.E., Bresseleers, C.M. and Vanderborght, B.M. *Analytical Chemistry,* **49**, 1326 (1977).
48 Van Espen, P., Nullers, H. and Adams, F.C. *Fresenius Zeitschrift für Analytische Chemie,* **285**, 215 (1977).
49 Figura, P. and McDuffie, B. *Analytical Chemistry,* **52**, 1433 (1980).
50 Clanet, F., Delangele, R. and Popoff, G. *Water Research,* **15**, 59 (1981).
51 Everaets, F.M., Verbeggen, M., Reijenga, J.C. *et al. Journal of Chromatography,* **320**, 263 (1985).
52 Pakalns, P. *Analytica Chimica Acta,* **120**, 289 (1980).
53 Corsini, A., Chiang, S. and DiFricia, R. *Analytical Chemistry,* **54**, 1433 (1982).
54 Wan, C., Chiang, S. and Corsini, A. *Analytical Chemistry,* **57**, 719 (1985).
55 Dehairs, F., De Bandt, M., Baeyens, W., Van den Winkel, F. and Hoenig, M. *Analytica Chimica Acta,* **196**, 33 (1987).
56 Frigieri, P., Trucco, R., Ciaccolini, I. and Pampurini, G. *Analyst (London),* **105**, 541 (1980).
57 Janzer, V.J. *Journal of Research of the US Geological Survey,* **1**, 113 (1973).
58 Yamamoto, O. *Analytical Abstracts,* **14**, 6669 (1967).
59 Nowicki, J.L., Johnson, K.S., Coale, K.H., Elrod, V.A. and Liberman, S.H. *Analytical Chemistry,* **66**, 2732 (1994).
60 Boyle, E.A., Handy, B. and Van Green, A. *Analytical Chemistry,* **59**, 1499 (1987).
61 Pai, S.C., Whung, P.Y. and Lai, R.L. *Analytica Chimica Acta,* **211**, 257 (1988).
62 Pai, S.C. *Analytica Chimica Acta,* **211**, 271 (1988).
63 Sturgeon, R.E., Bermann, S.S., Desauliniers, A. and Russell, D.S. *Talanta,* **27**, 85 (1980).
64 Rasmussen, L. *Analytica Chimica Acta,* **125**, 117 (1981).
65 Greenberg, R.R. and Kingston, H.M. *Journal of Radioanalytical Chemistry,* **71**, 147 (1982).
66 Kingston, H.M., Barnes, I.L., Brady, I.J., Rains, T.C. and Champ, M.A. *Analytical Chemistry,* **50**, 2064 (1978).

67 Bruland, K.W., Franks, R.P., Knauer, G.A. and Martin, J.G. *Analytica Chimica Acta*, **105**, 233 (1979).
68 Hirata, S. and Handa, K. *Bunseki Kagaku*, **36**, 213 (1987).
69 Bafti, F., Cardinale, A.M. and Bruzzone, R. *Analytica Chimica Acta*, **270**, 79 (1992).
70 Rasmussen, L. *Analytica Chimica Acta*, **117**, 125 (1981).
71 Paulson, A.J. *Analytical Chemistry*, **58**, 183 (1986).
72 Bender, M.L., Klinkhammer, G.P. and Spencer, D.W. *Deep Sea Research*, **24**, 799 (1977).
73 Klinkhammer, G.P. and Bender, M.L. *Earth Planet Science Letters*, **46**, 361 (1980).
74 Batley, G.E. and Farrah, Y.J. *Analytica Chimica Acta*, **99**, 283 (1978).
75 Gardner, J. and Yates, J. In *International Conference, Management and Control of Heavy Metals in the Environment*, Water Research Centre, Stevenage Laboratory, Stevenage, UK; CEP Consultants, Edinburgh, pp. 427–430 (1979).
76 Boniforti, R., Ferraroli, R., Frigieri, P., Heltai, D. and Queirezza, G. *Analytica Chimica Acta*, **162**, 33 (1984).
77 Olsen, S., Pessanda, L.C.R., Ruzicka, J. and Hansen, E.H. *Analyst (London)*, **108**, 905 (1983).
78 Jorgensen, S.S. and Petersen, K. In Paper presented at 9th Nordic Atomic Spectroscopy and Trace Element Conference, Reykjavik, Iceland (1983).
79 Malamas, F., Bengtsson, M. and Johansson, G. *Analytica Chimica Acta*, **169**, 1 (1984).
80 Krug, F.J., Reis, B.F. and Jorgensen, S.S. In *Proceedings of the Workshop on Locally Produced Laboratory Equipment for Chemical Education*. Copenhagen, Denmark, p.121 (1983).
81 Berman, S.S., McLaren, J.W. and Willie, S.N. *Analytical Chemistry*, **52**, 488 (1980).
82 Sturgeon, R.E., Berman, S.S. and Desauliniers, J.A.U. *Analytical Chemistry*, **52**, 1585 (1980).
83 Reimer, P.A. and Myazaki, A. *Journal of Analytical Atomic Spectroscopy*, **7**, 1238 (1992).
84 Mykytiuk, A.P., Russell, D.S. and Sturgeon, R.E. *Analytical Chemistry*, **52**, 1281 (1980).
85 Blake, W.E., Bryant, M.W.R. and Waters, A. *Analyst (London)*, **94**, 49 (1969).
86 Lee, C., Kim, N.P., Lee, I.C. and Chung, K.S. *Talanta*, **24**, 241 (1977).
87 Lieser, K.H., Calmano, W., Heuss, E. and Neitzert, V. *Journal of Radioanalytical Chemsitry*, **37**, 717 (1977).
88 Lee, C., Kim, N.B., Lee, I.C. and Chung, K.S. *Talanta*, **24**, 241 (1977).
89 Lo, J.M., Wei, J.C. and Yeh, S.J. *Analytical Chemistry*, **49**, 1146 (1977).
90 Murthey, R.S.S. and Ryan, D.E. *Analytical Chemistry*, **55**, 682 (1983).
91 Huhn, C.A. and Bacon, M.P. *Analytical Chemistry*, **57**, 2065 (1985).
92 Stiller, M., Mantel, M. and Rappaport, M.S. *Journal of Radioanalytical and Nuclear Chemistry*, **83**, 345 (1984).
93 Matthews, A.D. and Riley, J.P. *Analytica Chimica Acta*, **51**, 287 (1970).
94 Isshiki, K. and Nakayama, E. *Analytical Chemistry*, **59**, 291 (1987).
95 Nakahara, T. and Chakrabarti, C.L. *Analytica Chimica Acta*, **104**, 99 (1979).
96 Subramanian, K.S., Meranger, J.C., Langfond, C.H. and Allen, C. *International Journal of Environmental Analytical Chemistry*, **16**, 33 (1983).
97 Kempster, B.L. and Van Vliet, H.R. *Water South Africa*, **4**, 125 (1978).
98 Morrison, G.M.P. *Environmental Technology Letters*, **8**, 393 (1987).

Chapter 26

Cations: Adsorption on metal oxides and metals

26.1 Non saline waters

26.1.1 Manganese dioxide

Manganese dioxide supported on glass fibres has been used for the preconcentration of lead from non saline waters [1]. Up to 75mg of lead are adsorbed per g of manganese dioxide.

Lead is recovered by heating the glass fibre filters in 6M hydrochloric acid and the solution analysed by atomic absorption spectrometry using the 217nm lead line. In actual river water analysis there was no pH adjustment or other pre-treatment of the sample. The presence of relatively high concentrations of manganese during lead analyses caused no interferences or aspirating problems.

The adsorption capacity of the manganese dioxide was investigated by determining the amount of lead adsorbed from varying amounts in 2L non saline water solutions. Adsorption is quantitative up to 75mg lead, with a maximum adsorption of approximately 190mg Pb g^{-1} lead from 500mg in solution. Samples of river water collected before and after a heavy rainfall gave the following results: before rain – Pb$_{(aq)}$ 1.6µg L^{-1}, Pb solid, 5.6µg L^{-1}; after rain – Pb$_{(aq)}$ 6.3µg L^{-1}, Pb solid 19.8µg L^{-1}. The increase after rain was presumably due to lead-bearing run-off from city streets which drain into the river.

Hashimoto et al. [2] preconcentrated lead–214 and bismuth–214 on millepore filters, pre-impregnated with manganese dioxide, lead–214 and bismuth–214 were then measured in the filter material by direct alpha spectrometric analysis. Michel et al. [3] preconcentrated radium–226 and radium–228 from 1L samples of water on to manganese dioxide impregnated on to acrylic fibres. Radium was then leached from the fibres and co-precipitated with barium sulphate prior to γ counting. Elsinger et al. [4] used a similar technique employing 100L samples. Butts et al. [5] and Rama et al. [6] respectively, have preconcentrated and determined radium–226 and radium–224 in non saline waters by a similar technique.

Crespo *et al.* [7] preconcentrated uranium, thorium and americium from river and lake waters onto manganese dioxide prior to the determination of these elements.

26.1.2 Titanium dioxide, zirconium dioxide, alumina and hydrated ferric oxide

Lieser *et al.* [8] showed that heavy metals, iron, cobalt, nickel, copper, zinc, lead and uranium, are bound to titanium dioxide, zirconium dioxide and alumina in varying degrees. Remarkably high fractions of zinc, lead and uranium are taken up by titanium dioxide and by aluminium oxide. The comparison between the sorbent fraction and the fraction present in suspended matter shows that these fractions are comparable for iron, cobalt, nickel, and copper, but not for zinc, lead and uranium.

Zhang *et al.* [9] preconcentrated lead by depositing it on a microcolumn of activated alumina and eluting with nitric acid before determination by flame atomic absorption spectrometry. The limit of detection for a 25ml sample was 0.36µg L $^{-1}$. The relative standard deviation for concentrations of 40 and 4µg L $^{-1}$ were 1.4 and 12%, respectively. Schell *et al.* [10] preconcentrated americium and plutonium from 4000L water by passage down a column of 0.3µm Millepore filters and sorption beds of alumina.

Indium has been preconcentrated from rain water by adsorption on hydrated ferric oxide which is then irradiated with neutrons [11]. Beta particles from the indium–116 were counted with a gas flow proportional counter supplemented with X–ray counting with a Ge(Li) detector.

Beinrohr *et al.* [12] used alumina to preconcentrate chromium from non saline waters.

Nikitina and Stephanov [13] used laser-induced luminescence in two methods for the determination of uranium in non saline waters. One method concentrated uranium on titanium dioxide and the other was direct measurement of the luminescence in a solution of sodium polysilicate. The detection limit of the direct method was approximately 0.02µg uranium L $^{-1}$.

Titanium dioxide has been used to gather the following elements from non saline water samples: bismuth, cadmium, cobalt, chromium, copper, iron, germanium, indium, manganese, nickel, lead, antimony, tin, tellurium, thallium, vanadium and zinc [46].

26.1.3 Gold, silver and copper

Non saline waters

Gold has been used for the preconcentration of mercury [14–24]. Neske *et al.* [18] determined mercury at the µg L $^{-1}$ level ng L $^{-1}$ level. The method is based on the preconcentration using sample sizes of up to 1L, using direct contact with a gold collector followed by high sensitivity atomic absorption spectrometry. Determinations of organomercury compounds were also possible. Wittman [22] heated water samples in an electric furnace and collected the mercury released on a gold wire. The preconcentrated mercury is then vaporised and carried through an absorption cell, so that the mercury atomic absorption sequel at 253.7nm increases sharply to a maximum. Mercury in water has been preconcentrated from lake water by reduction to elemental mercury, purging out mercury with an inert gas, and collecting the mercury on a gold trap [21]. Mercury was then released from the trap as a sharp pulse by heating and determined by atmospheric pressure helium microwave induced plasma emission spectrometry.

The preconcentration of mercury on copper or silver foil has been studied [14,15,25–27]. In this procedure the mercury present in a large volume sample is swept on to the silver or copper and subsequently released by heating to 800°C. The concentrated pulse of mercury is swept with argon into cold vapour atomic absorption spectrometer. Down to 0.1µg L $^{-1}$ mercury has been determined this way.

26.1.4 Lead sulphide

Gold has been preconcentrated by adsorption on lead sulphide [28].

26.1.5 Magnesium oxide

A magnesium oxide filter has been used to preconcentrate cobalt from non saline waters [29].

26.2 Seawater

26.2.1 Manganese dioxide

Harvey and Dutton [30] determined nanogram amounts of cobalt in seawater after preconcentration on manganese dioxide formed by photochemical oxidation of divalent manganese in a photochemical reactor. The sample (1L) containing 100µg manganese was irradiated in a Hanovia reactor fitted with a 2W low-pressure Hg discharge lamp radiating mainly at 254 and 185nm. The manganese dioxide deposit that adhered to the silica jacket of the reactor was dissolved in 0.15M hydrochloric acid containing a trace of sulphur dioxide, the solution was evaporated to dryness and the residue, was dissolved in 4ml of 0.625M hydrochloric acid; 1ml 5M aqueous ammonia and 0.1ml of 0.1% dimethylgloxime in ethanol were added, and cobalt was determined by pulse polarography. The polarograph was operated in the derivative mode, starting at –1.0V; a 50mV pulse height and 1s mercury drop life were used.

Bacon and Anderson [31] preconcentrated thorium–234, thorium–230 and thorium–228 in seawater on manganese dioxide coated on plastic netting prior to standard radiochemical counting techniques.

26.2.2 Manganese dioxide fibres

Oceanographers use different techniques for measuring radium–226 in seawater. Some workers store the sample in a 20L glass bottle and extract successive generations of radon–222 [35,36]. Others quantitatively extract the radium on to manganese fibre and measure radon–222 directly emanating from the manganese fibre [37] or in a hydrochloric acid extract from the fibres [38]. The radon–222 activity is then determined by an α-scintillation counting. All of these techniques give high levels of reproducibility and accuracy as determined by the oceanographic consistency of the results [35,36].

The introduction of high–resolution, high-efficiency γ-ray detectors composed of lithium-drifted germanium crystals has revolutionised γ-measurement techniques. Thus, γ-spectrometry allows the rapid measurement of relatively low-activity samples without complex analytical preparations. A technique described by Michel et al. [3] uses Ge(Li) γ-ray detectors for the simultaneous measurements of radium–228 and radium–226 in non saline waters. This method simplifies the analytical procedures and reduces the labour while improving the precision, accuracy and detection limits. In this method the radium isotopes are preconcentrated in the field from 100 to 1000L water samples on to manganese impregnated acrylic fibre cartridges, leached from the fibre and co-precipitated with barium sulphate. Lower limits of detection are controlled by the volume of water processed through the manganese fibres. In a 1d count, samples as low as 10dpm are measured to ±10% uncertainty. This manganese fibre-γ-ray technique is shown to be more accurate than the actinium–228 methods recommended by the US Environmental Protection Agency [39] and as accurate but more rapid than the thorium–228 ingrowth procedure.

Table 26.1 shows the results for samples from the Environmental Protection Agency Radium in Water Crosscheck Program using both the thorium–228 ingrowth and the Ge(Li) γ-ray techniques. Both types of analyses were run on 3.8L samples. The thorium–228 ingrowth samples were stored for 3–11 months to accumulate sufficient activity and then counted by α-spectrometry for 2d. The Ge(Li) samples were stored for 3 weeks and counted for 400–1000min. Both techniques required a separate analysis for the absolute radium–226 activity whereas the thorium–228

Table 26.1 Comparison of the known values of the EPA radium in water cross-check samples with both the 228Th ingrowth/222Rn emanation and Ge(Li) γ-ray methods*

EPA no.	Ra/²²⁸Ra activity ratio		²²⁸Ra (decays min⁻¹)	
	Th ingrowth	Known	Th ingrowth	Known
Dec 78	1.04 ± 0.06	0.96 ± 0.20	21.1 ± 0.8	19.8 ± 2.9
Mar 79	1.59 ± 0.31	1.66 ± 0.34	28.6 ± 5.5	30.2 ± 4.4
Apr 79	1.05 ± 0.05	1.05 ± 0.22	13.6 ± 0.5	13.8 ± 2.0
	Ge(Li)	Known	Ge(Li)	Known
Sept 80	1.07 ± 0.05	1.10 ± 0.24	29.1 ± 1.5	31.9 ± 1.8
Oct 80	0.66 ± 0.03	0.72 ± 0.09	19.8 ± 1.0	20.4 ± 1.8
Dec 80	0.76 ± 0.06	0.79 ± 0.17	22.0 ± 1.1	23.3 ± 3.6

*Measurements are reported as weighted means with weighted standard deviations. The known values are reported by EPA with expected laboratory standard deviations

Source: Reproduced with permission from Michel, J. et al. [3] American Chemical Society

ingrowth method requires an additional analysis for the total radium–226 in the large volume sample to obtain the ^{228}Ra/^{226}Ra activity ratio.

Table 26.2 shows the results for seawater samples collected by the manganese dioxide fibre technique and analysed by both the Ge(Li) and thorium–228 ingrowth techniques. Although the bulk activities of each isotope vary for the different techniques, the radium–228/radium–226 activity ratios are in close agreement.

26.2.3 Barium sulphate

Oceanographers have developed methods to measure the radium–226 content of seawater as it is a useful tracer of mixing in the ocean. These procedures are based on concentrating radium from a large volume of seawater, removing all thorium–228 from the sample and ageing the sample while a new generation of thorium–228 partially equilibrates with radium–228. After storage periods of 6–12 months, the sample is spiked with thorium–230 and after ion exchange and solvent exchange separations, the thorium isotopes are measured in a γ-ray spectrometer system utilising a silicon surface barrier detector. Early work was based on concentrating the radium from the seawater sample by adding barium and coprecipitating with barium sulphate. This concentration procedure has been replaced by one involving the extraction of radium from seawater on acrylic fibre coated with manganese dioxide [33,40] (Mn fibres, see section 26.2.2). By use of this technique, volumes of 200–2000L may be sampled routinely.

Table 26.2 Comparison of ^{226}Ra, ^{228}Ra and ^{228}Ra/^{226}Ra activity ratio in seawater of the same sampling using two different methods

Sample	^{226}Ra (decays min $^{-1}$)*	^{228}Ra (decays min $^{-1}$)*	^{228}Ra/^{226}Ra	Method*
451–99	260 ± 7	210 ± 10	0.81 ± 0.04	1
	305 ± 2	223 ± 5	0.73 ± 0.02	2
452–99	235 ± 10	125 ± 13	0.53 ± 0.06	1
	229 ± 12	153 ± 3	0.67 ± 0.04	2
453–99	154 ± 1	56 ± 7	0.36 ± 0.04	1
	161 ± 9	53 ± 2	0.33 ± 0.02	2
454–99	246 ± 13	94 ± 11	0.38 ± 0.05	1
	291 ± 8	94 ± 3	0.32 ± 0.01	2

*Method 1, Ge(Li) γ-ray spectrometry. Results reported as weighted means and weighted standard deviations of both ISD counting and efficiency uncertainties. Method 2, ^{226}Ra by radon emanation; ^{228}Ra by ^{228}Th ingrowth. Results reported as ISD counting uncertainties

Source: Reproduced with permission from Michel, J. *et al.* [3] American Chemical Society

26.2.4 *Gold*

Olafsson [41–43] described a procedure for the preconcentration and determination of mercury in seawater in which the sample (450ml) is acidified with nitric acid, aqueous stannous chloride is added, and the mercury is entrained in a stream of argon into a silica tube wound externally with resistance wire and containing pieces of gold foil, on which the mercury is retained. The tube and its contents are then heated electrically to about 320°C and the vaporised mercury is swept by argon into a 10cm silica absorption cell in an atomic absorption spectrophotometer equipped with a recorder. The absorption (measured at 253.7nm) is directly proportional to the amount of mercury in the range 0–24ng per sample.

Gill and Fitzgerald [44] isolated mercury from seawater using tin(II) chloride (0.5ml of 50% solution) reduction and gas phase stripping with collection and concentration on gold columns. The gas flow system used two gold-coated bead columns (the collection and the analytical columns) to transfer mercury into the gas cell of an atomic absorption spectrometer. By careful control and estimation of the blank, a detection limit of 0.21pM was achieved using 2L of seawater. The accuracy and precision of this method were checked by comparison with aqueous laboratory and National Bureau of Standards (NBS) reference materials spiked into acidified non saline water samples at picomolar levels.

Various other workers have studied various amalgamation methods for the preconcentration of mercury [14,15,17,45].

26.3 Potable water

26.3.1 Gold

Preconcentration of mercury on gold or silver wire or foil is an interesting idea which has been pursued by other workers. Thus Temmerman *et al.* [23] evaluated and optimised reduction aeration/amalgamation as a method for the analysis of mercury in potable water by cold vapour atomic absorption spectrometry. Parameters investigated were: aeration time, aeration gas flow rate, type of gas distributor, amount of reductant and mercury concentration. The aerated mercury was preconcentrated on a gold-coated quartz sand absorber and subsequently thermally desorbed. This enabled interference by water vapour to be avoided and calibration to be simplified. The detection limit was 0.6ng L^{-1} and reproducibility was better than 10%.

Kunert *et al.* [24] preconcentrated mercury in potable water on gold or silver foil prior to its determination by cold vapour graphite furnace atomic absorption spectrometry and cold vapour atomic absorption spectrometry. Using adsorption on gold or silver wire Temmerman *et al.* [23] were able to determine down to 6.6µg L^{-1} mercury in potable water.

26.4 Ground water

26.4.1 Manganese dioxide fibres

Radium is extracted from ground waters with high efficiency on the manganese dioxide fibre [32]. Radium is removed from the manganese fibre by reducing the manganese dioxide with hot hydrochloric acid or desorbing the radium into cold dilute nitric acid. The hydrochloric acid treatment removes the radium quantitatively from the fibres but generates a considerable quantity of chlorine. The nitric acid treatment is much easier and safer but only removes about 70% of the radium [33]. Measurements of radium–226 are simpler than those for radium–228 and are more precise. These measurements are generally made by concentrating the radium from up to a few litres via barium sulphate precipitation followed by thick source counting or by radon–222 extraction following dissolution of barium sulphate [34].

References

1 Matthews, K.M. *Analytical Letters*, **16**, 633 (1983).
2 Hashimoto, I., Satal, K. and Aoyage, M. *Journal of Radioanalytical and Nuclear Chemistry Articles*, **92**, 407 (1985).
3 Michel, J., Moore, W.S. and King, P.T. *Analytical Chemistry*, **53**, 1885 (1981).
4 Elsinger, R.J., King, P.T. and Moore, W.S. *Analytica Chimica Acta*, **144**, 277 (1982).
5 Butts, J., Todd, J.F., Berch, I., Moore, W.S. and Moore, D.G. *Marine Chemistry*, **25**, 349 (1988).

6 Rama, S., Todd, J.F., Butts, J.L. and Moore, W.S. *Marine Chemistry*, **22**, 43 (1987).
7 Crespo, M.T., Gascon, J.L. and Acena, M.I. *Science of the Total Environment*, **130**, 383 (1993).
8 Lieser, K.H., Quandt, S. and Gleitsmann, B. *Fresenius Zeitschrift für Analytische Chemie*, **298**, 378 (1979).
9 Zhang, Y., Riby, P., Cox, A.G. et al. *Analyst (London)*, **113**, 125 (1988).
10 Schell, W.R., Nevissi, A. and Huntamer, D. *Marine Chemistry*, **6**, 143 (1978).
11 Bhatki, K.S. and Dingle, A.N. *Radiochemical and Radioanalytical Letters*, **3**, 71 (1970).
12 Beinrohr, E., Manova, A. and Dzurov, J. *Fresenius Journal of Analytical Chemistry*, **355**, 528 (1996).
13 Nikitina, S.A. and Stephanov, A.V. *Radiokhimlya*, **25**, 606 (1986).
14 Fishman, M.J. *Analytical Chemistry*, **42**, 1462 (1970).
15 Muscat, V.I., Vickers, T.J. and Audrery, A. *Analytical Chemistry*, **44**, 218 (1972).
16 Matsunaga, K. *Mizushori-gijutsu*, **15**, 341 (1974).
17 Olafsson, J. *Analytica Chimica Acta*, **68**, 207 (1974).
18 Neske, P., Hellwig, A., Dornheim, L. and Triene, B. *Fresenius Zeitschrift für Analytische Chemie*, **318**, 498 (1984).
19 Alder, J.F. and Hickmann, D.H. *Analytical Chemistry*, **49**, 338 (1977).
20 Siemer, D.L. and Hageman, L. *Analytical Chemistry*, **52**, 105 (1980).
21 Chau, Y.K. and Saitoh, H. *Environmental Science and Technology*, **4**, 839 (1970).
22 Wittman, Z. *Talanta*, **28**, 271 (1981).
23 Temmerman, E., Dumarey, R. and Daws, R. *Analytical Letters (London)*, **18**, 203 (1985).
24 Kunert, I., Komarek, J. and Summiri, J. *Analytica Chimica Acta*, **106**, 285 (1979).
25 Doherty, P.E. and Dorsett, R.S. *Analytical Chemistry*, **43**, 1887 (1971).
26 Kaiser, I., Schock, P. and Togl, G. *Talanta*, **23**, 889 (1975).
27 Kalb, G.W. *Atomic Absorption Newsletter*, **9**, 54 (1970).
28 McHugh, J.B. *Applied Spectroscopy*, **4**, 66 (1983).
29 Lo, J.M. and Lee, J.D. *Analytica Chimica Acta*, **318**, 391 (1996).
30 Harvey, B.R. and Dutton, J.W.R. *Analytica Chimica Acta*, **67**, 377 (1973).
31 Bacon, M.P. and Anderson, R.F. In *Trace Metals in Sea Water*, Proceedings of a NATO Advanced Research Institute on Trace Metals in Sea Water, 30/3–3/4/81, Sicily, Italy, eds C.S. Wong et al. Plenum Press, New York, p. 368 (1981).
32 Moore, W.S. and Cook, L.M. *Nature (London)*, **253**, 30 (1975).
33 Moore, W.S. *Deep Sea Research*, **23**, 647 (1976).
34 US Environmental Protection Agency. In Radiochemical Methodology for Drinking Water Regulations EPA 600/4–75–005 (1975).
35 Ku, T.L., Huh, C.A. and Chen, P.S. *Earth Planet Science Letters*, **49**, 293 (1980).
36 Chung, Y. *Earth Planet Science Letters*, **49**, 319 (1980).
37 Moore, W.S. *Estuarine, Coastal Shelf Science*, **12**, 713 (1982).
38 Reid, D.F., Key, R.M. and Schink, D.R. *Earth Planet Science Letters*, **43**, 223 (1979).
39 US Environmental Protection Agency. In *National Interium Primary Drinking Water Regulations*, EPA–57019–79–003 (1976).
40 Moore, W.S. and Reid, D.F. *Journal of Geophysical Research*, **78**, 8880 (1973).
41 Olafsson, J. In *Trace Metals in Seawater*, Proceedings of a NATO Advanced Research Institute on Trace Metals in Seawater, 30/3–3/4/81, Sicily, Italy, eds. C.S. Wong et al. Plenum Press, New York, p. 476 (1981).
42 Olfasson, J. *Analytica Chimica Acta*, **68**, 207 (1974).
43 Olafsson, J. *Marine Chemistry*, **11**, 129 (1982).
44 Gill, G.A. and Fitzgerald, W.F. *Marine Chemistry*, **20**, 227 (1987).
45 Carr, R.A., Hoover, J.B. and Wilkniss, P.E. *Deep Sea Research*, **19**, 747 (1972).
46 Vassileva, E., Proinova, I. and Hadjivanou, K. *Analyst (London)*, **121**, 607 (1996).

Chapter 27

Cations: Inorganic coprecipitation techniques

A further method of preconcentration involves adding to the sample a solution of a suitable metal such as iron, zirconium or indium and following this by a precipitating agent which precipitates not only the added metal but also coprecipitates the metals which it is required to determine in the sample. The precipitate isolated from a large volume of original water sample is then isolated and dissolved in a small volume of acid to provide a concentrate for analysis.

27.1 Non saline waters

27.1.1 Ferric hydroxide

Pande and Sarin [2] have described a technique which can be used to enhance the sensitivity of the sodium diethyldithio-carbamate method for arsenic which involves coprecipitation of arsenic (arsenic(III) and arsenic(V)) with ferric hydroxide at pH8.9 to 9.1 in ammonia medium, which preconcentrates the arsenic. The filtered ferric hydroxide is dissolved in a small volume of hydrochloric acid and the standard sodium diethyldithiocarbamate spectrophotometric method is applied.

Ferric hydroxide has also been used for the preconcentration of selenium in non saline water [3]. Selenium ($0.2-5\mu g\ L^{-1}$) is coprecipitated with hydrated iron(III) oxide, dissolved in hydrochloric acid, reprecipitated as the element, and determined by molecular emission cavity analysis. For a 250ml water sample, the detection limit is $0.2\mu g\ L^{-1}$.

Chakravatry and Van Grieken [4] used ferric hydroxide coprecipitation when determining traces of manganese, nickel, copper, zinc and lead in non saline waters and seawater. Prior to analysis of the concentrate by energy dispersive X-ray fluorescence spectrometry, the optimum preconcentration procedure involved adding 2mg iron to a 200ml water sample, adding dilute sodium hydroxide up to pH9, filtering off on a Nuclepore membrane after a 1h equilibration time and analysing. Quantitative recoveries could then be obtained for nickel, copper, zinc and lead at the $10\mu L^{-1}$ level in waters of varying salinity while manganese

Table 27.1 Influence of sample matrix on the coprecipitation (2.5µg metal spikes; 200ml sample; pH9; 1h equilibration)

Element	% coprecipitation⁻ ± standard deviation per measurement			
	Bidistilled water	*3% NaCl*	*North Sea water*	*Synthetic seawater*
Initial Fe amount: 0.8mg				
Mn	86 ± 9	93 ± 6	37 ± 6	
Ni	104 ± 7	100 ± 7	73 ± 4	
Cu	92 ± 4	95 ± 3	89 ± 9	
Zn	109 ± 7	106 ± 7	103 ± 13	
Pb	98 ± 5	91 ± 4	99 ± 10	
Initial Fe amount: 2.5mg				
Mn	98 ± 7	93 ± 3	48 ± 7	50 ± 10
Ni	112 ± 3	108 ± 8	95 ± 7	110 ± 3
Cu	101 ± 8	98 ± 4	99 ± 7	103 ± 2
Zn	104 ± 9	103 ± 4	106 ± 2	109 ± 4
Pb	98 ± 4	102 ± 4	99 ± 4	10 ± 4

Source: Reproduced with permission from Chakravartry, R. and Van Grieken, R. [4] Gordon AC Breach

was partially collected. The precision is 7–8% at the 10µg L^{-1} level and the detection limits are in the 0.5–1µg L^{-1} range.

While traces of lead and zinc collect quantitatively beyond pH 6.0 and 8.0 respectively, copper and nickel exhibit a nearly quantitative coprecipitation only at pH9.0 and above. In the case of manganese, the collection yield is only about 55–59% at pH9.0 and above. At the natural pH (8.0) of seawater zinc and lead are coprecipitated totally while copper and nickel are collected up to about 80% and manganese only about 11% with an initial iron concentration of 10mg L^{-1}. As an optimum pH for the efficient collection of all the five trace elements, a pH of 9.0 was chosen.

Parallel precipitation experiments carried out on natural and synthetic seawater, on 3% sodium chloride solutions, and on bidistilled water samples, show that the strongly adsorbed ions, zinc and lead are not affected by the changing matrix. Manganese and to a lesser extent copper and nickel appear to be collected significantly better in bidistilled water and sodium chloride solution than in seawater. Probably the metal speciation is a most important factor in this context. However, Table 27.1 proves that no inhibitory effect due to the highly ionic complex matrix exists at low iron concentrations. A remarkably high effective preconcentration factor of 15000 is achieved in this method.

Table 27.2 Application of ferric hydroxide coprecipitation to the preconcentration of metals in non saline waters

Metal	Type of water sample	Dissolving agents[a]	Analytical finish	Detection limit*	Ref.
Ga(III)	Trade effluents	None	Scintillation spectrometry	–	[9]
Cr total and Cr(III)	Non saline	Hydrochloric acid	Atomic absorption spectrometry	0.2µM	[10]
Cr(III)	Non saline	Hydrochloric acid	Atomic absorption spectrometry	Cr 40ng L⁻¹	[11]
Be, Cu, Zn, Pb, Bi, Co, Ni, Cd	Non saline	Hydrochloric acid	Emission spectrography	1–10µg L⁻¹	[12]
Se(IV)	Non saline	–	Molecular emission cavity analysis	0.2µg L⁻¹	[3,13,14]
Zr, Hf	Non saline	–	Isotope dilution mass spectrometry	–	[15]
Pb, Cu, Zn, Cd	Non saline	–	Differential pulse polarography	–	[16]
V	Non saline	–	Spectrophotometric	–	[17]
Zr	Non saline	–	Spectrophotometric	–	[18]
Cu	Non saline	Nitric acid	AAS	<2.4µg L⁻¹	[19]
Th, U	Non saline	–	Spectrophotometric	0.07µg L⁻¹	[20]
Cr	Non saline	–	X-ray fluorescence spectroscopy	0.4–0.7µg	[21]
Cr(III), Cr(VI)	Non saline	–	AAS	–	[11,22]
Cr(III)	Non saline	Hydrochloric acid	AAS	–	[23]

[a]If no dissolving agent for the ferric hydroxides is mentioned then the sample is analysed directly in the solid state
*µg L⁻¹ unless otherwise stated

Source: Own files

Table 27.3 Determination of lead in non saline water samples*

Lead added (μg)	Lead found (μg)	Relative error (%)
1.00	0.92	−8.0
1.50	1.52	+1.3
2.00	2.06	+3.0
2.50	2.45	−2.0
3.00	3.06	+2.0
3.50	3.40	−2.9
5.00	4.90	−2.0

*Volume of sample = 200ml

Source: Reproduced with permission from Shrivastava, A.C. and Tandon, S.G. [24] *International Journal of Environmental Analysis*

Ferric hydroxide coprecipitation coupled with an X-ray fluorescence spectroscopic finish has also been used for the preconcentration and analysis of copper [5]; iron, zinc and lead [6]; cadmium, lead and copper [7], and iron, zirconium and tin [8] in non saline waters. Other applications of ferric hydroxide coprecipitation are reviewed in Table 27.2.

27.1.2 Zirconium hydroxide

Shrivastava and Tandon [24] preconcentrated lead in unpolluted non saline or seawaters by coprecipitation with zirconium hydroxide. This is done by adding zirconyl oxychloride to water samples that have been brought to pH8.2–9.5 with ammonia. The precipitate is redissolved in 50% v/v nitric acid and its absorbance is measured at 283.3nm. The relative error of the method with made up seawater samples was about 3% at the 10μg L^{-1} level. With polluted water samples the method was reliable for lead determination in the range 26–600μg L^{-1}.

To check the efficiency of this method for the determination of lead in seawater, Shrivastava and Tandon [24] added spiked known amounts of lead into the non saline water. The results are summarised in Table 27.3. The excellent agreement between the amount of lead added and found establishes the reliability of the method.

Other applications of zirconium hydroxide are reviewed in Table 27.4.

27.1.3 Hafnium hydroxide

Ueda and coworkers [30,31] applied methods based on coprecipitation with hafnium hydroxide to the determination of copper [30] and

Table 27.4 Applications of zirconium hydroxide coprecipitation to the determination of metals in non saline waters

Metal	Type of water sample	Dissolving agents	Analytical finish	Detection limit*	Ref.
Cu, Ni, Co	River and potable	–	–	<1µg L^{-1}	[25]
Cd	Non saline	–	–	<1µg L^{-1}	[26]
Cu, Mn, Pb	Non saline	Hydrochloric acid	AAS	–	[27]
Misc.	Non saline	–	–	–	[28]
As, Sb	Non saline	–	X-ray fluorescence spectroscopy	As 0.3µg Sb 6.1µg (100ml sample)	[29]

*µg L^{-1} unless otherwise stated

Source: Own files

indium(III) and gallium(III) [31] in non saline waters. Detection limits achieved by atomic absorption spectrometry were less than 4µg L^{-1} copper and less than 8µg L^{-1} (gallium and indium).

27.1.4 Lead phosphate and lead sulphate

Obiols *et al.* [33] describe an analytical method for determination of chromium speciation in non saline waters. The chromium is coprecipitated with lead phosphate to remove chromium(III) and chromium(VI). The chromium(VI) is coprecipitated in a separate determination with lead sulphate. The chromium in the fractions is determined by electrothermal atomic absorption spectrometry.

27.1.5 Barium sulphate

Radium–228 has been preconcentrated from non saline waters by coprecipitation with barium and lead sulphate. Radium–226 is then estimated in the precipitate by beta counting or by alpha counting [34]. Barium sulphate impregnated alumina has been used to preconcentrate radium isotopes prior to their determination by X-ray spectrometry [35]. Acena and Crespo [36] preconcentrated radium–226 by coprecipitation with barium sulphate. De Jong and Wiles [37] preconcentrated thorium–230 by coprecipitation with barium sulphate. Mikac and Branica [38] preconcentrated inorganic lead from solutions also containing alkyl–lead species by coprecipitation with barium sulphate following the addition of barium chloride and sodium sulphate. Alkyl–lead compounds did not interfere in this procedure.

27.1.6 Miscellaneous inorganic coprecipitating agents

These are summarised in Table 27.5.

27.2 Seawater

27.2.1 Ferric hydroxide

Spectrophotometric methods

Manganese–54 has also been determined by a method [54] using coprecipitation with ferric hydroxide. The precipitate is boiled with hydrogen peroxide and the iron is removed by extraction with isobutyl methyl ketone. Zinc is separated on an anion exchange column and the manganese is separated by oxidising it to permanganate in the presence of tetraphenylarsonium chloride and extracting the resulting complex with chloroform prior to a spectrophotometric finish. Both ^{65}Zn and ^{54}Mn

Table 27.5 Miscellaneous inorganic coprecipitating agents in non saline waters

Element pre-concentrated	Coprecipitating agent	Carrier	Comments	Finish technique	Detection limit*	Ref.
[137]Cs	Ferrocyanide	Caesium	–	β-counting		[39]
[90]Sr	Carbonate	Carbonate	–	β-counting		[39]
[214]Bi	Bismuth	Bismuth	–	Scintillation counting		[40]
Au	Tellurium	–	–	AAS		[41]
Pb	Manganese dioxide	–	Acid solution, lead converted to plumbane	Hydride AAS		[42]
[214]Pb	–	Lead	–	Scintillation counting		[40]
[33]P	Molydophosphate	–	–	Radiochemical		[43]
[90]Sr	Carbonate	–	–	–		[44,45]
As, Sb, Bi, Se, Te	Lanthanum hydroxide	–	–	AAS		[46]
[60]Co	Cobaltic hydroxide	–	Converted to hexane amino cobalt(III) complex which is extracted into methyl isobutyl ketone	Liquid scintillation counting		[47]
Te(IV), Te(VI)	Magnesium hydroxide	–	–	Hydride generation AAS	0.5pmol L⁻¹	[48]
Sn	Magnesium hydroxide	–	Coprecipitated in alkaline medium	Anodic scanning voltammetry at −1.0V	0.02	[49]
20 metals	Iron sulphide and ferric hydroxide	–	–	Neutron activation analysis	–	[50]
Au	Lead sulphide	–	–	X-ray spectrometry	0.2	[51]
Sr		Carbonate	No interference from Ca or Mg	X-ray fluorescence spectroscopy	<2	[52]
As	Selenium	–	–	X-ray fluorescence spectroscopy	200	[53]

*μg L⁻¹ unless otherwise stated

Source: Own files

are counted with a 512-channel analyser with a well-type NaI(T1) crystal (7.6 × 7.6cm). Recoveries of known amounts of ^{65}Zn and ^{54}Mn were between 74% and 84% and between 69% and 74% respectively.

Shigematsu *et al.* [55] determined cerium fluorometrically at the 1µg L $^{-1}$ level in seawater. Quadravalent cerium is coprecipitated with ferric hydroxide and the precipitate is dissolved in hydrochloric acid and interfering ions are removed by extraction with isobutyl methyl ketone. The aqueous phase is evaporated almost to dryness with 70% perchloric acid, then diluted with water and passed through a column of *bis*-(2–ethylhexyl) phosphate on poly(vinyl chloride), from which Ce(IV) is eluted with 0.3M perchloric acid. The eluate is evaporated, then made 7M in perchloric acid and treated with Ti(III), and the resulting Ce(III) is determined spectrofluorometrically at 350nm (excitation at 255nm).

Elderfield and Greaves [56] have described a method for the mass spectrometric isotope dilution analysis of rare earth elements in seawater. In this method the rare earth elements are concentrated from seawater by coprecipitation with ferric hydroxide and separated from other elements and into groups for analysis by anion exchange [57–62] using mixed solvents. Results for synthetic mixtures and standards show that the method is accurate and precise to ±1%; and blanks are low (eg 10 $^{-12}$moles La and 10 $^{-14}$moles Eu). The method has been applied to the determination of nine rare earth elements in a variety of oceanographic samples.

Atomic absorption spectrometry

Cranston and Murray [10,63] took the samples in polyethylene bottles that had been precleaned at 20°C for 4d with 1% distilled hydrochloric acid. Total chromium, Cr(IV) + Cr(III) + Cr$_p$ (particulate chromium), was coprecipitated with iron(II) hydroxide and reduced chromium (Cr(III) + Cr$_p$) was coprecipitated with iron(III) hydroxide. These coprecipitation steps were completed within minutes of sample collection to minimise storage problems. The iron hydroxide precipitates were filtered through 0.4µm Nuclepore filters and stored in polyethylene vials for later analyses in the laboratory. Particulate chromium was also obtained by filtering unaltered samples through 0.4µm filters. In the laboratory the iron hydroxide coprecipitates were dissolved in 6M distilled hydrochloric acid and analysed by flameless atomic absorption. The limit of detection of this method is about 0.1–0.2nmol L $^{-1}$. Precision is about 5%.

Nakayama *et al.* [64] have described a method for the determination of chromium(III), chromium(VI) and organically bound chromium in seawater. They found that seawater in the sea of Japan contained about 9 × 10 $^{-9}$M dissolved chromium. This is shown to be divided as about 15% inorganic Cr(III), about 25% inorganic Cr(IV), and about 60% organically bound chromium. These workers studied the coprecipitation behaviours of chromium species with hydrated iron(III) and bismuth oxides.

Mullins [11] has described a procedure for determining the concentrations of dissolved chromium species in seawater. Chromium(III) and chromium(VI) were separated by coprecipitation with hydrated iron(III) oxide and total dissolved chromium were determined separately by conversion to chromium(VI), extraction with ammonium pyrrolidinediethyldithiocarbamate into methyl isobutyl ketone and determination by atomic absorption spectroscopy. The detection limit is 40ng L $^{-1}$ Cr. The dissolved chromium not amenable to separation and direct extraction is calculated by difference. In the waters investigated, total concentrations were relatively high (1–5µg L $^{-1}$) with chromium(VI) the predominant species in all areas sampled with one exception, where organically bound chromium was the major species. A standard contact time of 4h was found to be necessary for the quantitative coprecipitation of chromium on ferric oxide. The rsd values for the determination of chromium(III), chromium(VI) and total dissolved chromium were generally 10.0, 5.0 and 5.0% respectively. From these results, the rsd for the calculated concentration of the bound species was 20%.

Coprecipitation as iron hydroxide in the case of seawaters is also discussed in section 27.1.1.

Neutron activation analysis

The neutron activation method for the determination of arsenic and antimony in seawater has been described by Ryabinin et al. [65]. After coprecipitation of arsenic acid and antimony in a 100ml sample of water by addition of a solution of ferric iron (10mg iron per litre) followed by aqueous ammonia to give a pH of 8.4 the precipitate is filtered off and, together with the filter paper, is wrapped in polyethylene and aluminium foil. It is then irradiated in a silica ampoule in a neutron flux of 1.8×10^{13} neutrons cm $^{-1}$ for 1–2h. Two days after irradiation, the γ-ray activity at 0.56MeV is measured with use of a NaI(Tl) spectrometer coupled with a multichannel pulse–height analyser, and compared with that of standards.

Gas chromatography

Siu and Berman [66] determined selenium in seawater by gas chromatography after coprecipitation with hydrous iron(III) oxide. Following a brief stirring and settling period the coprecipitate was filtered and dissolved in hydrochloric acid, derivatised to 5–nitropiazselenol and extracted into toluene. The selenium was determined by gas chromatography–electron capture detection. The detection limit was 1pg injected or 5ng selenium per litre of seawater using a 200ml sample. Precision was 6% at 25pg selenium per litre.

Radiochemical methods

Silant'ev *et al.* [67] have described a procedure for the determination of strontium–90 in small volumes of seawater. This method is based on the determination of the daughter isotope yttrium–90. The sample is acidified with hydrochloric acid, heated, and, after addition of iron, interfering isotopes are separated by double coprecipitation with ferric hydroxide. The filtrate is acidified with hydrochloric acid, yttrium carrier added, the solution set aside for 14d for ingrowth of yttrium–90, and $Y(OH)_3$ precipitated from the hot solution with carbon dioxide free aqueous ammonia. Then $Y(OH)_3$ is reprecipitated from a small volume in the presence of hold-back carrier for strontium, the precipitate dissolved in the minimum amount of nitric acid, the solution heated, and yttrium oxalate precipitated by adding precipitated oxalic acid solution. The precipitate is collected and ignited at 800–850°C to Y_2O_3. The cooled residue is weighed to determine the chemical yield, then sealed in a polyethylene bag and the radioactivity of the saturated yttrium measured on a low-background β-spectrometer.

Wong [68] has described a method for the radiochemical determination of plutonium in seawater. This procedure permits routine determinations of ^{239}Pu activities (dpm) down to 0.004dpm per 100L of seawater. The plutonium is separated from seawater by coprecipitation with ferric hydroxide. After further treatment and purification by ion exchange, the plutonium is electrodeposited on to stainless steel discs for counting and resolution of the activity by α-spectrometry. For 30 samples the average deviation was generally well within the 1so counting error. For seawater the average recovery was 52 ± 18%. The most serious interference is from ^{228}Th, which is present in most samples and is also a decay produce of the ^{236}Pu tracer.

Ballestra *et al.* [69] described a radiochemical measurement for determination of technetium–99 in rain, rivers and seawater which involved reduction to technetium(IV), followed by iron hydroxide precipitation and oxidation to the heptavalent state. Technetium(VII) was extracted with xylene and electrodeposited in sodium hydroxide solution. The radiochemical yield was determined by gamma counting on an anti-coincidence shield GM–gas flow counter. The radiochemical yield of 50 to 150L water samples was 20–60%.

Cabezon *et al.* [70] proposed a simultaneous separation of copper, cadmium and cobalt from seawater by co-flotation with octadecylamine and ferric hydroxide as collectors. They studied and optimised the experimental parameters. The drawbacks arising from the low solubility of octadecylamine and the corresponding sublates in water were avoided by use of a 6M hydrochloric acid–methyl isobutyl ketone–ethanol (1:2:2 v/v) mixture. The results obtained by means of this method were

Table 27.6 Application of ferric hydroxide coprecipitation to the preconcentration of metals in seawaters

Metal	Type of water sample	Dissolving agents	Analytical finish	Detection limit $\mu g\ L^{-1}$	Ref.
Zn–65, Mn–54	–	Hydrogen peroxide	γ-spectrometry	–	[71]
Zr	–	–	Spectrophotometric	–	[18]
Al, Pb, V	North Atlantic seawater	–	–	–	[72]
Mo, Vi		Hexadecyl triamethyl ammonium bromide with octylamine used as surfactant	Coflotation then electrochemical	0.7–5.7	[73]
Pu		Acid	Ion exchange chromatography	–	[74]

Source: Own files

compared with those given by the standard ammonium pyrrolidine-dithiocarbamate methyl isobutyl ketone extraction method.

Other applications of ferric hydroxide in preconcentration are reviewed in Table 27.6.

27.2.2 Indium hydroxide

Inductively coupled plasma atomic emission spectrometry

Hiraide *et al.* [75] developed a multi-element preconcentration technique for chromium(III), manganese(II), cobalt, nickel, copper(II), cadmium and lead in artificial seawater using coprecipitation and flotation with indium hydroxide followed by inductively coupled plasma atomic emission spectrometry. The metals are simultaneously coprecipitated with indium hydroxide adjusted to pH9.5 with sodium hydroxide, ethanolic solutions of sodium oleate and dodecyl sulphate added, and then floated to the solution surface by a steam of nitrogen bubbles. Cadmium may be completely recovered without the coprecipitation of magnesium. Concentrations of heavy metals (chromium(III), manganese(II), cobalt, nickel, copper(II), cadmium and lead) in 1200ml of artificial seawater were increased 240-fold, while those of sodium and potassium were reduced to 2–5% and those of magnesium, calcium and strontium to 50%. Down to 1μg L $^{-1}$ of the above-mentioned heavy metals can be determined by this procedure. However, it is emphasised that real seawater samples were not included in this study.

27.2.3 Zirconium hydroxide

Inductively coupled plasma atomic emission spectrometry

Akagi *et al.* [76] used zirconium coprecipitation for simultaneous multi-element determinations of trace metals in seawater by inductively coupled plasma atomic emission spectrometry. The coprecipitation procedure, ageing and washing of coprecipitates, and optimal pH conditions are described, together with spectral interferences. Recoveries of most metals increased with increase in pH except for hexavalent chromium and hexavalent molybdenum. Improved detection limits for 17 metals are reported including aluminium, arsenic, cadmium, cobalt, chromium, copper, iron, lanthanum, manganese, molybdenum, nickel, lead, antimony, titanium, vanadium, yttrium and zinc.

Yoshimura and Uzawa [77] coprecipitated cadmium in seawater with zirconium hydroxide prior to determination by square wave polarography. The precipitate was dissolved in hydrochloric acid and cadmium concentration was determined from the peak height of the polarogram at

–0.64V. The calibration curve was linear for concentrations ≤5.0μ of cadmium.

See also section 27.1.2.

27.2.4 Calcium carbonate

Nozaki and Tsunogai [78] determined lead–210 and polonium–210 in seawater. The lead–210 and polonium–210 in a 30–50L sample were coprecipitated with calcium carbonate together with lead and bismuth and were then separated from calcium by precipitation as hydroxides. The precipitate was dissolved in 0.5M hydrochloric acid, and polonium–210 was deposited spontaneously from this solution on to a silver disc and determined by α-spectrometry. Chemical yields of lead and bismuth were determined in a portion of the solution from which the polonium was deposited; hydroxides of lead and other metals were precipitated from the remainder of this solution and after a period exceeding 3 months, the polonium–210 produced by decay of lead–210 was determined as before. The activity of lead–210 was calculated from the activity of polonium–210. The method was used to determine the vertical distribution of lead–210 and polonium–210 activities in surface layers of the Pacific Ocean.

Tsunogai and Nozaki [79] analysed Pacific Ocean surface water by consecutive coprecipitations of polonium with calcium carbonate and bismuth oxychloride after addition of lead and bismuth carriers to acidified seawater samples. After concentration, polonium was spontaneously deposited onto silver planchets. Quantitative recoveries of polonium were assumed at the extraction steps and plating step. Shannon et al. [80] who analysed surface water from the Atlantic Ocean near the tip of South Africa, extracted polonium from acidified samples as the ammonium pyrrolidinedithiocarbamate complex into methyl isobutyl ketone. They also autoplated polonium onto silver counting disks. An average efficiency of 92% was assigned to their procedure after calibration with ^{210}Po–^{210}Pb tracer experiments.

27.2.5 Caesium chloride

Mason [81] has described a rapid method for the separation of caesium–137 from a large volume of seawater. In this procedure the sample (50L) was adjusted to pH1 with nitric acid and ammonium nitrate (100g), and caesium chloride (30mg) was added as carrier. A slurry is prepared of ammonium molybdophosphate (7.5g) and Gooch crucible asbestos (715g) with 0.01M ammonium nitrate and deposited by centrifugation on a filter paper fitted in the basket of a continuous–flow centrifuge. The sample was centrifuged at 600–3000rpm and the deposit washed on the filter with 1M ammonium nitrate (60–70ml) and 0.01M nitric acid (30–40ml). The caesium collected on the filter was then prepared for counting by the method of Morgan and Arkell [82]. With this method the caesium can be extracted in less than 1h.

27.2.6 Gallium hydroxide

Akaji and Haraguchi [83] performed simultaneous multielement determination of trace metals in 10ml samples of seawater by gallium coprecipitation followed by inductively coupled plasma atomic emission spectrometry. The techniques employed were coprecipitation with gallium hydroxide and microsample introduction. A preconcentration factor of 200 times was achieved by dissolving gallium precipitate with 50µL of nitric acid. The solution of one drop size (50µL) was introduced into the plasma via a cross-flow nebuliser. Integration time and sample volume were examined in detail for the optimisation of the measurement conditions of the microsampling technique. With internal standard-isation using gallium, the concentrations of aluminium, titanium, chromium, manganese, iron, cobalt, nickel, copper, zinc, yttrium and lead were determined. The detection limits of these elements ranged from 500ng L^{-1} (lead) to 10ng L^{-1} (copper and zinc). The precisions of most elements for unacidified coastal seawater are about 10%, and accuracy for manganese, which is the only element measurable for an acidified seawater standard, is fairly good. The analytical procedure is so simple that about 50 samples could be preconcentrated and analysed in 1h. This method was applied to the analysis of some natural seawaters, and it was found that this method was effective, especially for the analysis of seawaters where the available sample volume was very small.

27.2.7 Lead phosphate and sulphate

Zhang et al. [32] preconcentrated various chromium species (total Cr, Cr(IV), Cr(III)) in potable and estuarine and seawater samples by coprecipitation with lead sulphate or lead phosphate prior to determination by neutron activation analysis and gamma spectrometry. Lead phosphate will collect both trivalent and hexavalent chromium while lead sulphate collects hexavalent chromium only. The procedure had a detection limit of 0.1µg L^{-1} for chromium in seawater when 800ml samples were used.

Recoveries of both chromium(III) and chromium(VI) were excellent for both sample types with lead phosphate. Chromium(VI) was quantitatively recovered by the lead sulphate procedure from potable water but its recovery from seawater was incomplete (87%) because of the considerable amount of competing species, especially sulphate, present. Under the

Table 27.7 Applications of coprecipitating agents to the preconcentration of cations in seawater

Cation	Coprecipitating agent	Analytical finish	Detection limit*	Ref.
Cd, Zn, Mn, Co, Cu, Ni, V, Cr; Si, B, Be, Ca, Mg, Mg, Li, lanthanides	Magnesium hydroxide	Inductively coupled plasma atomic emission spectrometry	<4 (200ml sample)	[84]
Te(VI)	Magnesium hydroxide	AAS	0.5pmol L^{-1}	[51]
Hg, Pb, Cd	Copper sulphide	AAS	–	[85]
Rb	$Ni_3K_2[Fe(CN)_6]$	X-ray fluorescence spectroscopy	2–4µg absolute	[86]
Cd, Cu, Pb	Palladium salts	AAS	–	[87]
Misc. trace metals	Maleic acid/ammonium hydroxide buffer system	–	–	[88]
U	$Th(OH)_4$	Spectrofluorometry	200	[89]
Ag(I), Cd(II), Cr(III) Cu(II), Mn(II), Th(IV) U(VI), Zr(IV)	Lead phosphate	Neutron activation analysis	Ag 0.2 / Cd 3.2–3.8 / Cr 7 / Cu 2.7 / Mn 0.07–1.7 / Th 0.3–0.9 / U 0.007–0.018 / Zr 40	[90]
^{237}Np	Fluoride	α-spectrometry	– (100ml sample)	[91]

*µg L^{-1} unless otherwise stated

Source: Own files

conditions used, the detection limit based on 2(background)$^\text{¼}$ was 0.08µg chromium, ie 0.1µg L $^{-1}$ chromium for 800ml samples. Coprecipitation as the phosphate has also been used to preconcentrate lead in potable water at levels down to 50µg L $^{-1}$ prior to polarographic analysis.

27.2.8 Other inorganic coprecipitating agents

It is seen in Table 27.7 that other coprecipitating agents (magnesium hydro-oxide, palladium salts, copper salts $Ni_3K_2[Fe(CN)_6]$, malic acid, thorium hydroxide and lead phosphate) have been applied to a limited number of preconcentrations of cations.

27.3 Estuary waters

Zhang and Smith [92] preconcentrated various chromium species (total Cr, Cr(VI), Cr(III)) in estuarine and seawater samples by coprecipitation with lead sulphate or lead phosphate prior to determination by neutron activation analysis and gamma spectrometry. Lead phosphate will collect both trivalent and hexavalent chromium while lead sulphate collects hexavalent chromium only. The procedure had a detection limit of 0.1µg L $^{-1}$ for chromium in seawater when 800ml samples are used.

Recoveries of both chromium(III) and chromium(VI) were excellent for both sample types with lead phosphate. Chromium(VI) was quantitatively recovered by the lead sulphate procedure from potable water but its recovery from seawater was incomplete (~87%) because of the considerable amount of competing species, especially sulphate, present. Under the conditions used, the detection limit based on 2(background)$^\text{¼}$ was 0.08µg chromium, ie 0.1µg L $^{-1}$ chromium for 800ml samples.

27.4 Potable waters

Vijan and Sadana [42] observed that copper and nickel present in potable waters interferes with the determination of lead by the hydride generation atomic absorption spectrometric technique. They eliminated this interference by coprecipitating lead with manganese and nitric acid and the evolved plumbane was analysed by an automated hydride–atomic absorption method.

Table 27.8 reports results on potable water samples obtained by this technique, also by direct injection atomic absorption spectroscopy, graphite furnace atomic absorption spectroscopy, and differential pulse anodic scanning voltammetry. The graphite furnace method employed did not allow for matrix interference and this is seen in the poor agreement of lead contents obtained between this and the other methods. The results obtained by conventional flame atomic absorption

Table 27.8 Determination of lead in drinking water by four methods ($\mu g\ L^{-1}$)

Sample	Atomic absorption spectrometry	Hydride atomic absorption spectrometry	Graphite furnace atomic absorption spectrometry		Differential pulse anodic scanning voltammetry	
			As received	Coprecipitated	As received	Digested with nitric acid
1	64	49	38	54	45	55
2	41	35	44	30	27	33
3	69	80	48	54	50	54
4	67	78	60	88	48	60
5	76	75	48	110	57	75

Source: Reproduced with permission from Vijan, P.N. and Sadana, R.S. [42] Elsevier Science, UK

Table 27.9 Preconcentrations on potable waters

Cations	Preconcentrating agent	Analytical finish	Detection limit*	Ref.
Mo(VI)	Co-flotation with Fe(OH)$_3$ and octadecylamine	Differential pulse polarography	–	[93]
As	Fe(OH)$_3$	Spectrophotometric	1	[94]
Cd, Cu, Pb, Mn	Zr(OH)$_4$	–	–	[95]
Pb	Phosphate	Polarography	50	[96]

*µg L^{-1} unless otherwise stated

Source: Own files

spectrometry are included in Table 27.8. The method requires at least a 10-fold preconcentration in order for reliable signals to be obtained for the low lead levels. The accuracy of measurement at these concentrations is limited and necessitates the use of high damping and scale expansion in addition to background correction.

The results obtained by differential pulse anodic scanning voltammetry when nitric acid digestion is employed, are in good agreement with those obtained by graphite furnace atomic absorption spectrometry with manganese dioxide coprecipitation and hydride graphite furnace atomic absorption spectrometry with manganese dioxide coprecipitation.

Iron and zirconium hydroxides have also been used in pre-concentrations in potable waters (Table 27.9).

Coprecipitation of arsenic with ferric hydroxide (added as ferric chloride) has been used for the preconcentration of mg L^{-1} amounts of arsenic in potable water [1,2].

References

1 Nakashima, S. *Analyst (London)*, **103**, 1031 (1978).
2 Pande, S.P. and Sarin, R. *Indian Journal of Environmental Health*, **22**, 189 (1980).
3 Kouimtzis, T.A., Safoniou, M.C. and Papadoyannis, T.T. *Analytica Chimica Acta*, **123**, 315 (1981).
4 Chakravartry, R. and Van Grieken, R. *International Journal of Environmental Analytical Chemistry*, **11**, 67 (1982).
5 Naito, W., Yoneda, A. and Azumi, T. *Water Purification and Liquid Wastes Treatment*, **20**, 329 (1979).
6 Bruninx, E. and Meyl, E.V. *Analytica Chimica Acta*, **80**, 85 (1975).
7 Laxen, D.P.H. *Chemical Geology*, **47**, 321 (1984–5).

8   Naito, W., Takahata, N., Yaneda, A. and Azumi, T. *Water Purification and Liquid Wastes Treatment*, **20**, 529 (1979).
9   Rafaeloff, R. *Radiochemistry and Radioanalytical Letters*, **9**, 373 (1972).
10 Cranston, R.E. and Murray, J.W. *Analytica Chimica Acta*, **99**, 275 (1978).
11 Mullins, R.L. *Analytica Chimica Acta*, **165**, 97 (1984).
12 Lebedinskaya, M.P. and Chuiko, V.T. *Zhur. Analit. Khim.*, **28**, 863 (1975).
13 Belcher, R., Kouimtzis, T.A. and Townshend, A. *Analytica Chimica Acta*, **68**, 297 (1974).
14 Belcher, R., Bogdanski, S.L., Heyden, E. and Townshend, A. *Analytica Chimica Acta*, **113**, 13 (1980).
15 Boswell, S.M. and Elderfield, H. *Marine Chemistry*, **25**, 197 (1988).
16 Frimmel, F.H. and Gewwicz, H. *Journal of Science of the Total Environment*, **60**, 57 (1987).
17 Shcherbinina, S.D. and Petrova, S. Yu. Energetik (8), 21. Ref: *Zhur. Analit. Khim.* 19GD(23) Abstract No. 23G135 (1972).
18 Sastry, V.N., Krishnamoorthy, T.M. and Sarma, T.P. *Current Science (Bombay)*, **38**, 279 (1969).
19 Nishoika, H., Maeda, Y. and Azumi, T. *Nippon. Kalsui Gakkalshi*, **40**, 100 (1986).
20 Quiang, Y., Tian, Z., Jiang, T. *et al. He Huaxue Yu Fangshe Huaxue*, **8**, 230 (1986).
21 Inoue, N., Yoneda, A., Maeda, Y. and Azumi, T. *Kenkhu Hokoku–Hime Kogyo Daigaku*, **40A**, 100 (1987).
22 Suranova, Z.P., Oleinck, G.M. and Morozov, A.A. Izv. Yyssh. Lcheb. Zavod Khim. Tekhnol., 12 149 (1969). Ref. *Zhur. Khim.* 19GD(14) Abstract No. 14892 (1969).
23 Sugimoto, F., Maeda, Y. and Azumi, T. *Nippon Kalsui Gakkaishi*, **42**, 22 (1988).
24 Shrivastava, A.C. and Tandon, S.G. *International Journal of Environmental Analysis*, **12**, 169 (1982).
25 Nakashima, S. and Yagi, M. *Analytica Chimica Acta*, **147**, 213 (1963).
26 Nakashima, S. and Yagi, M. *Analytical Letters*, **17**, 1693 (1984).
27 Abe, K., Ito, M., Kikuchi, H. *et al. Eisel Kagku*, **33**, 258 (1987).
28 Nakamura, T., Oka, H., Ishii, M. and Sata, J. *Analyst (London)*, **119**, 1397 (1994).
29 Tanaka, S., Nakamura, M. and Hashimoto, V. *Bunseki Kagaku*, **36**, 114 (1987).
30 Ueda, J. and Yamazaki, H. *Analyst (London)*, **112**, 283 (1987).
31 Ueda, J. and Mizui, C. *Analytical Science*, **4**, 417 (1988).
32 Zhang, H.F., Holzbecher, J. and Ryan, D.E. *Analytica Chimica Acta*, **149**, 385 (1983).
33 Obiols, J., Devesa, R., Garcia-Berro, J. and Serra, J. *International Journal of Environmental Analytical Chemistry*, **30**, 197 (1987).
34 Moran, M.C., Garcia-Tenorio, R., Garcia-Mantano, E., Garcia-Leon, M. and Madunga, G. *International Journal of Applied Radiation and Isotopes*, **37**, 383 (1986).
35 Perkins, R.W. In *Report of Atomic Energy Commission*. US BNWL 1051. Part 2. Radium and Radiobarium measurement in seawater and freshwater by sorption and direct multidimensional X-ray spectrometry (1979).
36 Acena, M.L. and Crespo, M.T. *Journal of Radioanalytical and Nuclear Chemistry*, **126**, 77 (1988).
37 De Jong, I.G. and Wiles, D.R. *Water, Air and Soil Pollution*, **23**, 197 (1984).
38 Mikac, N. and Branica, M. *Analytica Chimica Acta*, **212**, 349 (1988).
39 Alekson'Yan, O.M. *Gidrokim Mater.*, **53**, 163 (1972).
40 Michaelis, M.L. *Chemikerzeiture Chemical Apparatus*, **93**, 883 (1969).
41 McHugh, J.R. *Applied Spectroscopy*, **4**, 66 (1983).
42 Vijan, P.N., Sadana, R.S. *Talanta*, **27**, 321 (1980).
43 Palagyi, S. *International Journal of Applied Radiation and Isotopes*, **34**, 755 (1983).

44 Gusev, N.G., Ya, U. and Margulis, A.N. In *Dosimetric and Radiometric Methods* Atomizdat, Moscow, p. 444 (1966).
45 Barelta, E.J. and Knowles, F.R. *Journal of the Association of Official Analytical Chemists*, **69**, 540 (1986).
46 Thompson, M., Pahlavonpur, B., Walton, S.J. and Kirkbright, G.F. *Analyst (London)*, **103**, 568 (1978).
47 Claassen, H.C. *Analytica Chimica Acta*, **52**, 229 (1970).
48 Andreae, M.O. *Analytical Chemistry*, **56**, 2064 (1984).
49 Portratuyi, U.P., Malyuta, V.F. and Chuiko, U.T. *Zhur. Analig. Khim.*, **28**, 1337 (1973).
50 Bart, G. and Von Gunten, H.R. *International Journal of Environmental Analytical Chemistry*, **6**, 25 (1979).
51 Petit, L. *Revues, Internationale d'Oceanographic Medicale*, **19**, 79 (1985).
52 Nishioka, H., Yoneda, A., Maeda, Y. and Azumi, T. *Nippon Kaisui Gakkaishi*, **39**, 393 (1986).
53 Hemens, C.M. and Elson, C.M. *Analytica Chimica Acta*, **188**, 311 (1986).
54 Stah, S.M. and Rao, S.R. *Current Science (Bombay)*, **41**, 659 (1972).
55 Shigematsu, T., Nishikawa, Y., Hiraki, K., Goda, S. and Tsujimatsu, Y. *Japan Analyst*, **20**, 575 (1971).
56 Elderfield, H. and Greaves, H.J. *Trace Metals in Seawater*. In Proceedings of a NATO Advanced Research Institute on Trace Metals in Seawater. 30/3–3/4/81, Sicily, Italy. Eds L.S. Wong *et al.*, Plenum Press, New York, p. 427 (1981).
57 Hooker, P.J. BA Thesis, University of Oxford (1974).
58 Hooker, P.J., O'Nions, P.K. and Pankhurst, R.J. *Chemical Geology*, **16**, 189 (1975).
59 Korkisch, J. and Arrhenius, G. *Analytical Chemistry*, **36**, 850 (1964).
60 Faris, J.P. and Warton, J.W. *Analytical Chemistry*, **34**, 1077 (1962).
61 Desai, H.P., Krishnamoorthy, I.R. and Sandar-Das, M. *Talanta*, **11**, 1249 (1964).
62 Brunfelt, A.O. and Steinnes, E. *Analyst (London)*, **94**, 979 (1969).
63 Cranston, R.E. and Murray, J.W. *Limnology and Oceanography*, **25**, 1104 (1980).
64 Nakayama, E., Kuwamoto, T., Takoro, H. and Fujinaka, F. *Analytica Chimica Acta*, **131**, 247 (1981).
65 Ryabinin, A.I., Romonov, A.S., Khatawov, S.O., Kist, A.A. and Khamidova, R. *Zhur. Analit. Chim.*, **27**, 94 (1972).
66 Siu, K.W.H. and Berman, S.S. *Analytical Chemistry*, **56**, 1086 (1984).
67 Silant'ev, A.N., Chumichev, U.B. and Vakulouski, S.M. (1970). Trudy Inst. eksp. Met. glav. uprav. gidromet, Sluzhty Sov. Minist. SSSR, 15, (2) Ref. Zhur. Khim. 19GD (1) Abstract No. 1 G209 (1971).
68 Wong, K.M. *Analytica Chimica Acta*, **56**, 355 (1971).
69 Ballestra, S., Barci, G., Holm, E., Lopez, J. and Gastand, J. *Journal of Radioanalytical and Nuclear Chemistry*, **115**, 51 (1987).
70 Cabezon, L.M., Caballero, M., Cele, R. and Perez-Bustamante, J.E. *Talanta*, **31**, 597 (1984).
71 Shah, S.M. and Rao, S.R. *Current Science (Bombay)*, **41**, 659 (1972).
72 Wesel, C.P., Duce, R.A. and Fasching, J.L. *Analytical Chemistry*, **56**, 1050 (1984).
73 Hidalgo, J.L., Gomez, M.A., Caballero, M., Cela, R. and Perez-Bustamante, J.A. *Talanta*, **35**, 301 (1988).
74 Chen, Q., Aarkrog, A., Nielson, S.P. *et al. Journal of Radioanalytical and Nuclear Chemistry*, **172**, 281 (1993).
75 Hiraide, M., Ito, T. and Baba, M. *Analytical Chemistry*, **52**, 804 (1980).
76 Akagi, T., Noriri, Y., Matsui, M. and Haraguchi, H. *Applied Spectroscopy*, **39**, 662 (1985).

77 Yoshimura, W. and Uzawa, Z. *Bunseki Kagaku*, **36**, 367 (1987).
78 Nozaki, Y. and Tsunogai, M.I. *Analytica Chimica Acta*, **64**, 209 (1973).
79 Tsunogai, S. and Nozaki, Y. *Geochemical Journal*, **5**, 165 (1971).
80 Shannon, L.V., Cherry, R.D. and Orren, M.F. *Geochimica and Cosmochimica Acta*, **34**, 701 (1970).
81 Mason, W.J. *Radiochemical and Radioanalytical Letters*, **16**, 237 (1974).
82 Morgan, A. and Arkell, O. *Healthy Physics*, **9**, 857 (1963).
83 Akaji, T. and Haraguchi, H. *Analytical Chemistry*, **62**, 81 (1990).
84 Buchanan, A.S. and Hannaker, P. *Analytical Chemistry*, **56**, 1379 (1984).
85 Patini, A.M. and Morozov, N.P. *Zhur. Analitches Koi Khimmi*, **31**, 282 (1976).
86 Lebedev, V.A., Alvares, U.V., Krasnyanskii, A.V., Shurupuva, T.I. and Golubtsov, I.V. Vesin. Mosk. University, Ser 2, Khim., **28**, 174 (1987).
87 Zuang, Z., Yang, C., Wang, X., Yang, P. and Huang, B. *Fresenius Journal of Analytical Chemistry*, **355**, 277 (1996).
88 Pai Su-Cheng, Chen Tsai-Chu, Wong, G.T.F. *Analytical Chemistry*, **62**, 774 (1990).
89 Leung, G., Kim, Y.S. and Zeitlin, H. *Analytica Chimica Acta*, **60**, 229 (1972).
90 Halzbecher, J. and Ryan, D.E. *Journal of Radioanalytical Chemistry*, **74**, 25 (1982).
91 Holm, E., Aarkrog, A. and Ballestra, S. *Journal of Radioanalytical Chemistry*, **115**, 5 (1987).
92 Zhang, O. and Smith, R.S. *Analytical Chemistry*, **55**, 100 (1983).
93 Hidalgo, J.L., Gomez, M.A., Cabbalero, M. Cela, R. and Perez-Bustramante, S.A. *Talanta*, **35**, 301 (1988).
94 Senften, H. *Mitt. Geb. Lebensmittel*, **64**, 152, (1973).
95 Chiba, I. and Takakura, Y. Fukushimaken Eisel Kogal Kenkyusho Nenpo, 1986, 70–75 (1987).
96 Spasojevic, B.D. and Tovanovic, D.A. *Acta Pharm. Jugoslavia*, **21**, 103 (1971).

Chapter 28

Cations: Organic coprecipitation techniques

In some methods of analysis, eg X-ray fluorescence spectrometry, preconcentrated samples are required in the solid not the aqueous phase. In such methods the metal ion or ions to be determined are reacted with a suitable organic chelating agent under appropriate conditions to produce an insoluble complex. The small amount of precipitate thereby obtained from a large volume of water sample might then be left to 'age' prior to a careful filtration under controlled conditions and dried. The solid thereby obtained is suitable for metals analysis by X-ray fluorescence spectrometry of the solid, or after dissolution in a suitable reagent by liquid methods such as atomic absorption spectrometry.

By far the greatest amount of work has been carried out using dithiocarbamic acid derivatives as coprecipitates.

28.1 Non saline waters

28.1.1 Dithiocarbamic acid derivatives

Atomic absorption spectrometry

Tris-pyrrolidinedithiocarbamate–cobalt(III) has been used to coprecipitate lead(II) prior to its determination by atomic absorption spectrometry [1].

X-ray fluorescence spectrometry

Vanadium(VI) and vanadium(V) have been determined by energy dispersive X-ray fluorescence spectrometry following preconcentration by precipitation as the diethyldithiocarbamate [2]. Pentavalent vanadium was precipitated at pH1.8 and tetravalent vanadium was precipitated with the same reagent at pH4. The precipitates were collected by vacuum filtration on a membrane filter.

Experimental conditions for the preconcentration by precipitation of various metal complexes from water are discussed below [3].

1   Sodium diethyldithiocarbamate (NaDDC). Samples (100ml) were adjusted to pH4 ± 0.05 and buffered with 2ml of 0.1M potassium hydrogen phthalate solution (pH4) then 5ml of aqueous 0.1% (w/v) solution of sodium diethyldithiocarbamate was added with stirring. After the precipitate had aged for 15min, the solution was filtered under vacuum through a 25mm 0.45μm pore membrane filter and the filter was air dried and mounted between Mylar films.

2   1-Pyrrolidinecarbodithioic acid, ammonium salt (APDC). This procedure was as for sodium diethyldithiocarbamate with the exceptions that 1ml of aqueous 1% (w/v) APDC solution was added and the precipitates were aged for 20min before filtration.

3   Sodium dibenzyldithiocarbamate (DBDTC). This procedure was as for sodium diethyldithiocarbamate with the exceptions that 1ml of methanolic 1% (w/v) solution was used and precipitates were aged for a total of 30min.

4   Complexation with 8-quinolinol (oxine) and adsorption of complexes on activated carbon. Samples (100ml) were adjusted to pH8 ± 0.1 and buffered with ammonium chloride/ammonia solution (10ml, pH8). An amount of oxine solution (8mg ml^{-1} in acetone) was added followed by 100mg of activated carbon. The resulting suspension was then rotated in a glass vessel on a pot mill for 1h, after which time the activated carbon was collected by filtration of the solution through a membrane filter. The filter with the collected activated carbon was mounted between Mylar films while still moist. The lower Mylar support was perforated and the sample dried under vacuum.

5   Precipitation with thionalide and poly(vinylpyrrolidone) (PVP). To a 100ml sample were added 5ml of an aqueous 0.1% (w/v) solution of poly(vinylpyrrolidone) and 5ml of a 0.1% (w/v) solution of thionalide (mercaptoacetic acid naphthylamide) in glacial acetic acid. The pH was adjusted to 4 ± 0.05 and the solution was filtered through a membrance filter (47mm diameter, 0.45cm pore size) after 15min stirring and 5min ageing. Then 25mm diameter sections of these filters were punched out, air dried, and mounted between Mylar films for X-ray analysis.

Following these preconcentration procedures, Ellis and Leyden [3] determined elements by X-ray fluorescence spectrometry.

On the basis of an extensive study of these various preconcentration methods based on precipitation, Ellis and Leyden [3] found that methods based on precipitation by a dithiocarbamate were optimal. These workers have also reported on the effect of interfering species in their methods [4]. The agents most likely to cause interference in actual sample analysis were found to be chromium(III), manganese(II), iron(III), cobalt(II), copper(II), zinc(II), lead(II), cadmium(II), sodium chloride and calcium. It was found that for dithiocarbamate preconcentration methods, over half of the interference problems that were encountered were spectral in nature and compensation can be made for these. On the other hand the other preconcentration methods suffered from mainly chemical interferences, for which corrections cannot be made.

Leyden et al. [5] used the following preconcentration procedures. The diethyldithiocarbamates were formed by adding 5ml of the 0.1% solution of the diethyldithiocarbamate sodium salt to a sample that has been adjusted to pH4 with 2ml of phosphate buffer. The solution was filtered through a Gelman filter (Metricet, Ga–6, 0.45µm) after standing for about 15min. The ammonium pyrrolidinedithiocarbamates were prepared similarly, but they were stirred for 2–3min before ageing at least 20min. The formation of the oxinates was accomplished by adjusting the pH value of the water samples to 8 ± 0.1 with ammonium chloride–ammonia buffer and adding a predetermined amount of oxine solution according to the recommendations given by Vanderborght et al. [6,7], which take into account the amount of oxine needed to complex calcium and magnesium and trace ions plus an excess of $5mg\ L^{-1}$ of reagent.

Recoveries obtained by energy dispersive X-ray spectrometry for several elements are tabulated in Table 28.1. The unsatisfactory nature of these and other data presented by Leyden et al. [5] indicate that much more work was required in the development of precipitation methods. Later work by Ellis et al. [3,4] would appear to be the more reliable.

Preconcentration by coprecipitation with iron dibenzyldithiocarbamate

Caravajal et al. [8] have described a method for the determination of uranium based on the coprecipitation of dissolved uranium in non saline waters at pH4 using an iron dibenzyl-dithiocarbamate carrier complex. The precipitate is collected as a thin film and measured by wavelength dispersive X-ray fluorescence spectrometry. Ultraviolet irradiation was used prior to coprecipitation to alleviate the effect of filter-clogging colloids such as humic acids. The iron carrier significantly improved the filtration time, while irradiation improved both the filtration time and the recovery from solutions containing below 10µg of uranium. The results are precise and accurate. A detection limit of $0.4µg\ L^{-1}$ was achieved for 500ml water samples.

The use of iron dibenzyldithiocarbamate as a coprecipitant for uranium in non saline waters appears to be quite promising. Both ultraviolet irradiation and the addition of a carrier ion are essential if the applicability is to be extended to water samples where the concentration of uranium present is low $(0.1–10µg\ L^{-1})$ to provide for reasonable filterability and recovery. Although the iron carrier when absent did not

Table 28.1 Recovery 10μg L^{-1} relative to recovery at 200μg L^{-1} (%)

	Mn	Fe	Co	Cu	Zn	As	Hg	Pb
Sodium diethyl dithiocarbamate	–	51 ± 12	67 ± 16	115 ± 22	30 ± 28	–	61 ± 12	21 ± 9
Ammonium pyrrolidine-dithiocarbamates	–	72 ± 41	82 ± 14	84 ± 20	–	81 ± 15	46 ± 12	–
Activated carbon and oxinates	94 ± 27	–	109 ± 28	–	–	–	123 ± 22	110 ± 18

Source: Reproduced with permission from Leyden, D.E. et al. [5] Gordon AC Breach

Table 28.2 Preconcentration of metals in non saline waters using dithiocarbamic acid derivatives

Cations determined	Organic coprecipitant	Analytical finish	Detection limit*	Ref.
Se	Polyvinyl pyrrolidine thionalide, diethyldithiocarbamate	–	1	[9,10]
Se	Ammonium pyrrolidine dithiocarbamate in presence of iron(III)	X-ray fluorescence spectroscopy	0.6–5	[11]
Co	Ti(IV) and diethyldithiocarbamate	X-ray fluorescence spectroscopy	0.4	[12]
Misc. metals	Pyrrolidinedithiocarbamate	–	–	[13]
Au	Lead diethyldithiocarbamate	Neutron activation analysis	–	[14]

*μg L^{-1} unless otherwise stated

Source: Own files

appear to influence the precision, it had a dramatic effect on the filtration time of the sample; in contrast, irradiation improved both the filtration time and the recovery from solutions containing below 10μg of uranium. It is apparent from extrapolation data that the linearity of the method extends past 50μg. Although this extension may exist, the need for preconcentration becomes less essential. Thus, the preconcentration of uranium from non saline waters with an iron dibenzyldithiocarbamate carrier complex prior to determination by X-ray spectrometry is precise, applicable to river water samples and free from interferences from major ions normally present in non saline water. Judging from the results obtained for spiked non saline water samples, the method would appear to give accurate as well as precise results.

Various other workers have studied the use of dithiocarbamic acid derivatives for the preconcentration of metals (Table 28.2).

28.1.2 8-hydroxyquinoline

This has been used as a coprecipitant in the determination of aluminium, vanadium, copper, molybdenum, zinc and uranium in lake waters, underground waters, and seawater [15]. The final isolated solid was examined by neutron activation analysis.

Akatsuka and Atsuya [16] preconcentrated copper and manganese from river and potable water using 8-hydroxyquinoline and magnesium

Table 28.3 Miscellaneous coprecipitants used in preconcentration of cations in non saline waters

Cations determined	Organic coprecipitant	Comments	Analytical finish	Detection limit*		Ref.
Co, Cd, Ca, Cr Mn, U, Zn	1-(2-pyridylazo) naphthol	–	Neutron activation analysis	Co Cd Cu Cr Mn U Zn	0.04 0.8 0.3 0.2 0.0006 0.006 0.3	[17]
U	Oxinate	On bed of phenolphthalein	Neutron activation analysis	<0.5		[18]
Sn	1.10 phenanthroline/ tetraphenylboron		AAS	1ng		[19]
Ag, As, Cd, Co, Fe, Hg, Mo, Sb, Se, W, Zn	Thionalide (2-mercapto–N–2- naphthyl acetamide)	pH9.1	Neutron activation analysis	–		[20]
As	Thionalide	–	Neutron activation analysis			[21]
Misc. metals	Thionalide	Co and Mg do not interfere	X-ray fluorescence spectroscopy	μg L⁻¹ level		[9]
Mo, U	Uron and induron loaded cellulose	–	–	–		[22]

*μg L⁻¹ unless otherwise stated

Source: Own files

as carrier. Copper and manganese ions were coprecipitated at pH7.0–8.5 (dilute ammonia solution) and were placed after filtering and drying in a mini-cup to be inserted in a graphite furnace atomic absorption spectrometer. Detection limits for copper and manganese were 12 and 14ng L^{-1}, respectively, for 300ml portions of tap and river waters.

28.1.3 Other organic coprecipitants

Other organic coprecipitants which have received a limited amount of study include 1(2-pyridylazo)naphthol, oxine, 1,10 phenanthroline/tetraphenylboron, thionalide (2-mercapto–N–2-naphthyl(acetamide)) and uron loaded cellulose. See Table 28.3.

28.2 Seawater

28.2.1 Dithiocarbamic acid derivatives

Atomic absorption spectrometry

Copper, nickel, and cadmium have been preconcentrated and determined at the ng L^{-1} level by coprecipitation with cobalt pyrrolidinedithiocarbamate followed by dissolution of the precipitate in an organic solvent and analysis by graphite furnace atomic absorption spectrometry [23]. Excellent results for the distribution of nickel and cadmium in the ocean were obtained by this technique.

X-ray fluorescence spectrometry

Stiller *et al.* [24] have described the determination of cobalt, copper and mercury in the Dead Sea by neutron activation analysis followed by X-ray spectrometry and magnetic deflection of β-ray interference. The metals were coprecipitated with lead ammonium pyrrolidinedithiocarbamate and detected by X-ray spectrometry following neutron activation. Magnetic fields deflect the β-rays while the X-rays reach the silicon (lithium) detector unaltered. The detectors have low sensitivity to γ rays. The concentration of cobalt found by this method was 1.3μg L^{-1}, about one-fifth of that measured previously, while that of copper, 2.0μg L^{-1} agreed with results of some previous workers. The concentration of mercury was 1.2μg L^{-1}.

Other applications of dithiocarbamic acid derivatives to the preconcentration of cations in seawater are listed in Table 28.4.

28.2.2 1,10 phenanthroline/tetraphenylboron

Dogan and Haerdi [19] determined iron in seawater by graphite furnace

Table 28.4 Applications of dithiocarbamic acid derivatives to the preconcentration of cations in seawater

Cations determined	Organic coprecipitant	Comments	Analytical finish	Detection limit*	Ref.
Hg, Au, Cu	Lead diethyldithiocarbamate	–	Neutron activation analysis		[25]
Cd, Co, Cu Hg, Mn, Th, Th, U, V, Zn	(1) (1-(2-thiazolylazo)-2-naphthol (2) Pyrrolidinedithiocarbamate (3) N-nitrosophenyl-hydroxylamine	–	Neutron activation analysis	µg L⁻¹ level	[26]
Mo	Sodium diethyldithiocarbamate	–	X-ray fluorescence spectroscopy	0.3	[27]
Heavy metals	Ammonium pyrrolidinedithiocarbamate	–	X-ray fluorescence spectroscopy	–	[28]
Ni	Sodium diethyldithiocarbamate	Precipitate redissolved in nitric acid	AAS	0.5	[29]

*µg L⁻¹ unless otherwise stated

Source: Own files

atomic absorption spectrometry. These workers added 0.1–1.0ml 25M 1, 10-phenanthroline and 0.1–1.0ml 0.2M tetraphenyl boron (both reagents were freshly prepared) to 50–1000ml of water sample which had been previously filtered through a 0.45µm millepore filter. The pH of this solution was adjusted to 5.0 before addition of coprecipitating reagents. The precipitate thus obtained was either filtered or centrifuged and dissolved in a 1–5ml aliquot of ammoniacal alcohol (methanol, ethanol or isopropanol) solution, pH = 8–9 or in Lumatom®. For large volumes of water, the dissolution of the coprecipitate must be carried out with Lumatom® since a precipitate is formed due to other ions present with ammoniacal alcohol solution.

28.2.3 1(–2-pyridylazo)–2 naphthol

Atsuya and Itoh [30] showed that 1–(2-pyridylazo)–2–naphthol used in conjunction with dimethyl glyoxime and nickel as a carrier ion is a very useful method for coprecipitating cadmium, copper, manganese, lead and zinc prior to analysis by graphite furnace atomic absorption spectrometry. Coprecipitation with this complex increased the analytical sensitivity of the conventional graphite furnace method by 1000-fold. Detection limits ranged from 6µg L^{-1} (cadmium) to 50ng L^{-1} (lead) in lake and seawater.

28.2.4 5,7-dibromo–8-hydroxyquinoline

Riley and Topping [31] used this coprecipitant to preconcentrate lead, cerium, copper, zinc, iron, cobalt, manganese and chromium in seawater. 85% recovery was obtained for manganese while low recoveries were obtained for lead, silver and cerium.

28.3 Potable water

28.3.1 Tri(pyrrolidine dithiocarbamate cobalt(III))

This coprecipitant was used to preconcentrate lead from potable water prior to its determination by atomic absorption spectrometry [32].

References

1 Krishnamurty, K.V. and Reddy, M.M. *Analytical Chemistry*, **49**, 222 (1977).
2 Hirayama, K. and Keyden, D. *Analytica Chimica Acta*, **188**, 1 (1986).
3 Ellis, A.T. and Leyden, D.E. *Analytica Chimica Acta*, **142**, 73 (1982).
4 Ellis, T.A., Leyden, D.E., Wegscheider, W., Jablonski, B.B. and Bodnar, W.B. *Analytica Chimica Acta*, **142**, 89 (1982).

5 Leyden, D.E., Wegscheider, W. and Bodnar, W.B. *International Journal of Environmental Analytical Chemistry*, **7**, 85 (1979).
6 Vanderborght, B.M., Verbeeck, J. and Van Grieken, R.E. *Bulletin of the Belgian Chemical Society*, **86**, 23 (1977).
7 Vanderborght, B.M. and Van Grieken, R.E. *Analytical Chemistry*, **49**, 311 (1977).
8 Caravajal, G.S., Mahan, K.L. and Leyden, D.E. *Analytica Chimica Acta*, **135**, 205 (1982).
9 Panayappan, R., Venezky, D.L., Gilfrich, J.V. and Birks, L.C. *Analytical Chemistry*, **50**, 1125 (1978).
10 Birks, C. private communication.
11 Pradzynzki, A.H., Henry, R.E. and Stewart, J.L.S. *Radiochemical and Radioanalytical Letters*, **21**, 273 (1975).
12 Nishoika, H., Maeda, Y. and Azumi, T. *Kenkyu Hokoku-Himeji Kogyo Daiggaku*, **39A**, 56 (1986).
13 Beasley, P.I., Rao, R.R. and Chatt, A. *Journal of Radioanalytical and Nuclear Chemistry*, **179**, 267 (1994).
14 Byrko, V.M., Vizhenskii, V.A. and Molchanova, T.P. *Zh. Anal. Khim.*, **42**, 15767 (1987).
15 Fujinaga, T., Kusaka, R., Koyama, H. *et al.* *Journal of Radioanalytical Chemistry*, **13**, 301 (1973).
16 Akatsuka, K. and Atsuya, I. *Analytica Chimica Acta*, **202**, 223 (1987).
17 Bem, H. and Ryan, D.E. *Analytica Chimica Acta*, **166**, 189 (1984).
18 Carmella Crespi, V., Genova, N., Melonis, N. and Oddone, M. *Journal of Radioanalytical and Nuclear Chemistry*, **114**, 303 (1987).
19 Dogan, S. and Haerdi, W. *International Journal of Environmental Analytical Chemistry*, **8**, 249 (1980).
20 Zmijewska, W., Polkowska Motrenko, H. and Stakowska, H. *Journal of Radioanalytical and Nuclear Chemistry Articles*, **84**, 319 (1984).
21 Ray, B.J. and Johnson, D.E. *Analytica Chimica Acta*, **62**, 196 (1972).
22 Burba, P. and Willmer, P.G. *Fresenius Zeitschrift für Analytische Chemie*, **324**, 298 (1986).
23 Boyle, I.A. and Edmond, J.M. *Analytica Chimica Acta*, **91**, 189 (1977).
24 Stiller, M., Mantel, M. and Rappaport, M.S. *Journal of Radioanalytical and Nuclear Chemistry*, **83**, 345 (1984).
25 Lo, J.M., Wei, J.C. and Yeh, S.J. *Analytical Chemistry*, **49**, 1146 (1977).
26 Rao, R.R. and Chat, A. *Journal of Radioanalytical and Nuclear Chemistry*, **168**, 439 (1993).
27 Kimura, A., Yoneda, A., Maeda, Y. and Azumi, T. *Nippon Kalsui Gakkaishi*, **40**, 141 (1985).
28 Civici, N. *Journal of Radioanalytical and Nuclear Chemistry*, **186**, 303 (1994).
29 Nisioka, H., Assadamongkol, S., Maeda, Y. and Azumi, T. *Nippon Kalsui Gakkaishi*, **40**, 286 (1987).
30 Atsuya, I. and Itoh, K. *Fresenius Zeitschrift für Analytische Chemie*, **329**, 750 (1988).
31 Riley, J.P. and Topping, G. *Analytica Chimica Acta*, **44**, 234 (1969).
32 Department of the Environment. Methods for the Examination of Waters and Associated Materials, Lead in Potable Waters by Atomic Absorption Spectrometry. HMSO, London (1976).

Chapter 29

Cations: Miscellaneous preconcentration methods

There are a number of additional techniques which have been adapted for the preconcentration of cations as a means of increasing the sensitivity of analysis for these substances. Many of these techniques are of value in particular applications and are discussed here under various broad headings. Techniques that have been used for cations include: electrolytic; gas evolution; flotation; evaporation/distillation; precipitation/ crystallisation and dialysis/osmosis.

29.1 Non saline waters

29.1.1 Electrolytic methods

Frick and Tallman [1] have described a flow cell for the determination of mercury in water by electrodeposition followed by atomic absorption spectrometry. They use a commercially available non-coated graphite furnace tube ($0.2222 \pm 0.0051n$ id) as the working electrode in a thin-layer, flow-through configuration. The thin-layer design enhances deposition efficiency and hence sensitivity compared to deposition carried out with an open tubular electrode. The accuracy of the electrodeposition–atomic absorption method was tested by analysing an Environmental Protection Agency reference sample (accepted mercury concentration of $7.6\mu g\ L^{-1}$). The method yielded mean results of 7.3 to $7.8\mu g\ L^{-1}$. A river water sample, after $0.45\mu m$ filtration was also analysed by the electrodeposition atomic absorption spectrometric system and by cold vapour method, with each technique giving the same value of $2.8\mu g\ L^{-1}$.

Muhlbaier et al. [2] determined cadmium in Lake Michigan water by electrodeposition on an amalgamated gold wire followed by electrothermal atomic absorption spectrometry or by mass spectrometric isotope dilution analysis used as a reference method. This process achieved a 5000-fold enrichment in cadmium content. The sample container – electrolysis cell used by these workers is shown in Fig. 29.1. Prior to use, the entire apparatus was cleaned with a phosphate free detergent, rinsed, leached with 10% v/v aqueous Ultrex ultrapure nitric

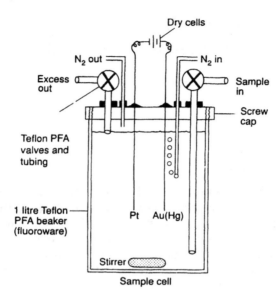

Fig. 29.1 Schematic representation of combination sample container and electrodeposition cell
Source: Reproduced with permission from Muhlbaier, J. et al. [2] American Chemical Society

acid and then rinsed and filled with Nanopure water and placed in double plastic bag. After connection to the outlet of the sampling device the container was flushed with several times its volume of sample before a subsample was entrained by closing the stopcocks. As shortly thereafter as feasible, the sample was treated with 1ml of 10% v/v nitric acid.

The cell's anode consisted of a 30ml diameter high purity platinum wire 10cm long. The cathode was 30ml gold wire of the same length, amalgamated by brief immersion in triple distilled mercury. Because the mercury is incorporated into the concentrate during stripping, its purity must be ascertained in advance to avoid introducing spurious cadmium. Three 1.5V dry cells connected in series provided the EMF for deposition. This arrangement resulted in a current of 2–3mA.

After electrodeposition, the cathode and anode were separately stripped with 200μL of 10% v/v Ultrex nitric acid contained in a micro test tube made by sealing one end of a length of 1mm capillary tubing. The strippings were analysed by atomic absorption spectrometry or by the mass spectrometric reference method.

Muhlbaier et al. [2] conclude that recoveries obtained by atomic absorption spectrometry are appreciably lower than those obtained by

Table 29.1 Cadmium contents of Lake Michigan samples by two techniques

	Cadmium content ± SD (ng L⁻¹)	
	Sample A	Sample B
Electrodeposition–atomic absorption spectrometry	24 ± 8	25 ± 2
Electrodeposition–mass spectrometric isotope dilution analysis	38 ± 8	31 ± 8
% recovery obtained in atomic absorption method relative to mass spectrometric referee method	63.2	80.6

Source: Reproduced with permission from Muhlbaier, J. et al. [2] American Chemical Society

the mass spectrometric reference method (Table 29.1). The latter method was suitable for determining nanogram levels of cadmium in fresh water samples. Once the proper precautions are taken, the procedure yields a concentrate that is free of significant interferences and which gives steady ion currents from samples and blanks containing 25–100ng total cadmium. In practice the lower working limit is determined by the blank value, rather than inherent sensitivity and it is the variability introduced by contamination that appears to control the overall precision. Clean room conditions would definitely be required if metal concentrations below 10µg L⁻¹ were being determined.

Wang and Mahmoud [3] described preconcentration procedure for ultra trace quantities of iron(II) based on the effective interfacial accumulation of iron(III)/Solochrome Violet RS (SVRS) chelate on the hanging mercury drop electrode. After degassing, a preconcentration potential of –0.40V was applied while the solution was stirred. After concentration, the stirring was stopped and the voltammogram recorded to a final potential of –1.10V. The height of the chelate peak which occurred at a potential of 0.28V more negative than the peak of the free dye was proportional to the concentration of the iron. The limit of detection at a solution pH of 5.1, after 1min preconcentration was 0.04µg L⁻¹ and the relative standard deviation at the 0.1µM level was 4.7%. The effects of interferences due to coexisting metal ions or organic surfactants were evaluated. Iron(III) could be determined in the presence of iron(II).

Prabhu et al. [4] studied the preconcentration of copper(I) at a chemically modified carbon paste electrode containing 2,9–dimethyl–1, 10-phenanthroline. Deposition was rapid with 80–90% of the final response generated within 1min of exposure to copper(I) solution. Quantification of copper(I) from the area of the anodic curve for a 1s deposition time gave a limit of detection of 0.3µM and a linear response

Table 29.2 Applications of electrochemistry to the preconcentration of cations in non saline water

Cations	Comments	Analytical finish	Detection limits*	Ref.
Pb, Cd, Tl	Cathodic deposition	Isotope dilution mass spectrometry	µg L^{-1} to ng L^{-1}	[5]
Pb	Electrolysis on silanised graphite rod	X-ray fluorescence spectroscopy	15pg	[6]
Cr(VI)	Deposited on tungsten wire electrode coated with tri–n–octyl phosphine oxide then Cr dissolved in hydrochloric acid	AAS	0.3	[7]
Pb	Electrodeposition on mercury film tungsten electrode	AAS	0.0008	[8]
Cd, Cr, Cu, Mn, Ni, Pb, Zn	Electrochemical deposition on mercury film electrode	Inductively coupled plasma atomic emission spectrometry	Zn 100 Cr 5200	[9]
Ga	Adsorption on Ga Solochrome Violet RS, then voltammetry. Chelate adsorbed on hanging mercury drop electrode	–	0.08	[10]
Sn	Adsorptive stripping voltammetry of tin tropalone complex with preconcentration on to hanging mercury drop electrode	–	0.028	[11]
Se	Selenium extracted mesh 3,3'diamino-benzidene and concentrated as SeHg on mercury electrode at –0.45V	–	0.01	[12]
Cr	Chromium preconcentrated as HgCrO$_4$ at 0.45V and determined from the stripping peak at 0.35V	–	<3 × 10^{-9}M	[13]

Table 29.2 continued

Cations	Comments	Analytical finish	Detection limits*	Ref.
Hg(II)	Mercury preconcentrated as anionic forms in chloride medium onto poly(4-vinyl–pyridine) gold film electrode	–	–	[14]
Cd	Cadmium complexed with dithiocarbamate extracted into Freon then back-extracted into acidic solution then anodic stripping voltammetry	–	0.025	[15]
Zr	Adsorption of Zr–Solochrome RS Violet complex onto hanging mercury drop electrode	–	$2.3 \times 10^{-10}M$	[16]

*µg L^{-1} unless otherwise stated

Source: Own files

up to 10μM. Relative standard deviation was 5%. A small preconcentration signal was observed for silver(I), but there were no direct responses for copper(II), zinc(II), cobalt(II), lead(II), nickel(II) or iron(III), although concentrations of these suppressed the copper(I) response.

Further applications of electrochemistry to the preconcentration of cations are reviewed in Table 29.2.

29.1.2 Dialysis and osmosis

Non saline waters

Dialysis has been applied to the determination of the chemical form of trace elements in non saline water [17,18]. Benes and Steinnes [18] have described a method in which the dialysis bag filled with pure water is immersed in the non saline water sample, allowing the dialysis and adsorption equilibrium to be established.

Hart and Davies [17] combined dialysis with ion exchange and this combination enabled the determination to be completed in a much shorter time (5h) than in the conventional dialysis methods (24h). They showed that about 100% of cadmium, copper, lead, zinc and iron were dialysed in the 5h period. The results gave information on the concentration of ionic species present in the non saline water sample. The experimental set up is illustrated in Fig. 29.2. The dialysis unit was from a Technicon Auto Analyzer II: 'Cuprophanel' membranes were used. This membrane consists of regenerated cellulose with a pore size of approximately 0.8nm.

Analyses were mainly performed using atomic absorption spectrophotometry – Varian Techtron AA with a laminar flow burner (Techtron type AB 51) or carbon rod furnace (CRA–63). The satisfactory transition metal recoveries obtained following a 5h dialysis of a synthetic non saline water sample containing typical levels of sodium, potassium, calcium, magnesium and anions expected in non saline water samples were recorded.

Wilson and Di Nunzio [19] used dialysis to preconcentrate nickel and cobalt by factors of up to 20 at the 20μL $^{-1}$ level.

Kasthurikrishnan and Koropchak [20] have developed a pre-concentration method based on the Donnan dialysis to provide rapid extraction of cations from non saline waters in which detection limits in the ppt range are achievable.

Stec et al. [21] have investigated the preconcentration of transition metals ions from dilute aqueous solution using osmosis. They describe the manufacture, separation and performance characteristics of the osmotic preconcentration cell. Cellulose acetate membranes were used as semi-permeable barriers between sample solutions and saturated salt

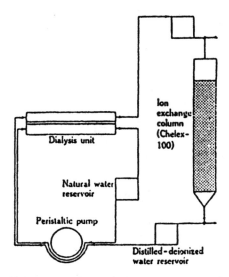

Fig. 29.2 Dialysis ion exchange apparatus
Source: Reproduced with permission from Hart, R.T. and Davies, S.H.R. [17] Marine and Freshwater Research

solutions. These membranes were composed of dense microporous surface layers with highly porous substructure. After assembly of the cell, saturated salt solutions and water were cycled through the outer and inner channels, respectively, for approximately 15min to assure a steady state permeation rate. Analyte recovery data were obtained on a multi-element inductively coupled plasma emission spectrometer. The stability of permeation rates was dependent on the three parameters: temperature, salt concentration, and the flow rate of salt solution through the cell. Complexation of metal ions with EDTA prior to osmotic preconcentration was used to prevent loss of analytes by permeation through the membrane. A comparison of chelated and non-chelated recoveries for cadmium, copper, manganese, nickel and zinc was made and the results tabulated. The reproducibility for unchelated ion recoveries was variable (64–95%) but the major limitation of this technique was the back-permeation of sodium chloride from the saturated salt solution.

29.1.3 Solvent sublation

Solvent sublation is a flotation process in which the material adsorbed on the surface of gas bubbles is collected on a layer of immiscible liquid instead of as a layer of foam over an aqueous phase as occurs in the conventional foam flotation process.

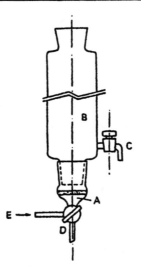

Fig. 29.3 Experimental device used for the solvent sublation process. (A) Detachable section with sintered glass plate; (B) variable size column; (C) stopcock; (D) three way stopcock and (E) gas inlet
Source: Reproduced with permission from Cervera, J. *et al.* [22] Royal Society of Chemistry

Cervera *et al.* [22] studied the feasibility of using this process for the separation and preliminary concentration of copper as its dithizonate. Sodium lauryl sulphate was used as the collector with dithizone solutions in methyl isobutyl ketone. Cervera *et al.* [22] established optimum experimental conditions. Sample volumes of several litres can be handled making this a very effective process with the possibility of pre-concentration factors of up to 100. The experimental conditions are described below.

The experimental device used is shown in Fig. 29.3. A water cooled high efficiency reflux condenser was attached on the upper part of the flotation column to prevent evaporation losses from the upper organic layer, which was important in long-duration experiments (more than 1h). The column had a capacity of about 3.5L and a cross-section of 70cm² and was fitted at the lower end with a detachable section furnished with a sintered glass place of low porosity (no. 3). A Pye Unicam SP 9800 atomic absorption spectrophotometer using an air–acetylene flame or equivalent was used for determining concentrations of copper(II) in aqueous layer during sublation process.

An air stream was used as the carrier gas, presaturated with water vapour, using a membrane pump and the flow was regulated by means of a needle valve. The gas flow rate was controlled by means of a bubble rotameter. An optimum flow rate of carrier gas (air presaturated with water vapour to minimise carriage of vapour from the aqueous phase by the ascending gas bubbles) at a limit of 80ml min $^{-1}$ was adopted. The optimum flow rate of the carrier gas depends on a number of parameters directly related to the porosity and mechanical quality of the sintered glass plate used, to the column size (especially its cross-section), and to the salinity of the aqueous phase (an increase of which favours mixing of the two phases). Care should be taken to keep the flow rate reasonably constant, which is fairly easy as the samples to be taken from the aqueous phase during the process are small enough not to bring about a significant variation in flow rate owing to the resulting charge losses of the liquid column, which produce slight increases in the flow rate.

Under the experimental conditions used, the total decrease in height of the aqueous phase column arising from the periodic extraction of samples for analysis did not exceed about 3% of the initial height thus making it unnecessary to make flow corrections following removal of samples for analytical monitoring.

Attention must be paid however, to the eventual transport of organic phase by the gas bubbles, which might reach a considerable extent owing to the small volume of organic phase used (25ml) especially when the process is run for long periods (up to several hours). This problem can be greatly reduced by attaching a suitable water cooled glass condenser of high efficiency to the upper end of the column combined with the use of organic solvents of low volatility (methyl isobutyl ketone, octan–2–ol, etc). Provided that the experiments are carried out within a reasonable time (about 1h or less) there is no need to take special precautions with the dithizone solutions, although it is preferable to isolate the system from heat and light.

Cervera et al. [22] demonstrated that their solvent sublation process was not applicable to the determination of trace metals in estuarine and seawater. They concluded that the solvent sublation technique allows a highly effective separation and preconcentration of trace amounts of copper from aqueous solution as the dithizonate within a narrow pH range. The yield of the process is 100% reaching this value for sodium lauryl sulphate to copper(II) ratios greater than 20:1 in not more than 20min.

In order to obtain more accurate results for comparative purposes, a separate extraction process was carried out using 250ml extraction funnels and keeping the same organic to aqueous layer volume ratios used in the solvent sublation method. The kinetics and yields of the corresponding processes (using in all instances an initial copper(II) concentration of 1mg L $^{-1}$) are compared in Table 29.3, which demonstrates clearly the difference between the two processes and the outstanding concentration features exhibited by the solvent sublation process.

Table 29.3 Comparative results of copper(II): dithizone solvent sublation and solvent extraction obtained under the same experimental conditions

Experiment No.	NaLS:Cu(II) molar ratio	Solvent extraction		Solvent sublation	
		Extraction (%)	Time to reach equilibrium (min)	Yield (%)	Time necessary (min)
1	40:1	73.3	~30	100.0	~20
2	15:1	75.6	~20	95.6	~45
3	7:1	78.4	<3	78.3	—*
4	1:1	78.2	<2	28.4	—*
5	0:1	78.5	<1	—	—

*Yield obtained after 60min

Source: Reproduced with permission from Cervera, J. et al. [22] Royal Society of Chemistry

29.1.4 Direct evaporation

Although this is an obvious method for preconcentrating water samples prior to analysis, this technique has several disadvantages, namely the risk of sample contamination and the lengthy nature of the evaporative process. Nevertheless some work has been carried out on the application of evaporative methods [23–25] to the determination of trace elements [24] including copper [25]. Preconcentration of dilute solutions of 12 metals at the 10^{-12}g L^{-1} level by non-boiling evaporation followed by atomic absorption spectrometry has been applied [24].

Thompson et al. [26] attempted to increase the sensitivity of the technique by carrying out a rapid evaporative 10:1 preconcentration on the sample before instrumental analysis. They studied the effects of background interference and its on-peak correction on realistic detection limits of 30 elements, including high concentrations of calcium and magnesium. Recoveries of 24 and out of 32 elements during preconcentration were found to be adequate.

In order to investigate the performance and likely sources of error, Thompson et al. [26] concentrated synthetic trial solutions resembling fresh waters and analysed them by the procedure outlined above. Realistic estimates of the effect on detection limits of increasing background corrections were obtained by concentrating and analysing ten replicate samples of deionised water and ten of the synthetic fresh waters containing high levels of calcium (200mg L^{-1}) and magnesium (30mg L^{-1}) only. The resulting detection limits are given in columns 4 and 5 of Table 29.4).

The calcium and magnesium levels used in this study are approximately 20 times higher than those in average river waters. The interference effects in river water analysis will therefore be proportionally reduced.

The precision of the method measured using the 10 replicate samples expressed as twice the coefficient of variation and averaged over 20 analytes was 8.0% at the $50\mu g$ L^{-1} level and 7.0% at the $500\mu g$ L^{-1} level. The presence of 200mg L^{-1} of calcium and 30mg L^{-1} of magnesium increased these figures to 8.8 and 7.8% respectively after interference correction.

In order to study the recovery of 21 elements during sample preconcentration, two level element spikes were added to tenfold replicates of pure water. At the $500\mu g$ L^{-1} level only silver (33% low) and antimony (17% low) showed recoveries that are significantly low at 95% confidence limits. At the $50\mu g$ L^{-1} level manganese (17% low), arsenic (16% low) and molybdenum (8% low) also showed values significantly below the spike added. Three other elements studied separately showed low results: titanium (up to 70% low), zirconium (up to 16% low) and beryllium (up to 16% low). For these elements with high ionic potentials the low results probably represent loss of analyte by chemisorption or hydrolysis. For other elements low returns could be due to statistical inaccuracy and not losses (eg manganese).

Recovery of the principal ionic constituents (aluminium, calcium, iron(III), potassium, magnesium, sodium, phosphate and sulphate) was found to be quantitative in separate experiments with the preconcentration of non saline water samples. No significant loss of hydrochloric acid occurred during the evaporation.

The effect of high levels of calcium and magnesium on the practical detection limit can be seen by a comparison of columns 4 and 5. The detection limits in column 5 are calculated in a way that includes both inaccuracy and imprecision in the background correction. To the normal two standards deviations of noise over 10 samples has been added the absolute value of the correction bias. For example, sulphur has a detection limit of $70\mu g$ L^{-1} by direct nebulisation, which is improved to $8\mu g$ L^{-1} by the preconcentration method in the absence of interfering elements. The interference, mainly from calcium, produced $1505\mu g$ L^{-1} of apparent sulphur with a standard deviation of $20\mu g$ L^{-1}. The total calculated interference correction is $1488\mu g$ L^{-1} leaving an uncorrected residual of $+17\mu g$ L^{-1} of sulphur. The detection limit is therefore recorded as $(2 \times 20) + 17 = 57\mu g$ L^{-1}. A deterioration of the detection limit due to this cause is evident in a number of elements, notably sulphur and molybdenum. Although the uncertainties in the detection limits make rigorous interpretation difficult, there appears generally to be an increase in the detection limit equal to approximately 10% of the total background interference.

Table 29.4 Applicability of ICP to water analysis

Element	(2) Wavelength/ nm (order)	(3) Detection limits ($\mu g\,L^{-1}$) Direct	(4) By pre-concentration Soft	(5) Hard	(6) Interference ($\mu g\,L^{-1}$)	(7) Recovery	(8) Average river water	(9) Concentrations ($\mu g\,L^{-1}$) GL	(10) EEC MAC	(11) Applicability Average river	(12) EEC
Ag	328.1 × 2	2	0.3	0.5	1.3	a	0.3		10	c	c
Al	308.2 × 2	50	15	6	0		400	50	200		d
As	193.8 × 2	30	2	—	0	b	2		50	c	
Ba	455.4 × 1	4	0.4	0.3	0		10	100	c		
Be	313.0 × 2	0.1	0.02	0.03	0.04	a	0.47				c
Bi	223.1 × 2	30	2	4	7.7		0.005[b]			d	
Ca	317.9 × 2	60	5	—	—		1.5×10^4	1×10^5		c	c
Co	228.6 × 3	5	0.6	0.5	0.87		0.2				
Cd	226.5 × 3	2	0.2	0.3	0.52		0.03[a]				
Cr	267.7 × 2	3	0.2	0.8	3.0		—		5	d	c
Cu	324.8 × 2	2	0.2	0.3	2.0		7		50	c	c
Fe	259.9 × 2	40	8	5	0		100[a]	100		c	c
Hg	194.2 × 1	4	0.6	1.5	3.4		0.07	50	200	c	c
K	766.4 × 1	100	9	9	0		2300	1×10^4	1.2×10^4	c	
Li	670.8 × 1	1	0.1	0.1	0		—		—		c
Mg	279.0 × 2	100	100	—	—		4100	3×10^4	5×10^4		c
Mn	257.6 × 2	10	1	2	0	b	7	20	50		c
Mo	281.6 × 2	5	0.6	4	31	b	—				
Na	589.0 × 1	50	30	20	0		6300	2×10^4	1.8×10^5	d	
Ni	231.6 × 2	8	0.8	0.9	2.5		1.5a		50	c	c
Pb	220.3 × 2	30	4	4	12.3		3		50	c	c
Sb	206.8 × 2	80	5	6	8.3	a	—		10		c

Table 29.4 continued

Element (2)	Wavelength/ nm (order)	Detection limits ($\mu g\ L^{-1}$) Direct (3)	By pre-concentration Soft (4)	Hard (5)	Inter-ference ($\mu g\ L^{-1}$) (6)	Recovery (7)	Average river water (8)	Concentrations ($\mu g\ L^{-1}$) GL (9)	EEC MAC (10)	Applicability Average river (11)	EEC (12)
Se	196.1	80	8	12	45		0.2		10		
Sn	190.0 × 2	7	0.6	0.4	8.1		?				
Sr	407.8 × 1	2	0.2	0.1	1.6		50			c	
Te	214.3 × 2	30	2	5	5.8		?				
Ti	337.3 × 2	60	5	7	0		3				
V	311.1 × 2	2	0.2	0.1	2.8	a	0.9			c	
Zn	202.5 × 3	7	0.8	1.4	4.8	a	20	100		c	c
Zr	349.6 × 2	3	0.2	0.9	4.6		?				c

Column 3: detection limit (2σ) from 10 readings of blank solution by direct nebulisation.

Column 4: detection limit (2σ) from 10 replicate blank preparations by pre-concentration.

Column 5: detection limit (2σ) from 10 replicate samples with 200 mg L^{-1} of calcium and 30 mg L^{-1} of magnesium prepared by pre-concentration.

All values of detection limits are approximate and can vary by 100% by random fluctuations.

Column 6: the background interference from 200 mg L^{-1} of calcium and 30 mg L^{-1} of magnesium expressed in $\mu g\ L^{-1}$ analyte; 0 signifies no measurable interference.

Column 7: a elements giving low recoveries on spikes at 50 and 500 $\mu g\ L^{-1}$; b elements giving low recoveries on spikes at 50 $\mu g\ L^{-1}$ only.

Column 8: average of median reported river concentrations. Question marks signify uncertain or unknown values.

Columns 9 and 10: EEC guide levels (GL) and maximum admissible concentrations (MAC) of 1980.

Column 11: elements for which determination at average river levels is c suitable or d marginal.

Column 12: elements for which determination below EEC levels is c suitable or d marginal.

Source: Reproduced with permission from Thompson, M. et al. [26] Royal Society of Chemistry

Table 29.5 Relative mean values and standard deviations obtained by freeze drying

Element	Concentration ($\mu g\ L^{-1}$)	Relative mean value and standard deviation, all values included
As	1	1.81 ± 2.4
Au	0.01	0.36 ± 0.20
Ba	100	0.67 ± 0.037
Br	10	1.08 ± 0.14
Ca	10^5	1.04 ± 0.17
Cd	1	4.11 ± 3.12
Ce	0.1	4.84 ± 0.91
Co	1	1.10 ± 0.11
Cr	1	1.04 ± 0.34
Cu	10	2.75 ± 3.0
Eu	0.1	1.21 ± 0.55
Fe	100	0.72 ± 0.35
Hg	0.1	31.2 ± 106
K	1000	1.16 ± 0.58
La	0.1	1.14 ± 0.32
Mo	1	1.08 ± 0.26
Na	5000	0.87 ± 0.09
Sb	1	0.90 ± 0.33
Sc	0.02	1.29 ± 0.18
Se	1	1.04 ± 0.20
U	1	1.17 ± 0.15
Zn	10	0.98 ± 0.09

Source: Reproduced with permission from Lieser, L.H. *et al.* [30] Elsevier Sequoia, Lausanne

29.1.5 Freeze drying

Jørstad and Selbu [27] used freeze-drying and irradiation, then they concentrated ^{75}Se by controlled potential electrolysis and obtained a $0.68\mu g\ L^{-1}$ detection limit. Favourable sensitivities are obtained when a large volume of water, eg 200ml is freeze-dried after addition of graphite and the pelletised residue is analysed by an X-ray energy spectrometric procedure in which the matrix corrections are based on the scatter peaks in the spectrum, but the detection limit of $1\mu g\ L^{-1}$ is still well above natural concentrations.

Crocker and Merritt [28] preconcentrated water by freeze drying prior to the determination of $0.1\mu g\ L^{-1}$ amounts of elements of atomic weight 7 to 238 in rain water by mass source mass spectrometry. Miscellaneous elements (30) in lake water have been preconcentrated by freeze drying prior to determination at the $0.01\mu g\ L^{-1}$ level by spark source mass spectrometry [29].

Lieser *et al.* [30] used freeze drying as a means of preconcentrating potable and lake water samples prior to the determination of trace metals by neutron activation analysis of the dried solid. Some results obtained by this procedure are given in Table 29.5. These show the sensitivity, reproducibility and accuracy of the procedure. Unfortunately, the authors give no details of the freeze drying procedure they adopted.

To preconcentrate ground water samples prior to the determination of uranium–239 by epithermal neutron activation analysis Juleli and Parry [31] freeze dried 50ml aliquots of filtered and acidified groundwater samples. Uranium–239 peaks of gamma-ray energy at 74.7keV were measured to calculate uranium concentrations applying the total peak area method against that of a uranium standard. Results of this technique were compared against a set of data obtained by conventional instrumental neutron activation using neptunium–239 and a set of data from laser induced fluorometry. The three data sets obtained for ground water samples and for US Geological Survey standard reference samples were in good agreement especially when the total dissolved solids of the samples was low.

29.1.6 Aspiration methods

Harsanyi *et al.* [32] examined the preconcentration method of Topping and Pirie [33] in which the mercury in a 4L sample is reduced to metallic mercury with stannous chloride and aspirated into 50ml of a potassium permanganate–sulphuric acid solution then, after further reduction with stannous chloride, determined by the cold vapour atomic absorption method. Down to 8ng of mercury per litre could be determined.

29.1.7 Cocrystallisation

This technique has been subject to a limited amount of investigation in the case of molybdenum organic chelate complexes [34–37].

29.1.8 Reductive precipitation

Skogerboe *et al.* [38] preconcentrated trace elements from non saline water samples by precipitation as the reduced element or boride. Trace iron was added as a carrier to improve recovery.

29.1.9 Flotation methods

Hidalgo *et al.* [39] preconcentrated molybdenum(VI) in potable water and seawater by coflotation on iron(III) hydroxide. Coflotation was achieved by means of surfactants; octadecylamine for tap water samples, hexadecyltrimethylammonium bromide for non saline waters, and both hexadecyltrimethylammonium bromide and octadecylamine for seawater samples. Differential pulse polarography using the catalytic wave caused by molybdenum(VI) in nitrate medium was applied to the preconcentrate. Good reproducibility was obtained with mean values of 0.7µg L $^{-1}$ and 5.7µg L $^{-1}$ for molybdenum(VI) in tap water and seawater respectively.

The technique has also been used for the determination of ng L $^{-1}$ levels of cadmium in fresh water [40]. In this method cadmium in 1L of sample is coprecipitated with zirconium hydroxide, the precipitate is separated by dissolved air flotation, then dissolved in a small volume of dilute hydrochloric acid prior to analysis by atomic absorption spectrometry.

To investigate the applicability of this method to the separation and determination of cadmium in non saline fresh water, Nakashima and Yagi [41] checked recoveries of known amounts of cadmium added to fresh water samples. For this purpose, 1L aliquots of clear uncontaminated stable and river water were filtered through 0.4µm nuclepore filters after the addition of 3ml of hydrochloric acid per litre of sample immediately after collection. The analytical system could be successfully applied to the separation and determination of cadmium at nanogram levels in fresh waters. The cadmium concentrations in the potable and river water samples were low: 7.8 and 18.7µg L $^{-1}$ respectively.

29.1.10 Gas evolution methods

Hydride formation

In this procedure, the element of interest is converted to a hydride which is then preconcentrated in a cold trap, then released for estimation by a suitable technique; usually atomic absorption spectrometry or gas chromatography.

A number of elements in the fourth, fifth, and sixth groups of the periodic system form hydrides upon reduction with sodium borohydride, which are stable enough to be of use for chemical analysis (Ge, Sn, Pb, As, Sb, Se, Te). Of these elements Andreae [42] has investigated in detail arsenic, antimony, germanium and tin. The inorganic and organometallic hydrides are separated by a type of temperature-programmed gas chromatography. In most cases it is optimal to combine the functions of the cold trap and the chromatographic column in one device. The hydrides are quantified by a variety of detection systems which take into account the specific analytical chemical properties of the elements under investigation. For arsenic, excellent detection limits (about 40pg) can be obtained with a quartz tube cuvette burner which is positioned in the beam of an atomic absorption spectrophotometer. For some of the methylarsines, similar sensitivity is available by an electron capture detector. The quartz-burner–atomic absorption spectrometry system has a detection limit of 90pg for tin; for this element much lower limits (about 10pg) are possible with a flame photometric detection system, which uses the extremely intense emission of the SnH molecule at 609.5nm. The formation of GeO at the temperatures of the quartz tube furnace makes this device quite insensitive for the determination of germanium. Excellent detection limits (about 140pg) can be reached for this element by the combination of the hydride generation system with modified graphite furnace atomic absorption spectrometry.

Many of the methods make use of the condensation of the hydrides in a cold trap at liquid nitrogen temperature. In these methods a packed cold trap is used to serve both as a substrate to collect the hydrides at liquid nitrogen temperature, and to separate arsine and the methylarsines gas chromatographically by controlled heating of the trap [43]. As(III) and As(V) can be differentiated by a prereduction step and by control of the pH at which the reduction takes place. The lowest detection limits were achieved by cold-trapping of the hydrides and subsequent introduction into either a quartz cuvette furnace or into a commercial graphite furnace [44].

The determination of the hydride element species consists of five steps:

1 the reduction of the element species to the hydrides;
2 the removal of interferent volatiles from the gas stream;
3 the cold-trapping of the hydrides;
4 the separation of the substituted and unsubstituted hydrides from each other and from interfering compounds; and
5 the quantitative detection of the hydrides.

Only when the very contamination-sensitive electron-capture detector is used is it necessary to provide separate gas streams, one for the reaction and stripping part of the system, the other for the carrier gas stream of the column and detector. Otherwise, the same gas stream can be used to strip the hydrides from solution and to carry them into the detector, which greatly simplifies the apparatus. Initially, column packings of glass beads and wool [45] were used; however, these packings produce poor separation of the methylated species from each other and badly tailing peaks. Andreae [42] therefore used a standard gas chromatographic packing (15% OV–3 on Chromosorb W/AWDMCS, 60–80mesh) in U-tubes for the separation of the inorganic and alkyl species of arsenic, antimony and tin. This packing is quite insensitive to water and produces sharp and well-separated peaks.

Depending on the detector used, some volatile compounds which are formed or released during the hydride generation step may interfere with the detection of the hydrides of interest. Most prominent among them are water, carbon dioxide, and, in the case of anoxic water samples, hydrogen sulphide. The atomic absorption detector is insensitive toward these compounds; thus no precautions need to be taken when this detector is used. It has been found convenient in some applications however, to remove most of the water before it enters the cold trap/column which serves to condense and separate the hydrides. This can be accomplished by passing the gas stream through a larger cold trap cooled by a dry ice–alcohol mixture or by an immersion cooling system [46]. This method was also used with water-sensitive detectors, eg the electron-capture detector for methylarsines [47] or with plasma discharge detectors (eg Crecelius [48]).

Cutter [49] proposed a method for selenium in which various forms of selenium in a large volume of sample are reduced with sodium borohydride. The volatile selenium is swept off with helium and collected in a liquid nitrogen trap. The methyl selenium species are separated by gas chromatography and measured in a quartz tube atomic absorption furnace at levels corresponding to 5ng L $^{-1}$ in the original sample. Other workers have applied a similar technique [50,51].

Orivini et al. [52] determined arsenic(III) arsenic(V), and total inorganic arsenic using selective hydride generation coupled with neutron activation analysis. Shaikh and Tallman [53] determined arsenic with nanogram sensitivity by reducing arsenic species to hydrides and collecting them in a liquid nitrogen trap. They are selectively vaporised from the trap and directly injected into a graphite furnace atomic absorption spectrometer.

Hanson et al. [54] generated the gaseous hydrides of antimony(III), arsenic(III) and tin(II) by sodium hydride reduction. The hydrides were swept from a large volume of sample on to a Poropak column where they were separated and detected at a gold gas-porous electrode by measurement of their electro-oxidation currents. Detection limits were 0.2µg L $^{-1}$ for arsenic(III) and antimony(III) and 0.8µ L $^{-1}$ for tin(II).

Vien and Fry [55] employed a hydride generation gas chromatographic system with an intermediate cold trap that required no drying agents or carbon dioxide scrubbers in their preconcentration of arsenic, selenium, tin and antimony from non saline waters. The combination of a photoionisation detection and cryogenic entrapment was also tolerant to water vapour, volatile boranes and other spurious byproducts from a borohydride reactor, besides improving sensitivity by several orders of magnitude. Detection limits for arsenic, selenium, antimony and tin were as low as 0.025ng in a 28ml sample.

29.2 Seawater

29.2.1 Electrolytic methods

For most tasks in the trace chemistry of non saline waters, voltammetric

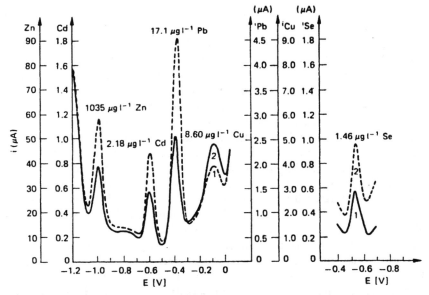

Fig. 29.4 Voltammogram of the simultaneous determination of Cu, Pb, Cd and Zn with DPASV at the HMDE and subsequent determination of Se(IV) by DPCSV in the same run in rainwater at an adjusted pH of 2. Preconcentration time for DPASV is 3min at −1.2V, for DPCSV 5min at −0.2V. 1 = Original analyte; 2 = after first standard addition. Total analysis time with two standard additions, 30–40min
Source: Own files

determination requires preconcentration, because in a group of simultaneously determined ecotoxic heavy metals one usually has levels below 0.1µg L $^{-1}$. Electrochemical preconcentration can be attained in the following two ways, depending whether differential pulse stripping voltammetry (differential pulse anodic scanning voltammetry) or adsorption differential pulse voltammetry has been applied.

Heavy metals capable of forming amalgams, ie Cu, Pb, Cd, Zn, etc., are plated at a stationary mercury electrode consisting of a hanging mercury drop electrode with adjustment of a rather negative potential of −1.2V versus the Ag/AgCl reference electrode for several minutes. To speed up mass transfer, the solution is stirred with a magnetic bar at 900rpm. Their preconcentration is achieved by the accumulation of the heavy metals in the mercury drop. Subsequently the stirring is terminated, and after a quiescent period of 30s the potential is scanned into the anodic direction in the differential pulse mode. At the respective redox potential the plated heavy metal is reoxidised and the corresponding current is recorded (Fig. 29.4). The voltammetric peak heights obtained are proportional to the bulk concentrations of the respective trace metals in the analyte. The hanging drop mercury electrode can usually be applied down to trace levels of 0.1–0.05µg L $^{-1}$.

At lower ultra-trace levels the less voluminous mercury film electrode has to be used. It consists of a mercury film of only several hundred Å thickness on a glassy carbon electrode as support. The fabrication of this glassy carbon electrode is critical for obtaining an optimal mercury film electrode suitable to perform determinations down to 1ng L $^{-1}$ or below.

Certain trace substances such as Se(IV) can be determined in a similar manner by differential cathodic stripping voltammetry. For selenium a rather positive preconcentration potential of –0.2V is adjusted. Se(IV) is reduced to Se^{2-} and Hg from the electrode is oxidised to Hg^{2+} at this potential. It forms, with Se^{2-} on the electrodes a layer of insoluble HgSe and in this manner the preconcentration is achieved. Subsequently the potential is altered in the cathodic direction in the differential pulse mode. The Hg(II) resulting peak produced by the Hg(II) reduction is proportional to the bulk concentration of Se(IV) in the analyte. This procedure has made rapid progress and could now be used to determine lead, cadmium, copper zinc, uranium, vanadium, molybdenum, nickel and cobalt in seawater, with great sensitivity and specificity, allowing study of metal speciation directly in the unaltered sample. The technique used preconcentration of the metal at a higher oxidation state by adsorption of certain surface active complexes, after which its concentration was determined by reduction.

Examples of extraction of trace metals from seawater as chelates with subsequent determination by electrothermal atomic absorption spectrometric procedures have been described but these, and similar methods, are seldom effective and satisfactory when the matrix is very complex and the analyte concentration very low. In contrast, the coupling of electrochemical and spectroscopic techniques, ie electrodeposition of a metal followed by detection by atomic absorption spectrometry, has received limited attention. Wire filaments, graphite rods, pyrolytic graphite tubes and hanging drop mercury electrodes have been tested for electrochemical preconcentration of the analyte to be determined by atomic absorption spectrometry. However, these *ex situ* preconcentration methods are often characterised by unavoidable irreproducibility, contaminations arising from handling of the support, and detection limits unsuitable for lead detection at sub-ppb levels.

These drawbacks could be certainly avoided by performing *in situ* deposition. The sole attempt in this direction was made by Torsi [56] who set up an apparatus which permitted both *in situ* electrochemical preconcentration of the analyte from a flowing solution and almost complete suppression of matrix effects because the matrix could be removed by suitable washing. The feasibility of this approach was successfully tested with respect to lead determinations in different salt solutions (mainly ammonium acetate) and some preliminary results were reported for real samples, such as seawater and urine [56].

Torsi et al. [56,57] carried out a systematic investigation to establish the potentialities of such an apparatus. The apparatus is basically an electrothermal device in which the furnace (or the rod) is replaced by a small crucible made of glassy carbon. Fig. 29.5(a) represents an overall view of the apparatus. Fig. 29.5(b) shows a block diagram of the electrolysis circuit; the crucible (6) acts as a cathode while the anode is a platinum foil dipped into either the sample solution reservoir (1) or the washing solution reservoir (2). The pre-electrolysis was performed at constant current by a 500V DC variable power supply (5). Under these conditions, the cathode potential is not controlled so that other metals can be co-deposited with lead. There is no great need to control the deposition potential, because the spectral selectivity is sufficiently good to prevent interferences by other metals during the atomic absorption step.

A typical atomisation signal obtained in this way is shown in Fig. 29.6. As can be seen, the baseline increases smoothly with time as a consequence of an upward lift of the crucible caused by thermal expansion of the material.

A calibration curve for lead in seawater obtained by the standard addition method is shown in Fig. 29.6(b), a deviation from linearity is observed at higher lead concentrations. The estimated value for the original sample was found to be 0.51µg L $^{-1}$ with confidence limits at the 95% confidence level of ±0.036µg L $^{-1}$, compared with a value of 0.65 ± 0.08µg L $^{-1}$ obtained by anodic scanning.

Cowen et al. [58] showed that polonium can be electrodeposited on to carbon rods directly from acidified seawater, stripped from the rods and autoplated onto silver counting discs with an overall recovery of 85 ± 4%, and an electrodeposition time of 16h [59]. These workers compared two procedures for concentrating polonium–210 from seawater:

1   Coprecipitation upon partial precipitation of the natural calcium and magnesium with sodium hydroxide.
2   Electrodeposition of polonium directly from acidified seawater on to carbon rods.

Polonium thus concentrated was autoplated on to silver counting discs held in spinning Teflon holders. A comparison of results obtained by the two methods is shown in Table 29.6. Recoveries of polonium–208 tracer in the precipitation method were 77 ± 7% ($n + 8$) compared with 40 ± 2% ($n = 2$) for the electrodeposition method with 16h plating time, 64 ± 1% ($n = 2$) in 24h, and 85 ± 4% ($n = 2$) in 48h. Even though the electrode-position method requires less attention, it requires long plating times for high recoveries. Thus the recovery of polonium–210 by direct plating appears to be rate limited by diffusion of polonium to the cathodes since the

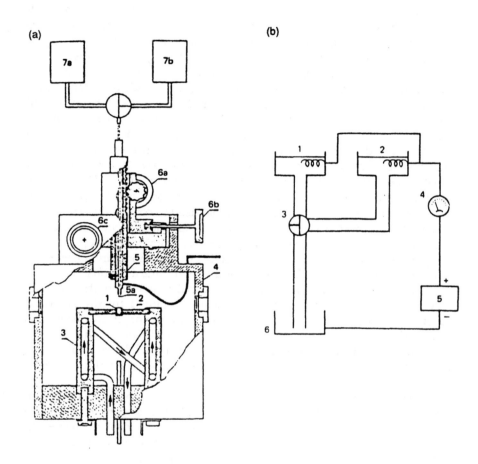

Fig. 29.5 Overall view of the apparatus

(a) (1) Vitreous carbon crucible; (2) graphite rod; (3) water-cooled, steel column electrical leads; (4) plexiglas cover; (5) feeder; (5a) feeder tip; (6a–c) slide knobs; (7a,b) washing and sample solution reservoirs. The glassy carbon crucible (1) was 8mm high, 5mm od, 3mm id, 6mm deep (Le Carbon Lorraine, Paris), graphite rods (2) keep the crucible firmly in position. Water-cooled stainless steel columns (3) press the graphite rods against the crucible by two screws hidden inside and act also as electrical leads. The plexiglas box (4) allowed the use of the controlled inert atmosphere necessary to avoid drastic reduction of the absorption signal caused by oxygen. The solution feeder (5) can be moved up and down by means of a knob (6a) into a metal block attached to the upper part of the plexiglas box. A three-way stopcock at the cylinder top connects, by Teflon tubing (11.5mm od, 0.8mm id) the feeder tip (5a) to the washing and sample solution reservoirs. Other knobs (6b and c) enable the feeder to be moved in the horizontal plane. The three knobs permit a micrometric spatial adjustment of the feeder tip. (b) Electrolysis circuit layout. (1,2) sample and washing solution compartment; (3) three-way stopcock; (4) ammeter; (5) 500V DC variable power supply; (6) crucible

Source: Reproduced with permission from Torsi, G. [56] Publishers of *Analytica Chimica (Rome)*

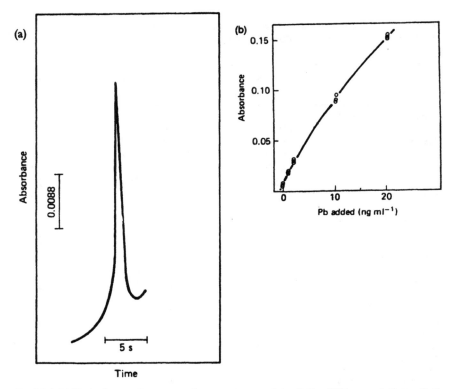

Fig. 29.6 (a) Typical recorder trace of seawater containing 2.8ng Pb^{2+} per ml after a 2min electrolysis time. (b) Calibration curve for lead in seawater, pH 1.9
Source: Reproduced with permission from Torsi, G. [56] Publishers of *Analytica Chimica (Rome)*

applied potential difference is far in excess of that required to reduce Po(IV) to polonium.

Breyer and Gilbert [12] used a differential pulse cathodic stripping voltammetry in an attempt to improve the sensitivity of the voltammetric determination of selenium after extraction as the 3,3'–diaminobenzidine piazselenol. After formation of the piazselenol, the selenium was deposited on a mercury electrode at –0.45V. The limit of detection of selenium by this method was 0.01µg L $^{-1}$. Interferences could be avoided by extraction of the piazselenol into toluene followed by back-extraction into 0.5M hydrochloric acid.

Donat and Bruland [60] preconcentrated cyclohexane–1,2–dione dioxime (nioxime) complexes of cobalt(II) and nickel(II) from 10ml seawater samples on to a hanging mercury drop electrode by controlled adsorption. Cobalt(II) and nickel(II) reduction currents were measured by

Table 29.6 Comparison of sodium hydroxide precipitation and carbon rod plating methods for concentrating ^{210}Po from aliquots of acidified seawater removed from 50L parent samples

Method	Date sampled	^{210}Po activity (pCi L^{-1})[a]	Recovery (%)
Sample 1 (0.5M Hcl)[b]			
NaOH	12/5/75	0.115 ± 0.009	71
NaOH	12/5/75	0.113 ± 0.009	66
Carbon rod	12/5/75	0.116 ± 0.007	81[c]
Carbon rod	12/5/75	0.110 ± 0.005	89[c]
NaOH	17/5/75	0.104 ± 0.008	74
NaOH	17/5/75	0.097 ± 0.008	80
Sample 2 (0.5M HCl)[d]			
NaOH	20/5/75	0.031 ± 0.002	86
NaOH	20/5/75	0.025 ± 0.002	74
Carbon rod	20/5/75	0.034 ± 0.002	63[e]
Carbon rod	20/5/75	0.035 ± 65[e]	
Carbon rod	27/5/75	0.034 ± 0.003	38[f]
Carbon rod	27/5/75	0.028 ± 0.002	41[f]
NaOH	27/5/75	0.040 ± 0.003	73
NaOH	27/5/75	0.034 ± 0.002	90

[a] ±1 standard counting error
[b] Collected on 25/4/75 at Scripps Pier
[c] Electroplating time, 48h
[d] Collected on 20/5/75 at Scripps Pier
[e] Electroplating time, 24h
[f] Electroplating time, 16h

Source: Reproduced with permission from Cowen, J.P. et al. [58] American Chemical Society

differential pulse cathodic stripping voltammetry. Detection limits for cobalt and nickel were 6pM and 0.45mM, respectively.

29.2.2 Freeze drying

Gordon and Larson [61] used photon activation analysis to determine strontium–87 in seawater. Samples (2ml, acidified to pH1.67 or 2.54 for storage) were filtered and freeze dried. The residues, together with strontium standards, were wrapped in polyethylene and aluminium foil and irradiated in a 30MeV bremsstrahlung flux of γ-radiation. After irradiation, the samples were dissolved in 50ml of acidified water and strontium–87m was separated by precipitation as strontium carbonate for counting (γ-ray spectrometer, Ge(Li) detector, and multi-channel pulse–height analyser). The standard deviation at the 7ppm strontium level was ±0.47. Freeze drying followed by neutron activation analysis enabled Jørstad and Selbu [27] to determine to 0.7µg L^{-1} selenium in seawater.

29.2.3 Flotation methods

Coflotation

Hiraide *et al.* [62] used continuous flow coprecipitation–flotation for the radiochemical separation of cobalt–60 from seawater. The cobalt–60 activity was measured by liquid scintillation counting with greater than 90% yield and a detection limit of 5fCi L^{-1} seawater.

Coflotation with octadecylamine and ferric hydroxide as collectors has been used to separate copper, cadmium and cobalt from seawater [80]. The method was based on the coflotation or adsorbing colloid flotation technique. The substrates were dissolved in an acidified mixture of ethanol, water and methyl isobutyl ketone to increase the sensitivity of the determination of these elements by flame atomic absorption spectrophotometry. The results were compared with those of the usual ammonium pyrrolidinedithiocarbamate–methyl isobutyl ketone extraction method. While the mean recoveries were lower, they were nevertheless considered satisfactory, between 79 and 98%.

Froth flotation

Tseng and Zeitlin [63] studied the applicability of an intrinsically rapid technique namely adsorption colloid flotation. This separation procedure uses a surfactant-collector–inert gas system, in which a charged surface inactive species is adsorbed on a hydrophobic colloid collector of opposite charge; the colloid with the adsorbed species is floated to the surface with a suitable surfactant and inert gas, and the foam layer is removed manually for analysis by a methylene blue spectrometric procedure. The advantages of the method include a rapid separation, simple equipment and excellent recoveries. These workers used the flotation unit that was devised by Kim and Zeitlin [64]. The recovery of the selenium was found to be 100 ± 10% (at the 95% confidence level). Using this method Tseng and Zeitlin [63] found 0.40 ± 0.12µg L^{-1} selenium in seawater. See also section 29.1.9.

Colloid flotation

Adsorbing colloid flotation has been used to separate uranium from seawater [65]. To the filtered seawater (500ml; about 1.5µg U) is added 0.05M ferric chloride (3ml), the pH is adjusted to 6.7 ± 0.1 and the uranium present as $(UO_2(CO_3)_3)^{4-}$ is adsorbed on the colloidal ferric hydroxide which is floated to the surface as a stable froth by the addition of 0.05% ethanolic sodium dodecyl sulphate (2ml) with an air-flow (about 10ml min^{-1}) through the mixture for 5min. The froth is removed and dissolved in 12M hydrochloric acid – 16M nitric acid (4:1) and the uranium is salted out with a solution of calcium nitrate containing EDTA, and determined spectrophotometrically at 555nm by a modification of a Rhodamine B method. The average recovery of uranium is 82%; co-adsorbed WO_4^{2-} and MoO_4^{2-} do not interfere.

Adsorbing colloid flotation has also been used by Williams and Gillam [66]. Voyce and Zeitlin [69] have used adsorption colloid flotation to determine mercury in seawater. The sample (500ml) is treated with concentrated hydrochloric acid, an aqueous solution of cadmium sulphate and a fresh aqueous solution of sodium sulphate are added. The pH is adjusted to pH1.0 and solute poured into a flotation cell with a nitrogen flow of 10ml min^{-1}. Ethanolic octadecyltrimethylammonium chloride is injected and the froth dissolved in aqua regia in a flameless atomic absorption cell. Following the reduction of mercury with stannous chloride the mercury vapour is flushed from the system. To determine organically bound mercury, the sample (500ml) is treated with 0.5M sulphuric acid and aqueous potassium permanganate and set aside for 24h. Aqueous hydroxylammonium chloride is added and the determination completed as above. The amounts of mercury in the samples are calculated by reference to the standard absorptions. Average recoveries of 0.05µg mercury were 88%.

Murthey and Ryan [70] used colloid flotation as a means of preconcentration prior to neutron activation analysis for arsenic, molybdenum and vanadium. Hydrous iron(III) is floated in the presence of sodium dodecyl sulphate with small nitrogen bubbles from 1L of seawater at pH5.7. Recoveries of arsenic, molybdenum and vanadium were better than 95% while that of uranium was about 75%. Adsorption colloid flotation using dodecylamine as surfactant has been used to separate zinc with 95% efficiency from seawater [65].

29.2.4 Gas evolution methods

Amankwah and Fasching [44] have discussed the determination of arsenic(V) and arsenic(III) in seawater by solvent extraction–absorption spectrometry using the hydride generation technique. Bertine and Lee [71] have described hydride generation techniques for determining Sb(V) and Sb(III), Sb–S species and organoantimony species in frozen seawater samples.

Sturgeon et al. [72] have described a hydride generation atomic absorbtion spectrometric method for the determination of antimony in seawater. The method uses formation of stibene using sodium borohydride. Stibine gas was trapped on the surface of a pyrolytic graphite coated tube at 250°C and antimony determined by atomic absorption spectrometry. An absolute detection limit of 0.2ng was obtained and a concentration detection limit of 0.04µg L $^{-1}$ obtained for 5ml sample volumes. Preconcentration in a toluene cold trap following arsine generation was successful for determining monomethyl arsenate and dimethyl arsenite but non-quantitative recoveries of dimethyl arsinite have been reported [73]. Molecular rearrangements occurring during arsine generations are, however, reported to be minimised if sodium borohydride is introduced as a pellet [74]. To simplify the determination of the different species, improved separation of the arsines using gas chromatography may be necessary.

Carbonyl formation

Lee [75] described a method for the preconcentration of nanogram or sub–nanogram amounts of nickel in seawater. Dissolved nickel is reduced by sodium borohydride to its elemental form which combines with carbon monoxide to form nickel carbonyl. The nickel carbonyl is stripped from solution by a helium–carbon monoxide mixed gas stream, collected in a liquid nitrogen trap, and atomised in a quartz tube burner of an atomic absorption spectrophotometer. The sensitivity of the method is 0.05ng of nickel. The precision for 3ng nickel is about 4%. No interference by other elements is encountered in this technique. Between 0.3 and 0.6µg L $^{-1}$ nickel was found by this method, in a vertical profile of water samples taken down to 1200m in the Santa Catalina Basin.

Fission track method

Fission track method has also been used by Hashimoto [67]. In this method, the uranium is first coprecipitated with aluminium phosphate [68]; the precipitate is dissolved in dilute nitric acid, and an aliquot of the solution is transferred to a silica ampoule into which small pieces of muscovite are inserted before sealing. The uranium is then determined by measuring the density of fission tracks formed on the muscovite during irradiation of the ampoule for 15min at 80°C in a reactor. The muscovite is etched with hydrofluoric acid for 1h before the photomicrography; the density is referred to that obtained with standard solution of uranium. There is no interference from thorium, and no chemical separations are required. An average concentration of 3–40 ± 12µg uranium per litre was obtained, in good agreement with the normally accepted value.

Table 29.7 Methods for the preconcentration of cations in potable water

Cations determined	Method	Analytical finish	Detection limit*	Ref.
Fe	Interfacial accumulation of iron Solochrome Violet RS chelate on hanging mercury drop electrode	–	0.04	[3]
Ge	–	Hydride generation AAS	0.004	[76]
Hg	–	Cold vapour AAS	<4	[77]

*μg L⁻¹ unless otherwise stated

Source: Own files

29.3 Potable water

Methods for the preconcentration of cations in potable water are reviewed in Table 29.7.

29.4 Ground water

29.4.1 Freeze drying

Juleli and Parry [31] determined very low levels of uranium–239 in ground waters using a rapid neutron activation method known as the cyclic activation system. The method involved freeze drying 50ml aliquots of filtered and acidified groundwater samples followed by epithermal neutron activation analysis of the residue. Uranium–239 peaks of gamma ray energy at 74.7keV were measured to calculate uranium concentrations applying the total peak area method against that of a uranium standard. Results of this technique were compared against a set of data obtained by conventional instrumental neutron activation analysis using neptunium–239 and a set of data from laser induced fluorometry. The three data sets obtained for ground water samples and for US Geological Survey standard reference samples were in good agreement especially when the total dissolved solids of the samples was low.

29.5 Waste water

29.5.1 Electrolytic methods

Trace amounts of tetravalent thorium were determined in wastewaters by complexation with tetravalent thorium in an acetic acid–sodium hydroxide solution as reported by Chen and Wei [78]. The thorium–tetravalent thorium chelate was preconcentrated by adsorption onto a mercury electrode and determined by cathodic stripping voltammetry. The method had a detection limit of 10^{-8} mol L^{-1} and a relative standard deviation of 2.5%.

29.6 Sewage effluents

29.6.1 Dialysis

Morrison [79] estimated bioavailable metal uptake rates for storm water and treated sewage effluents using dialysis with receiving resins. The metal chelating resin, Chelex–100, in the calcium form was incorporated within a dialysis bag sealed at both ends. The bag was immersed in the sampled solution for 1–4d before releasing the chelating resin to a separating column. Metals were eluted from the resin with 1M nitric acid followed by ultra-pure water and concentrations determined by graphite furnace and flame atomic absorption spectrometry. Results were expressed in terms of a metal uptake rate per unit surface area and time (pg mm^{-2}.h^{-1}). Background resin uptake rates from an unpolluted stretch of river for zinc (0.79), cadmium (0.01), lead (0.02) and copper (0.06) were exceeded at three storm water outfalls (Backebolsimotet, Bergsjon, and Floda) which provided the following range of resin uptake rates for zinc (2.5–67.1), cadmium (0.01–0.14), lead (0.15–2.4) and copper (0.26–41.3). Mean efficiencies for the reduction of bioavailable uptake rates between influent and effluent water of a sewage treatment plant were derived for zinc (42%), lead (49%) and copper (33%).

References

1 Frick, D.A. and Tallman, D.E. *Analytical Chemistry*, **54**, 1217 (1982).
2 Muhlbaier, J., Stevens, C., Graczy, K. and Tisue, T. *Analytical Chemistry*, **54**, 496 (1982).
3 Wang, J. and Mahmoud, J. *Fresenius Zeitschrift für Analytische Chemie*, **327**, 789 (1987).
4 Prabhu, S.V., Baldwin, R.P. and Kryger, L. *Analytical Chemistry*, **59**, 1074 (1987).
5 Trettenbach, J. and Henmann, K.C. *Fresenius Zeitschrift für Analytische Chemie*, **322**, 306 (1985).
6 Rigin, V.I. and Rigina, I.V. *Analytical Chemistry*, **34**, 1211 (1979).
7 Jin, L., Yang, L., Xu, T. and Fang, Y. *Fenxi Huaxue*, **16**, 410 (1988).
8 Xu, B., Xu, T., Shen, M. and Fang, Y. *Fenxi Huaxue*, **14**, 623 (1986).
9 Malinski, T., Fish, J. and Matusiewicz, H. Proceedings Water Technology Conference, 14 1986. (Adv. Water Anal. Treat.), 347–359 (1987).
10 Wang, J. and Zadell, J.M. *Analytica Chimica Acta*, **185**, 229 (1986).
11 Wang, J. and Zadell, J. *Talanta*, **34**, 909 (1987).
12 Breyer, P. and Gilbert, B.P. *Analytica Chimica Acta*, **201**, 33 (1987).
13 Yang, J. and Tang, S. *Xiangtan Daxue Ziran Kexue Xuebao*, **4**, 64 (1986).

14 Zen, J.M. and Chung, M.J. *Analytical Chemistry*, **67**, 3571 (1995).
15 Adelson, S.B. and Brown, K.A. *Analyst (London)*, **112**, 221 (1987).
16 Wang, J., Peng, T. and Varughese, K. *Talanta*, **34**, 561 (1987).
17 Hart, B.T. and Davies, S.H.R. *Australian Journal of Marine and Freshwater Research*, **28**, 105 (1977).
18 Benes, P. and Steinnes, E. *Water Research*, **8**, 947 (1974).
19 Wilson, R.L. and Di Nunzio, J.C. *Analytical Chemistry*, **53**, 692 (1981).
20 Kasthurikrishnan, N. and Koropchak, J.A. *Analytical Chemistry*, **65**, 857 (1993).
21 Stec, R.J., Kortyyahann, S.P. and Taylor, H.E. *Analytical Chemistry*, **58**, 3240 (1986).
22 Cervera, J., Cella, R. and Perez-Bustamanate, J.A. *Analyst (London)*, **107**, 1425 (1982).
23 Bontron, C. and Martin, S. *Analytical Chemistry*, **51**, 140 (1979).
24 Bontron, C. *Analytica Chimica Acta*, **61**, 140 (1972).
25 Department of the Environment Water Council Standing Committee of Analysts. In *Methods for the examination of waters and associated materials. Copper in potable waters by atomic absorption spectrometry.* RP 22 B HMC, HMSO, London (1982).
26 Thompson, M., Ramsey, M.J. and Pahlavanpour, B. *Analyst (London)*, **107**, 1330 (1982).
27 Jørstad, K. and Selbu, B. *Analytical Chemistry*, **52**, 672 (1980).
28 Crocker, I.H. and Merritt, W.F. *Water Research*, **6**, 285 (1972).
29 White, M.N. and Lisk, D.J. *Journal of the Association of Official Analytical Chemists*, **53**, 1055 (1970).
30 Lieser, K.H., Heuss, W.C. and Neitzen, V. *Journal of Radioanalytical Chemistry*, **37**, 717 (1977).
31 Juleli, Y.M. and Parry, S.J. *Journal of Radioanalytical and Nuclear Chemistry*, **102**, 337 (1986).
32 Harsanyi, E., Polos, L. and Pungor, F. *Analytica Chimica Acta*, **67**, 229 (1973).
33 Topping, T. and Pirie, E. *Analytical Abstracts*, **25**, 1307 (1973).
34 Kulathilake, A.I. and Chatt, A. *Analytical Chemistry*, **52**, 828 (1980).
35 Kim, Y.S. and Zeitlin, H. *Analytica Chimica Acta*, **46**, 1 (1969).
36 Head, P. and Burton, J.D. *Journal of Marine Biological Association, UK*, **50**, 439 (1970).
37 Weiss, H.V. and Lai, M.G. *Talanta*, **8**, 72 (1961).
38 Skogerboe, R.K., Hanagan, W.H. and Taylor, H.E. *Analytical Chemistry*, **57**, 2815 (1985).
39 Hidalgo, J.L., Gomez, M.A., Caballero, M., Cela, R. and Perez-Bustamate, A. *Talanta*, **35**, 301 (1988).
40 Shi-Yu, L. and Ping, G. *Talanta*, **8**, 844 (1984).
41 Nakashima, S. and Yagi, M. *Analytica Chimica Acta*, **147**, 213 (1983).
42 Andreae, M.O.P.I. In *Trace Metals in Seawater*, Proceedings of a NATO Advanced Research Institute on Trace Metals in Seawater, 30/3/81–3/4/81, Sicily, Italy, eds. C.S. Wong *et al.* Plenum Press, New York (1981).
43 Braman, R.S. and Forebach, C.C. *Science*, **182**, 1247 (1973).
44 Amankwah, S.A. and Fasching, J.L. *Talanta*, **32**, 111 (1985).
45 Andreae, M.O. and Froelich, P.N. *Analytical Chemistry*, **53**, 287 (1981).
46 Foreback, C.C. In *Some Studies on the Detection and Determination of Mercury, Arsenic and Antimony in Gas Discharges*, Thesis, University of South Florida, Tampa (1973).
47 Tomkins, M.A. In *Environmental Analytical Studies in Antimony, Germanium and Tin*, Thesis, University of South Florida, Tampa (1977).

48 Crecelius, E.A. *Analytical Chemistry*, **50**, 826 (1978).
49 Cutter, G.A. *Analytica Chimica Acta*, **98**, 59 (1978).
50 McDaniel, M., Shendrikar, A.D., Reiszner, K.D. and West, P.W. *Analytical Chemistry*, **48**, 2240 (1976).
51 Knudson, E.J. and Christian, G.D. *Analytical Letters*, **36**, 1039 (1973).
52 Orvini, E., Delfanti, R., Gallorini, M. and Speziali, M. *Analytical Proceedings (London)*, **18**, 237 (1973).
53 Shaikh, A.N. and Tallman, D.E. *Analytica Chimica Acta*, **98**, 251 (1978).
54 Hanson, E.H., Ruzicka, J. and Reitz, B. *Analytica Chimica Acta*, **89**, 241 (1977).
55 Vien, S.H. and Fry, R.C. *Analytical Chemistry*, **60**, 465 (1988).
56 Torsi, G. *Analytica Chimica (Rome)*, **67**, 557 (1977).
57 Torsi, G., Oesimoni, E., Palimisano, F. and Sabbatini, L. *Analytica Chimica Acta*, **124**, 143 (1981).
58 Cowen, J.P., Hodge, V.F. and Folson, T.R. *Analytical Chemistry*, **49**, 494 (1977).
59 Reid, D.F., Key, R.M. and Schink, D.R. *Earth Planet Science Letters*, **43**, 223 (1979).
60 Donat, J.R. and Bruland, K.W. *Analytical Chemistry*, **60**, 240 (1988).
61 Gordon, C.M. and Larson, R.E. *Radiochemical and Radioanalytical Letters*, **5**, 369 (1970).
62 Hiraide, M., Sakurai, K. and Mizuike, A. *Analytical Chemistry*, **56**, 2851 (1984).
63 Tseng, J.H. and Zeitlin, H. *Analytica Chimica Acta*, **101**, 71 (1978).
64 Kim, Y.S. and Zeitlin, H. *Separation Science*, **6**, 505 (1971).
65 Kim, Y.S. and Zeitlin, H. *Analytical Chemistry*, **43**, 1390 (1971).
66 Williams, W.J. and Gillam, A.H. *Analyst (London)*, **103**, 1239 (1979).
67 Hashimoto, T. *Analytica Chimica Acta*, **56**, 347 (1971).
68 Smith, J. and Grimaldi, O. *Bulletin of the US Geological Survey*, **1006**, 125 (1957).
69 Voyce, D. and Zeitlin, H. *Analytica Chimica Acta*, **69**, 27 (1974).
70 Murthey, R.S.S. and Ryan, D.E. *Analytical Chemistry*, **55**, 682 (1983).
71 Bertine, K.K. and Lee, D.S. In *Trace Metals in Seawater*. Proceedings of a NATO Advanced Research Institute on Trace Metals in Seawater. 30/3–3/4/81, Sicily, Italy, (ed. Wang *et al.*), Plenum Press, New York (1981).
72 Sturgeon, R.E., Willie, S.N. and Berman, S.S. *Analytical Chemistry*, **57**, 2311 (1985).
73 Pierce, F.D. and Brown, H.R. *Analytical Chemistry*, **49**, 1417 (1977).
74 Talmi, Y. and Bostick, D.T. *Analytical Chemistry*, **47**, 2145 (1975).
75 Lee, D.S. *Analytical Chemistry*, **54**, 1182 (1982).
76 Tao, G. and Fang, Z. *Journal of Analytical Atomic Spectroscopy*, **8**, 577 (1993).
77 Bertenshaw, M.P. and Wagstaffe, K. *Analyst (London)*, **107**, 664 (1982).
78 Chen, Q. and Wei, R. *Tongi Daxue Xuebao*, **15**, 355 (1987).
79 Morrison, G.M.P. *Environmental Technology Letters*, **8**, 393 (1987).
80 Cabezon, L.M., Caballero, H., Cole, M. and Perez-Bustamante, J.A. *Talanta*, **31**, 597 (1984).

Chapter 30

Preconcentration of multication mixtures

Techniques discussed in chapters 20 to 29 are concerned with the preconcentration of either single cations or, in some cases, with several cations. Enough has been said in these chapters to demonstrate the applications of these various techniques. There is however, much more published work available on the applications of preconcentration techniques which handle several cations as opposed to a single cation. These methods are not discussed in the preceding chapters but are summarised briefly below. Summaries of further methods for preconcentration of multications in non saline waters, sea waters and potable waters are given, respectively, in Tables 30.1 to 30.3.

Table 30.1 Summary of further methods available for preconcentration of multications in non saline waters

Elements	Chelating agent	Solvent	Detection limit*	Comments	Analytical finish	Ref.
		Chapter 2 Chelation–solvent extraction methods				
V, Cr, Fe, Co, Ni, Cu, Zn, Mo, Cd, Pb	Ammonium pyrrolidone dithiocarbamate	2,6-dimethyl-4-heptanone	–	–	AAS	[1]
Cd, Co, Cu, Ni, Pb, Zn	Ammonium pyrrolidone dithiocarbamate	Methyliso-butyl ketone	–	Study of effect of sample matrix and pH	AAS	[2]
Cd, Cu, Pb, Ni, Zn	Sodium diethyl dithiocarbamate	Chloroform	–	Interlaboratory study	AAS	[3]
Cd, Cu, Pb	Pyrrolidine dithiocarbamate	Methyliso-butylketone	–		AAS and ICAPES	[4]
Cd, Zn, Cu Re, Ni	Ammonium Pyrrolidine dithiocarbamate-diethyldithiocarbamate mixture	4-methyl pentane-2-one	0.1	–	AAS	[5]
Cu, Zn, Cd, Pb	Dithizone, quinolinal, acetylacetone	–	0.1–2	–	AAS ASV	[6]
Cd, Ag, Bi, Co, Au Ni, Pb, Tl, Zn	Hexamethylene ammonium hexamethylene dithiocarbamate	Xylene-di-isopropyl ketone	–	Interlaboratory study	AAS GFAAS	[7]

Table 30.1 continued

Elements	Chelating agent	Solvent	Detection limit*	Comments	Analytical finish	Ref.
Co, Ni, Cu, Zn, Cd, Pb	Ammonium pyrrolidine dithiocarbamate	Methyl isobutyl ketone	–	In brackish water	AAS	[8]
Misc.	Ammonium pyrrolidine dithiocarbamate	2,4-dimethyl heptanone	–	–	AAS	[9]
Cu, Ni, Fe, Co, Cd, Zn, Pb	Ammonium pyrrolidine dithiocarbamate	Misc.	–	–	AAS	[10]
Cd, Pb, Co, Ag, Ni	Sodium diethyl dithiocarbamate	Methyl isobutyl ketone	–	–	AAS	[11]
Cd, Fe, Zn, Cu, Mn, Pb	Sodium diethyl dithiocarbamate	Isoamyl alcohol	–	–	AAS	[12]
Au, Ag	Dithizone	Methyl isobutyl ketone	–	–	AAS	[13]
Cu, Fe, Co, Cd, Pb, Zn	Mixed chelates	Hexane	1–3	–	AAS	[14]
Fe, Zn, Cu, Cd, Pb	Benzylamine-pelargonic acid	Decane		Effect of pH studied	AAS	
Cd, Co, Cu, Fe, Ni, Pb	Sodium diethyl dithiocarbamate	Carbon tetrachloride	–	–	AAS	[15]
Au, Ag	Polyorgs XI–N	–	Ag 0.002 Au 0.005	–	AAS	[16]
Cd, Cu, Mn, Pb, Zn	Dimethylglyoxime/ Ni/I-(2-pyridyl–azo) 2–naphthol	–	0.006– 005	–	AAS	[17]

*μg L⁻¹ unless otherwise stated

Table 30.1 continued

Elements	Chelating agent	Solvent	Detection limit*	Comments	Analytical finish	Ref.
Cd, Co, Cr, Fe, Mn, Mo, Ni, Pb, V, Zn	Ammonium pyrrolidine dithiocarbamate hexa methylene ammonium dithiocarbamate thiocarbamate	Xylene	0.02–05	–	ICPAES	[18]
22 elements	Dibenzylammonium dibenzyl thiocarbamate	2-ethyl hexyl acetate	–	–	ICPAES	[19–21]
Sb, As	Ammonium pyrrolidine dithiocarbamate	–	10	–	Neutron activation analysis	[22]
Misc.	Ammonium pyrrolidine dithiocarbamate	–	–	–	X-ray spectrometry	[23]
Heavy metals	Bis(trifluoroethyl)-dithiocarbamate	Carbon tetrachloride	–	–	GLC	[24]
Co, Cr	Di(trifluoroethyl) dithiocarbamate	Toluene	Co 0.05 Co 0.2	–	GLC	[25]
Sb, As	Ammonium pyrrolidine carbodithioate	Chloroform	Sb 1ng L^{-1} As 1ng L^{-1}	Tri and penta valent species determined	Neutron activation analysis	[22]

*μg L^{-1} unless otherwise stated

Table 30.1 continued

Elements	Chelating agent	Solvent	Detection limit*	Comments	Analytical finish	Ref.
Mn, Cu, Pb, Co	Monothenoyltrifluoro-acetone and triphenyl-phosphine oxide	Cyclohexane	—	—	AAS	[26]
Pb, Zn, Cd	8-quinolinol	—	Cd 0.6ng L^{-1} Zn 1.9ng L^{-1} Pb 2.3ng L^{-1} (400 ml sample)	—	AAS	[27]
Cd, Co, Cu, Fe, Ni, Pb	Diethyldithio carbamate	Carbon tetrachloride	Cd 10 pg Co 150 pg Cu 125 pg Fe 100 pg Ni 250 pg Pb 100 pg		AAS	[15]
Cd, Cu, Pb, Mo, Ni	Pyrrolidine dithiocarbamate	—	—	—	ICPAES	[28]
Cd, Cu, Ni, Pb	Ammonium pyrrolidine dithiocarbamate	Di isobutyl ketone	—	—	ICPAES	[29]
Cu, Ni, Pb, Cd	Ammonium pyrrolidine dithiocarbamate	Me Cl₃	Cu 0.3 Pb 0.7 Ni 0.5 Cd 0.02	—	AAS	[30]
Th, La Yt, Sr	Crown ethers Crown ethers	— —	— —	— —	— —	[31] [32]

Table 30.1 continued

Metal	Solid adsorbent	Solvent	Detection limit*	Comments	Analytical finish	Ref.
	Chapter 3 Adsorption on immobilised solid chelating adsorbents					
Fe, Co, Ni, Cu, Zn, Cd, Mo, Cr, V Ga, In	Chelating functional groups on silica gel Quinolin-8-ol on glass beads	–	–	–	ICPAES	[33]
		–	–	–	–	[34]
Lanthanides	8-hydroxyquinoline on silica gel	Supported organophosphorus extractant	–	1 L sample	Isotope dilution ICPMS	[35]
V, Cr, Co, Ni, Cu, Zn, Se, Mo, Ce, Te, Hg, Tl, Pb, Bi	Metal chelates of bis(ethoxyethyl) dithio carbamate adsorbed on phenyl modified silica column	Quarternary solvent mixture for elution on column	–	–	Reverse phase liquid chromatography ultraviolet detection	[36]
Misc.	8-hydroxyquinoline on silica gel	–	–	–	AAS	[37]
Cu, Pb, Ni, Zn, Cd, Co	2,2'dipyridyl-4-amino 3-hydrazino-5-mercapto-1,2,4 triazole hydrazone on silica gel	–	–	–	AAS	[38,39]
Cu, Pb, Cd, Zn	2-mercapto benzthiazole on glass beads	Methyl isobutyl ketone	–	Selective precon-centration of copper and lead possible at pH6.5	AAS	[40]
Ag, As, Co, Cr, Cu, Fe, Hg, Mn, Pb, Zn	Sodium diethyl dithiocarbamate on glass beads	–	–	Inconclusive results	–	[41]

*μg L⁻¹ unless otherwise stated

Table 30.1 continued

Chapter 4 Adsorption on non chelating solid adsorbents

Metal	Solid adsorbent and complexing agent	Solvent	Detection limit*	Comments	Analytical finish	Ref.
18 Misc metals	Cellulose, piperazine dithiocarbamate	–	–	–	Neutron activation analysis	[42]
Transition elements Zn, Cd, Mn	2,2'diaminodiethyl-amine cellulose	–	0.1–0.5 by X-ray fluorescence spectroscopy	Collection yields 85%, good recovery at pH up to 5	X-ray fluorescence spectroscopy and neutron activation analysis	[43,44]
Pb, Cu, Ni, Cd, Zn	Cellulose hyphan (Cellex P)	1 M nitric acid for elution from column	–	pH not critical over range 5–8	–	[45]
Misc.	Natural cellulose	–	–	Study of sorbtive properties	–	[46,47]
Cu, Pb, Zn, Cd, Ni, Co	Polyarlyonitrile with attached thioamide groups	–	–	High selectivity for named metals Cr(II) and Cr(V) have very different sorbtive properties	AAS	[48]
Na, K, Ca, Mg, Cd, Cr, Mn, Ni, Pb, Zn, Co, Fe, Cu	Ion exchange column containing nicotinium molybdophosphate	Selective elution with 4M ammonia	–	Recovery 98–102%	AAS	[49]
Misc.	Cellulose immobilised triethylene tetramine pentaacetic acid	–	–	–	–	[50]

Table 30.1 continued

Metal	Solid adsorbent and complexing agent	Solvent	Detection limit*	Comments	Analytical finish	Ref.
Misc.	Carbamate loaded polyurethane	–	–	–	–	[51,52]
Misc.	Poly(amido–amine) loaded polyurethone foam	–	–	–	–	[53]
Misc.	Tributyl phosphate plasticised dibenzoyl methane loaded polyurethone foam	–	–	–	–	[54]
Misc.	Sulphonated styrene– divinylbenzene resin	–	–	–	–	[55]
Misc.	Silica immobilised algae cells	–	–	–	–	[56]
Misc.	Silica immobilised lichen and seaweed biomass	–	–	–	–	[57]

Chapter 5 Adsorption on active carbon

Metals	Adsorbent	Detection limits	Comments	Analytical finish	Ref.
Mo,W	Active carbon	Mo 0.4 W 1.2 by ICPAES Mo 0.06 W 0.06 by ICPMS	–	ICPAES and and ICPMS	[58]
As, Sb	Active carbon	–	–	Hydride generation AAS	[59]

*µg L^{-1} unless otherwise stated

Table 30.1 continued

Metals	Adsorbent	Detection limit*	Comments	Analytical finish	Ref.
Al, As, K, Mg, Mn, Na, Sr, U, V, Zn	Active carbon	–	–	–	[60]
Au, Ag, Ga, In, Tl, Cd, La, Mo, Ni, Cu	Active carbon	–	–	AAS	[61]
Cd, Zn	Active carbon	–	–	–	[62]

Metal	Resin	Method of desorption from resin	Detection limit*	Comments	Analytical finish	Ref.
Chapter 6 Adsorption of anion and cation exchange resins						
Zn, Cd, Cu, Ni, Mo, Co, Pb	Chelex–100	–	–	Study of interferences by non ionic detergents, detergent additives and soaps, 92% recovery	–	[63,64]
Cd, Co, Cu, Mn, Ni, Pb, Cr(III), Fe(III)	Chelex–100	1% nitric acid	<1	–	AAS	[65,66]
Zn, Cd, Pb, Cu	Chelex–100	–	–	Speciation study	Differential anodic stripping voltammetry	[67]
Zn, Cd, Hg, Pb	Chelex–100	–	–	–	–	[68]
Fe, Cu, Ni, Cd, Co, Zn, Pb, Mn	Chelex–100	–	–	–	–	[69]

Table 30.1 continued

Metal	Resin	Method of desorption from resin	Detection limit*	Comments	Analytical finish	Ref.
Al, Eu, Ti, Cd, Co, Cu, Cr, Fe, Mn, Mo, Ni, Si, Sn, Th, U, V, Zn	Chelex–100	–	–	–	–	[70]
Cu, Cd, Pb, Zn	Chelex–100	–	sub μg L^{-1} to μg L^{-1}	–	Anodic stripping voltammetry	[71]
Cd, Cu, Pb	Strongly basic anion exchange resin	Nitric acid	–	Preliminary treatment of sample with HBr–ascorbic acid	AAS	[72]
Co, Zn, Cd, Pb	Convert to 8-hydroxy-quinoline-5-sulphonate then strongly basic anion exchange resin	For Zn, Cd, Pb, 2M nitric acid For Co 12M hydrochloric acid	0.2–4.0	–	AAS	[73]
Heavy metals	Anion exchange resin with salicylic acid functional groups	–	0.05–0.5	–	Flow injection analysis	[74,75]
Zn, Cu, Ni, Pb, Co, Mn, Fe, Cr	Amberlite I R 120 cation exchange resin	Saturated sodium chloride	–	–	X-ray spectroscopy	[76]
Co, Ni, Cu, Zn, Pb	KU–2 cation exchange resin	Hydrochloric acid	–	–	Spectrography	[77]
Co, Cr, Cu, Fe, Ni, Zn	Acropane cation exchange resin	–	–	–	AAS	[78]
Cu, Fe, Mg	XAD–1 and XAD–2 cation exchange resins	–	–	–	X-ray fluorescence spectroscopy	[79]

*μg L^{-1} unless otherwise stated

Table 30.1 continued

Metal	Resin	Method of desorption from resin	Detection limit*	Comments	Analytical finish	Ref.
Fe, Co, Ni, Cu, Pb, Bi	XAD–2 cation exchange resin with pyrocatechol violet	–	–	–	AAS	[80]
Heavy metals	XAD–cation exchange resin treated with indium	–	–	–	AAS	[81]
Cu, Zn, Pb, Mn, Co, Ni, Cd	KU2 and KU23 cation exchange resins	–	–	–	AAS	[82]
Al, Ca, K, Mg, Mn, Na	Remove humic acids with cation exchange resin	–	–	–	Neutron activation analysis	[83]
Cr, Ni Misc	Anion exchange resin Metals complexed with butane 2,3 dione bis (N–pyridinoacetyl hydrazone), adsorbed on XAD resin	–	sub ppm low ng L^{-1}	–	ICPAES ICPAES	[84] [85]

Metal	Adsorbent	Reagent for dissolution from adsorbent	Detection limit	Comments	Analytical finish	Ref.
				Chapter 8 Adsorption on metal oxides and metals		
Na, Cd, Cr, Se, B	Tantalum or gold foil	–	–	100-fold preconcentration	Proton activation analysis	[86,87]

Table 30.1 continued

Metal	Adsorbent	Reagent for dissolution from adsorbent	Detection limit*	Comments	Analytical finish	Ref.
Fe, Co, Ni, Cu, Zn, Pb, U	TiO_2, ZrO_2, Al_2O_3	—	—	—	—	[88]
Cu, Pb, Cd	Fe oxides MNO_2	—	—	—	—	[89]
U, Th, Pb, Am	Al_2O_3	—	—	—	—	[90]
Bi, Cd, Co, Cr, Cu, Fe, Ge, In, Mn, Te, Tl, V, Zn	MnO_2, TiO_2	—	—	—	—	[91]
Cu, Ni, Co	$Zr(OH)_4$	—	1	94–102% recovery	—	[78]
Mn, Ni, Cu, Zn, Pb	$Fe(OH)_3$	—	0.5–1	—	X-ray fluorescence spectroscopy	[92]
Cu, Pb, Cd	$Fe(OH)_3$	—	—	—	—	[93]
Fe, Zn, Pb	$Fe(OH)_3$	—	—	—	X-ray fluorescence spectroscopy	[94]
Be, Cu, Zn, Pb, Bi, Co, Ni, Cd	$Fe(OH)_3$	—	—	—	Emission spectrography	[95]
Fe, Zr, Sn	$Fe(OH)_3$	—	—	—	X-ray fluorescence spectroscopy	[96]
Ni, Pb, Zn	$Fe(OH)_3$	—	—	—	—	[97]
Pb, Cu, Zn, Cd	$Fe(OH)_3$	—	—	—	Differential pulse polarography	[98]
U, Th	$Fe(OH)_3$	—	0.07	U and Th separated chromatographically	Spectrophotometric	[99]
Ga, In	$Ga(OH)_3$	—	0.5	—	AAS	[100]
Misc.	$Zr(OH)_4$	—	—	—	—	[101]
As, Sb	$Zr(OH)_4$	—	As 0.3µg Sb 6.1µg	—	X-ray fluorescence spectroscopy	[102]

*µg L⁻¹ unless otherwise stated

Table 30.1 continued

Metal	Adsorbent	Reagent for dissolution from adsorbent	Detection limit*	Comments	Analytical finish	Ref.
Cd, Cu, Mn, Pb	Zr(OH)$_4$	–	–	–	AAS	[103]
As, Sb, Bi, Se, Te	La(OH)$_3$	–	0.05	–	–	[104]

Chapter 10 Organic coprecipitation techniques

Metal	Coprecipitant	Reagent for dissolution of coprecipitate	Detection limit	Comments	Analytical finish	Ref.
Misc.	Thioanlide–polyvinyl–pyrrolidine	–	–	–	X-ray fluorescence spectroscopy	[105]
Fe, Co, Ni, Mn, Cu, Zn	8-mercapto quinoline bis(8-quinolyl)disulphide	–	–	–	–	[106]
Co, Cd, Cu, Cr, Mn, U, Zn	1-(2-pyridyl azo)-2-naphthol	–	Co 0.04 Cd 0.8 Cu 0.3 Cr 0.2 Mn 0.006 U 0.006 Zn 0.3	0.8L sample	Neutron activation analysis	[107]
Ag, As, Cd, Co, Fe, Hg, Mo, Mn, Sb, Sc, Se, W, Zn	Thionalide	–	–	Coprecipitation at pH9	–	[108]

Table 30.1 continued

Metal	Technique	Detection limit	Comments	Analytical finish	Ref.
Chapter 11 Miscellaneous coprecipitation techniques					
Ge, Sn, Pb, As, Sb, Se, Te	Cold trapping after reduction with sodium bromide	As, Se, Sb, Sn 0.025ng (30ml sample)	–	Gas chromatography with AAS detector	[109–112]
Cd, Cr, Cu, Mn, Ni,	Electrochemical ion film electrode	Zn 100 Cr 5000	–	ICPAES	[113]
Cr, Co, Cu, Hg, Ni, Zn, Ag, Bi, Cd, In, Pb, Se, Te, Tl	Electrolysis on silanised rod atomically heated graphite rod	Ag 15pg (Pb)	–	X-ray fluorescence spectroscopy	[114–117]
Misc.	Reductive precipitation	–	–	–	[118]
Misc. (wide range of elements up to uranium)	Freeze drying	0.1	–	Spark source mass spectrometry	[119]
Misc.	Dialysis	ppb	–	–	[120]
Cd, Pb	Adsorption on resin	Cd <0.1 Pb 2	–	AAS	[121]
Misc.	Solid phase techniques	–	Review	–	[122]
As, Se	Preconcentration on antimony coated inner surface of electro-thermal analyser	As 1ng Se 0.5ng	–	Hydride generation AAS	[123]

*$\mu g\ L^{-1}$ unless otherwise stated

Source: Own files

Table 30.2 Summary of further methods available for preconcentration of multications in seawaters

Elements	Chelating agent	Solvent	Detection limit*	Comments	Analytical finish	Ref.
As(III), As(V)	–	Misc.	–	–	Hydride generation AAS	[124]
				Chapter 2 Chelation–solvent extraction methods		
Misc. heavy metals	Diethyldithio-carbamates	Methyl isobutyl ketone	See below	Methods based on formation metal dithiocarbamate complexes in water sample then back extraction of complexes with oxygenated solvent, methyl isobutyl ketone or chlorinated solvent	AAS	[125–134]
Misc. heavy metals	Diethyldithio-carbamates	Methyl isobutyl ketone or chlorinated solvents	See below	Methods based on formation of complexes as above then back extraction of solution extract with dilute nitric acid for analysis of aqueous phase	AAS	[125–134]
Co, Fe, In, Zn	Trifluoroacetone	Acetone	–	–	–	[135]
Mn, Fe, Co, Ni, Zn, Pb, Cu	Hexahydro-azepine-l-carbodithioate	Butyl acetate	Ni 0.2, Zn 0.4 Fe 1.5, Pb 2.6 Co 0.6, Cu 0.5 Ni 0.6	–	AAS	[136]

Table 30.2 continued

Elements	Chelating agent	Solvent	Detection limit*	Comments	Analytical finish	Ref.
		Chapter 2 Chelation–solvent extraction methods				
Fe, Pb, Cd, Co, Ni, Cr, Mn, Zn, Cu	Diethyldithio-carbamate	MIBK or xylene	–	–	AAS	[137]
Fe, Cu	Ammonium pyrrolidine dithiocarbamate	MIBK	Cu <1 Fe <1	–	AAS	[138]
Cd, Zn, Pb, Ca, Ni, Cu, Ag	Dithizone	Chloroform	Ag 0.05, Cu 0.06, Cd 0.05, Ni 0.3, Zn 0.6, Co 0.04, Pb 0.04	–	AAS	[139]
Cd, Cu, Pb, Ni, Zn	(a) Ammonium dipyrrolidine dithiocarbamate (b) Ammonium dipyrrolidine dithiocarbamate plus diethyl dithiocarbamate	MIBK	Cu 10, Cd 2, Pb, 4, Ni 16, Zn 30	–	AAS	[140]
Cu, Ni, Cd	Ammonium pyrrolidine dithiocarbamate	MIBK	–	–	Graphite furnace AAS	[141]
Ag, Cd, Cr, Cu, Fe, Ni, Pb, Zn	Ammonium dipyrrolidine dithiocarbamate	MIBK	Ag 0.02, Ni 0.10, Cd 0.03, Pb 0.03, Cr 0.05, Zn 0.03, Cu 0.05, Fe 0.20	–	Graphite furnace AAS	[128]

*μg L^{-1} unless otherwise stated

Table 30.2 continued

Elements	Chelating agent	Solvent	Detection limit*	Comments	Analytical finish	Ref.
Cu, Cd, Zn, Ni	Diethyl dithiocarbamate plus ammonium pyrrolidine dithiocarbamate	Chloroform	Cu 1.0, Zn 2 Cd 0.2, Ni 10	–	Graphite furnace AAS	[134]
Cd, Pb, Ni, Cu, Zn	Ammonium pyrrolidine dithiocarbamate plus diethyl dithiocarbamate	Freon	–	–	Graphite furnace AAS	[142]
Cu	Ammonium pyrrolidine dithiocarbamate	MIBK	Cu 0.5	–	Graphite furnace AAS	[143]
Cd, Cu, Ni, Zn	Dithizone	Chloroform	Cu 0.006 Cd 0.004 Ni 0.032 Zn 0.016	–	Graphite furnace AAS	[144]
Cd, Cu, Fe	Ammonium pyrrolidine dithiocarbamate plus diethyl dithiocarbamate	Freon	–	–	Graphite furnace AAS	[145]

Table 30.2 continued

Elements	Chelating agent	Solvent	Detection limit*	Comments	Analytical finish	Ref.
Cd, Zn, Pb, Cu, Fe, Mn, Co, Cr, Ni	Ammonium pyrrolidine N carbodithioate plus 8-hydroxy-quinoline	MIBK	Fe 0.08, Zn 0.34 Cu 0.10, Pb 0.06 Cd 0.02	—	Graphite furnace AAS	[131]
Cd	Ammonium pyrrolidine dithiocarbamate	Carbon tetrachloride	Cd 0.006	—	Graphite furnace AAS	[146]
Cd, Zn, Pb, Fe, Mn, Cu, Ni, Co, Cr	Dithiocarbamate	MIBK			Graphite furnace AAS	[147]
Cd, Co, Cu, Fe, Mn, Ni, Pb, Zn	Ammonium pyrrolidine dithiocarbamate	Chloroform	Cd <0.0001 Pb <0.016 Cu <0.012 Zn <0.08 Fe <0.02 Mn <0.004 Ni <0.012	—	Graphite furnace AAS	[148]
Cd, Co, Cu, Fe, Mn, Ni, Pb, Zn	Ammonium pyrrolidine dithiocarbamate	Chloroform	Cd 0.02, Pb 0.04 Cu 0.24, Zn 1.0 Fe 0.24, Mn 0.02 Ni 0.08	—	Graphite furnace AAS	[149]
Mn, Cd	Ammonium pyrrolidine dithio-carbamate and diethylammonium diethyldithiocarbamate	Freon	Mu 0.01 Cd 0.027	—	Graphite furnace AAS	[150]

*µg L^{-1} unless otherwise stated

Table 30.2 continued

Elements	Chelating agent	Solvent	Detection limit*	Comments	Analytical finish	Ref.
Cd, Cu, Fe, Pb, Ni, Zn	Ammonium pyrrolidine dithiocarbamate and diethyl-ammonium diethyldithiocarbamate	Freon	–	–	Graphite furnace AAS	[151]
Cd, Co, Cu, Sn, As, Mo	Ammonium pyrrolidine dithiocarbamate	–	–	–	AAS	[152]
Cu, Bi, Pb, Cd, Zn	Diethyl and dibutyl dithiophosphate	Carbon tetrachloride	Cu 0.6, Bi 0.5 Cd 0.8, Zn 0.8 Pb 0.05	–	AAS	[153]
La, Y	–	Solvents, then back-extraction into aqueous phase	0.6–3pg mL $^{-1}$	1L sample	ICPAES	[154]
Al, V, Cu, Mo, Zn, U	8-hydroxy quinolates	Solvent extraction	–	–	Neutron activation analysis then γ ray spectrometry	[155]

Table 30.2 continued

Elements	Chelating agent	Solvent	Detection limit*	Comments	Analytical finish	Ref.
Se(V)	3,3 diamino-benzene piazselenol	Piazselenol extracted into toluene then back-extraction into 0.5M hydrochloric acid	0.01	After formation of piazselenol, selenium deposited on mercury electrode	–	[156]

Metal	Solid adsorbent	Solvent	Detection limit*	Comments	Analytical finish	Ref.
			Chapter 3 Adsorption onto immobilised chelators			
Se, Sb	Ammonium pyrrolidine dithiocarbamate adsorbed onto C$_{18}$ bonded silica gel	Methanol, then evaporation to dryness and dissolution in 1% nitric acid	Se(IV) 0.007 Sb(III) 0.05 Sb(V) 0.05	300ml sample conc. in factor of 200 achieved	AAS	[157]
Cd, Cu	C$_{18}$ column loaded with sodium diethyldithio-carbamate	–	0.004–0.024 (0.5ml sample)	–	AAS	[158]

*µg L^{-1} unless otherwise stated

584

Table 30.2 continued

Metal	Solid adsorbent	Solvent	Detection limit*	Comments	Analytical finish	Ref.
Cu, Ni, Cd	C_{12} column coated with ammonium pyrrolidine-1-yldithioformate cetyl trimethyl ammonium bromide ion pair; metal complexes separated on C_{18} column prior to HPLC	–	0.5 (10ml sample)	–	HPLC	[159]
Transition metals eg Cu in presence of wide range of other metals	Silica immobilised 8-quinolol	–	–	Alkaline earths do not interfere	Flow injection AAS	[37,161–163]
Cd, Cu, Pb, Zn	Mercapto-(3-mercaptopropyl) trimethyl-oxysilane modified silica gel	–	–	Eluting agents investigated, 95% recovery of metals reported	–	[164]
Cd, Pb, Zn, Cu, Fe, Mn, Ni, Co	Silica immobilised 8-hydroxyquinoline	–	Co0.0002 to Fe 0.04 (50–100ml sample)	–	AAS	[165]
Cu, Ni, Cd	8-quinolinol immobilised silica	–	0.016–0.07	–	ICPAES	[166]
Cd, Cu, Zn, Pb, Fe, Mn, Ni, Co	Silica immobilised 8-hydroxyquinoline	–	–	50–90-fold preconcentration	AAS	[163]

Table 30.2 continued

Metal	Solid adsorbent	Solvent	Detection limit*	Comments	Analytical finish	Ref.
Ag, Al, Bi, Cd, Cu, Fe, Ga, Mn, Ni, Pb, Tl	7-dodecenyl-8-quinolinol impregnated XAD–4 macroporous resin	Solvent extraction of resin with DDO (7-(1-vinyl-3,3,6,6 tetra-methylhexyl 8-quinolinol)	–	–	–	[167]
Misc.	Sodium bis (2-hydroxyethyl) dithiocarbamate	–	–	–	–	[168]
Transition metals	Chelating resin	–	–	–	Neutron activation analysis	[70]
Hg, Pb	Tubular membrane containing resin with dithiocarbamate groups	–	–	–	–	[169]
Ga, Tl, In	8-hydroxy-quinoline immobilised resin	–	0.0001–0.004	–	ICPAES	[180]
Cd, Cu, Hg, Pb	Dithiocarbamate cellulose derivative	–	–	–	Neutron activation analysis	[171]

*μg L⁻¹ unless otherwise stated

Table 30.2 continued

Chapter 4 Adsorption on non chelating solid adsorbents

Metals	Solid adsorbent and complexing agent	Solvent	Detection limit	Comments	Analytical finish	Ref.
Lanthanides	Complexation with bis (2-ethylhexyl) hydrogen phosphate and adsorption on to C$_{18}$ resin cartridge			1–5L samples; 200–1000Cd enrichments	ICPMS	[172]
Misc.	Adsorption on modified carbon	–	–	–	–	[173]
Misc.	Adsorption on chitosan	–	–	–	–	[174,175]
Misc.	Adsorption on p-dimethylamino-benzyldene rhodamine	–	–	–	–	[176]
Mo(V)	Adsorption on cellulose phosphate column	–	–	–	ICPAES	[177]
Cd, Cu, Mn, Ni, Pb, Zn	Adsorption on chelamine resin	–	–	90% recovery	–	[178]
Cd, Co, Cu, Fe, Mn, Ni, Pb, Zn	Adsorption on maleic acid–ammonium hydroxide	–	–	Adsorption is critically dependent on pH	–	[179]

Table 30.2 continued

Metals	Solid adsorbent and complexing agent	Solvent	Detection limit	Comments	Analytical finish	Ref.
Cd, Cu, Co, Ni, Pb, Zn	Sodium bis (2-hydroxy-ethyl) dithiocarbamate complexes formed then adsorbed on XAD–4 resin column	–	–	Automated analysis system	–	[168]
Fe(III), Cu(II), Cd(II), Pb(II), Zn(II)	Adsorption on poly-acrylamid-oxine column	Resin equilibrated with 1:1 hydrochloric acid to extract metals	–	–	AAS	[180]

Metal	Anion exchange resin	Method of desorption from resin	Detection limit	Comments	Analytical finish	Ref.
Chapter 6 Adsorption on anion exchange resins						
U	Strongly basic ion exchange resin in presence of azide ions	1M hydrochloric acid	<3	–	Spectro-photometry	[181]
V, Co, Cu, Zn, Cd	Adsorption on anion exchange resin in thiocyanate medium	2N nitric acid	–	–	AAS	[182]

*μg L^{-1} unless otherwise stated

Table 30.2 continued

Metal	Anion exchange resin	Method of desorption from resin	Detection limit	Comments	Analytical finish	Ref.
Cd, Pd, In, Au, Pu, Tc	Single lead of resin adsorption	–	–	–	Mass spectrometry	[183]
Pt, Ir	Single lead resin adsorption	–	–	–	AAS	[184]

Metal	Cation exchange resin	Method of desorption from resin	Detection limit	Comments	Analytical finish	Ref.
			Chapter 7 Adsorption on cation exchange resins			
Zn, Cd, Pb, Cu	Cation exchange resin in H+ form	–	–	Study of effect of pH	–	[185–188]
Cd, Mn, Cr, Cu, Ni, Fe, Co, Pb	Amberlite XAD–7 resin	1% nitric acid	Preconcentration factors of 40 achieved	Prior removal of humic materials at pH1–2 is necessary prior to preconcentration on Amberlite XAD–7 at pH8	AAS	[66,189]
Heavy metals Cr. Cu. Mn	Chelex–100	–	–	Optimum pH6.5	–	[190]
	Chelex–100 and Lewatit TP207	–	–	–	AAS	[191]
Cd, Cu, Co, Mn, Ni, Pb, Zn	Chelex–100	–	–	–	–	[192]

Table 30.2 continued

Metal	Cation exchange resin	Method of desorption from resin	Detection limit*	Comments	Analytical finish	Ref.
Cd, Pb, Cu, Zn	Chelex–100	2M nitric acid	–	Preconcentration factor of 10 achieved	AAS	[193]
Misc.	Chelex–100	–	–	–	FIA	[74,194–196]
Cd, Co, Cu, Fe, Mn, Ni, Pb, Zn	Chelex–100	2.5M nitric acid	low µg L^{-1} range	–	AAS	[197]
Cu, Cd, Zn, Ni	Chelex–100	2M nitric acid	Cu 0.006 Cd 0.006 Zn 0.015 Ni 0.015	–	AAS	[134]
Cd, Zn, Pb, Cu, Fe, Mn, Co, Cr, Ni	Chelex–100	2.5M nitric acid	–	–	AAS	[131]
Cd, Pb, Ni, Cu, Zn	Chelex–100	2.5M nitric acid	Cd 0.01 Pb 0.16–0.28 Ni 0.24–0.68 Cu 0.6 Zn 1.8	–	AAS	[142]
Co, Cr, Cu, Fe, Mn, Ni, Si, Th, U, V, Zn	Chelex–100	–	–	–	Neutron activation analysis	[198,199]
Mn, Fe, Co, Ni, Cu, Cr	Chelex–100	–	–	Comparison of chelation–solvent extraction co-precipitation with Fe(OH)$_3$ or Mg(OH)$_2$ methods by neutron activation analysis	AAS	[200]

*µg L^{-1} unless otherwise stated

Table 30.2 continued

Metal	Cation exchange resin	Method of desorption from resin	Detection limit*	Comments	Analytical finish	Ref.
Mn, Cd, Cu, Cr, Fe, Co, Ni, Zn, Pb, Th, V	Chelex–100	–	Cr 0.14, Th 0.004 Mn 0.16, V 0.06 Fe 1.2, Co 0.006 Cu 0.08, Zn 0.02	Study of effect of flow rate and pretreatment	AAS and neutron activation analysis	[201]
			Chapter 9 Inorganic coprecipitation methods			
Misc.	$Fe(OH)_3$	Nitric acid	200-fold pre-concentration achieved	–	AAS	[202]
Al, Pb, V	$Fe(OH)_3$	–	–	–	–	[203]
Hg, Pb, Cd	CuS	–	–	–	AAS	[204]
Se	$Fe(OH)_3$	Hydrochloric acid	0.005 200ml sample	–	Gas chromatography	[205]
Al, As, Cd, Co, Cr, Cu, Fe, La, Mn, Mo, Ni, Pb, Sb, Ti, V, Y, Zn	$Zr(OH)_4$	–	–	–	ICPAES	[202]
Cu, Cd, Co	Coflotation with octadecylamine and $Fe(OH)_3$	–	–	Substrate dissolved in acidified mixture of water, methanol and methyl isobutyl ketone	AAS	[206]
Al, Tl, Cr, Mn, Fe, Co, Ni, Pb, Cu, Zn, Y	$Ga(OH)_3$	–	Pb 0.5, Cu 0.01 Zn 0.001	–	ICPAES	[207]
Al, Cu, Pb	Palladium salts	–	–	–	ASS	[208]

Table 30.2 continued

Chapter 10 Organic coprecipitation methods

Metal	Coprecipitant	Reagent for coprecipitation of coprecipitate	Detection limit	Comments	Analytical finish	Ref.
Hg, Au, Cn	Lead diethyldithiocarbamate	–	–	–	Neutron activation analysis	[209]
Cd, Co, Cu, Hg, Mn, Th, U, V, Zn	(I-(2-thiazolylazo)-2-naphthol-ammonium pyrrolidine dithiocarbamate N-nitrosophenylhydroxyl-amine	–	–	–	Neutron activation analysis	[210]
Heavy metals	Ammonium pyrrolidine dithiocarbamate	–	–	–	X-ray fluorescence spectroscopy	[211]
Al, Co, Cu, Fe, Pb, Mn, Ni, V, Zn	Poly-5-vinyl-8-hydroxy-quinoline	–	–	Alkaline earths coprecipitated	–	[212]
Cu, Ni, Cd	Cobalt pyrrolidine dithiocarbamate	Organic solvent	ng L^{-1}	–	AAS	[141]

*μg L^{-1} unless otherwise stated

Source: Own files

Table 30.3 Summary of further methods available for preconcentration of multications in potable waters

Metal	Chelating agent	Solvent	Detection limit*	Comments	Analytical finish	Ref.
Chapter 2 Chelation–solvent extraction techniques						
Ag, Cd, Co, Cr, Cu, Fe, Mn, Ni, Pb	Ammonium pyrrolidine dithiocarbamate	Methyl isobutyl ketone	–	Effect of analytical parameters studied	AAS	[213]

Metal	Solid adsorbent	Solvent	Detection limit*	Comments	Analytical finish	Ref.
Chapter 3 Adsorption on immobilised solid chelating adsorbents						
Misc. incl. Cu	Silica immobilised 8-quinolinol	–	–	Effect of analytical parameters studied	Flow injection	[37]
Cu, Cd, Fe, Mn, Ni, Zn	8-quinolinol complexes on poly(chlorotrifluoro-ethylene) resin	–	–	–	–	[214]
Ag, Cd, Cu, Zn	Bismuththiol(II) complexes on poly (chlorotrifluoro) ethylene resin	–	–	–	–	[214]
Cd, Cu, Zn	8-quinolinol-5-sulphonic acid complexes on poly(chlorotrifluoro-ethylene) resin	–	–	–	–	[214]

Table 30.3 continued

Metal	Cation exchange resin	Method of desorption from resin	Detection limit*	Comments	Analytical finish	Ref.
	Chapter 7 Adsorption on cation exchange resins					
Cr, Mn, Fe, Co, Ni, Cu, Zn, Cd, Pb	Amberlite IR 120 (H form)	5M hydrochloric acid	Cr 3, Mn 0.5 Co 1, Cu 0.5, Zn 2 Cd 0.1 Pb 6	pH adjusted to 2.0 to 2.5, ascorbic acid added to eliminate iron interference	AAS	[215]
Bi, Cu, In, Cu, Ni	Amberlite XAD–2 loaded with pyro-catechol violet	Aqueous acid	–	–	AAS	[80]
Cd, Cu, Pb, Zn	Chelex–100	–	–	Low recovery of Pb and Cu due to humic acid interference	AAS	[216]

Metal	Coprecipitant	Reagent for dissolution of coprecipitate	Detection	Comments	Analytical finish	Ref.
	Chapter 9 Inorganic coprecipitation methods					
Cd, Cu, Pb, M	Zr(OH)$_4$	–	–	–	–	[217]

*µg L^{-1} unless otherwise stated

Source: Own files

References

1 Bone, K.M. and Hibbert, W.D. *Analytica Chimica Acta*, **107**, 219 (1979).
2 Tessir, A., Campbell, P.G.C. and Bisson, M. *International Journal of Environmental Analytical Chemistry*, **7**, 41 (1979).
3 Analytical Quality Control (Harmonised Monitoring) Committee Water Research Centre, Harlow, Bucks, UK. *Analyst (London)*, **110**, 109 (1985).
4 Rubio, R., Huguet, J. and Ranret, G. *Water Research*, **18**, 423 (1984).
5 Webster, T.B. *Water Pollution Control*, **79**, 511 (1980).
6 Ihnat, M., Gordon, A.D., Gaynor, L.D. *et al. International Journal of Environmental Analytical Chemistry*, **8**, 259 (1980).
7 Dorenemann, A. and Kleist, H. *Analyst (London)*, **104**, 1030 (1979).
8 British Standards Institution UK BS 6068 (Section 2.29). Determination of cobalt, nickel, copper, zinc, cadmium in flame atomic absorption spectrometric methods (1987).
9 Pakalns, P. *Water Research*, **15**, 7 (1981).
10 Sourova, J. and Capkova, A. *Vodni Hospodarstvi, Series B*, **30**, 133 (1980).
11 Tweeten, T.N. *Analytical Chemistry*, **48**, 64 (1976).
12 Chormann, F.H., Spencer, M.J., Lyons, W.B. and Mayewski, P.A. *Chemical Geology*, **53**, 25 (1985).
13 Savitsky, V.N., Peleshenko, V.I. and Osadchiy, C. *Hydrobiology Journal*, **1**, 60 (1986).
14 Savitsky, V.N., Peleshenko, V.I. and Osadchii, V.I. *Journal of Analytical Chemistry, USSR*, **42**, 540 (1987).
15 Chakraborti, D., Adams, F., Van Moal, W. and Irgolic, K.J. *Analytica Chimica Acta*, **196**, 23 (1987).
16 Shvoeva, O.P., Kuchava, G.P., Kubrakova, I.V. *et al. Zh. Anal. Khim*, **41**, 2186 (1986).
17 Atsuya, J. *Fresenius Zeitschrift für Analytische Chemie*, **329**, 750 (1988).
18 Tao, H., Miyazaki, A., Bansho, K. and Umezaki, Y. *Analytica Chimica Acta*, **156**, 159 (1984).
19 Moore, R.V. *Analytical Chemistry*, **54**, 895 (1982).
20 Sugi'yama, M., Fujino, O., Kihara, S. and Matsui, A. *Analytica Chimica Acta*, **181**, 159 (1986).
21 Smith, C.L., Matooka, J.M. and Willson, W.R. *Analytical Letters (London)*, **17**, 1715 (1984).
22 Mok, W.H. and Wai, C.M. *Analytical Chemistry*, **59**, 233 (1987).
23 Tisue, T., Suls, C. and Keel, R.T. *Analytical Chemistry*, **57**, 82 (1985).
24 Rigi, V.I. and Yurtaev, P.V. *Soviet Journal of Water Chemistry and Technology*, **8**, 77 (1986).
25 Schaller, H. and Neeb, R. *Fresenius Zeitschrift für Analytische Chemie*, **327**, 170 (1987).
26 Ueda, K., Kitahara, S., Kubo, K. and Yamamoto, Y. *Bunseki Kagaku*, **36**, 728 (1987).
27 Akatsuka, K., Nobuyama, N. and Atsuya, K. *Analytical Science*, **4**, 281 (1988).
28 Shan, X., Tie, J. and Xie, G. *Journal of Analytical Atomic Spectroscopy*, **3**, 259 (1988).
29 Wada, K., Matsuchita, T., Hizume, S. and Kojima, K. *Bunseki Kagaku*, **37**, 405 (1988).
30 Apte, S.C. and Gunn, A.M. *Analytica Chimica Acta*, **193**, 147 (1987).
31 Wood, D.J., Eishani, S., Du, H.S., Natole, N.R. and Wal, C.M. *Analytical Chemistry*, **65**, 1350 (1993).
32 Du, H.S., Wood, D.J., Elshani, S. and Wal, C.M. *Talanta*, **40**, 173 (1993).

33 Hirayama, K. and Unchara, N. *Nihon Dalgaku Kogakubu Kiyo Bunrul K*, **28**, 149 (1987).
34 Mohammed, B., Ure, A.M. and Littlejohn, D. *Journal of Analytical Atomic Spectroscopy*, **8**, 325 (1993).
35 Esser, B.K., Volpe, A., Kenneally, J.M. and Smith, D.K. *Analytical Chemistry*, **66**, 1736 (1994).
36 Munder, A. and Ballschmidter, K. *Fresenius Zeitschrift für Analytische Chemie*, **323**, 869 (1986).
37 Marshall, M.A. and Mottola, A. *Analytical Chemistry*, **57**, 129 (1985).
38 Samara, C. and Kouimtzis, T.A. *Chemosphere*, **16**, 405 (1987).
39 Samara, C. and Kouimtzis, T.A. *Analytica Chimica Acta*, **174**, 305 (1985).
40 Terada, K., Matsumoto, K. and Inaba, T. *Analytica Chimica Acta*, **170**, 225 (1985).
41 Leydon, D.E., Wegscheider, W. and Bodnar, W. *International Journal of Environmental Analytical Chemistry*, **7**, 85 (1979).
42 Imai, S., Muroi, M., Hamaguchi, A. and Kuyama, H. *Analytical Chemistry*, **55**, 1215 (1983).
43 Smits, J. and Van Grieken, R. *International Journal of Environmental Analytical Chemistry*, **9**, 81 (1981).
44 Reggers, G. and Van Grieken, R. *Fresenius Zeitschrift für Analytische Chemie*, **317**, 520 (1984).
45 Brajster, K. and Slonawka, K. *Analytica Chimica Acta*, **185**, 271 (1986).
46 Burba, P. and Willman, P.G. *Talanta*, **30**, 381 (1983).
47 Dingman, J., Siggia, S., Barton, C. and Hoscock, K.B. *Analytical Chemistry*, **44**, 1351 (1972).
48 Samchuk, A.I., Kazakevich, Y.E., Romonov, N.N. *et al*. *Soviet Journal of Water Chemistry and Technology*, **10**, 63 (1988).
49 Nyangababo, J.T. and Hamya, S.W. *Bulletin of Environmental Contamination and Toxicology*, **36**, 924 (1986).
50 Burba, P., Rocha, J.A. and Schulte, A. *Fresenius Journal of Analytical Chemistry*, **346**, 414 (1993).
51 Beasley, P.I., Rao, R.R. and Chatt, A. *Journal of Radioanalytical and Nuclear Chemistry*, **179**, 267 (1994).
52 Hsieh, T. and Liu, L.K. *Analytica Chimica Acta*, **282**, 221 (1993).
53 Pesavento, M., Soldi, T., Riolo, C., Profumo, M. and Barbucci, R. *Environmental Protection Engineering*, **16**, 49 (1991).
54 Aziz, M., Behair, G. and Shakir, K. *Journal of Radioanalytical and Nuclear Chemistry*, **172**, 319 (1993).
55 Wada, H., Matsuchita, M., Yasui, T. *et al*. *Journal of Chromatography*, **657**, 87 (1993).
56 Mahan, C.A. and Holcombe, J.A. *Spectrochimica Acta, Part B*, **47B**, 1483 (1992).
57 Ramelow, G.J., Liu, L, Himel, C. *et al*. *International Journal of Environmental Analytical Chemistry*, **53**, 219 (1993).
58 Hall, G.E.M., Jefferson, C.W. and Michel, F.A. *Journal of Geochemical Exploration*, **30**, 63 (1988).
59 Maggi, L., Meloni, S., Queirazza, G. and Genova, N. *Journal of Trace Micropr. Techn.*, **1**, 369 (1983).
60 Lamphun, N.A., Moebiuss, O.A. and Keller, C. *Science of the Total Environment*, **70**, 415 (1988).
61 Samchuk, A.I. *Soviet Journal of Water Chemistry and Technology*, **9**, 57 (1987).
62 Bhattaacharyya, D. and Cheng, Y.R. *Environmental Progress*, **6**, 110 (1987).
63 Pakalns, P., Batley, G.E. and Cameron, A.J. *Analytica Chimica Acta*, **99**, 233 (1978).
64 Pakalns, P., Batley, G.E. and Cameron, A.J. *Analytica Chimica Acta*, **99**, 333 (1978).

65 Corsini, A., Chiang, S. and Di Frucia, R. *Analytical Chemistry*, **54**, 1433 (1982).
66 Wan, C., Chiang, S. and Corsini, A. *Analytical Chemistry*, **57**, 719 (1985).
67 Duinker, J.C. and Kramer, C.J.M. *Marine Chemistry*, **5**, 207 (1977).
68 Clanet, F., Delangele, R. and Popoff, G. *Water Research*, **15**, 591 (1981).
69 Everaerts, F.M., Verbeggen, M., Reijenga, J.C. *et al. Journal of Chromatography*, **320**, 263 (1985).
70 Greenberg, R.R. and Kingston, H.M. *Analytical Chemistry*, **55**, 1160 (1983).
71 Figura, P. and McDuffie, B. *Analytical Chemistry*, **52**, 1433 (1980).
72 Korkische, J. and Sario, A. *Analytica Chimica Acta*, **76**, 393 (1975).
73 Berge, D.G. and Going, J.E. *Analytica Chimica Acta*, **123**, 19 (1981).
74 Fang, Z., Xu, S. and Zhang, S. *Analytica Chimica Acta*, **164**, 41 (1984).
75 Fang, Z., Ruzicka, J. and Hansen, E.H. *Analytica Chimica Acta*, **164**, 23 (1984).
76 Zhang, H.F., Holzbecher, J. and Ryan, D.E. *Analytica Chimica Acta*, **149**, 385 (1983).
77 Nakashima, A. *Analytical Chemistry*, **51**, 654 (1979).
78 Nakashima, S. and Yagi, M. *Analytical Letters (London)*, **17**, 1693 (1984).
79 Mackey, D.J. *Journal of Chromatography*, **236**, 81 (1982).
80 Brajter, K., Olbrych-Slezynska, E. and Staskiewicz, M. *Talanta*, **35**, 65 (1988).
81 Hiraide, M., Arima, Y. and Mizuike, A. *Analytica Chimica Acta*, **200**, 171 (1987).
82 Pilipenko, A.T., Safronova, V.G. and Zakrevskaya, L.V. *Soviet Journal of Water Chemistry and Technology*, **9**, 74 (1987).
83 Duffy, S.J., Hay, G.W., Micklethwaite, K. and Vanloon, G.W. *Science of the Total Environment*, **76**, 203 (1988).
84 Petrucii, F., Alimonti, A., Laszlity, A., Horvath, Z. and Caroli, S. *Canadian Journal of Applied Spectrosocpy*, **39**, 113 (1994).
85 Yang, H., Huang, K., Jiang, S., Wu, C. and Chau, C. *Analytica Chimica Acta*, **282**, 437 (1993).
86 Bankert, S.F., Bloom, S.D. and Sauter, G.D. *Analytical Chemistry*, **45**, 692 (1973).
87 Fiarman, S. and Schneier, G. *Environmental Science and Technology*, **6**, 79 (1972).
88 Leiser, K.H., Quandt, S. and Gleitsman, B. *Fresenius Zeitschrift für Analytische Chemie*, **298**, 378 (1979).
89 Analiitia, T.U. and Pickering, W.F. *Water, Air and Soil Pollution*, **35**, 171 (1987).
90 Crespo, M.T., Gascon, J.L. and Acena, M.I. *Science of the Total Environment*, **130**, 383 (1993).
91 Vassileva, E., Proinova, I. and Hadjivanov, K. *Analyst (London)*, **121**, 607 (1996).
92 Chakravarty, R. and Van Grieken, R. *International Journal of Environmental Analytical Chemistry*, **11**, 67 (1982).
93 Bruninx, E. and Meyl, E.V. *Analytica Chimica Acta*, **80**, 85 (1975).
94 Laxen, D.P.H. *Chemical Geology*, **47**, 321 (1984-5).
95 Lebedinskaya, M.P. and Chuiiko, V.T. *Zhur. Analit. Khim.*, **28**, 863 (1973).
96 Naito, W., Takahata, N., Yoneda, A. and Azumi, T. *Water Purif. Liquid Wastes Treatment*, **20**, 529 (1979).
97 Bowers, H.R. and Huang, C.P. *Water Research*, **21**, 757 (1987).
98 Frimmel, F.H. and Geywitz, J. *Science of the Total Environment*, **60**, 57 (1987).
99 Quiang, Y., Tian, Z., Jiang, T. *et al. Huaxue Yu Fangshe Huaxue*, **8**, 230 (1986).
100 Ueda, J. and Mizui, C. *Analytical Science*, **4**, 417 (1988).
101 Nakamura, T., Oka, H., Ishii, M. and Sato, J. *Analyst (London)*, **119**, 1397 (1994).
102 Tanaka, S., Nakamura, M. and Hashimoto, V. *Bunscki Kagaku*, **36**, 114 (1987).
103 Abe, K., Ito, M., Kikuchi, H. *et al. Eisel Kagku*, **33**, 258 (1987).
104 Thompson, M., Pahlavapour, B., Walton, S.J. and Kirkbright, G.F. *Analyst (London)*, **103**, 568 (1978).

105 Panayappan, R., Venezky, D.L., Gilfrich, J.V. and Birks, L.S. *Analytical Chemistry*, **50**, 1125 (1978).
106 Bankovsky, Y.A., Vircavs, M.V., Veveris, O.E., Pelve, A.R. and Vircava, D.E. *Talanta*, **34**, 179 (1987).
107 Bem, H. and Ryan, D.E. *Analytica Chimica Acta*, **166**, 189 (1984).
108 Zmijewska, W., Polkowska-Motrenko, H. and Stakowska, H. *Journal of Radioanalytical and Nuclear Chemistry Articles*, **84**, 319 (1984).
109 Braman, R.S. and Foreback, C.C. *Science*, **182**, 1247 (1973).
110 Foreback, C.C. Some studies on the detection and determination of mercury, arsenic and antimony in gas discharges. Thesis, University of South Florida, Tampa (1973).
111 Tompkins, M.A. Environmental analytical studies in antimony, germanium and tin. Thesis, University of South Florida, Tampa (1977).
112 Andreae, M.O.P.I. Trace metals in seawater. Eds. C.S. Wong *et al. Proceedings of a NATO Advanced Research Institute on Trace Metals in Seawater*. 30/3/81–3/4/81 Sicily, Italy, Plenum Press (1981).
113 Malinski, T., Fish, J. and Matusiewicz, H. Proceedings of Water Technology Conference (1986), 14 (Adv. Water Anal. Treat.) 347–59 (1987).
114 Vassos, B.H., Hirsch, R.F. and Letterman, H. *Analytical Chemistry*, **457**, 92 (1973).
115 Torsi, G. and Palmisano, F. *Science of the Total Environment*, **37**, 35 (1981).
116 Brandenberger, H. and Bader, H. *Atomic Absorption Newsletter*, **6**, 101 (1967).
117 Fairless, C. and Bard, A.J. *Analytical Letters (London)*, **5**, 433 (1972).
118 Skogerboe, R.K., Hanagan, W.H. and Taylor, H.E. *Analytical Chemistry*, **57**, 2815 (1985).
119 Crocker, J.H. and Merritt, W.F. *Water Research*, **6**, 285 (1972).
120 Kasthurikrishnan, N. and Koropchak, J.A. *Analytical Chemistry*, **65**, 857 (1993).
121 Isozaki, A., Ueki, K., Sazaki, H. and Utsumi, R. *Bunseki Kagaku*, **36**, 672 (1987).
122 Nickson, R.A., Hill, S.J. and Worsfold, P.J. *Analytical Proceedings (London)*, **32**, 387 (1995).
123 Brovko, I.A. *Zh. Anal. Khim*, **42**, 1637 (1987).
124 Amaukwah, S.A. and Fasching, J.L. *Talanta*, **32**, 111 (1985).
125 Brooks, R.R., Presley, B.J. and Kaplan, I.R. *Talanta*, **14**, 809 (1967).
126 Kremling, K. and Peterson, H. *Analytica Chimica Acta*, **70**, 35 (1974).
127 Kinrade, J.D. and Van Loon, J.C. *Analytical Chemistry*, **46**, 1894 (1974).
128 Jan, T.K. and Young, D.R. *Analytical Chemistry*, **50**, 1250 (1978).
129 Stolzberg, R.J. In *Analytical Methods in Oceanography* (ed. T.R.P. Gibb, Jr.) Advanced Chemistry, No. 147, American Chemical Society, Washington, DC, p.30 (1975).
130 Danielsson, L., Magnusson, B. and Westeriund, S. *Analytica Chimica Acta*, **98**, 45 (1978).
131 Sturgeon, R.E., Berman, S.S., Desauliniers, A. and Russell, D.S. *Talanta*, **27**, 85 (1980).
132 Magnusson, B. and Westerlund, S. *Analytica Chimica Acta*, **131**, 63 (1981).
133 Armansson, H. *Analytica Chimica Acta*, **88**, 89 (1977).
134 Bruland, K.W., Franks, R.P., Knauer, G.A. and Martin, J.H. *Analytica Chimica Acta*, **105**, 233 (1979).
135 Lee, M.G. and Burrell, D.C. *Analytica Chimica Acta*, **62**, 153 (1972).
136 Tsalev, D.L., Alimarin, I.P. and Neiman, S.I. *Zhur Analytical Khim*, **27**, 1223 (1972).
137 El-Enamy, F.F., Mahmond, K.F. and Varma, M.M. *Journal of the Water Pollution Control Federation*, **51**, 2545 (1979).
138 Pellenberg, R.E. and Church, T.M. *Analytica Chimica Acta*, **97**, 81 (1978).

139 Armannsson, H. *Analytica Chimica Acta*, **110**, 21 (1979).
140 Brugmann, L., Danielsson, L.R., Magnusson, B. and Westerlund, S. *Marine Chemistry*, **13**, 327 (1983).
141 Boyle, E.A. and Edmond J.M. *Analytica Chimica Acta*, **91**, 189 (1977).
142 Rasmussen, L. *Analytica Chimica Acta*, **125**, 117 (1981).
143 Ediger, R. D., Peterson, G. E. and Kerber, J. D. *Atomic Absorption Newsletter*, **13**, 61 (1974).
144 Smith, R.G. and Windom, H.L. *Analytica Chimica Acta*, **113**, 39 (1980).
145 Danielsson, L.G., Magnusson, B. and Westerlund, S. *Analytica Chimica Acta*, **98**, 47 (1978).
146 Sperling, K.R. *Fresenius Zeitschrift für Analytische Chemie*, **301**, 254 (1980).
147 Sturgeon, R.E., Berman, S.S., Desauliniers, J.A.H. *et al. Analytical Chemistry*, **52**, 1585 (1980).
148 Lo, J.M., Yu, J.C., Hutchison, J.C. and Wal. C.M. *Analytical Chemistry*, **54**, 2536 (1982).
149 Lo, J.M., Yu, J.C., Hutchinson, F.I. and Wal, C.M. *Analytical Chemistry*, **54**, 2538 (1982).
150 Statham, P.J. *Analytica Chimica Acta*, **169**, 149 (1985).
151 Danielsson, L.G., Magnusson, B. and Westerlund, S. *Analytica Chimica Acta*, **144**, 183 (1982).
152 Jin, L., Wu, D. and Ni, Z. *Huaxue Xuebao*, **45**, 808 (1987).
153 Radionova, T.V. and Ivanov, V.M. *Zh. Anal. Khim.*, **41**, 2181 (1986).
154 Shabani, M.B.S., Akagi, T., Shimizu, H. and Masuda, A. *Analytical Chemistry*, **62**, 2709 (1990).
155 Fujnaga, T., Kirsaka, R., Koyama, M. *et al. Journal of Radioanalytical Chemistry*, **13**, 301 (1973).
156 Breyer, P. and Gilbert, B.P. *Analytica Chimica Acta*, **201**, 33 (1987).
157 Sturgeon, R.E., Willie, S.N. and Berman, S.S. *Analytical Chemistry*, **57**, 6 (1985).
158 Wang, M., Yuzefousky, A.I. and Michel, R.G. *Microchemical Journal*, **48**, 326 (1993).
159 Comber, S. *Analyst (London)*, **118**, 505 (1993).
160 Sugawar, K.R., Weetall, H.H. and Schucker, G.D. *Analytical Chemistry*, **46**, 489 (1974).
161 Gudes da Mota, M.M., Romer, F.G. and Griepink, B. *Fresenius Zeitschrift für Analytische Chemie*, **287**, 19 (1977).
162 Moorhead, E.D. and Davis, P.H. *Analytical Chemistry*, **46**, 1879 (1974).
163 Sturgeon, R.E., Bermann, S.S., Willie, S.N. and Desaulniers, J.A.H. *Analytical Chemistry*, **53**, 2337 (1981).
164 Volkan, M., Ataman, O.Y. and Howard, A.G. *Analyst (London)*, **112**, 1409 (1987).
165 Nakashima, S., Sturgeon, R.E., Willie, S.N. and Bermann, S.S. *Fresenius Zeitschrift für Analytische Chemie*, **330**, 592 (1988).
166 Lau, C.R. and Yang, M. *Analytica Chimica Acta*, **287**, 111 (1994).
167 Isshiki, K., Tsuji, F., Kuwamoto, T. and Nakayama, E. *Analytical Chemistry*, **59**, 2491 (1987).
168 Van Geen, A. and Boyle, E. *Analytical Chemistry*, **62**, 1705 (1990).
169 Johansson, M., Emteborg, H., Glad, B., Reinholdsson, F. and Baxter, D.C. *Fresenius Journal of Analytical Chemistry*, **351**, 461 (1995).
170 Orians, K.J. and Boyle, E.A. *Analytica Chimica Acta*, **282**, 63 (1993).
171 Murthy, R.S.S. and Ryan, D.E. *Analytica Chimica Acta*, **140**, 163 (1982).
172 Masuda, A. *Analytical Chemistry*, **64**, 737 (1992).
173 Zharikov, V.F. and Senyavin, I.M. Trudy gos Okeanogr. Inst. 9710 (101), 128 Ref: Zhur. Khim 195D (7) Abstract No. 7G189 (1971).

174 Riccardo, A.A., Muzzarelli, R.A.A. and Tubertini, O. *Journal of Chromatography*, **47**, 414 (1970).
175 Muzzarelli, R.A.A. and Sipos, L. *Talanta*, **18**, 853 (1971).
176 Terada, K., Morimoto, K. and Kiba, T. *Analytica Chimica Acta*, **116**, 127 (1980).
177 Ogura, H. and Oguma, K. *Microchemical Journal*, **49**, 220 (1994).
178 Blain, S., Appriou, P. and Handel, H. *Analytica Chimica Acta*, **272**, 91 (1993).
179 Atienza, J., Herrero, M.A., Maquiera, A. and Puchades, R. *Critical Reviews of Analytical Chemistry*, **23**, 1 (1992).
180 Cabella, M.B. and Barnes, R.M. *Analytical Chemistry*, **52**, 2347 (1980).
181 Kuroda, R., Oguma, K., Mukai, N. and Iwamoto, M. *Talanta*, **34**, 433 (1987).
182 Kiriyama, T. and Kuroda, R. *Mikrochimica Acta*, **1**, 405 (1985).
183 Koide, M., Lee, D.S. and Stallard, M.O. *Analytical Chemistry*, **56**, 1956 (1984).
184 Hodge, V., Stallard, H.O., Koide, M. and Goldberg, E.D. *Analytical Chemistry*, **58**, 616 (1986).
185 Florence, J.M. and Batley, G.E. *Talanta*, **23**, 179 (1976).
186 Riley, J.P. and Taylor, D. *Analytica Chimica Acta*, **41**, 175 (1968).
187 Abdullah, M.J. and Royle, L.G. *Analytica Chimica Acta*, **58**, 283 (1972).
188 Figura, P. and McDuffie, B. *Analytical Chemistry*, **49**, 1950 (1977).
189 Mackay, D.J. *Marine Chemistry*, **11**, 169 (1982).
190 Pai, S.C., Whung, P.Y. and Lai, R.L. *Analytica Chimica Acta*, **211**, 257 (1988).
191 Bafti, F., Cardinale, A.M. and Bruzzone, R. *Analytica Chimica Acta*, **270**, 79 (1992).
192 Pai, S.C. *Analytica Chimica Acta*, **211**, 271 (1988).
193 Olsen, S., Persanda, L.C.R., Ruzicka, J. and Hansen, E.H. *Analyst (London)*, **108**, 905 (1983).
194 Jorgensen, S.S. and Petersen, K. Paper presented at 9th Nordic Atomic Spectroscopy and Trace Element Conference, Reykjavik, Iceland (1983).
195 Malamas, F., Bengtsson, M. and Johansson, G. *Analytica Chimica Acta*, **160**, 1 (1984).
196 Krug, F.J., Reis, B.F. and Jorgensen, S.S. Proceedings of the Workshop on Locally Produced Laboratory Equipment for Chemical Education, Copenhagen, Denmark, p. 121 (1983).
197 Kingston, H.M., Barnes, I.L., Brady, T.J., Rains, I.C. and Champ, M.A. *Analytical Chemistry*, **50**, 2064 (1978).
198 Greenberg, R.R. and Kingston, H.M. *Journal of Radioanalytical Chemistry*, **71**, 147 (1982).
199 Kingston, H.M. and Pella, P.A. *Analytical Chemistry*, **53**, 233 (1981).
200 Bonifort, R., Ferraroli, R., Frigieri, P., Heltai, D. and Queirazza, G. *Analytica Chimica Acta*, **162**, 33 (1984).
201 Paulson, A. *Analytical Chemistry*, **58**, 183 (1986).
202 Akagi, T., Noriji, Y., Matsui, M. and Haraguchi, H. *Applied Spectroscopy*, **39**, 662 (1985).
203 Weisel, C.P., Duce, R.A. and Fasching, J.L. *Analytical Chemistry*, **56**, 1050 (1984).
204 Patin, A.A. and Morozov, N.P. *Zhurnal Analiticheskoi Khimii*, **31**, 282 (1976).
205 Siu, K.W.H. and Bermann, S.S. *Analytical Chemistry*, **56**, 1806 (1984).
206 Cabezon, L.M., Cabellero, M., Cela, M. and Perez-Bustamante, J.A. *Talanta*, **31**, 597 (1984).
207 Akagi, T. and Haraguchi, H. *Analytical Chemistry*, **62**, 81 (1990).
208 Zuang, Z., Yang, C., Wang, X., Yang, P. and Huang, B. *Fresenius Journal of Analytical Chemistry*, **355**, 277 (1996).
209 Lo, J.M., Wei, J.C. and Yeh, S.J. *Analytical Chemistry*, **49**, 1146 (1977).

210 Rao, R.R. and Chat, A. *Journal of Radioanalytical and Nuclear Chemistry*, **168**, 439 (1993).
211 Civici, N. *Journal of Radioanalytical and Nuclear Chemistry*, **186**, 303 (1994).
212 Buono, J.A., Buono, J.C. and Fasching, J.L. *Analytical Chemistry*, **47**, 1926 (1975).
213 Subramanian, K.S. and Meranger, J.C. *International Journal of Environmental Analytical Chemistry*, **7**, 25 (1979).
214 Yamaguchi, T., Zhang, L., Matsumoto, K. and Terada, K. *Analytical Science*, **8**, 85 (1992).
215 Kempster, P.L. and Van Vliet, H.R. *Water South Africa*, **4**, 125 (1978).
216 Subramanian, K.S., Meranger, J.C., Langford, C.H. and Allen, C. *International Journal of Environmental Analytical Chemistry*, **16**, 33 (1983).
217 Chiba, I. and Takakura, Y. Fukushima-ken Eisel Kogal Kenkyusho Nenpo (1986), 70–75 (1987).

Chapter 31

Cations: On-line preconcentration techniques

Preconcentration techniques by their nature introduce an additional step into analytical procedures for trace analysis of water samples.

Quite an appreciable amount of effort has recently been put into developing on-line preconcentration procedures. As well as speeding up the analysis such methods increase the output and decrease analytical time and introduce the possibility of unattended overnight operation.

On-line methods have an additional advantage that analytical quality control methods can be incorporated into the method with the additional possibility of improved statistical parameters such as precision and accuracy.

31.1 Flow injection analysis methods

31.1.1 Seawater

Olsen et al. [1] have described an on-line flow analysis injection preconcentration system using a microcolumn of Chelex–100 allowing the determination of lead at the $10\mu g\ L^{-1}$ level and cadmium and zinc at the $1\mu g\ L^{-1}$ level in seawater. The sampling rate was 30–60 samples per hour, considerably greater than that achieved by manual preconcentration techniques. Readout was available in 60 to 100s after sample injection.

These workers devised a simple flow injection system to inject samples into a flame atomic absorption spectrometer.

The apparatus used by Olsen et al. [1] comprised an atomic absorption spectrometer (Varian, Model AA–1275) connected in parallel to a recorder (Radiometer Servograph REC 80, furnished with an REA 112 high sensitivity module) and via a homemade interface to a computer (PET Commodore Model 3032, combined with a Commodore CBM, Model 2040, dual–drive floppy disk) and a printer (Tractor, Model 3022). For the automated flow injection analysis system (Fig. 31.1) the contact to the computer was triggered by a micro-switch in the injection valve of the flow injector analysis system, which served to inform the computer of the sequence of events. The automated system in Fig. 31.1 additionally

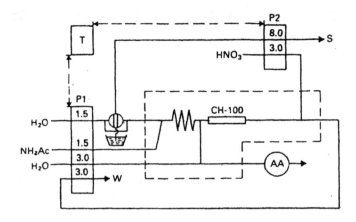

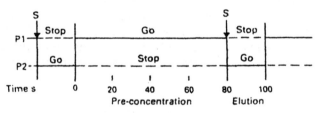

Fig. 31.1 Fully automated FIA–FAA system

Fully automated FIA–FAA system operated via two peristaltic pumps, the stop and go sequences of which are controlled by an electronic timer (T). The stated pumping rates are in ml mm^{-1} with sample volume 2ml. The operation consists of a preconcentration cycle from the column counter-currently.

The ammonium acetate buffer solutions used in the FIA system were prepared by aqueous dilutions of a 2M stock solution made by mixing 55.5ml of 99% acetic acid with 112.5ml of 25% ammonia solution: redistilled water was added to a total volume of 500ml. Any readjustment of the pH in the dilution solutions was done by the addition of ammonia.

The Chelex–100 cation exchange resin, 50–100mesh (sodium form), was purchased from Bio–Rad Laboratories.

Source: Reproduced with permission from Olsen *et al.* [1] Royal Society of Chemistry

included a timer that controlled the timing of the stop–go intervals of the two attached peristaltic pumps (Ismatec, Model Mini–S–840).

The flow injection system consisted of an FIAstar unit, details of which have been published by Ruzicka *et al.* [2]. All connecting tubes consisted of 0.5mm id Microline. As the FIAstar valve is furnished with an external loop in order to regulate the injected sample volume, change of volume was simply effectuated by changing the external loop. For small sample volumes (less than 200μL) the sample loop consisted of 0.5mm id tubing, while tubes of larger id were used for larger sample volumes (up to 1.57mm id, for sample volumes exceeding 1ml).

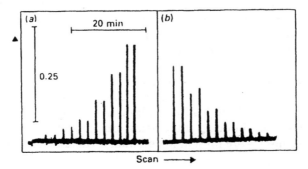

Fig. 31.2 Calibration runs for (a) lead (217.0nm) (20, 50, 100, 250, 350 and 500ppb); and (b) cadmium (228.8nm) (2, 5, 10, 20, 35 and 50ppb) as executed in the FIA–FAA system of Fig. 31.1.
Source: Reproduced with permission from Olsen, S. *et al.* [1] Royal Society of Chemistry

The fully automated apparatus in Fig. 31.1 incorporates a single valve two-pump system. This was equipped with an electronic timer capable of sequencing pump 1 and pump 2 in a stop–go mode, each preconcentration–elution cycle being initiated by the turn of the injection valve in Fig. 31.1. During the preconcentration cycle pump 1(P1) was going while the sample was injected by turning the valve (S). Afterwards, the injected zone was mixed with a buffer in the coil and passed through the microcolumn. In the next sequence, pump 1 was stopped and pump 2 (P2) started, permitting the eluting acid to move through the column in the opposite direction and thereby transport the eluted metal into the FAA spectrophotometer. Note that when pump 1 is stopped the liquids in the thus closed circuit cannot move in either direction. The sampling cycle is completed when the peak appears, whereupon pump 2 may be stopped and pump 1 reactivated, thus establishing the high pH inside the microcolumn and thereby making it ready for the next preconcentration step. Note also that, as in the previous system, the seawater matrix never enters the FAA spectrophotometer and that the microcolumn is carefully regenerated prior to each sampling cycle, while being operated in counter–current fashion. Typical calibration runs for series of lead and cadmium standards are shown in Fig. 31.2.

The manifold components shown in Fig. 31.1 were all incorporated into an integrated microconduit [3], which is a new way of fabricating flow through systems. In the integrated micro–conduit all connectors, mixing coil and the ion exchange column are imprinted within a plastic block, which is then permanently sealed by a second plate. Therefore, connecting tubing and Swagelock-type connectors are eliminated as the whole structure is embedded within a $7 \times 4 \times 1.5$cm thick plastic block with four inlets and two outlets. Such a microflow injection manifold can then be placed within and operated by a BIFOK–Tecator FIA 5020 instrument which has two pumps, injection valve with variable volume, and appropriate timing and sequencing facilities.

It is necessary to refer to a serious deficiency from which all methods for the determination of trace amounts of metals in non saline waters and seawater suffer. At the parts per billion level and below there may be a substantial fraction of the lead or cadmium present in forms that have been described as 'bound', thus being inaccessible to determination because these metals are present in colloidal form, insoluble in acid, or in the form of strong complexes that are not dissociated. As only the 'labile' and 'free' metals can be electrodeposited, extracted or adsorbed on Chelex–100, only these fractions are measurable by the techniques described by Olsen et al. [1] because the seawater matrix itself interferes in the direct measurement by flame or graphite furnace atomic absorption.

Shortly after the above work was published, Fang et al. [4] published further work describing an efficient flow injection system with on-line ion exchange preconcentration for the determination of copper, zinc, lead and cadmium in seawater at the µg L^{-1} and sub µg L^{-1} level by atomic absorption spectrometry.

The behaviour of the different chelating exchangers used was studied with respect to their preconcentration characteristics, with special emphasis on interferences encountered in the analysis of seawater.

These detection limits are a considerable improvement on those achieved in the earlier work [1] utilising Chelex–100 as adsorbent. Since the original works of Olsen et al. [1] valves designs have been improved; in addition to Chelex–100, other ion exchange resins have been tested, and different flow systems have been proposed to increase the efficiency of the process [5–8]. These developments indicate that on line methods not only increases the speed of the preconcentration process, but could also ultimately rival the sensitivity and speed of graphite furnace atomic absorption spectrometry. The designs suggested so far are summarised in Table 31.1 including that discussed by Fang et al. [4]. For comparison of the efficiency of the different systems, the degree of preconcentration attained in 1min (including sampling time and elution time) is taken as a measure for evaluation. Table 31.1 clarifies the main object of the method developed by Fang et al. [4], ie to combine the advantages of the previous designs while avoiding the shortcomings as much as possible. The system described here is more efficient than the previous designs, being capable of attaining 50–105-fold preconcentration at a sampling frequency of 60h^{-1}. A number of heavy metals can thus be determined at sub µg L^{-1} levels by flame atomic absorption spectrometry, the detection limits for copper, zinc, lead and cadmium being 0.07, 0.03, 0.5 and 0.05µg L^{-1} respectively, with precisions ranging between 1.2 and 3.2% at µg L^{-1} levels.

Table 31.1 The development of flow injection preconcentration systems

Basic configuration:

(P, pump; C, carrier; S, sample; A, eluent; Ch–100, Chelex column; AA, aas detector; W, waste)

Development	Concn. efficiency of system (x-fold min $^{-1}$)	Ref.
Improved design: (1) no discharge of effluent into nebuliser (prevents blocking); (2) counter–current elution (improves column operation);(3) confluenced buffer (improves mixing)	6	[1]
(1), (2) incorporated into design; (4) single valve design (simplifies construction and manipulation); (5) acid eluent flowing through nebuliser only during elution (lessens corrosion of nebuliser); (6) sampling by time and flow rate instead of by sample loop (increases efficiency, decreases sample dispersion)	9	[5,8]
(1), (2), (4), (5) incorporated into system with dual column operated alternately (increases efficiency)	13–18	[6]
(3), (5), (6) incorporated into system; (8) resin with surface bound immobilised functional groups (improves sensitivity)	20	[7]
(1–6), (8) incorporated into system; (9) parallel sampling, sequential elution dual-column system	50–105	[4]

Source: Reproduced with permission from Fang *et al.* [4] Elsevier Science, UK

Previous studies have shown that the type of ion exchange resin is important in the overall performances of the systems. Therefore, besides the study on the hardware of the system, Fang *et al.* [4] compared the different types of resins applied for preconcentrations in flow injection analysis with respect to their behaviour when used for determinations of heavy metals in water.

These include Chelex–100 (35mg) (in the ammonium form), or 10mg of 8-quinolinol-based resin [7] or 100mg of weakly acidic 122 resin (phenol formaldehyde with salicylate functional groups) [6]. Details of the two-layer 8-channel preconcentration value and the analysis manifold are given in Figs. 31.3 and 31.4.

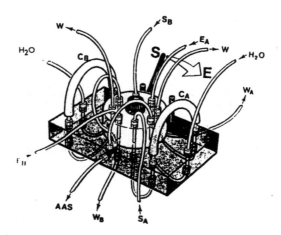

Fig. 31.3 Dual-column ion exchange preconcentration valve

SA, SB, samples A and B; CA, CB, ion exchange columns A and B; EA, EB, eluant (2M nitric acid) for columns A and B; WA, WB, waste lines for samples and eluants A and B; W, waste lines; AAS, atomic absorption spectrometer. The dimensions of the base plate of the valve are 70 × 15 × 10mm. For details of operation, see text and Fig. 31.4.

Source: Reproduced with permission from Fang, Z. *et al.* [4] Elsevier Science, UK

The manifold was set up as shown in Fig. 31.4. The pumps were connected to the double timer; the sampling time was normally fixed to 100s, while the elution time was set to 10s for both columns. The sampling flow rate was 6.0ml min $^{-1}$ and the elution flow rate 4.0ml min $^{-1}$.

With the valve in the elution position ((E) in Fig. 31.3 and (b) in Fig. 31.4) samples A and B in 0.025M ammonium acetate buffer (final pH9.5) were pumped by pump I through the pump tubes and valve to waste in order to wash out the previous sample. (This is a preparation step prior to the sampling stage and should be completed simultaneously with the elution of analyte from the previous sample on column A. If 10s is insufficient for the wash out of the previous sample, ie when carryover occurs, the elution period of column A should be lengthened. The pump tubes and connecting tube for the sampling should also be kept as short as possible to facilitate washout.) When the recorder returned to baseline following the elution of column A the valve was turned manually to the sampling position ((a) in Fig. 31.4), which actuated the microswitch initiating the monitoring of the sampling time. At the end of the sampling time, pump I was stopped automatically. The valve was at once returned manually to elution position (b) which actuated the second channel of the timer, starting pump II. The eluant, 2M nitric acid, was pumped through column B, eluting the analyte directly into the nebuliser of the

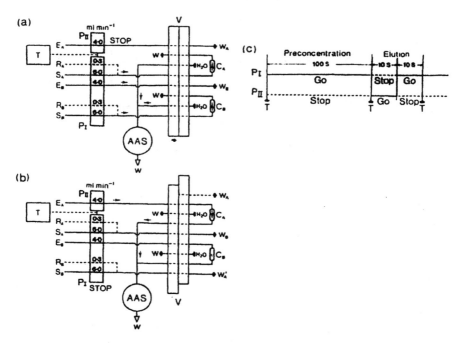

Fig. 31.4 Manifold for dual column on-line ion exchange preconcentration system with flame atomic absorption detection

(a) PI, PII pumps I and II; T, timer; V, valve; because of the circular arrangement of the valve channels, WA represents the same channel as WA; other symbols are the same as for Fig. 31.3. (c) Time sequencing program for valve operation. The points marked T indicate turn of valve. Note that the elution of both columns A and B takes place in position (b) of the valve, but with pumps PI and PII sequenced stop–go and go–stop respectively

Source: Reproduced with permission from Fang, Z. *et al.* [4] Elsevier Science, UK

spectrometer and producing a transient signal which was recorded. The elution was stopped after 10s, at which point pump I resumed its action, the 10s stop period of pump I being used to change samples. Then elution of column A was effected by means of pump I producing a second signal. After the recorder had returned to baseline, which was about 10s after the start of the elution, the preparative washout of the next two samples was completed and another turn of the valve commenced the next sampling sequence (cf. Fig. 31.4). Distilled water was aspirated into the nebuliser during the sampling period to wash it out. During elution the water flowed to waste. Before closing down, the entire system was washed by pumping distilled water through the lines. This is particularly important because a chelating resin could be seriously damaged by being left in 2M nitric acid overnight. When Chelex–100 is used, the column should be

Table 31.2 Performance of dual-column flow injection system with sequential elution/preconcentration and different chelating resins

Resin	Characteristic concentration (µg L⁻¹)				LOQ[a] (µg L⁻¹)				LOD[b] (µg L⁻¹)			
	Cu	Zn	Pb	Cd	Cu	Zn	Pb	Cd	Cu	Zn	Pb	Cd
Chelex-100	0.5	0.12	1.5	0.2	0.2	0.11	2.5	0.2	0.07	0.04	0.8	0.07
8-Quinolol	0.6	0.07	0.8	0.13	0.3	0.08	1.4	0.15	0.09	0.03	0.5	0.05
122 (weakly acidic)	0.5	0.13	1.0	0.2	0.2	0.12	1.6	0.2	0.07	0.04	0.6	0.07

Resin	RSD (%)				Recovery[c] (%)				Concentration efficiency[d] (fold min⁻¹)			
	Cu	Zn	Pb	Cd	Cu	Zn	Pb	Cd	Cu	Zn	Pb	Cd
Chelex-100	2.2	3.2	1.3	1.9	99	92	95	93	88	50	70	60
8-Quinolol	2.3	2.6	2.2	1.9	70	73	101	52	80	87	100	105
122 (weakly acidic)	2.2	1.8	1.2	2.3	97	94	101	44	88	47	83	60

[a] Limit of quantiation (10σ)
[b] Limit of detection based on 99.7% (3σ) confidence level
[c] From a seawater matrix with the composition of 3.1% NaCl, 1300mg L⁻¹ Mg, 400mg L⁻¹ Ca
[d] Concentration efficiency = [concentration (enrichment factor) × [sampling frequency (min⁻¹)]

Source: Reproduced with permission from Fang, Z. et al. [4] Elsevier Science, UK

washed in the H⁺form, as washing in the NH₄⁺ form will create excessive swelling and pressure in the column.

The results listed in Table 31.2 were obtained using the system shown in Fig. 30.3 with a buffer non-confluencing module at a sampling rate of 60h-1 (10ml samples), the buffer in the samples being adjusted to a level of 0.025M ammonium acetate at pH9.5. The table provides not only an evaluation of the performance of the system, but also a comparison of the behaviour of the different resins (see below).

The concentration efficiencies for the elements studied were significantly higher than in previous designs with the same resin, the improvement being due to the factors discussed below. The concentration efficiency was calculated by multiplying the concentration (enrichment) factor by the sampling frequency per minute. As long as the sample volume was kept within certain limits, so that there was no breakthrough of the analyte, the products were almost constant for a given experimental set-up. The concentration factor was obtained by comparing the peak height of the recording obtained during elution with the response of standard solutions of the same metal ion aspirated continuously directly into the nebuliser at the same flow rate as the eluant; the ratio of the concentrations which gave the same absorption signal is the concentration factor.

Resing and Measures [9] have described a highly sensitive fluorometric method for the determination of aluminium in seawater by flow injection analysis with on-line preconcentration. The method employs in-line preconcentration of aluminium on a column of resin immobilised 8-hydroxyquinolone. The column is subsequently eluted into the flow injection analysis system from the resin to form acidified seawater. The eluted aluminium reacts with lumogallion to form a chelate, which is detected by its fluorescence. The fluorescence is enhanced ~5-fold by the addition of a micelle forming detergent, Brij–35. The method has a detection limit of ~0.15nM and a precision of 1.7% at 2.4nM. The method has a cycle time of ~3 min and can be readily automated. The ease of use and relative freedom from contamination artefacts makes this method ideal for shipboard determination of aluminium in seawater.

Ke and Liu [10] have described a method based on laser-enhanced ionisation detection for the determination of lead in seawater by flow injection analysis with on-line preconcentration and separation. In this procedure the flow injection manifold is incorporated with a microcolumn packed with a C_{18} bonded silica. The chelating agent DDPA is used to form the Pb–DDPA complex, which may be sorbed in the microcolumn and then eluted with methanol. The preconcentrated lead is then detected by the laser enhanced ionisation technique with either single-step or two-step excitation. At 5 and 15mL volume–fixed sample loading, the detection limits of 11 and 3.3µg L⁻¹ and enrichment factors of 16 and 48 are achieved, respectively, using a two-step flow injection–laser enhanced ionisation.

The sensitivity of this system proves to be better by at least 1 order of magnitude than that of conventional laser enhanced ionisation method. The flow injection–laser enhanced ionisation also increases the tolerance of matrix interference. The laser enhanced signal is slightly reduced to 80% intensity as 10000mg L $^{-1}$ sodium and potassium matrixes are mixed in the lead solution. The resistance to the alkali matrixes is enhanced ~4 times that reported previously using a similar water immersed probe as a laser enhanced ionisation collector. Flow injection–laser enhanced ionisation is applied to detect the lead content in seawater, achieving a result of 0.0112 ± 0.0006µg L $^{-1}$ consistent with the certified value of 0.013 ± 0.005µg L $^{-1}$.

31.1.2 Non saline waters

Aggarwal et al. [11] have developed a method for the determination of rare earth elements in non saline waters at sub ppt levels by inductively coupled plasma mass spectrometry and flow injection inductively coupled mass spectrometry.

The method described by Aggarwal et al. [11] is capable of determining the rare earth elements at concentrations down to 0.1µg L $^{-1}$ (0.5pmol kg $^{-1}$) in aqueous samples using flow injection coupled to an inductively coupled plasma mass spectrometer. The method is an extension of a technique described by Shabani et al. [12] to allow application to a wider range of ionic strengths. The method is sufficiently rapid to enable samples to be processed through the purification stage in less than 90min. More than 99.5% of the barium in the sample is removed during processing, ensuring that isobaric interference of barium oxide on the rare earth elements is <2% of the rare earth element signal. The detection limits of this method show an improvement of up to 30 times better than the original method. This improvement was achieved by using flow injection techniques that allow the sample to be concentrated in a smaller volume for analysis. The technique has been successfully employed for the determination of rare earth elements in Icelandic hydrothermal fluids.

Zhang et al. [13] have described an on-line preconcentration of lead by a flow injection system prior to flame atomic absorption spectrometry. The analyte was deposited on an alumina microcolumn and subsequently eluted with nitric acid. The limit of detection was 0.36µg L $^{-1}$ and the relative standard deviations at 40 and 4µg L $^{-1}$ levels in non saline waters were 1.4% and 12% respectively.

31.1.3 Potable water

Tao and Fang [14] preconcentrated germanium in a graphite furnace using a flow injection hydride generation technique to determine down to 4ppt of the element using a 5ml sample.

Table 31.3 On-line preconcentrations by adsorption on various solids

Metal	Type of water sample	Solid adsorbent	Elution of metals from solid	Detection limit	Analytical finish	Ref.
Co	Sea	8-quinolinol on silica gel	–	–	Chemical luminescence	[15]
Fe(II)	Sea	C₁₈ impregnated with ferrozine	–	0.1mM	Spectrophotometric	[16]
Pb	Non saline	Modified cellulose	–	Sub ppb	AAS	[17]
Misc.	Non saline	Formation of metal dithiocarbamates and adsorption on C₁₈ bonded silica loaded cetyltrimethyl ammonium bromide dithiocarbamate ion pair	Gradient of acetonitrile and water	–	HPLC	[18]
Cu(II)	Potable	Silica immobilised 8-quinolinol	–	–	FIA–AAS	[19]

Source: Own files

31.2 On-line adsorption on solids and desorption

Some applications of this technique are given in Table 31.3.

31.3 Miscellaneous on-line methods

Agudo *et al.* [20] used liquid–liquid extraction combined with on-line monitoring to determine lead in non saline waters.

Martinez-Jiminez *et al.* [21] employed a continuous precipitation and filtration flow system coupled with an atomic absorption spectrometer for the preconcentration and determination of lead. Lead(II) forms a precipitate with ammonia which is retained on a stainless steel filter, then redissolved with nitric acid. The method was proposed for the determination of lead in non saline waters in the range of 1.2–1500µg L $^{-1}$ and the relative standard deviation was ≤3.6%.

Down to 2µg L $^{-1}$ aluminium has been measured in non saline waters by on-line concentration and inductively coupled plasma mass spectrometry. Speciation studies showed good agreement of this method with an established high performance liquid chromatographic method [22].

References

1 Olsen, S., Pessenda, L.C.R., Ruzicka, J. and Hansen, E.H. *Analyst (London)*, **108**, 905 (1983).
2 Ruzicka, J., Hansen, E.H. and Ramsing, A.U. *Analytica Chimica Acta*, **134**, 55 (1982).
3 Ruzicka, J., Janata, J. and Hansen, E.H. unpublished work.
4 Fang, Z., Ruzicka, A. and Hansen, E.H. *Analytica Chimica Acta*, **164**, 23 (1984).
5 Jørgensen, S.S. and Petersen, K. Paper presented at 9th Nordic Atomic Spectroscopy and Trace Element Conference, Reykjavik, Iceland (1983).
6 Fang, Z., Xu, L. and Zhang, S. *Analytica Chimica Acta*, **164**, 41 (1984).
7 Malamas, M., Bengtsson, M. and Johansson, G. *Analytica Chimica Acta*, **160**, 1 (1984).
8 Krug, F.J., Reis, B.F. and Jørgensen, S.S. Proceedings of the Workshop on Locally Produced Laboratory Equipment for Chemical Education, Copenhagen, Denmark, p. 121 (1983).
9 Resing, J.A. and Measures, C.I. *Analytical Chemistry*, **66**, 4105 (1994).
10 Ke, C.B. and Liu, K.C. *Analytical Chemistry*, **71**, 1561 (1999).
11 Aggarwal, J.K., Shabani, M.B., Palmer, M.R. and Ragnarsdottir, K.V. *Analytical Chemistry*, **68**, 4418 (1996).
12 Shabani, M.B., Akagi, T., Shimuzu, H. and Masuda, H. *Analytical Chemistry*, **62**, 2709 (1990).
13 Zhang, Y., Riby, P., Cox, A.G. *et al.* *Analyst (London)*, **113**, 125 (1988).
14 Tao, G. and Fang, Z. *Journal of Analytical Atomic Spectroscopy*, **8**, 577 (1993).
15 Hirata, S., Hashimoto, Y., Aikara, M. and Mallika, G.V. *Fresenius Journal of Analytical Chemistry*, **355**, 676 (1996).
16 Blain, S. and Terguer, P. *Analytica Chimica Acta*, **308**, 425 (1995).
17 Naghmush, A.M., Pyrzynska, K. and Trojanowicz, M. *Talanta*, **42**, 851 (1995).

18 Irth, H., De Jong, G.J., Brinkman, U.A.T. and Frei, R.W. *Analytical Chemistry*, **59**, 98 (1987).
19 Marshall, M.A. and Mottola, H.A. *Analytical Chemistry*, **57**, 729 (1985).
20 Agudo, M., Rios, A. and Valcarcel, M. *Analytical Chemistry*, **65**, 2941 (1993).
21 Martinez-Jiminez, P., Gallego, M. and Valcarcel, M. *Analyst (London)*, **112**, 1233 (1987).
22 Fairman, B., Sanz-Medel, A. and Jones, P. *Journal of Analytical Atomic Spectroscopy*, **10**, 281 (1995).

Chapter 32

Rationale, preconcentration of cations

Procedures for particular cations methods used in the various types of water samples are reviewed in Tables 32.1 to 32.5.

These tables list the type of cation, also the type of water sample, the analytical finish employed, the detection limits achieved by various workers and the section in the book in which further information can be obtained.

Keynotes to the abbreviations used are given at the top of the respective tables.

Table 32.1 Preconcentration of metals in non saline waters

Key to preconcentration method: chelation–solvent extraction (1); immobilised chelation (2); immobilised non chelation (3); active carbon adsorbtion (4); anion exchange resins (5); cation exchange resins (6); adsorption on metals and metal oxides (7); coprecipitation (8); on-line methods (9); gas evolution methods (10); electrolytic methods (11); electrochemical methods (12); cathodic deposition methods (13); osmosis (14); dialysis (15); flotation (16); stripping methods (17); solvent sublation (18); evaporation (19); aeration (20); co-crystallisation (21); freeze drying (22); ion exclusion (23)

   Key to analytical finish: atomic absorption spectrometry (1); spectrophotometric (2); inductively coupled plasma atomic emission spectrometry (3); inductively coupled plasma mass spectrometry (4); neutron activation analysis (5); X-ray fluorescence spectroscopy (6); emission spectrometry (7); supercritical fluid methods (8); anodic stripping voltammetry (9); gas chromatography (10); scintillation counting (11); X-ray spectrometry (12); high performance liquid chromatography (13); differential pulse polarography (14); radiochemical methods (15); flow injection analysis (16); electron spin resonance spectroscopy (17); spectrophotometric methods (18); voltammetry (19); isotope dilution (20); ultraviolet spectroscopy (21); mass spectrometry (22); γ ray spectrometry (23); molecular emission cavity analysis (24); titration (25); luminescence methods (26); spectrofluoro-metry (27); flame photometry (28); laser enhanced ionisation (29)

Metal	Preconcn. method	Analytical finish	Detection limit*	Section No.	[Ref.]
Al	1			20.1.2	[64]
	2	1		21.1.3	[15]
	2			21.1.3	[20]
	3	18		21.1.3	[21]
	3	5		24.2.3	[10]
	6	5		25.2.1.2	[10]
	8	5		28.1.2	[27]
	9	3	2	31.1.3	[22]
Am	7			26.1.2	[11]
				26.1.1	[7]
Sb	1	1		20.1.7	[143]
	1	2	200	20.1.5	[103]
	1	7		20.1.3	[71]
	1	8	1	20.1.1	[60,61]
	1	9		20.1.1	[62]
	1	5	0.001	20.1.1	[54]
	2	1		21.1.4	[85]
	2	1		21.1.1	[7]
	2	1		21.1.3	[16]
	8	6	6.1 µg absolute	27.1.2	[29]
	8	1		27.1.6	[46]
	8	5		28.1.3	[20]
	10		0.025 ng absolute	29.1.1	[55]
	10			29.1.10	[42,43]
	10		0.2	29.1.1	[54]
As	1	1		20.1.7	[143]
	1	8		20.1.1	[60,61]
	1	1	0.07	20.1.1	[19]
	8	6	0.02–0.1	20.1.1	[56]

*µg L $^{-1}$ unless otherwise stated

Metal	Preconcn. method	Analytical finish	Detection limit*	Section No.	[Ref.]
	1	5	10	20.1.1	[53]
	2	1		21.1.3	[19]
	3	6	0.5	22.1.3	[12]
	8	2		27.1.1	[1,2]
	8	1		27.1.6	[26]
	8	5		28.1.3	[20,21]
	8	6	0.13µg absolute	27.1.2	[29]
	8	6		28.1.1	[5]
	8	6	200	27.1.6	[53]
	10	1		29.1.1	[53]
	10	1			
	10	5		29.1.10	[42,43]
	10		0.025µg absolute	29.1.1	[55]
	10		0.2	29.1.1	[54]
Ba	1	6		20.1.1	[5]
Be	1	1		20.1.5	[109]
	1	10	1	20.1.6	[141]
	1	1		20.1.2	[64]
	2	1	<0.001	22.1.4	[27]
	3	1		21.1.4	[23]
	8	7	1–10	27.1.1	[12]
Bi	1	8	µg L^{-1}	20.1.1	[60,61]
	1	8	1	20.1.5	[125]
	1	7		20.1.3	[71]
	3	18		21.1.2	[10]
	7			26.1.1	[2]
	8	11		27.1.6	[40]
	8	1		27.1.6	[46]
Cd	1	9		20.1.5	[58]
	1	1	0.2	20.1.1	[4]
	1	1		20.1.7	[143]
	1	1	0.5	20.1.1	[24]
	1	3		20.1.1	[43]
	1	1		20.1.1	[12,17–19, 26,27]
	1	1		20.1.1	[37]
	1	1		20.1.1	[15,21, 25]
	1	1		20.1.5	[112]
	1	1		20.1.4	[74]
	1	1	0.02	20.1.1	[36]
	1	1	0.0066	20.1.5	[121]
	1	1	0.5–1.1	20.1.2	[63]
	1	1	10pg absolute	20.1.1	[16]
	1	1	1	20.1.5	[101]
	1	3		20.1.1	[44]
	1	3		20.1.1	[45]
	1	3	0.017	20.1.1	[39]

*µg L^{-1} unless otherwise stated

Metal	Preconcn. method	Analytical finish	Detection limit*	Section No.	[Ref.]
	1	6	0.02–1	20.1.1	[56]
	1	7		20.1.3	[71]
	1	12	0.025	20.1.1	[58]
	2	13	0.5	21.1.2	[13]
	2	1		21.1.1	[5,6]
	2	1		21.1.3	[18]
	2	1	10µg absolute	21.1.1	[21]
	2	1	1–5	21.1.4	[42]
	3		0.0015	21.1.3	[20]
	3			22.1.3	[15]
	3			21.1.3	[18]
	3			22.1.3	[11]
	4	1		23.1	[4]
	4	5		23.1	[7]
	4	6		23.1	[2]
	5	2		24.2.3	[12,15]
	5	2	µg L^{-1}	24.2.1	[1]
	5	3		24.2.3	[25]
	6	1		25.2.2.1	[42]
	6	1		25.2.1.1	[5]
	6	1	1–3	25.2.1.1	[6]
	6	5	0.1	25.2.1.2	[8]
	6	6		25.2.1.3	[11]
	6	9	sub µg L^{-1}	25.2.2.3	[49]
	6			25.2.2.4	[53]
	6			25.2.2.4	[50,51]
	6			25.2.2.2	[34]
	8	5		28.1.3	[20]
	8	5	0.8	28.1.3	[17]
	8	14		27.1.1	[16]
	8	7	1–10	27.1.1	[12]
	8			27.1.1	[7]
	8		<1	27.1.2	[26]
		9	0.025	29.1.1	[15]
	11	1		29.1.1	[2]
	12	3		29.1.1	[8]
	13			29.1.1	[4]
	14			29.1.2	[21]
	15	1		29.1.2	[17]
	16			29.1.9	[40]
Cs	1	6		20.1.1	[5]
	2			21.1.5	[38]
	3			22.1.5	[34]
	6	15		25.2.1.4	[21–24]
	8	15		27.1.6	[39]
Ca	1	3		20.1.1	[43]
	1	1		20.1.1	[1]

*µg L^{-1} unless otherwise stated

Metal	Preconcn. method	Analytical finish	Detection limit*	Section No.	[Ref.]
	1	1	2	20.1.5	[105]
	2	1 and 16	1–5	2.1.4	[41]
	2	16	<1	21.1.1	[1]
	6	5		25.2.1.2	[10]
	6		<1	25.2.2.4	[53]
	6	6		25.2.1.3	[12]
Ce	4	5		23.1	[7]
	6	1	0.05–0.5	25.2.1.1	[2,3]
	8	11		27.1.6	[47]
Cr	1	18	1	20.1.5	[122]
	1	1	5	20.1.5	[80]
	1	1	1	20.1.1	[3]
	1	1		20.1.1	[2,34]
	1	1	150pg absolute	20.1.1	[16]
	1	1	Cr(VI) 0.3 Cr(III) 0.2	20.1.1	[31,32]
	1	3	0.2–0.5	20.1.1	[39]
	2	1		21.1.3	[19]
	3	6	0.5	22.1.3	[12]
	4	5		23.1	[7]
	5	3	sub µg L^{-1}	24.2.3	[27]
	6	2		25.2.1.4	[15,25]
	6		<1	25.2.2.4	[53]
	6	6		25.2.1.3	[12]
	8	1	0.04	27.1.1	[11,22]
	8	1	0.2µM	27.1.1	[58]
	8	1		27.1.1	[23]
	8	1		27.1.4	[33]
	8	5	0.1	27.1.4	[32]
	8	5	0.2	28.1.3	[17]
	8	6	0.4–0.7	27.1.1	[21]
	7			26.1.2	[12]
	12	1	0.3	29.1.1	[7]
	12	3	5200	29.1.1	[9]
	17		<3 × 10^{-8}M	29.1.1	[13]
Co	1	1		20.1.1	[1,2,16,18,21]
	1	1		20.1.5	[112]
	1	1	30	20.1.5	[100]
	1	1	2	20.1.1	[4]
	1	1		20.1.4	[74]
	1	1	1	20.1.1	[3]
	1	7		20.1.3	[71]
	1	3		20.1.1	[43]
	1	3	0.02–0.5	20.1.1	[39]
	1	6		20.1.1	[5]
	1	8	µg L^{-1}	20.1.1	[60,61]
	1	10	4 × 10^{-11}g absolute	20.1.6	[140]

*µg L^{-1} unless otherwise stated

Metal	Preconcn. method	Analytical finish	Detection limit*	Section No.	[Ref.]
	1			20.1.2	[64]
	2	2	5	21.1.1	[4]
	2	1		21.1.3	[19]
	2	1		21.1.1	[5,6,19]
	2	1	preconc. 1000	21.1.3	[7]
	3	2		22.1.2	[9]
	3			22.1.3	[11]
	3	2	<1	21.1.1	[2]
	4	5		23.1	[7]
	4	6		23.1	[2]
	4			23.1	[1]
	5	1	µg L^{-1}	24.2.1	[1]
	5	2		24.2.3	[14]
	6	2		24.2.1.4	[17]
	6	1	0.05–0.5	25.2.1.1	[2,3]
	6	1		25.2.2.1	[42]
	6	1		25.2.1.1	[5]
	6	6		25.2.1.3	[11]
	6	6		25.2.1.3	[12]
	6			25.2.2.4	[51]
	6			25.2.2	[39]
	6			25.2.2.4	[53]
	8	5	0.04	28.1.3	[17]
	8	5		28.1.3	[20]
	8	6		28.1.1	[5]
	8	6	0.4	28.1.1	[12]
	8	7	1–10	27.1.1	[12]
	8		<1	27.1.2	[25]
	7			26.1.5	[29]
Cu	1	2	2	20.1.5	[86]
	1	1		20.1.1	[2,15,18,21, 26,27]
	1	1	1000–10000	20.1.1	[28]
	1	1		20.1.1	[37]
	1	3		20.1.1	[37]
	1	1		20.1.1	[3]
	1	1	0.5	20.1.1	[4]
	1	1	0.3	20.1.1	[36]
	1	1	125pg absolute	20.1.1	[16]
	1	1	0.5–1.6	20.1.2	[63]
	1	1	2	20.1.5	[101]
	1	1		20.1.5	[106,112]
	1	1	20	20.1.5	[100]
	1	1		20.1.7	[143]
	1	3		20.1.1	[44]
	1	6	0.02–1	20.1.1	[56]
	1	7		20.1.3	[71]

*µg L^{-1} unless otherwise stated

Metal	Preconcn. method	Analytical finish	Detection limit*	Section No.	[Ref.]
	1	10		20.1.6	[142]
	1			20.1.5	[127]
	2	1		21.1.1	[5,6]
	2	1	1–5	21.1.4	[41]
	2	1		21.1.3	[17,18]
	2	1	<1	21.1.3	[14]
	2	13	0.5	21.1.2	[13]
	2			21.1.4	[29]
	4	1		23.1	[4]
	4	6		23.1	[2]
	4			23.1	[3]
	3	1		22.1.1	[1]
	3	6	0.5	22.1.3	[12]
	3			22.1.1	[1]
	3			22.1.3	[16]
	6	2		25.21.4	[17]
	6	18		25.2.1.1	[1]
	6	18		25.2.1.4	[16]
	6	1		25.2.1.1	[5,7]
	6	1	0.05–0.5	25.2.1.1	[2,3]
	6	1		25.2.1.4	[7]
	6	1		25.2.2.1	[42]
	6	6		25.2.1.3	[12]
	6	9	sub µg L^{-1}	25.22.3	[49]
	6			25.2.2	[34,39,41]
	6			25.2.2.4	[51]
	8	1		27.1.2	[16,27]
	8	1	<4	27.1.3	[30,31]
	8	1	<2.4	27.1.1	[19]
	8	6		28.1.1	[5]
	8	6		27.1.1	[5]
	8	6	0.5–1	27.1.1	[4]
	8	5		28.1.2	[15]
	8	5	0.3	28.1.3	[17]
	8	6	1–10	27.1.1	[12]
	8	14		27.1.1	[16]
	8			27.1.1	[7]
	8		<1	27.1.2	[25]
	7			26.1.2	[8]
	11	19		29.1.1	[5]
	15	1		29.1.2	[17,18]
	14			29.1.2	[21]
	18			29.1.3	[22]
	13	3		29.1.1	[9]
	19			29.1.4	[23–25]
Eu	1	6		20.1.1	[5]
	4	5		23.1	[7]

*µg L^{-1} unless otherwise stated

Metal	Preconcn. method	Analytical finish	Detection limit*	Section No.	[Ref.]
Gd	1	7		20.1.3	[71]
Ga	1	5	0.001	20.1.1	[50,51]
	2			23.1	[4]
	8	1	<8	27.1.3	[30,31]
	8	11		27.1.1	[9]
		19	0.08	29.1.1	[10]
	2			21.1.3	[20]
Ge	9	16	4ppt	31.1.3	[14]
	10			29.1.10	[42,43]
	1	1		20.1.1	[30]
	1	1	5pmole absolute	20.1.2	[66]
	4	1		23.1	[4]
	4	1	0.010	23.1	[5]
	4	5		23.1	[7]
	5	5		24.2.2	[8]
	5	16	f molar	24.2.3	[28]
	8	1		27.1.6	[41]
	8	5		28.1.1	[14]
	8	12	0.2	27.1.6	[26]
	7			26.1.4	[28]
Hf	8	20		27.1.1	[15]
In	2			21.1.3	[20]
	3	1		22.1.1	[4]
	4	1		23.1	[4]
	8	1	<8	27.1.3	[30,31]
Fe	1	18	15	20.1.5	[127]
	1	1		20.1.1	[1–3,21,27]
	1	1	100pg absolute	20.1.1	[16]
	1	1	3	20.1.5	[101]
	1	1	40	20.1.5	[100]
	1	1		20.1.5	[112]
	1	3	0.02–0.5	20.1.1	[39]
	1	6	0.02–1	20.1.1	[56]
	1	8	µg L^{-1}	20.1.1	[61,62]
	1			20.1.2	[64]
	2	18		21.1.1	[10]
	2	1		21.1.3	[19]
	2			21.1.4	[29,32]
	2		<40	21.1.4	[34]
	3	1		22.1.1	[1]
	3	6	0.5	22.1.3	[12]
	3			22.1.4	[26,23–25]
	3			22.1.3	[11]
	4	6		23.1	[2]
	4	5		23.1	[7]
	4			23.1	[2]
	5	18	1–2ng absolute	24.2.3	[11]

*µg L^{-1} unless otherwise stated

Metal	Preconcn. method	Analytical finish	Detection limit*	Section No.	[Ref.]
	6	1	0.05–0.5	25.2.1	[2,3]
	6	6	0.5	25.2.1.3	[12]
	6	21	0–20	25.2.1.4	[30]
	6		<1	25.2.2.4	[53]
	6			25.2.2.4	[51]
	8	6		28.1.1	[5]
	8	6		27.1.1	[8]
	8	5		28.1.3	[20]
	8			27.1.1	[8]
	11	19		29.1.1	[3]
	15	1		29.1.2	[17,18]
	10			29.1.10	[42]
Ir	1	5	0.001	20.1.1	[50,51]
Lanthanides	2	3	0.0002–0.001	21.1.4	[40]
	2	4	1000 preconc.	21.1.1	[2]
	9	4	0.1	31.1.2	[11,12]
La	3	6	0.5	22.1.3	[12]
	4	5		23.1	[7]
	1			20.1.5	[132]
Pb	1	1		20.1.1	[3,15,17–19,21, 22,25–27,29]
	1	1	5	20.1.1	[33]
	1	1	0.7	20.1.1	[36]
	1	1	100pg absolute	20.1.1	[19]
	1	1	10	20.1.1	[24]
	1	1	2.5	20.1.1	[4]
	1	1	2.1–5.1	20.1.2	[63]
	1	1	5	20.1.5	[101]
	1	1	100	20.1.5	[100]
	1	1		20.1.5	[112]
	1	1	5	20.1.5	[107]
	1	18	0.1	20.1.5	[94]
	1	3	0.52	20.1.1	[39]
	1	3		20.1.1	[2,6,7,43–45]
	1	1		20.1.1	[37]
	1	5		20.1.1	[52]
	1	6	0.02–1	20.1.1	[56]
	1			20.1.2	[64]
	2	1		21.1.3	[18,19]
	2	1		21.1.1	[5,6]
	2	1	1–5	21.1.4	[41]
	2	1	sub ng L^{-1}	21.1.4	[21]
	2		<2	21.1.4	[39]
	3	1		22.1.1	[4]
	3	1	0.01	22.1.4	[36]
	3	6	0.5	22.1.3	[12]
	3			22.1.3	[11,16]

*µg L^{-1} unless otherwise stated

Metal	Preconcn. method	Analytical finish	Detection limit*	Section No.	[Ref.]
	4			23.1	[3]
	6	2		25.2.1.4	[17]
	6	1		25.2.1.1	[5]
	6	1		25.2.2.1	[42]
	6	6		25.2.1.3	[11,12]
	6	9	sub µg L^{-1}	25.2.2.3	[49]
	6			25.2.2.2	[34]
	6			25.2.2.4	[50,51]
	6		<1	25.2.2.4	[53]
	5	1	µg L^{-1}	24.2.1	[1]
	8	1		27.1.6	[42]
	8	1		27.1.2	[27]
	8	1		28.1.1	[1]
	8	7	1–10	27.1.1	[12]
	8	6		28.1.1	[15]
	8	6	0.5	27.1.1	[4]
	8	6		27.1	[8]
	8	14		27.1.1	[16]
	8			27.1.5	[38]
	8		26	27.1.2	[24]
	8			27.1.1	[7]
	8	11		27.1.6	[40]
	7			26.1.1	[1]
	7			26.1.2	[8]
	7	1	0.36	27.1.2	[9]
	9	1	1.2	31.3	[21]
	9	16	0.36	31.1.2	[13]
	9			31.3	[20]
	9		sub µg L^{-1}	31.2.1	[17]
	15	1		29.1.2	[17,18]
	10			29.1.10	[42,43]
	11	6		29.1.1	[6]
	12	3		29.1.1	[9]
	12	4	0.00008	29.1.1	[8]
	13	22		29.1.1	[5]
Mg	2	1	1–5	21.1.4	[41]
	3	1		22.1.1	[1]
	3	5		25.2.13	[10]
Mn	1	1		20.1.1	[2,27,35]
	1	1	0.1	20.1.5	[91]
	1	1	6	20.1.5	[100]
	1	3	0.02–0.5	20.1.1	[39]
	1	6		20.1.1	[5]
	1	6	0.02–1	20.1.1	[56]
	1	7		20.1.4	[72]
	1			20.1.4	[19]
	2	1		21.1.3	[19]

*µg L^{-1} unless otherwise stated

Metal	Preconcn. method	Analytical finish	Detection limit*	Section No.	[Ref.]
	2	1	1–5	21.1.4	[41]
	3	6	0.5	22.1.3	[12]
	3			22.1.3	[11]
	4	6		23.1	[2]
	4			23.1	[2]
	6	1		25.2.1.1	[5]
	6	1		25.2.2.1	[42,43]
	6	18	2µg absolute	25.2.1.4	[13]
	6	5		25.2.1.3	[10,11]
	6			25.2.2.4	[43,51]
	8	1		28.1.2	[16]
	8	6	0.5–1	27.1.1	[4]
	8	6		28.1.1	[5]
	8	5	0.006	28.1.3	[17]
	8	1		27.1.2	[27]
	12	3		29.1.1	[9]
	14			29.1.2	[21]
Hg	1	1		20.1.5	[104]
	1	1		20.1.2	[67]
	1	1	4ng absolute	20.1.2	[68]
	1	5	1	20.1.1	[48,49]
	1	8	µg L^{-1}	20.1.1	[60,61]
	1			20.1.5	[95]
	2	1	0.1	21.1.1	[9]
	2	1		21.1.3	[19]
	2	1	<1	21.1.4	[31]
	2	1	0.0002	21.1.4	[30]
	2	1		21.1.4	[24,25]
	3	6	0.5	22.1.3	[12]
	3			22.1.4	[27]
	3			22.1.3	[18]
	4	5		23.1	[7]
	4	1		24.2.1	[2]
	4	5	0.005µg absolute	24.2.2	[9]
	4	5	0.02	24.2.3	[30]
	6	6		25.2.1.3	[12]
	6			25.2.2.4	[50]
	8	6		28.1.1	[5]
	8	5		28.1.3	[20]
	7	1	<1	26.1.3	[18]
	7	1	6.6	26.1.3	[23]
	7	1	0.1	26.1.3	[14,15,25–27]
	7	1		26.1.3	[22]
	7	3		26.1.3	[21]
	7			26.1.3	[14–27]
	13			29.1.1	[14]
	20			29.1.6	[32]

*µg L^{-1} unless otherwise stated

Metal	Preconcn. method	Analytical finish	Detection limit*	Section No.	[Ref.]
	11	1		29.1.1	[1]
Mo	1	18	0.1	20.1.5	[83]
	1	1		20.1.1	[4]
	1	3	0.02–0.5	20.1.1	[39]
	1	1		20.1.5	[114–118]
	1	1		20.1.1	[3]
	1	7		20.1.3	[71]
	1			20.1.5	[113]
	2		<1	20.1.4	[20]
	3			20.1.4	[21,32]
	4	1		23.1	[4]
	4	3	0.4	23.1	[6]
	4	4	0.06	23.1	[6]
	4	5		23.1	[7]
	4			23.1	[1]
	5	1	0.2	24.1.2	[6]
	5	1	10	24.2.3	[24]
	5			24.2.3	[25]
	6			25.2.1.4	[32]
	8	5		28.1.2	[15]
	8	5		28.1.3	[20]
	8			28.1.3	[22]
	16			29.1.9	[39]
	21			29.1.7	[34–37]
Ni	1	1	1	20.1.1	[1–3,16,18, 21,23,36]
	1	1	1.5	20.1.1	[4]
	1	1	0.5	20.1.1	[36]
	1	1		20.1.2	[69]
	1	1	20	20.1.5	[100]
	1	3		20.1.1	[44]
	1	3	0.02–0.5	20.1.1	[39]
	1	14	1	20.1.5	[19,82]
	1	8		20.1.1	[60,61]
	1			20.1.2	[64]
	2	1		20.1.1.1	[5,6]
	2	1	1–5	20.1.4	[42]
	2	1	factor 1000	21.1.3	[17]
	2	1	100	21.1.4	[36]
	2	13	0.5	21.1.2	[13]
	3			22.1.3	[16]
	3	1		23.1	[4]
	3	6		23.1	[2]
	5		sub µg L^{-1}	24.2.3	[27]
	5	18	0.5mg absolute	24.2.3	[17]
	6	2		25.1.4	[17]
	6	18		25.2.1.4	[26]

*µg L^{-1} unless otherwise stated

Metal	Preconcn. method	Analytical finish	Detection limit*	Section No.	[Ref.]
	6	1		25.2.2.1	[42]
	6	1		25.2.1.1	[5]
	6	1	0.05–0.5	25.2.1.1	[2,3]
	6	1	100	25.2.1.4	[31]
	6	6		25.2.1.3	[11,12]
	6			25.2.2.4	[51]
	6			25.2.2	[39]
	6		<1	25.2.2.4	[53]
	8	6	0.57	27.1.1.1	[4]
	8	7	1–10	27.1.1	[12]
	8		<1	27.1.2	[25]
	14			29.1.2	[21]
	13	3		29.1.1	[9]
Nb	1	7		20.1.3	[71]
	1	18	0.1	20.1.5	[126]
Pd	1	18	0.1	20.1.5	[123]
	1	1	0.04	20.1.2	[70]
Pu	2			21.1.4	[33]
	5	23		24.2.3	[22]
	7			26.1.2	[10]
K	1	18		20.1.5	[111,120]
	6	5		25.2.12	[10]
Ra	6	11	0.03Ci L^{-1}	25.2.1.4	[24]
	7			26.1.1	[5,6]
	8	12		27.1.5	[35]
	8	15		27.1.5	[34]
	8			27.1.5	[36]
Sc	4	5		23.1.1	[7]
	8	5		28.1.3	[20]
Se	1	1	<100	20.1.5	[108]
	1	1		20.1.7	[143]
	2	1	0.007	21.1.1.3	[12]
	2	12	0.05	21.1.4	[26,27]
	3	6	0.5	22.1.3	[12]
	4	5		23.1	[7,10]
	4	12	0.05	23.1	[9]
	5	14	<1	24.2.3	[29]
	5	3	sub µg L^{-1}	24.2.3	[26]
	6	6	9	25.2.1.3	[12]
	1	6	10	20.1.1	[55]
	1	10	0.002	20.1.6	[82,124–139]
	1			20.1.5	[98]
	8	5		28.1.3	[20]
	8	1		27.1.6	[46]
	8	6		28.1.1	[11]
	8	24	0.2	27.1.1	[3,13,14]
	8		1	28.1.1	[9,10]

*µg L^{-1} unless otherwise stated

Metal	Preconcn. method	Analytical finish	Detection limit*	Section No.	[Ref.]
	12		0.01	29.1.1	[12]
	10		0.025ng absolute	29.1.1	[55]
	10			29.1.10	[42,43]
	10		0.005	29.1.1	[49–51]
Ag	1			20.1.1	[1,2,26]
	1	1	0.2	20.1.1	[38]
	1	1	2p mole	20.1.2	[66]
	1	1	0.0005	20.1.2	[19]
	1	1		20.1.2	[64]
	1	1	0.0002	20.1.5	[38]
	2	1		21.1.3	[19]
	4	5		23.1	[7]
	4	1		23.1	[4]
	5	1		24.2.1	[3]
	5	18		25.2.1.4	[29]
	5	5		28.1.3	[20]
Na	6	5		25.2.1.2	[10]
	6	6		25.2.1.3	[11]
	6	15		25.2.1.4	[18,19]
Sr	1	1	7–13	20.1.5	[130]
	1			20.1.4	[75,76]
	3		0.02pCi L^{-1}	22.1.4	[28,29]
	6	15		25.2.1.4	[21–23]
	6		5fCi L^{-1}	25.2.1.4	[18,19]
	8	1		27.1.6	[44,45]
	8	6	<2	27.1.6	[52]
Tc	1	15		20.1.5	[88]
Te	8	1		27.1.6	[46,48]
Tl	5	1	0.003	24.2.1	[5]
	5	9		24.2.1	[4]
	4			23.1	[4]
	13			29.1.1	[5]
Th	1			20.1.5	[132]
	5	18		24.2.3	[16]
	8	18	0.07	27.1.1	[20]
	8			27.1.5	[37]
Sn	1	1		20.1.5	[102]
	1	7		20.1.3	[71]
	3	18		22.1.2	[6]
	6	2		25.2.1.4	[17]
	8	1	1µg absolute	28.1.3	[19]
	8			27.1.1	[8]
	8	15		27.1.6	[10]
	10		0.025ng absolute	29.1.1	[55]
	19		0.028	29.1.1	[11]
	10		0.2	29.1.1	[54]
	10		90pg absolute	29.1.10	[42,43]

*µg L^{-1} unless otherwise stated

Metal	Preconcn. method	Analytical finish	Detection limit*	Section No.	[Ref.]
Ti	2		<1000	21.1.4	[35]
W	1	1		20.1.5	[84]
	1	7		20.1.3	[7]
	3	6	0.5	22.1.3	[12]
	4	3	1.2	23.1	[6]
	4	4	0.06	23.1	[6]
	8	5		28.1.3	[20]
U	1	18		20.1.5	[90]
	1	25		20.1.5	[99]
	1	11		20.1.5.1	[87]
	3	6	0.5	22.1.3	[12]
	3	1	0.1	21.1.3	[16]
	3	6	0.1	21.1.3	[16]
	3	3	0.1	21.1.3	[16]
	3	26	0.02	22.1.4	[32]
	3			22.1.4	[33]
	7	26		20.1.2	[14]
	7			26.1.2	[8]
	4	5		23.1	[7]
	5	27	0.3	24.2.3	[18]
	5	27	0.04	24.2.3	[19–21]
	6	27	0.3	25.2.1.4	[28]
	6		µg amount	25.2.1.4	[27]
	6	5		25.2.2.2	[44]
	6			25.2.2.4	[52]
	8	18	0.07	27.1.1	[20]
	8	5	0.006	28.1.3	[17]
	8	5	<0.5	28.1.3	[18]
	8	6		28.1.2	[8]
V	1	18		20.1.5	[89,93,96]
	1	1	10	20.1.5	[97]
	1	1		20.1.1	[3]
	1	3	0.02–0.5	20.1.1	[39]
	1	7		20.1.3	[71]
	2		<1000	21.1.4	[35]
	2		<1	21.1.4	[22]
	2			21.1.2	[11]
	3	6	0.5	22.1.3	[12]
	5	5	0.1	25.2.1.2	[9]
	5			24.2.3	[25]
	8	18		27.1.1	[17]
	8	5		28.1.2	[15]
	8	6		28.1.1	[2]
	8			28.1.3	[22]
Y	2	15		21.1.1	[3]
	3		0.02pCi L^{-1}	22.1.4	[28,29]
	1	2		20.1.5	[85]

*µg L^{-1} unless otherwise stated

Metal	Preconcn. method	Analytical finish	Detection limit*	Section No.	[Ref.]
	1			20.1.5	[131]
Zn	1	2		20.1.5	[78,79,86]
	1	18		20.1.5	[92–94]
	1	1	0.5	20.1.1	[4]
	1	1		20.1.1	[1,3,6,18,21,27]
	1	1		20.1.5	[77,112]
	1	1	0.03	20.1.5	[123]
	1	1	0.0019	20.1.5	[121]
	1	1	3	20.1.5	[101]
	1	1	0.3–1.1	20.1.2	[63]
	1	1	0.02	20.1.5	[81]
	1	1	10	20.1.5	[100]
	1	3		20.1.1	[45]
	1	3	0.02–0.5	20.1.1	[39]
	1	6		20.1.1	[5]
	1	6	0.02–1	20.1.1	[56]
	1	7		20.1.3	[7]
	1	8		20.1.1	[60,61]
	1			20.1.2	[64]
	2	1	factor 1000	21.1.1	[5,6]
	2	1	factor 1000	21.1.3	[17]
	2	1		21.1.3	[18,19]
	2	1		21.1.4	[29]
	2	1	1.5	21.1.4	[41]
	3	1		22.1.1	[1]
	3	6	0.5	22.1.3	[12]
	3			22.1.3	[16]
	4	5		23.1	[7]
	4	6		23.1	[2]
	5	18		24.2.3	[13]
	5	1	µg L⁻¹	24.2.1	[1]
	6	2		25.2.1.4	[17]
	6	1		25.2.2.1	[42]
	6	1	1–3	25.2.1.1	[51]
	6	1	0.05–0.5	25.2.1.1	[2,3]
	6			25.2.2	[34]
	6			25.2.2.4	[50,51]
	6	6		25.2.1.3	[12]
	6	6		25.2.1.3	[11]
	6	18	2	25.2.2.4	[14]
	6	9	sub µg L⁻¹	25.2.2.3	[49]
	7			26.1.2	[26]
	8	5		28.1.3	[20]
	8	5		28.1	[15]
	8	5	0.3	28.1.3	[17]
	8	6	0.5–1	27.1.1	[4]
	8	6		28.1.1	[5]

*µg L⁻¹ unless otherwise stated

Metal	Preconcn. method	Analytical finish	Detection limit*	Section No.	[Ref.]
	8	7	1–10	27.1.1	[12]
	8	14		27.1.1	[16]
	12	13	100	29.1.1	[9]
	14			29.1.2	[21]
	15	1		29.1.2	[17,18]
Zr	8	18		27.1.1	[18]
	8	20		27.1.1	[15]
	8			27.1.1	[8]
	12		2.3×10^{-10}M	29.1.1	[16]

Source: Own files

Table 32.2 Preconcentration of metals in seawaters

(Key as in Table 32.1)

Metal	Preconcn. method	Analytical finish	Detection limit*	Section No.	[Ref.]
Al	1	10		20.2.8	[210]
	1	18		20.2.7	[209]
	8	3		27.2.6	[83]
	9	16		31.1.1	[9]
	8			27.2.3	[76]
	8			27.2.1	[72]
Sb	2	1	0.05	21.2.2	[12]
	3	1	0.05	22.2.1	[40,42]
	6	5		25.3.2.5	[46,88]
	8	5		27.2.1	[65]
	8			27.2.3	[76]
	10	1		29.2.4	[72]
As	1	5		20.2.1	[179]
	1	1		20.2.1	[168]
	6	5		25.3.2.5	[46,65,88,94]
	8	3		27.2.3	[76]
	8	5		27.2.1	[65]
	10	1		29.2.4	[44]
	16			29.2.3	[70]
Ba	6	1		25.3.1	[55]
	6	5		25.3.2.5	[90]
Be	4	1	0.0006	23.2	[11]
	8	3	<4	27.2.7	[84]
Bi	1	1	0.003	20.2.1	[167]
	1	1		20.2.1	[170]

*μg L^{-1} unless otherwise stated

Metal	Preconcn. method	Analytical finish	Detection limit*	Section No.	[Ref.]
	3	13	sub μg	22.2.1	[41]
B	8	3	<4	27.2.7	[84]
Cd	1	1		20.2.1	[11,152,156,158, 160,163–165, 168,171,226]
	1	1	<0.001	20.2.1	[154]
	1	1	0.03	20.2.1	[8]
	1	1	0.02	20.2.1	[154]
	1	1	0.02	20.2.1	[11]
	1	1	0.003	20.2.1	[8]
	1	1	2ng L^{-1}	20.2.1	[151]
	1	1	0.05	20.2.2	[14]
	1	1	2	20.2.1	[156]
	1	3	0.02	20.2.1	[173]
	1	3	0.25	20.2.1	[147]
	1	3	<4	20.2.1	[174]
	1	1	0.0004	20.2.2	[187]
	1	1	0.027	20.2.1	[148]
	1	1	0.02	20.2.1	[148]
	1			20.2.1	[155]
	2	1	0.004–0.024	21.2.7	[62,72]
	2	1		21.2.1	[47]
	2	3		21.2.7	[66,82]
	2	6		21.2.6	[58,86]
	2			21.2.7	[63,76]
	3	1		22.2.3	[52]
	3	13	sub ng	22.2.1	[26,41]
	3			22.2.4	[55,56]
	4	5		23.2	[7]
	5	1		24.3.2	[43]
	5	14		24.3.2	[42]
	5			24.3.2	[44]
	6	1	0.006–0.01	25.3.2.1	[63,65,66,67]
	6	1	0.05	25.3.2.1	[3]
	6	1	0.0001	25.3.2.1	[71]
	6	1	preconcn. 40	25.3.1	[54]
	6	1	<1	25.3.2.1	[77]
	6	3		25.3.2.2	[81]
	6	5		25.3.2.5	[65,90]
	6	22		25.3.2.3	[63,84]
	6			25.3.2.1	[70,75]
	6			25.3.2	[62]
	8	1		27.2.7	[85,87]
	8	1		28.2.1	[23]
	8	1		28.2.3	[30]
	8	3	<1	27.2.2	[75]
	8	3	<4	27.2.7	[84]

*μg L^{-1} unless otherwise stated

Metal	Preconcn. method	Analytical finish	Detection limit*	Section No.	[Ref.]
	8	3		27.2.3	[76]
	8	5	I	28.2.1	[26]
				27.2.7	[90]
	8	14		27.2.3	[77]
	8			27.2.1	[23]
	16	I		29.2.3	[80]
Cs	I	I		20.2.8	[217]
	2	I		21.2.7	[65]
	2	23		21.2.7	[68]
	6	15	10pCi L⁻¹	25.3.1	[57]
	6	15		27.2.5	[81]
	6	I		25.3.1	[56]
Ca	I	I	10	20.2.1	[156]
	I	I		20.2.1	[170,178]
	I	6	<0.14µg absolute	20.2.3	[192]
	2	I		21.2.3	[54]
	6	5		25.3.2.5	[88–90]
	6	28		25.3.2.4	[85]
	8	I		28.2.3	[30]
	8	3	<1	27.2.2	[71]
	8	3	<4	27.2.7	[84]
	8	5	I	28.2.1	[26]
Ce	4	5		23.2	[7]
	6	5		25.3.2.5	[89,90]
	8	27		27.2.1	[55]
	8			28.2.4	[31]
Cr	I	I		20.2.1	[11,155,158,165]
	I	I	0.04	20.2.2	[14]
	I	I	0.05	20.2.1	[8]
	I	6	<0.14µg absolute	20.2.3	[192]
	I	10		20.2.8	[211–213]
	I			20.2.1	[155]
	2	I		20.2.1	[53–55]
	2	3		21.2.7	[82]
	2	18		21.2.1	[48–50]
	3	19	10pmole L⁻¹	22.2.4	[62]
	4	5		23.2	[7]
	6	I	0.05	25.3.2.1	[71]
	6	5	0.14	25.3.2.1	[71]
	6	I	factor ×40	25.3.1	[54]
	6	I		25.3.2.1	[76]
	6	3		25.3.2.1	[76]
	6	I		25.3.2.1	[63]
	6	3		25.3.2.2	[81]
	6	5		25.3.2.5	[65,87,90]
	6			25.3.2.1	[69]
	8	I	0.1–0.2nmole L⁻¹	27.2.1	[10,63]

*µg L⁻¹ unless otherwise stated

Metal	Preconcn. method	Analytical finish	Detection limit*	Section No.	[Ref.]
	8	1	40	27.2.1	[11]
	8	3		27.2.6	[83]
	8	3	<1	27.2.2	[75]
	8	3	<4	27.2.7	[84]
	8	3		27.2.3	[76]
	8	5	7	27.2.7	[90]
	8			27.2.1	[64]
Co	1	18	0.04µmol L⁻¹	20.2.4	[197,198]
	1	18		20.2.6	[191,202–208]
	1	1		20.2.1	[11,154,155, 157,158,168, 169]
	1	6		20.2.1	[184]
	1	6		20.2.3	[191]
	1	6	<0.14µg absolute	20.2.3	[192]
	1	13	5pmole L⁻¹	20.2.1	[183]
	1	12		20.2.1	[182]
	1	22		20.2.8	[216]
	1			20.2.1	[155]
	2	1		21.2.1	[47]
	2	3		21.2.7	[82]
	2	26		21.2.7	[67]
	2	23		21.2.7	[69]
	2			21.2.6	[58]
	3	1		22.2.2	[50]
	3	27		22.2.1	[44]
	3			22.2.4	[26]
	3			22.2.3	[52]
	4	5		23.2	[7]
	5	1		24.3.2	[43]
	6	1	0.04	25.3.2.1	[71]
	6	5	0.006	25.3.2.1	[71]
	6	1		25.3.2.1	[63,66]
	6	1	factor ×40	25.3.1	[54]
	7	14		26.2.1	[30]
	6	3		25.3.2.2	[81]
	6	3		25.3.2.1	[76]
	6	1		25.3.2.1	[76]
	6	5		25.3.2.5	[63,65,87,90,95]
	6	26	5pmole L⁻¹	25.3.1	[60]
	6			25.3.2	[62]
	8	3		27.2.6	[83]
	8	3		27.2.3	[76]
	8	3	<4	27.2.7	[26]
	8	3	<1	27.2.2	[75]
	16	1		29.2.3	[80]
	16	11	5fCi L⁻¹	29.2.3	[62]

*µg L⁻¹ unless otherwise stated

Metal	Preconcn. method	Analytical finish	Detection limit*	Section No.	[Ref.]
	9	26		31.2.1	[15]
	12	14		29.2.1	[52]
	8	5	1	28.2.1	[26]
	8	6		28.2.1	[24]
	8			27.2.1	[70]
	8			28.2.4	[31]
Cu	1	1		20.2.1	[11,152,154,156, 158,160,165, 166,168,169]
	1	1	0.5	20.2.1	[157]
	1	1	<1	20.2.1	[159]
	1	1	0.24	20.2.1	[154]
	1	1	0.1	20.2.1	[11]
	1	1	0.05	20.2.1	[8]
	1	1	0.05	20.2.1	[36]
	1	1	0.05	20.2.1	[8]
	1	1	$0.03\mu mole\ L^{-1}$	20.2.1	[151]
	1	1	0.06	20.2.2	[14]
	1	1	0.006	20.2.2	[187]
	1	3		20.2.1	[173]
	1	3	0.5	20.2.1	[147]
	1	3	<4	20.2.1	[174]
	1	6	13	20.2.1	[184]
	1	6		20.2.3	[191]
	1	17	2	20.2.3	[193]
	1	17		20.2.3	[194]
	1			20.2.1	[155]
	2	1		21.2.1	[1,42–47]
	2	1		21.2.7	[62]
	2	1		21.2.7	[72]
	2	3		21.2.7	[61,82]
	2			21.2.6	[58,86]
	2			21.2.7	[63,76]
	3	1	<1	22.2.4	[58,59]
	3	13	sub µg	22.2.1	[47]
	3			22.2.4	[55,56]
	3			22.2.2	[51]
	3			22.2.1	[43]
	3			22.2.3	[52]
	5	1		22.3.2	[45,46]
	6	1	0.006–0.6	25.3.2.1	[63,65,66]
	6	1		25.3.2.1	[76]
	6	3		25.3.2.1	[76]
	6	1		25.3.2.1	[3]
	6	1	<1	25.3.2.1	[17]
	6	1	factor ×40	25.3.1	[54]
	6	3		25.3.2.2	[81]

*µg L^{-1} unless otherwise stated

Metal	Preconcn. method	Analytical finish	Detection limit*	Section No.	[Ref.]
	6	5		25.3.2.5	[9,63,65, 88,90]
	6	5	0.08	25.3.2.1	[71]
	6	1	0.006	25.3.2.1	[71]
	6	19		25.3.2.5	[34]
	6	22		25.3.2.3	[63,84]
	6			25.3.2.	[62]
	6			25.3.2.1	[69,70]
	8	1		27.2.7	[87]
	8	1		28.2.1	[23]
	8	3	<4	27.2.7	[84]
	8	3		27.2.6	[83]
	8	5		28.2.1	[25]
	8	5		27.2.7	[90]
	8			28.2.4	[31]
	8			28.2.1	[24]
	8			27.2.1	[70]
	16	1		29.2.3	[80]
Eu	4	5		23.2	[7]
	6	5		25.3.25	[65]
Ga	2	3	0.0001–0.0004	21.2.7	[79]
Ge		1		29.3.4	[76]
Au	2			21.2.7	[64]
	4	5		23.2	[7]
	5	22		24.3.2	[44]
	8	5		28.2.1	[25]
	6	5		25.3.2.5	[65,91]
In	2	6	0.0001–0.0004	21.2.7	[79]
	1	10		20.2.8	[216]
	6	5		25.3.2.5	[85]
Ir	5	22		24.3.2	[44]
Fe	1	1		20.2.1	[152,154,155 166]
	1	1	0.2nmole L^{-1}	20.2.1	[151]
	1	1	0.2	20.2.1	[8]
	1	1	<1	20.2.1	[159]
	1	1	1.5	20.2.1	[157]
	1	1	0.20	20.2.1	[8]
	1	1	0.08	20.2.1	[11]
	1	1	<0.02	20.2.1	[154]
	1	1	0.24	20.2.1	[154]
	1	1	0.5	20.2.3	[151,190]
	1	3	<4	20.2.1	[174]
	1	3	0.25	20.2.1	[147]
	1	3		20.2.1	[173]
	1	6		20.2.3	[191]
	1	6	16	20.2.1	[184]

*μg L^{-1} unless otherwise stated

Metal	Preconcn. method	Analytical finish	Detection limit*	Section No.	[Ref.]
	1	6	<0.14µg absolute	20.2.3	[192]
	1	22		20.2.8	[216]
	1			20.2.1	[155]
	2	18	0.1nM	21.2.7	[69]
	2	1		21.2.3	[54]
	2	1		21.2.1	[47]
	2	1	40	21.2.7	[62]
	2	3		21.2.7	[82]
	2	26	0.05nM L^{-1}	21.2.7	[80]
	2	23		21.2.7	[68]
	3			22.2.4	[55]
	3			22.2.2	[51]
	3			22.2.3	[52]
	4	5		23.2	[7]
	5	18		24.3.1	[31]
	5	1		24.3.2	[42]
	6	1		25.3.2.1	[63,66]
	6	1	factor ×40	25.3.1	[54]
	8	3		27.2.3	[76]
	8	3		27.2.6	[83]
	8	3		27.2.4	[31]
	9	18	0.01nM	31.2.1	[16]
	12			29.2.4	[3]
	6	3		25.3.2.2	[81]
	6	3		25.3.2.1	[76]
	6	1		25.3.2.1	[76]
	6	5		25.3.2.5	[65,87,88,90]
	6	22		25.3.2.3	[63,84]
Lanthanides	1	1	0.0006–0.003	20.2.1	[177]
	1	4		21.2.7	[71]
	8	13		27.2.1	[56–62]
	8	3		27.2.7	[84]
La	4	5		23.2	[7]
	6	5		25.3.2.5	[90]
	8	3		27.2.3	[76]
	6	5		25.3.2.5	[46]
Pb	1	1		20.2.1	[11,154,156, 158,165,166, 168–170]
	1	1		20.2.2	[186]
	1	1	0.03	20.2.1	[8]
	1	1	0.04	20.2.1	[154]
	1	1	2.6	20.2.1	[157]
	1	1	<0.016	20.2.1	[154]
	1	1	0.06	20.2.1	[11]
	1	1	0.04	20.2.2	[14]
	1	1	4	20.2.1	[156]

*µg L^{-1} unless otherwise stated

Metal	Preconcn. method	Analytical finish	Detection limit*	Section No.	[Ref.]
	1	3	2.5	20.2.1	[147]
	1	3	0.6	20.2.1	[173]
	1	6	40	20.2.1	[184]
	2	19		20.2.1	[178]
	2	15		20.2.1	[180]
	2			20.2.1	[155]
	2	29		21.2.2	[53]
	2	1		21.2.1	[47]
	2	1		21.2.7	[62,63,74,76]
	2			21.2.6	[58,86]
	3	1		22.2.1	[47]
	3	3	0.0006	22.2.4	[63]
	3	13	sub ng	22.2.1	[41]
	3			22.2.3	[52]
	3			22.2.4	[55,56]
	5	14		24.3.2	[42]
	6	1	factor ×40	25.3.1	[54]
	6	1	0.016	25.3.2.1	[71]
	6	5	0.016	25.3.2.1	[71]
	6	1	0.16–0.28	25.3.2.1	[63,65,66]
	6	1	<1	25.3.2.1	[77]
	6	1	0.5	25.3.2.1	[3]
	6	3		25.3.2.2	[81]
	6	5		25.3.2.5	[87]
	6	9		25.3.2.5	[34]
	6	22		25.3.2.3	[63,84]
	6	3		25.3.2.2	[83]
	6			25.3.2.1	[70]
	6			25.3.2	[62]
	8	1	0.05	28.2.3	[30]
	8	1		27.2.7	[85,87]
	8	3	<1	27.2.2	[75]
	8			28.2.4	[31]
	8			27.2.3	[76]
	8			27.2.1	[72]
	8	15	15	27.2.4	[78]
	9	16	0.011	31.1.1	[10]
	9	16	1	31.1.1	[1–8]
	11			29.2.1	[56.57]
Li	8	3	<4	27.2.7	[84]
Mg	8	3	<4	27.2.7	[84]
Mn	1	1		20.2.1	[11,149,154,155, 158,163]
	1	1	<0.004	20.2.1	[154]
	1	1	30	20.2.3	[188,189]
	1	1	0.02	20.2.1	[157]
	1	1	0.02	20.2.1	[154]

*µg L⁻¹ unless otherwise stated

Metal	Preconcn. method	Analytical finish	Detection limit*	Section No.	[Ref.]
	1	1	0.07	20.2.1	[148]
	1	3	<4	20.2.1	[174]
	1	3	0.063	20.2.1	[147]
	1	6		20.2.3	[191]
	1	6	15	20.2.1	[184]
	1	6	<0.14µg absolute	20.2.3	[192]
	1			20.2.1	[12]
	1			20.2.1	[155]
	1	18		20.2.8	[218]
	2	1		21.2.3	[54]
	2	1		21.2.7	[62,76]
	2	1		21.2.1	[47,54]
	3			22.2.3	[49]
	3			22.2.4	[55]
	6	1		25.3.2.7	[63,66]
	6	1	factor ×40	25.3.1	[54]
	6	1		25.3.2.1	[66,72,73,76]
	6	1	0.004	25.3.2.1	[71]
	6	5	0.016	25.3.2.1	[71]
	6	3		25.3.2.2	[81]
	6	5		25.3.2.5	[65,90]
	6			25.3.2	[62,69]
	8	18		27.2.1	[54]
	8	1		28.2.3	[30]
	8	3	<4	27.2.7	[84]
	8	3	<1	27.2.2	[75]
	8	3		27.2.6	[83]
	8	3		27.2.3	[76]
	8	5	1	28.2.1	[26]
	8	5	0.07–1.7	27.2.7	[90]
	8			28.2.4	[31]
	8	13		27.2.1	[71]
Hg	1	1	5	20.2.1	[153]
	1	1	0.05	20.2.2	[14]
	2			21.2.6	[58,86]
	2			21.2.7	[74]
	3	1	0.016	22.2.1	[45]
	3	1		22.2.2	[48]
	3	13	sub ng	22.2.1	[41]
	4	5		23.2	[7]
	6	5		25.3.2.5	[46,63,91]
	7	1	6.6	26.3.1	[23]
	7	1		26.2.4	[41–43]
	7	1	0.21pmole	26.2.4	[44]
	7			26.2.4	[14,15,45]
	8	1		27.2.7	[85]
	8	5		28.2.1	[25]

*µg L^{-1} unless otherwise stated

Metal	Preconcn. method	Analytical finish	Detection limit*	Section No.	[Ref.]
	8	5	1	28.2.1	[26]
	8	6		28.2.1	[24]
		1		29.3	[77]
	16	1		29.2.3	[69]
Mo	1	1		20.2.1	[168]
	1	3		20.2.1	[173]
	2	1		21.2.6	[59–61]
	2	3		21.2.7	[82]
	2			21.2.7	[75]
	3	3		22.2.4	[53]
	4			23.2	[1]
	5	18		24.3.1	[34,36]
	5			24.3.2	[46]
	5			24.3.1	[38]
	6	5		25.3.2.5	[46,65,94]
	8		0.7–5.7	27.2.1	[73]
	8	3		27.2.3	[76]
	8	6	0.3	28.2.1	[27]
	5			24.3.1	[36,39]
	16			29.2.3	[70]
Ni	1	1		20.2.1	[11,36,154,156, 158,160,164, 166,169]
	1	1	0.032	20.2.2	[187]
	1	1	<0.012	20.2.1	[154]
	1	1	0.03	20.2.2	[14]
	1	1	0.10	20.1.1	[8]
	1	1	0.6	20.2.1	[157]
	1	1	0.5	20.2.4	[199]
	1	1	16	20.2.1	[156]
	1	18		20.2.4	[14]
	1	18	0.5	20.2.4	[197,198]
	1	3	0.5	20.2.1	[147]
	1	3		20.2.1	[173]
	1	6		20.2.3	[191]
	1	6	<0.14µg absolute	20.2.3	[192]
	1	6	8	20.2.1	[184]
	1			20.2.1	[155]
	2	1		21.2.7	[62,77]
	2	1		21.2.1	[47]
	2	3		21.2.7	[71]
	2			21.2.7	[76]
	3	1		22.2.4	[57]
	3	13	sub ng	22.2.1	[41]
	3			22.2.3	[52]
	3			22.2.4	[55]
	6	1	factor ×40	25.3.1	[54]

*µg L^{-1} unless otherwise stated

Metal	Preconcn. method	Analytical finish	Detection limit*	Section No.	[Ref.]
	6	1	0.015	25.3.2.1	[66,67]
	6	1		25.3.2.1	[63,65,71,77]
	6	3		25.3.2.2	[81]
	6	5		25.3.2.5	[65]
	6	22		25.3.2.3	[63,84]
	6			25.3.2	[62,70]
	8	1		28.2.1	[23]
	8	1	0.5	28.2.1	[29]
	8	3	<4	27.2.7	[84]
	8	3	<1	27.2.2	[75]
	8			27.2.3	[76]
	12	14		29.2.1	[60]
	10	1		29.2.4	[75]
Pd	2			21.2.7	[64,71]
	5	22		24.3.2	[44]
Pu	3	15		22.2.2	[49]
	5	22		24.3.2	[44]
	8	15	0.004dpm	27.2.1	[68]
	8	13		27.2.1	[74]
Po	8			27.2.4	[79]
	8	15		27.2.4	[78,80]
	11			29.2.1	[58]
	1		10dpm	26.2.2	[3]
	7	23		26.2.3	[33,40]
	7	11		26.2.2	[35–38]
Rh	5	5		24.3.2	[49]
Rb	8	6	2–4	27.2.7	[86]
Sc	6	5		25.3.2.5	[65]
Se	1	10		20.2.8	[138]
	3	1	0.007	22.2.1	[40,42]
	4	5		23.2	[7]
	3	19		22.2.4	[54]
	5	27		24.3.1	[41]
	6	5		25.3.2.5	[7,46,90]
	8	3	<4	27.2.7	[84]
	8	10	0.005	27.2.1	[66]
	8	15		27.2.1	[67]
	16		<0.4	29.2.3	[63]
Ag	1	1	2µmole L^{-1}	20.2.1	[151]
	1	1	0.02	20.2.1	[8]
	2			21.2.7	[64]
	3	1		22.2.3	[52]
	4	5		23.2	[7]
	5	5		24.3.2	[50]
	6	5		25.3.2.5	[65]
	8	5	0.2	27.2.7	[83]
Sr	2	1	0.007–0.073	21.2.7	[81]

*µg L^{-1} unless otherwise stated

Metal	Preconcn. method	Analytical finish	Detection limit*	Section No.	[Ref.]
	6	5		25.3.2.5	[87,88]
	22	5	0.7	29.2.1	[27]
Tc	5	22		24.3.2	[44]
	8	23		27.2.1	[69]
Te	6	1	0.02	25.3.2.1	[71]
	6	5	1.2	25.3.2.1	[71]
	8	1	5pmole L^{-1}	27.2.7	[51]
Tl	8	3		27.2.6	[83]
Th	7	15		26.2.1	[31]
	6	1	0.004	25.3.2.1	[71]
	6	5	0.004	25.3.2.1	[71]
	6	5		25.3.2.5	[84]
	8	5	1	28.2.1	[26]
	8	5	0.3–0.9	27.2.7	[90]
Sn	1	1		20.2.1	[168]
	5	18		24.3.1	[32]
	8	1		28.2.2	[19]
Ti	1	18		20.2.1	[185]
	2	3	0.0001–0.0004	21.2.7	[65,79]
	8			27.2.3	[76]
U	1	1		20.2.1	[144]
	1	6		20.2.3	[191]
	1	18		20.2.8	[219,220]
	1	27		20.2.8	[219,220]
	2			21.2.6	[58]
	4	5		23.2	[7]
	5	18		24.3.1	[40]
	6	5		25.3.2.5	[46,65,87,90,94]
	6	22		25.3.2.3	[84,63]
	8	27	200	27.2.7	[89]
	8	5		28.2.1	[26]
	8	5	0.007–0.018	27.2.7	[90]
	16			29.2.3	[66]
	16	2		29.2.3	[65]
V	1	1	<1	20.2.5	[201]
	1	6	<0.14µg absolute	20.2.3	[192]
	1	3		20.2.1	[173]
	1	3	0.38	20.2.1	[147]
	1	18	0.025	20.2.5	[200]
	2	3		21.2.7	[78]
	3	3		22.2.4	[53]
	5	1		24.3.2	[43,47]
	5	5		24.3.2	[47]
	5	18		24.3.1	[33]
	6	5	0.06	25.3.2.1	[71]
	6	5		25.3.2.5	[46,63,65,94]
	8	3	<4	27.2.7	[84]

*µg L^{-1} unless otherwise stated

Metal	Preconcn. method	Analytical finish	Detection limit*	Section No.	[Ref.]
	8	5		28.2.1	[26]
	8		0.7–5.7	27.2.1	[73]
	8			27.2.1	[72]
	8			27.2.3	[76]
	16			29.2.3	[70]
Y	8	3		27.2.6	[83]
	8			27.2.3	[76]
Zn	1	1		20.2.1	[8,154,156,158, 165,166,169, 170]
	1	3	<4	20.2.1	[174]
	1	3		20.2.1	[173]
	1	3	0.13	20.2.1	[147]
	1	6	0.14μg absolute	20.2.3	[192]
	1	6	13	20.2.1	[184]
	1	19		20.2.1	[158]
	1	10		20.2.8	[216]
	1			20.2.1	[155]
	2	1		21.2.7	[62]
	2	3		21.2.7	[82]
	2	23		21.2.7	[68]
	2	1		21.2.1	[47]
	2	1	2.4ng dm^{-3}	21.2.7	[70]
	2			21.2.7	[76]
	3	1	2.4ng dm^{-3}	22.2.1	[46]
	3	19		22.2.4	[60]
	3			22.2.3	[52]
	3			22.2.4	[55,56]
	3			22.2.2	[50]
	4	5		23.2	[7]
	5	1		24.3.2	[43,44]
	6	1	0.5	25.3.2.1	[68]
	6	1	<1	25.3.2.1	[77]
	6	1	0.015–1.8	25.3.2.1	[63,65–67]
	6	1	0.03	25.3.2.1	[3]
	6	1	0.016	25.3.2.1	[71]
	6	1		25.3.2.1	[76]
	6	3		25.3.2.1	[76]
	6	3		25.3.2.2	[81]
	6	5		25.3.2.5	[46,65,90
	6	5	0.20	25.3.2.1	[71]
	6	22		25.3.2.3	[63,84]
	6	9		25.3.2.5	[34]
	6		0.1n mole	25.3.1	[59]
	6			25.3.2	[62]
	1	1	1.0	20.2.1	[14]
	1	1	0.4	20.2.1	[157]

*μg L^{-1} unless otherwise stated

Metal	Preconcn. method	Analytical finish	Detection limit*	Section No.	[Ref.]
	1	1	0.6	20.2.2	[14]
	1	1	0.34	20.2.1	[11]
	1	1	0.03	20.2.1	[8]
	1	1	30	20.2.1	[156]
	1	1	0.016	20.2.2	[187]
	1	1	<0.08	20.2.1	[154]
	8	1		28.2.3	[30]
	8	3		27.2.6	[83]
	8	3		27.2.7	[84]
	8	5		28.2.1	[26]
	8	23		27.2.1	[71]
	8			28.2.4	[31]
	8			27.2.3	[76]
Zr	8	18		27.2.1	[18]
	8	5	40	27.2.7	[90]
	2	23		21.2.7	[13]

*μg L⁻¹ unless otherwise stated

Source: Own files

Table 32.3 Preconcentration of metals in estuary waters

For key see Table 32.1

Metal	Preconcn. method	Analytical finish	Detection limit*	Section No.	[Ref.]
Cd	1	1		20.3.1	[155,166,221, 222]
	1	1		20.3.2	[223]
	1	1	0.5	20.3.1	[36]
Cr	8	5	0.1	27.3	[26]
Co	1	1		20.3.1	[155,221,222]
Cu	1	1		20.3.1	[20,23,37,38, 71,155,166, 221–228]
	1	1	0.3	20.3.1	[46]
Fe	1	1		20.3.1	[155,166,222]
Pb	1	1		20.3.1	[155,166]
	1	1		20.3.2	[223]
	1	1	sub ng L⁻¹	20.3.1	[221,222]
	1	1	0.7	20.3.1	[36]
Mn	1			20.3.1	[155,221,222]
Ni	1	1		20.3.1	[155,166]

*μg L⁻¹ unless otherwise stated

Table 32.3 continued

Metal	Preconcn. method	Analytical finish	Detection limit*	Section No.	[Ref.]
	I	I	sub ng L $^{-1}$	20.3.1	[221,222]
	I	I	0.02	20.3.1	[36]
Zn	I	I		20.3.1	[155,166]
	I	I		20.3.2	[223]
	I	I	sub ng L $^{-1}$	20.3.1	[221,222]

*µg L $^{-1}$ unless otherwise stated

Source: Own files

Table 32.4 Preconcentration of metals in potable waters

For key see Table 32.1

Metal	Preconcn. method	Analytical finish	Detection limit*	Section No.	[Ref.]
Al	5			24.4	[10]
As	8	18		27.4	[94]
Bi	I	I		20.4.1	[224]
Cd	I	I	0.05	20.4.1	[225]
	I	I		20.4.1	[20,224]
	3			22.4.1	[28]
	5	I		24.4	[51]
	6	I	0.1	25.4	[97]
	6	I		25.4	[96]
	8			27.4	[95]
Cr	6	I	3	25.4	[97]
Co	I	I		20.4.1	[224]
	6	I	I	25.4	[97]
Cu	I	I		20.4.1	[224]
	I	I	0.01	20.4.1	[225]
	I	16,11		31.2.1	[19]
	5	I		24.4	[51]
	6	I		25.4	[96]
	6	I	0.5	25.4	[97]
	8			27.4	[95]
	8	I		27.4	[42]
Fe	6	I		25.4	[97]
Pb	I	I		20.4.1	[224,226]
	I	I	0.9	20.4.1	[225]
	6	I		25.4	[96]

*µg L $^{-1}$ unless otherwise stated

Table 32.4 continued

Metal	Preconcn. method	Analytical finish	Detection limit*	Section No.	[Ref.]
	6	I	6	25.4	[97]
	5	I		24.4	[51]
	8	I 4	50	27.4	[96]
	8	I		28.3.1	[35]
Mn	6	I	0.5	25.4	[97]
	3			22.4.1	[67]
	8			27.4	[95]
Hg	7	I		26.1.3	[24]
	8	I 4		27.4	[93]
Ni	I	I		20.4.1	[23,224]
	6	I		25.4	[97]
	8	I		27.4	[42]
Ag	I	I		20.4.1	[224]
Tl	I	I		20.4.1	[224]
V	I	I 8		20.4.1	[226]
Zn	I	I	6	20.4.1	[225]
	I	I		20.4.1	[224]
	3			22.4.1	[67]
	6	I	2	25.4	[97]
	6	I		25.4	[96]

*μg L⁻¹ unless otherwise stated

Source: Own files

Table 32.5 Preconcentration of metals in miscellaneous waters

For key see Table 32.1

Metal	Preconcn. method	Analytical finish	Detection limit*	Section No.	[Ref.]
Ground waters					
Bi	3			22.6	[72]
Hg	3	18		22.6	[7]
Ra	7			26.2.2	[32]
U	2	18		21.3	[83]
	22	23		29.4.1	[31]
Sewage effluents					
Cd	15	I		29.6.1	[79]
	I			20.5	[227]
	6	I		25.5	[98]

*μg L⁻¹ unless otherwise stated

Metal	Preconcn. method	Analytical finish	Detection limit*	Section No.	[Ref.]
Cu	6	1		25.5	[98]
	15	1		29.6.1	[79]
Fe	1		1	20.5	[227]
Pb	1		1	20.5	[227]
	6	1		25.5	[98]
	15	1		29.6.1	[79]
Ni	1		1	20.5	[227]
Zn	1	1		20.5	[227]
	5			24.4	[13]
	6	1		25.5	[98]
	15	1		29.6.1	[79]
Trade effluents					
Sb	1			20.6	[71]
B	1			20.6	[71]
Cd	1			20.6	[71]
	3			22.5	[69]
	1			20.6	[71]
Gd	1			20.6	[71]
Hg	3			22.5	[70]
Mo	1			20.6	[71]
	3			22.5	[23–25,59,68,69]
Nb	1			20.6	[71]
	3			22.5	[69]
Ru	3			22.5	[69]
Sn	1			20.6	[71]
W	1			20.6	[71]
V	1			20.6	[71]
Zn	1			20.6	[71]
Zr	3			22.5	[71]
Waste waters					
Cd	2	1		21.4	[21]
Co	2	1		21.4	[84]
Pb	2	1		21.4	[84]
Ni	2	1		21.4	[85]
Th		19		29.5	[75]
Zn	2	1		21.4	[84]

*µg L^{-1} unless otherwise stated

Source: Own files

Chapter 33

Detection limits achievable for cations

33.1 Preconcentration methods

Detection limits achieved for non saline and seawaters by various methods of preconcentration are summarised in Table 33.1 (based on information given in Tables 32.1 to 32.2).

The methods of preconcentration used are discussed under five headings:

(a) Preconcentration by extraction of water with a relatively small volume of an organic solvent. For further details see Chapter 20.
(b) Preconcentration by adsorption of cations in water by various solids followed by either

    (i) desorption in solid with a relatively small volume of aqueous extractant or organic solvent; or

    (ii) direct analysis of the solid by techniques such as X-ray fluorescence spectroscopy or neutron activation analysis.

    The adsorbent solids used in this procedure include immobilised chelators, solid adsorbents, (eg silica gel, alumina), active carbon, metal oxides an metals. For further details see Chapters 21–23 and 26.
(c) Adsorption of cations from water on anionic or cation exchange resin, followed by desorption with a relatively small volume of aqueous reagent. For further details see Chapters 24 and 25.
(d) Coprecipitation methods. In this method of preconcentration an addition is made to a relatively large volume of water sample of a soluble salt such as hafnium or gallium, an alkali is then added which precipitates the hydroxide of this metal on which is coprecipitated the cation which it is required to preconcentrate. This solid is then isolated, for example by filtration or centrifugation and subject to a suitable analytical finish. For further details see Chapters 27 and 28.
(e) Miscellaneous preconcentration procedures. These include various preconcentration procedures which have been studied only to a small extent. For further details see Chapter 29.

The major techniques used for analytical finish for each of the five categories of preconcentraiton procedure are discussed below in more detail.

33.2 Analytical finishes

33.2.1 Atomic absorption spectrometry

Table 33.2 (based on Table 33.1) reports the ranges of detection limits achieved for non saline and seawaters by various workers when atomic absorption spectrometry is applied to the concentrated cation extracts of nine heavy metals obtained using the five methods of preconcentration discussed above.

It is seen that solvent extraction preconcentration techniques are very attractive achieving, at the best, detection limits as low as 2ng L $^{-1}$ for zinc in non saline waters and 6ng L $^{-1}$ for copper in seawater which are appreciably lower than those obtained by methods based on preconcentration on solids (10–500ng L $^{-1}$) ion exchange resins (16–500ng L $^{-1}$) or coprecipitation (5–500ng L $^{-1}$).

The detection limits obtained by solvent extraction preconcentration, followed by atomic absorption spectrometry are considerably lower than those achieved by direct atomic absorption spectrometry for heavy metals (viz 5–50µg L $^{-1}$) or graphite furnace atomic absorption spectrometry (0.02–2µg L $^{-1}$ in non saline waters).

Best available detection limit data for metals other than the heavy metals using an atomic absorption spectrometric finish are reported in Table 33.3. Again, extremely low detection limits are achievable, the 1–10ng L $^{-1}$ range being commonplace.

33.2.2 Inductively coupled plasma atomic emission spectrometry

The best available detection limits achievable by this technique are listed in Table 33.4. These results are comparable to those achievable using an atomic absorption spectrometric finish, ie in many instances in the ng L $^{-1}$ range. Again, detection limits achievable using coprecipitation preconcentration are the least sensitive, although, in many instances they will be adequate.

33.2.3 X-ray fluorescence spectroscopy

Although not quite as sensitive as the two previously discussed techniques, X-ray fluorescence spectroscopy combined with various preconcentration techniques is capable of giving detection limits in the 20ng L $^{-1}$ range (solvent extraction) and 500ng L $^{-1}$ (adsorption on solids ion exchange methods and coprecipitation) Table 33.5.

Table 33.1 Summary of detection limits achieved µg L⁻¹ for cations by various preconcentration methods

Key to analytical finishes: atomic absorption spectrometry (1); spectrophotometric method (2); inductively coupled plasma atomic emission spectrometry (3); inductively coupled plasma mass spectrometry (4); neutron activation analysis (5); X-ray fluorescence spectroscopy (6); emission spectrometry (7); supercritical fluid chromatography (8); anodic stripping voltammetry (9); gas chromatography (10); scinatillation counting (11); X-ray spectrometry (12); high performance liquid chromatography (13); differential pulse polarography (14); radiochemical methods (15); flow injection analysis (16); electron spin resonance spectroscopy (17); spectrophotometric methods (18); voltammetric methods (19); isotope dilution methods (20); ultraviolet spectroscopy (21); mass spectrometry (22); γ ray spectrometry (23); molecular emission cavity analysis (24); titration (25); luminescence analysis (26); spectrofluorometry (27); flame photometry (28); laser enhanced ionisation methods (29)

Cation	(a) Solvent extraction (analytical finish)		(b) Adsorption on solids (analytical finish)		(c) Ion exchange methods (analytical finish)		(d) Coprecipitation methods (analytical finish)		(e) Misc. preconcentration methods (analytical finish)	
	Non saline water	Seawater	Non saline water	Seawater	Non saline water	Seawater	Non saline water	Seawater	Non saline water	Seawater
Al	2 (on line)									
Sb	0.01 (5) 200 (2) 0.2 (10)		0.05 (1)						0.2 (gas evolution)	
As	0.01 (5) 0.02–1 (6) 0.07 (1)		0.5 (6)				200 (6)		0.2 (gas evolution)	
Be	0.02–1 (10)		<0.001 (1)	0.0006 (1)			1–10 (7)			
B								<4 (3)		
Bi	1	0.003 (1)						<4 (3)		
Cd	0.006–1.1 (1) 0.017 (3) 0.02 (6) 0.025 (12)	<0.0001–2(1) 0.02–<4 (3)	0.5 (13) 0.0015 1–5 (1)	0.004–0.24(1)	1–3 (1) 0.1 (5)	0.0001–1(1)	0.8 (5) 1–10 (7)	<1–<4 (3) 1 (5)		
Co	1–30 (1) 0.02–0.5 (3)	0.04–0.05(1)	<1 (2)		0.05–0.5 (1)	0.04 (1) 0.006 (5)	0.04 (5) 0.4 (6) 1–10 (7)	1–<4 (3) 1(5)		

Table 33.1 continued

Cation	(a) Solvent extraction (analytical finish)		(b) Adsorption on solids (analytical finish)		(c) Ion exchange methods (analytical finish)		(d) Coprecipitation methods (analytical finish)		(e) Misc. preconcentration methods (analytical finish)	
	Non saline water	Seawater	Non saline water	Seawater	Non saline water	Seawater	Non saline water	Seawater	Non saline water	Seawater
Cu	0.3–20 (1) 2 (2) 0.02–1 (6)	0.006–<1(1) 0.5–<4 (3) 13 (6) 2 (17)	1–5 (1) 0.5 (6) 0.5 (13)	<1 (1)	0.05–5 (1)	0.006–1 (1) 0.08 (5)	<2–<4 (1) 0.5–10 (6) 0.3 (5)	<4 (3)		
Ga	0.001 (5)			0.0001–0.0004 (3)						
Ge	0.004 (6)		0.01 (1)				<8 (1)			
In				0.0001–0.0004 (6)			0.2 (12) <8 (1)			
Fe	3–40 (1) 0.02–0.5 (3) 0.2–1 (6) 0.001 (5)	<0.02–1.5 (1)	0.5 (6)		0.05–0.5 (1) 0.5 (6) 0–20 (21)					
Ir	0.001 (5)	0.25–<4 (3) 16 (16)		40 (1)						
Lanthanides	0.0006–0.003 (1)		0.0002–0.001 (3) 0.5 (6)							0.1 (3) (on-line)
La										
Pb	0.7–5 (1) 0.02–1 (6)	0.02–4 (1) 0.006–2.5 (3) 40 (6)	0.01–5 (1) 0.5 (6)	0.0006 (3)	<1	0.016–<1(1) 0.016 (5)	1–10 (7) 0.5 (6)	0.05 (1) <1 (3) 15 (15) <4 (3)	0.00008 (4)	0.011–1 (16) (on-line)
Mg			1–5 (1)							
Mn	0.1–6 (1) 0.02–1.5 (3) 0.02–1 (6)	<0.004–30 (1) 30 (1)	1–5 (1) 0.5 (6)			0.004 (1)	0.5–1 (6) 0.0006 (5)	<1–<4 (3) 0.07–1.7 (5)		

Table 33.1 continued

Cation	(a) Solvent extraction (analytical finish)		(b) Adsorption on solids (analytical finish)		(c) Ion exchange methods (analytical finish)		(d) Coprecipitation methods (analytical finish)		(e) Misc. preconcentration methods (analytical finish)	
	Non saline water	Seawater	Non saline water	Seawater	Non saline water	Seawater	Non saline water	Seawater	Non saline water	Seawater
Hg	1 (8)	0.016–6.6(1)	0.1–6.6 (1) 0.0002–<1(1) 0.5 (6) 0.02 (5)	0.016–6.6(1)				1 (5)		
Mo	0.1 (8)		<1 0.4 (3) 0.06 (4)		0.2–10 (1)			0.7–5.7 0.3 (6)		
Ni	0.05–20 (1) 0.02–0.5(3) 1 (14)	<0.012–16(1) 0.5 (18) 0.5 (3) 8 (6)	1–100(1) 0.5 (13)	0.015(1)	0.05–0.5(1)		0.57(6) 1–10 (7)	0.5(1) <1–<4 (3)		
Nb	0.1 (18)									
Pd	0.1 (18) 0.04 (1)									
Rb	10(6) 0.002 (10)	0.007 (1)	0.5 (6) 0.007 (1) 0.05 (12)		<1 (14) 9 (6)		0.2 (24)	2–4 (6) <4 (3) 0.005 (10)	0.01 (electro- chemical) 10 (gas evolution)	<0.4 (flotation)
Se										
Ag	0.0002 –0.2(1) 7–13 (6)	0.02 (1)		0.007 –0.073(1)			<0.2 (6)	0.2 (5)		
Sr										
Te							0.02 (1) 1.2 (5)			

Table 33.1 continued

Cation	(a) Solvent extraction (analytical finish)		(b) Adsorption on solids (analytical finish)		(c) Ion exchange methods (analytical finish)		(d) Coprecipitation methods (analytical finish)		(e) Misc. preconcentration methods (analytical finish)	
	Non saline water	Seawater	Non saline water	Seawater	Non saline water	Seawater	Non saline water	Seawater	Non saline water	Seawater
Tl					0.003(1)					
Th						0.004 (1) 0.004 (5)	0.07 (18)	0.3–0.9 (5)		
Sn									0.028 (evaporation)	
Ti				0.0001–0.0004 (3)						
W	0.5 (6) 0.2 (3) 0.06 (4)									
U	0.1–0.5 (6) 0.1 (3) 0.02 (26)				0.04 0.03 (27)		0.07 (18) 0.006 (5)	0.007–0.018(5)		
V	10 (10) 0.02–0.5 (3)	<1 (1) 0.38 (3) 0.025 (18)	<1 0.5 (6)		0.1 (5)	0.06 (5)		<4 (3)		
Zn	0.0019–10(1) 0.02–0.5 (3) 0.02–1 (6)	0.13–<4(3) 13 (6) 0.03–30 (1)	1.5(1) 0.5 (5)		0.05–3(1) 2 (18)	0.015–18(1) 0.2 (5)	0.3 (5) 0.5–10 (6)			
Zr								40 (5)		

Source: Own files

Table 33.2 Preconcentration of heavy metals in non saline and seawaters (best reported detection limits µg L⁻¹) obtained by various methods of preconcentration and finish by atomic absorption spectrometry for non saline and sea waters

Metal	(a) Solvent extraction		(b) Adsorption on solids		(c) Ion exchange resins		(d) Coprecipitation		Direct conventional atomic absorption spectrometry (ie no preconcentration)	Direct GFAAS (ie no preconcentration)
	Non saline	Sea	Non saline	Sea	Non saline	Sea	Non saline	Sea		
Cd	0.006–1.1	<0.0001–2			1–3	0.0001–1			5	0.01
Cr	0.2–0.5	0.04–0.05					0.04		30	
Co	1–30	0.04–0.105			0.05–0.5	0.005			30	
Cu	0.3–2.0	0.006–<1	<1–5	<1	0.05–0.5	0.006–1	<2–<4		30	
Fe	3–40	<0.02–1.5	0.01–5		0.05–5				30	
Pb	0.7–5	0.02–4	1–5		0.16–1	0.004	0.016–<1	0.05	30	
Mn	0.1–6	<0.004–30	1–100						30	5
Ni	0.5–20	0.5–20	1.5	0.015	0.05–5			0.5	30	50
Zn	0.0019–10				0.05–3	0.015–18			50	0.4

Source: Own files

Table 33.3 Preconcentration of metals other than heavy metals (best reported detection limits) ($\mu g\ L^{-1}$) by various preconcentration methods and finish by atomic absorption spectrometry for non saline and sea waters

Metal	(a) Solvent extraction		(b) Absorption on solids		(c) Ion exchange		(d) Coprecipitation	
	Non saline	Sea	Non saline	Sea	Non saline	Sea	Non saline	Sea
Bi		0.003						
Ca	2	1						
Ce					0.05			
Ga								
Ge			0.01				<8	
In							<8	
Ir				4.0				
Lanthanides	0.0006							
Mg			1					
Hg		0.016	0.0002					
Mo			3		2			
Pd	0.04							
Se	0.007	0.007						
Ag	0.0002							
Sr				0.007				
Tl						0.004		
Th					0.003			
V	10	<1						

Source: Own files

Table 33.4 Preconcentration of metals (best reported detection limits, µg L^{-1}) by various preconcentration methods and finish by inductively coupled plasma atomic emission spectrometry for non saline and seawaters

Metal	(a) Solvent extraction		(b) Adsorption on solids		(c) Coprecipitation	
	Non saline	Sea	Non saline	Sea	Non saline	Sea
Be						<4
B						<4
Cd	0.017	0.02				<1
Cr	0.2					<1
Co	0.02					<1
Cu		<0.5				<4
Ga			<0.0004	<0.0004		
Fe	0.02					
Ir		0.25				
Lanthanides			0.0002			
Pb		0.006		0.0006		
Mg						<4
W	0.2					
Ti				0.0001		
U	0.1					
V	0.02	0.38				<4
Zn	0.02	0.13				

Source: Own files

33.2.4 Neutron activation analysis

Coupled to various preconcentration techniques this is an intrinsically sensitive technique for the determination of very low levels of a wide range of cations (generally 1–20ng L^{-1}, see Table 33.6).

Table 33.5 Preconcentration of metals (best reported detection limits, µg L^{-1}) by various preconcentration methods and finish by X-ray fluorescence spectroscopy for non saline and seawaters

Metal	(a) Solvent extraction		(b) Absorption on solids		(c) Ion exchange		(d) Coprecipitation	
	Non saline	Sea	Non saline	Sea	Non saline	Sea	Non saline	Sea
As	0.02						200	
Cd	0.02	0.02						
Cr			0.5					
Co							0.4	
Cu	0.02		0.5				0.4	
In				0.001				
Fe	0.02		0.5					
Pb	0.02	40	0.5		0.5		0.5	
Mn	0.02		0.5				0.5	
Hg			0.5					
Ni		8	0.5		9		0.5	0.3
Se	10		0.5		<2		0.5	
Sr	7							
W	0.5							
U	0.5							
V			0.5					
Zn	0.02	13	0.5					

Source: Own files

Table 33.6 Preconcentration of metals (best reported detection limits, µg L^{-1}) by various preconcentration methods and finish by neutron activation analysis for non saline and seawaters

Metal	(a) Solvent extraction		(b) Absorption on solids		(c) Ion exchange		(d) Coprecipitation	
	Non saline	Sea	Non saline	Sea	Non saline	Sea	Non saline	Sea
Sb	0.01							
As	0.01							
Cd					0.1		0.8	1
Ca						0.14		7
Cr							0.1	
Co						0.006	0.04	1
Cu						0.08		
Ga	0.0001							
Ir	0.0001							
Pb					0.016	0.016	0.07	
Mn								
Hg			0.02					
Se	10				9			
Ag								0.2
Te						1.2		
Th						0.004		
U					0.1			0.07
V						0.2		7
Zn								40

Source: Own files

Chapter 34

Anions: Chelation–solvent extraction techniques

34.1 Non saline waters

34.1.1 Antimonate, trivalent antimony, arsenate and trivalent arsenic

Metzger and Braun [1] preconcentrated antimonate (pentavalent antimony) by conversion to its chelate with n-benzoyl-n-phenylhydroxyamine and extraction into chloroform. Trivalent antimony was converted to its chelate with ammonium pyrrolidinedithiocarbamic acid and extracted into methyl isobutyl ketone. Antimony was determined in extracts of rain, snow and non saline water by anodic stripping voltammetry.

To preconcentrate arsenate and antimonate, Mok and Wai [2] and Mok et al. [3] first reduced them to their trivalent forms with thiosulphate and iodide at pH1.0 then adjusted to pH3.5–5.5, and, in the presence of citrate buffer and EDTA, extracted the trivalent complexes with a chloroform solution of ammonium pyrrolidinedithiocarbamate. Final analysis at the 1μg L⁻¹ level was achieved by neutron activation analysis.

Stary et al. [4] preconcentrated arsenate in non saline water by adding tungstate labelled with tungsten–185 and molybdate ions, and extracting the complex formed with a 1:2 dichloroethane solution of tetraphenylarsenium chloride. Down to 0.2μg L⁻¹ arsenate was then determined in the extract by radiochemical analysis. Dix et al. [5] preconcentrated arsenite, plus arsenate, monomethyl arsenate, and dimethyl arsenite by extraction with a cyclohexane solution of thioglycollic acid methyl ester. The methylthioglycollate derivatives were then determined by temperature programmed capillary column gas chromatography.

Nasu and Kau [6] preconcentrated arsenate, arsenite and phosphate by a procedure in which a floated (between aqueous and organic phase) ion pair of malachite green with molybdophosphate was dissolved by the addition of methanol to the organic layer. Phosphate and arsenate were determined by measuring absorbance of the organic phase. An oxidative (potassium dichromate) or a reductive (sodium thiosulphate) reaction was used for the determination of phosphate, arsenate and arsenite. A positive interference effect was observed in the presence of large amounts of silicon. This was overcome by acidification with concentrated hydrochloric acid. The method was applied to samples of hot spring water, seawater and ground water with almost complete recovery of added amounts.

34.1.2 Chromate and dichromate

Hexavalent chromium (ie chromate and dichromate) is reduced by diethyldithiocarbarmate to trivalent chromium [7,8], with which it forms an isobutyl methyl ketone soluble complex. Preconcentrated chromate is then determined in the solvent extract by atomic absorption spectrometry at 357.9nm.

Subramanian [9] studied the factors affecting the determination of trivalent and hexavalent chromium (chromate) by direct complexation with ammonium pyrrolidinedithiocarbamate, extraction of the complex into methyl isobutyl ketone, and determination by graphite furnace atomic absorption spectrometry. Factors studied included the pH of the solution, concentration of reagents, period required for complete extraction, and the solubility of the chelate in the organic phase. Based on the results, procedures were developed for selective determination of trivalent and hexavalent chromium without the need to convert the trivalent to the hexavalent state. For both states the detection limit was 0.2µg L $^{-1}$.

Schaller and Neeb [10] extracted di(trifluoroethyl)dithiocarbamate chelates of hexavalent and trivalent chromium and cobalt from aqueous solution at pH3 using a carbon–18 column. Adsorbed chelates were eluted with toluene before gas chromatographic analysis with electron capture detection. Broadening of the peaks of cobalt and chromium was reduced by arranging that the end of the capillary was 1cm within the detector body, the base of which was isolated with glass wool and aluminium foil. The last 0.2cm of the column was also protected with a sleeve of braided wire. The detection limit for chromium(VI) was 0.05µg L $^{-1}$.

34.1.3 Halides

To preconcentrate fluoride Miyazaki and Brancho [11] converted fluoride into the ternary lanthanum–alizarin complexone fluoride and extracted it into hexanol containing N,N-diethylaniline. The extract was analysed directly by inductively coupled plasma atomic emission spectrometry for the determination of fluoride. Measurement of the lanthanum(II) 333.75nm emission line and comparison with a calibration graph enabled fluoride concentrations as low as 0.59µg L $^{-1}$ to be determined in Japanese river water (polluted and unpolluted), coastal seawater, and potable water samples.

Miyazaki and Brancho [12] preconcentrated iodide by conversion to iodine and extraction into xylene. The extract was determined at 172.28nm using inductively coupled plasma atomic emission spectrometry. Iodate was reduced to iodide which was then treated in the same way to give a total iodide plus iodate content. Detection limits were, respectively, 8.3 and 21µg L^{-1} for iodate and iodide. Large concentrations of bromide interfere.

34.1.4 Orthophosphate

Most spectrophotometric methods of phosphate are based on the formation of a heteropoly acid with molybdate. The heteropoly acid formed (molybdophosphate) and its reduction product (so called molybdenum blue) have been used, as has the yellow vanadmolybdo-phosphoric acid. The protonated forms of these species have also been extracted into organic solvents for spectrophotometry. It well known that molybdophosphate reacts with cationic dyes and organic bases to form ion pairs. The ion pair formed can be separated as a precipitate from the aqueous solution [13–16]. Shida and Matsuo [17] reported a very sensitive spectrophotometric method based on formation of the ion associate of molybdophosphate with methylene blue, flotation of this ion pair between the aqueous phase and cyclohexane-4-methylpentan-2-one, and displacement of methylene blue from the ion pair by tetradecyldimethyl-benzylammonium ion.

Several preconcentrative extractions of molybdophosphate with cationic dyes have been studied. Safranine T has been used with a mixed acetophenone 1,2-dichlorobenzene solvent [18,19]. Crystal violet or iodine green forms an ion pair extractable with butanol–cyclohexane [20]; the methylene blue ion pair can be extracted with 4-methylpentan-2-one [21]. The ion pair with rhodamine B has been extracted into chloroform–butanol for fluorometry [22]. Most of these procedures are troublesome because pre-extractions of molybdophosphate or other measures are needed to avoid large reagent blanks.

In an attempt to improve the extractability of ion pairs of molydophosphate and thus the sensitivity of the determination of phosphate, Motomizu et al. [23] examined several cationic dyes and extracting solvents. They found that procedures based on ethyl violet and a cyclohexane-4-methylpentan-2-one mixture enabled 10µg L^{-1} concentrations of phosphate to be determined spectrophotometrically. In samples of river water and seawater, the phosphorus content is often in the µg L^{-1} range. Thus procedure requires small volumes of sample (below 10ml) and very simple vessels (25ml test tube) which are easily heated in order to hydrolyse any condensed phosphate. The absorption spectrum of the ion pair formed between molybdophosphate and ethyl violet in the organic phase obtained by the procedure shows a maximum absorbance at 602nm where the absorbance of the reagent blank is about 0.1. The calibration graphs obtained are linear in the range 0–0.6µg of phosphorus and the molar absorptivity calculated from the slope of the curve was 2.7×10^5 L $^{-1}$ mol $^{-1}$ cm $^{-1}$.

Silicate, vanadate and tungstate, which may react with molybdate to form heteropoly acids, do not interfere with the determination of phosphorus by the above method even when present at concentrations of 5×10^{-5}M, 10^{-5}M and 5×10^{-6}M, respectively. Arsenic(V) causes positive errors because it reacts with molybdate to form the heteropoly acid, which is quantitatively extracted into the organic phase. In non saline waters such as river water and seawater, the arsenic content is very much smaller than the phosphorus content, Tin(II) and (IV) ions at concentrations more than 10^{-6}M interfere; large amounts of tin(II) make it impossible to determine phosphate and tin(IV) causes negative errors.

Phosphorus occurs in non saline waters as orthophosphate, condensed phosphate (pyro-, meta- and polyphosphate) and organically bound phosphorus, all of which may be present in soluble forms and in suspension. Only orthophosphate can be determined directly by molybdophosphate procedures. Pyro-, tripoly- and poly-phosphate are completely hydrolysed to orthophosphate by acidification and heating. If samples stored in glass containers are not acidified, the phosphorus content decreases gradually, the decrease becoming significant after 1d in most cases. River water acidified with 1ml L $^{-1}$ of 5M sulphuric acid usually showed a constant content of phosphorus for about 2d. Thus if analyses cannot be completed within a few hours of collection, samples should be acidified.

In later work, Motomizu et al. [24] point out that in the determination of phosphate at µg L $^{-1}$ levels in waters the above method has certain disadvantages. First the absorbance of the reagent blank becomes too large for the concentration effect achieved by the solvent extraction to be of much use; for example when 20ml of sample solution and 5ml of organic solvent were used, the absorbance of the reagent blank was 0. 14. Second, the shaking time needed was long and the colour of the extract faded gradually if the shaking lasted more than 30min. In the course of attempts to improve on this method, they observed that malachite green gave a stable dark–yellow species in 1.5M sulphuric acid (probably a protonated one), whereas ethyl violet became colourless within 30min even in only 0.5M sulphuric acid.

Malachite green has several advantages over ethyl violet.

1   The absorbance of the reagent blank is very small and 20-fold concentration of phosphate by solvent extraction is possible.

2   The reagent solution, which consists of malachite green, molybdate and 1.5M sulphuric acid is stable at least for a month.

Table 34.1 Determination of phosphorus in river and seawater, and recovery test

Sample		Phosphorus found (ng ml^{-1})	Recovery test				
			Sample taken (ml)	P in sample (ng)	P added (ng)	P found (ng)	Recovery (%)
Asahi river	A	6.0	8	48	248	299	101
	B	17.9	8	143	248	387	99
Yoshii river	A	17.5	8	140	248	392	101
	B	16.4	8	131	248	390	103
	C	27.0	8	216	248	459	99
	D	39.4	5	197	248	446	100
Seashore of Seto Inland Sea							
Kojima		46.0	5	230	248	477	100
Shibukawa		17.4	8	139	248	391	101
Tamano		15.3	8	122	248	381	103
Ushimado		12.8	8	102	248	351	100
Nishiwaki		20.9	8	167	248	419	101

Source: Reproduced with permission from Motomizu *et al.* [24] Elsevier Science, UK

3 The method is less troublesome and shaking for 5min is enough to complete the extraction.
4 Colour fading in the organic phase does not occur during shaking and standing.

Most cations and anions commonly found in non saline waters do not interfere in this procedure, but arsenic(V) causes large positive errors; arsenic(V) at a concentration of 10µg L^{-1} produces an absorbance of 0.070 but can be masked with tartaric acid (added in the reagent solution). When arsenic(V) was present at concentrations of 50µg L^{-1}, it was masked with 0.1ml of 10^{-4}M sodium thiosulphate added after the sulphuric acid. Table 34.1 shows the results for determination of phosphorus in samples. Recovery tests were done by adding known amounts of phosphate. The results are also shown in Table 34.1. The recovery of phosphorus was good, 99–103%. The relative standard deviation for phosphorus was 0.6% for 21.0µg L^{-1} in seawater (12 replicates) and 1.1% for 4.3µg L^{-1} in potable water (10 replicates).

Motomizu and Oshima [25] have used the ion associate formed between molydophosphate and malachite green as the basis of a flow injection method for the determination of orthophosphate in river and potable waters in amounts down to 0.1µg L^{-1}. The ion associate was extracted with benzene-4-methylpentane-2-one solvent mixture to

Table 34.2 Instrumental operating conditions

ICP source	Shimadzu ICPS–2H
Operating frequency	27MHz
Load coil	2-turn copper tubing with teflon coating
Nebulizer	concentric
Spectrometer	Shimadzu GEW 170, 1.7m Ebert
Grating	2160 line mm $^{-1}$
Entrance slit width	30µm
Exit slit width	50µm
Reciprocal linear dispersion	0.26nm mm $^{-1}$ (1st order)
Recording console	Shimadzu RE–7
Pre-integration time	5s
Integration time	20s
Attenuation for photomultiplier high voltage	33 divisions

	DIBK extract	*Aqueous solution*
Operating power	1.6kW	1.2kW
Argon flow rates:		
coolant	13L min $^{-1}$	12L min $^{-1}$
plasma	1.1L m in $^{-1}$	0.9L min $^{-1}$
carrier	0.7L min $^{-1}$	0.8L min $^{-1}$
Observation height above load coil	14mm	16mm
Sample uptake rate	1.8ml min $^{-1}$	1.9ml min $^{-1}$

Source: Reproduced with permission from Motomizu and Oshima [25] Royal Society of Chemistry, UK

achieve the necessary preconcentration. Molybdenum blue solvent extraction procedures have also been applied to polluted waters [26].

Other molybdenum blue solvent extraction–preconcentration procedures include the use of isoamyl alcohol with a pulse polarographic finish [27] and methyl isobutyl ketone with an inductively coupled plasma atomic emission finish [28,29]. Inductively coupled plasma procedures have the advantage of being applicable to estuarine as well as river waters. Many of the earlier atomic absorption procedures had the disadvantage that arsenic, silicon and germanium cause positive errors because these elements also form reduced heteropoly acids. Muyazaki *et al.* [30,31] overcome this problem by determining phosphorus at the P' 214.91nm line as the Mo(II) 213.61nm line interferes with the P' 213.62nm line. To achieve the µg L $^{-1}$ sensitivity they required they incorporated a 100-fold preconcentration stage into the method.

An inductively coupled plasma atomic emission spectrometer instrument and its operating conditions are described in Table 34.2. The spectrometer was equipped with 14 fixed slits and a moving slit. It was

Table 34.3 Effect of other ions on phosphorus results (sample, 100ml; DIBK, 5ml; P, 10µg)

Other ion	Amount added	Recovery (%)	Other ion	Amount added	Recovery (%)
None	–	100	Fe(III)	1mg	101
As(V)	10µg	100	Cl⁻	2g	101
	50µg	102	I⁻	10mg	102
	100µg	105	Br⁻	10mg	99
As(III)	1mg	104	NO₂⁻	10mg	91
Si	10mg	102		10mg	104*
Ge	10mg	101	S₂O₈⁻	1g	103

*20mg of sulphamic acid added

Source: Reproduced with permission from Motomizu et al. [24] Elsevier Science, UK

also capable of selecting different wavelengths by changing the angle of the grating. The wavelength profile of the analytical line was obtained by moving the entrance slit over a small distance.

The effects of other ions are summarised in Table 34.3. No interference from silicon and germanium (which cause positive errors in the spectrophotometric method) was observed even at 1000-fold excess. Arsenic(V) was permissible up to 10 times the weight of phosphorus. Above this amount, a slight upward background shift was observed which caused a positive error exceeding 5%. However, non saline waters containing arsenic(V) in amounts more than 10 times that of phosphorus are rare. Most of the other anions did not show any interference and that of nitrite was decreased by addition of sulphamic acid. There was no interference from peroxidisulphate indicating that pretreatment with peroxidisulphate as used for the determination of total phosphorus in water is also applicable to this method.

The calibration graph obtained by the extraction method was linear for at least three orders of magnitude of concentration above the detection limit. The detection limit, defined as the concentration of phosphorus equivalent to three times the standard deviation of the background signal (3α) was 37µg L^{-1} phosphorus in the 5ml of extract. Hence, the detection limit for the original samples (500ml) was 0.37µg L^{-1} phosphorus. Measurements of 5µg of phosphorus, extracted from 500ml portions of aqueous solution gave a relative standard deviation of 2.1%.

The values obtained by the inductively coupled plasma method for river water samples are compared with those obtained by the spectrophotometric method in Table 34.4. They are in good agreement with each other. The precision of the values obtained by inductively coupled plasma atomic emission spectrometry is better than that achieved

Table 34.4 Phosphorus determination in river and seawater samples (500ml samples, 5ml di-isobutyl ketone

Sample	Phosphorus concentration (ng ml $^{-1}$)			
	ICPAES		Spectrophotometry	
	Range°	Average	Range°	Average
River water				
A	0.3–0.4	0.4	0.2–0.4	0.3
B	5.1–5.3	5.2	5.0–5.3	5.2
C	4.3–4.5	4.4	4.5–4.8	4.7
D	21.2–22.2	21.9	20.4–22.6	21.5
E	7.4–8.1	7.7	7.8–8.5	8.3
F	22.2–23.8	23.1	23.0–26.2	25.0

°Range of four measurements
ᵇ100ml of sample; 10ml of di-isobutyl ketone

Source: Reproduced with permission from Motomizu *et al*. [24] Elsevier Science, UK

spectrophotometrically, especially at low phosphorus concentrations where the absorbance is less than 0.05

Muyazaki *et al*. [29] have extended the detection limit of the inductively coupled plasma emission spectrometric technique down to the sub µg L $^{-1}$ level by measurements of the reduced molybdo-antimonyphosphoric acid, not at the phosphorus P^1 214.91nm line, as in the previous method, but at the more sensitive molybdenum(II) 202.03nm or the antimony(I) 206.83nm lines. This method is simple, sensitive and precise. Washing of the organic phase is not necessary because of the low solubility of di-isobutyl ketone in water.

The detection limits of the revised method defined as the concentration of phosphorus equivalent to three times the standard deviations of the background signal, were 5.2 and 45ng L $^{-1}$, respectively, for molybdenum and antimony measurements, when 500ml water samples and 5ml of di-isobutyl ketone were used. These values are about 100 and 10 times better than that for the direct phosphorus measurement. The significant improvements in the detection limits arise because the reduced molybdo-antimonyphosphoric acid has the composition $PSb_2 Mo_{10}O_{40}$ and the Mo(II) 202.03nm and Sb(I) 206.83nm emission lines are more sensitive than the P1 214.98nm emission line in inductively coupled plasma atomic emission spectrometry. The relative standard deviations ($n = 10$) of the complete procedure for 1µg of phosphorus are 2.0 and 2.55% for molybdenum and antimony measurements, respectively. The relative standard deviation of the background signal is 0.3%. The signals for the blank solution are not distinguishable from the background.

Table 34.5 Phosphorus determination in river samples

Sample	Phosphorus concentration (ng ml^{-1})					
	Via Mo		Via Sb		Via P	
	Range[a]	Mean[a]	Range[a]	Mean[a]	Range[a]	Mean[a]
River water						
A	0.49–0.50	0.49	0.4–0.6	0.5	0.4–0.5	0.5
B	0.48–0.50	0.49	0.4–0.5	0.4	0.3–0.4	0.4
C	5.80–6.01[b]	5.91	5.4–6.1	5.8	5.6–5.9	5.7
D	4.55–4.84[b]	4.61	4.0–4.5	4.4	4.3–4.5	4.4
E	4.60–4.94[b]	4.89	4.2–4.8	4.7	5.1–5.3	5.2

[a]Three measurements
[b]200ml sample, 10ml of DIBK

Source: Reproduced with permission from Muyazaki et al. [29] Elsevier Science, UK

Arsenic(V) caused serious interference, but germanium, silicon and arsenic(II) did not interfere in 100-fold amounts. Nitrite seriously interferes in spectrophotometric methods for molybdenum blue. Molybdenum measurements by the inductively coupled plasma method did not suffer from interference up to 5mg of nitrite (2500-fold excess over phosphorus). In the antimony measurement, nitrite was permissible up to 0.2mg (100-fold amount). Above that amount, the interference decreased with increasing amount of antimony in the molybdenum–antimony reagent, although the interference was not completely suppressed. The addition of sulphamic acid decreased the interference from nitrite. The difference in the interference mentioned above may be related to the mechanism of the reduction of molybdoantimonyphosphoric acid. No interference was observed from peroxidisulphate. Pretreatment with peroxidisulphate as used for the determination of total phosphorus in water, therefore, may be applicable to this method. The results obtained by the molybdenum and antimony measurements method for river samples are compared with those obtained by direct phosphorus measurements in Table 34.5. The results are in good agreement.

Bet-Pera et al. [30] have described an alternative phosphate pre-concentration procedure in which phosphate is converted to 12–molybdophosphoric acid which is then extracted into isobutyl acetate. After evaporation of the organic solvent the complex is dissolved in alkaline solution and after acidification molybdenum(VI) is reduced to molybdenum(III) using a Jones reductor. The resulting molybdenum(III) is reoxidised with iron(III) to molybdate and the resulting iron(II) is determined spectrophotometrically at 562nm as the iron(II) ferrozine complex. Phosphate in non saline water samples was determined by this method in amounts as low as 4µg L $^{-1}$ in the final solution with a relative precision of 12% at 2α value.

34.1.5 Silicate

Silicate can be preconcentrated [31] from water in microgram amounts by conversion to molydosilicic acid in a medium 0.1N sulphuric acid or hydrochloric acid containing ammonium molybdate. The heteropoly acid is extracted into butanol or isoamyl alcohol and reduced to molybdenum blue using stannous chloride prior to spectrophotometric determination.

34.2 Seawater

34.2.1 Antimonate, trivalent antimony, arsenate, arsenite and trivalent arsenic

The procedure described in section 34.1.1 is applicable to seawater [6].

34.2.2 Halides

The procedure described in section 34.1.3 is applicable to seawater [11].

34.3 Trade effluents

34.3.1 Thiosulphate

Chakraborty and Das [32] have described an indirect atomic absorption spectrometric method for determining thiosulphate in photographic processing effluents based on the formation of a stable ion association complex between lead, thiourea and thiosulphate in alkaline medium. The complex is extracted into n-butylacetate:n-butanol 2:1 and analysed directly by flame atomic absorption spectrometry. Down to 0.2µg L $^{-1}$ thiosulphate were determined.

References

1 Metzger, M. and Braun, H. *Analytica Chimica Acta*, **189**, 263 (1986).
2 Mok, W.M. and Wai, C.M. *Analytical Chemistry*, **59**, 233 (1987).
3 Mok, W.M., Shah, N.W. and Wai, C.M. *Analytical Chemistry*, **58**, 110 (1986).
4 Stary, J., Zeman, A., Kratzer, K. and Prasilova, J. *International Journal of Environmental Analytical Chemistry*, **8**, 49 (1980).
5 Dix, K., Cappon, C.J. and Toribara, T.H. *Journal of Chromatographic Science*, **25**, 164 (1987).
6 Nasu, T. and Kau, M. *Analyst (London)*, **113**, 1685 (1988).
7 Fukamachi, K., Morimoto, M. and Yanagawa, M. *Japan Analyst*, **21**, 26 (1972).

8 Yanagisawa, M., Suzuki, M. and Takequichi, T. *Mikrochimica Acta*, **3**, 475 (1973).
9 Subramanian, K.S. *Analytical Chemistry*, **60**, 11 (1988).
10 Schaller, H. and Neeb, R. *Fresenius Zeitschrift für Analytische Chemie*, **327**, 170 (1987).
11 Miyazaki, A. and Brancho, K. *Analytica Chimica Acta*, **198**, 297 (1987).
12 Miyazaki, A. and Brancho, K. *Spectrochimica Acta*, **423**, 277 (1987).
13 MacDonald, A.M.G. and Rivero, A.M. *Analytica Chimica Acta*, **37**, 414 (1967).
14 Kirkbright, G.F., Narayanaswamy, R. and West, T.S. *Analyst (London)*, **97**, 174 (1972).
15 Babko, A.K., Yu.F., Shkaravskii, F. and Ivanshkovich, E.M. *Ukranian Chemical Journal*, **33**, 30 (1967).
16 Pilipenko, A.T. and Shkaravskii, Yu.F., *Zhur. Anal. Khim.*, **29**, 716 (1974).
17 Shida, J. and Matsuo, T. *Bulletin of the Chemical Society of Japan*, **53**, 2868 (1980).
18 Ducret, L. and Drouillas, M. *Analytica Chimica Acta*, **21**, 86 (1959).
19 Sudakov, F.P., Klitina, V.I., Ya, T. and Dan'shova, H. *Zhur. Anal. Khim.*, **21**, 1333 (1966).
20 Babko, A.K., Shkaravskii, Yu.F. and Kulik, V.I. *Zhur Anal. Khim.*, **21**, 196 (1966).
21 Matsuo, T., Shida, J. and Kurihara, W. *Analytica Chimica Acta*, **91**, 385 (1977).
22 Kirkbright, G.F., Narayanasamy, R. and West, T.S. *Analytical Chemistry*, **43**, 1434 (1971).
23 Motomizu, S., Wakimoto, T. and Toei, K. *Analytica Chimica Acta*, **138**, 329 (1982).
24 Motomizu, S., Wakimoto, T. and Toei, K. *Talanta*, **31**, 235 (1984).
25 Motomizu, S. and Oshima, M. *Analyst (London)*, **112**, 295 (1987).
26 Chambe, A. and Gupta, V.K. *Analyst (London)*, **108**, 1141 (1983).
27 Fogg, A.G. and Yoo, K.S. *Analytical Letters*, **9**, 1035 (1976).
28 Muyazaki, A., Kimura, A. and Umezaki, Y. *Analytica Chimica Acta*, **127**, 93 (1981).
29 Muyazaki, A., Kimura, A. and Umezaki, Y. *Analytica Chimica Acta*, **138**, 121 (1982).
30 Bet-Pera, F., Srivasstava, A.K. and Jaselskis, B. *Analytical Chemistry*, **53**, 561 (1981).
31 Pavlova, M.W., Podal'skaya, B.L. and Shafran, I.G. Trudy uses Naucho issled. Inst. Khim. Reakt. osobo Christ. Khim. Veshchesto, 34, 185. Ref: Zhur. Khim. 19GD (1973) (11) Abstract No. 11G200 (1972).
32 Chakraborty, D. and Das, A.K. *Atomic Spectroscopy*, **9**, 115 (1988).

Chapter 35

Anions: Adsorption on solids

35.1 Sephadex C18 bonded silica

35.1.1 Non saline waters

35.1.1.1 Borate

Yoshimura *et al.* [1] have described a highly sensitive method for pre-concentrating borate based on adsorption on Sephadex G–25 gel in alkaline medium and reversible desorption in acid medium. The borate is then determined spectrophotometrically by the azomethine-4-method.

35.1.1.2 Chromate

Hexavalent chromium (chromate) has been preconcentrated from lake water on to C18-bonded silica prior to analysis by high performance liquid chromatography utilising an atomic absorption spectrophotometric detector [2]. Non saline pond water samples were adjusted to pH3.2 with acetic acid. The samples were analysed and 100% recovery of chromate was obtained even in the presence of chromium(III) ions. Chromate was eluted from the column (methanol/water 50/50) directly to the atomic absorption spectrometry detector.

Schaller and Neeb [3] extracted di(trifluoroethyl) dithiocarbamate chelates of hexavalent and trivalent chromium and cobalt from aqueous solution at pH3 using a carbon–18 column (see section 34.1.2). Various ion exchange resins have been used to preconcentrate chromate ions including Dowex AG 1–X4 anion exchange resin [4].

35.2 Polyurethane foam

35.2.1 Non saline waters

35.2.1.1 Chromate

Forag *et al.* [5] preconcentrated chromium(VI) in water using 1,5-diphenylcarbazide loaded polyurethane foam.

35.2.1.2 Iodide

Palagni [6] used the pulsed column bed technique to selectively separate and preconcentrate low levels of iodide–127 in non saline water. The pulsed bed consisted of a syringe containing a scintillation detector and a gamma ray spectrometer. The adsorbent consisted of open cell polyurethane foam impregnated with long chain tri-*n*-alkylamine which formed a complex with iodide.

35.2.1.3 Phosphate

Khan and Chow [7] have described a method in which the phosphate is converted into phosphomolybdate which is then extracted into polyurethane foam and analysed for molybdenum by X-ray fluorescence spectrometry directly on the foam. A polyether type polyurethane foam disc is squeezed for an hour in a mixture of phosphate solution, sodium molybdate and hydrochloric acid, spiked with phosphorus–32. After washing and drying, the foam disc is placed on plastic foam and stretched across the X-ray source holder. The method is simple and rapid and the precision is 5% for $0.25\mu g$ L^{-1} and 2%for 2.5mg L^{-1} phosphate. Equimolar amounts of silicon, germanium and arsenic(IV) appear to interfere with the determination.

35.3 Active carbon

35.3.1 Non saline waters

35.3.1.1 Phosphate

In a method developed to preconcentrate orthophosphate from non saline water Hashitani *et al.* [8] used activated carbon loaded with zirconium. A 0.1–1.0L water sample at pH1.5 was passed down a column of this material on which phosphate was adsorbed quantitatively and instantly below pH8 and desorbed above pH13.5. Pyrophosphate, tripoly-phosphate and metaphosphate behaved as orthophosphate.

35.3.1.2 Selenate and selenite

Selenate and selenite can be determined by energy dispersive X-ray spectroscopy after preconcentration of elementary selenium on activated carbon, ascorbic acid being used to reduce selenium to its elemental form [9–11].

35.4 Cellulose derivatives

35.4.1 Non saline water

35.4.1.1 Phosphate

A preconcentration technique based on nitrocellulose or acetyl cellulose membranes has been described for phosphate [12]. Phosphorus was collected as phosphomolybdenum on a nitrocellulose or acetylcellulose membrane in the presence of n-dodecyltrimethylammonium bromide. The membrane was dissolved in dimethylsulphoxide and the absorbance of dimethylsulphoxide solution measured. Moderate concentrations of silicate, anionic and non ionic surfactants and high concentrations of sodium chloride did not interfere. Arsenate interference could be eliminated by reducing arsenate to arsenite. Determination of condensed and organic phosphates was possible following their conversion to orthophosphoric acid. The limit of determination was 0.02µg L $^{-1}$ phosphate.

35.4.1.2 Selenate and selenite

2,2'–Diethylaminocellulose filters enabled selenate (SeO_4^{2-}) and selenite (SeO_3^{2-}) to be preconcentrated from non saline water [13,14] at pH3–6 with a detection limit of 0.05µg L $^{-1}$.

35.5 Immobilised diphenyl carbazone

35.5.1 Seawater

35.5.1.1 Chromate

Use of immobilised chelating agents for sequestering trace metals from aqueous and saline media presents several significant advantages over chelation–solvent extraction approaches to this problem. [15,16]. With little sample manipulation, large preconcentration factors can generally be realised in relatively short times with low analytical blanks. As a consequence of these considerations Willie et al. [17] developed a new approach to the determination of total chromium. This involves preliminary concentration of dissolved chromium from sea water by means of an immobilised diphenyl–carbazone chelating agent, prior to determination by atomic absorption spectrometry. A Perkin–Elmer Model 500 atomic absorption spectrometer fitted with a HGA–500 furnace with Zeeman background correction capability was used for chromium determinations. Chromium was first reduced to Cr(III) by addition of 0.5ml aqueous sulphur dioxide and allowing the sample to stand for several minutes. Aliquots of seawater were then adjusted to pH9.0±0.2

using high purity ammonium hydroxide and gravity fed through a column of silica at a nominal flow rate of 10ml min $^{-1}$.

The sequestered chromium was then eluted from the column with 10.0ml 0.2mol L $^{-1}$ nitric acid. More than 93% of chromium was recovered in the first 5ml of eluate by this method. Extraction of 80ng spikes of Cr(III) from 200ml aliquots of seawater was semi-quantitative.

References

1 Yoshimura, K., Kariya, R. and Torntani, T. *Analytica Chimica Acta*, **109**, 115 (1979).
2 Syty, A., Christenson, R.G. and Rains, T.C. *Atomic Spectroscopy*, **7**, 89 (1986).
3 Schaller, H. and Neeb, R. *Fresenius Zeitschrift für Analytische Chemie*, **327**, 170 (1987).
4 Aoyama, M., Habo, T. and Suzuki, S. *Analytica Chimica Acta*, **129**, 237 (1981).
5 Forag, A.G., El-Waki, A.M. and El-Shahawi, M.S. *Analyst (London)*, **106**, 809 (1981).
6 Palagni, S. *International Journal of Applied Radiation and Isotopes*, **34**, 55 (1983).
7 Khan, A.S. and Chow, A. *Analytical Letters (London)*, **16**, 265 (1983).
8 Hashitani, H., Okumira, M. and Fuginaza, K. *Fresenius Zeitschrift für Analytische Chemie*, **326**, 540 (1987).
9 Roblerecht, H. and Van Grieken, R. *Talanta*, **29**, 823 (1982).
10 Massee, R., Van der Sloot, H.A. and Das, H.A. *Journal of Radioanalytical Chemistry*, **38**, 157 (1977).
11 Orvini, E., Ladola, L., Gallorini, M. and Zerbia, T. In *Heavy Metals in the Environment*, Elsevier, Amsterdam, p. 657 (1981).
12 Taguchi, S., Ito-oko, E., Matsuyama, K. and Kashara, I. *Talanta*, **32**, 391 (1985).
13 Smits, J. and Van Grieken, R. *Analytica Chimica Acta*, **123**, 9 (1981).
14 Smits, J. and Van Grieken, R. *International Journal of Environmental Analytical Chemistry*, **9**, 81 (1981).
15 Myasoedova, G.V., Savvin, S.G. and Zhu, R. *Analytica Chimica*, **37**, 499 (1982).
16 Leyden, D.W. and Wegschneider, W. *Analytical Chemistry*, **63**, 1059A (1981).
17 Willie, S.N., Sturgeon, R.E. and Berman, S.S. *Analytical Chemistry*, **55**, 981 (1983).

Chapter 36

Anions: Adsorption on ion exchange resins

36.1 Anion exchange resins

36.1.1 Non saline waters

36.1.1.1 Borate

Inductively coupled plasma atomic emission spectrometry was used by Takahashi [1] for the determination of borate in non saline waters. The borate was preconcentrated on a column of Amberlite XE–243, then eluted with hydrochloric acid, and introduced directly into the nebuliser. Interference by iron and carbonate was noted.

Duchateau *et al.* [2] used isotope dilution mass spectrometry for the determination of borate in non saline waters. The borate was preconcentrated on an Amberlite IRA–743 borate selective ion exchange column.

Jun *et al.* [3] preconcentrated borate from non saline waters as its complex with chromotropic acid and octyltrimethyl ammonium chloride on an anion exchange column (TSK gel, I.C. Anion PW). The eluted concentrate was analysed by high performance liquid chromatography.

36.1.1.2 Chromate

Parkow *et al.* [4] have described a procedure for the preconcentration and differential analysis of traces of chromium(VI) and chromium(III) in non saline waters. The sample is filtered, acidified and divided into three portions, one of which is left untreated while the others are passed through a cation exchange resin (Dowex 50W X4 cation exchange resin) and an anion exchange resin respectively; the three aliquots are then treated with nitric acid, evaporated, and analysed by atomic absorption to give the concentrations of cationic, anionic and non ionic chromium. The concentration of chromium(III) is probably closely related to the sum of the cationic and non ionic fractions and the concentration of chromium(VI) corresponds to the anionic portion.

Various other ion exchange resins have been used to preconcentrate chromate ions including Dowex AG 1–X4 anion exchange resin [5],

Amberlite La–1 liquid anion exchanger [6] and Dowex Ag 1–X4 anion exchange resin [7].

Parkow and Januer [5] acidified water samples containing chromate to pH5 and then passed them upwards through a Dowex Ag 1–X4 anion exchange resin, so that the chromate was adsorbed in a narrow zone at the lower end of the resin bed. The chromate was eluted rapidly with small volumes of an acidic reductant solution which reacts with chromate on the column to form trivalent chromium during elution, thus producing very high concentration factors.

36.1.1.3 Halides

Iodide in non saline waters has been determined [8] by ion exchange high performance liquid chromatography using an iodide ion selective electrode as detector. On-line preconcentration of iodide on an anion guard cartridge allowed determinations down to 1nM.

Carlsson et al. [9] preconcentrated bromide and iodide on an anion exchanger, oxidised by peroxodisulphate to bromate, treated with iodide and measured the absorbance of the resulting tri-iodide. The sum of bromate and iodate produced in the oxidation was determined by treating the oxidised sample with iodide in hydrochloric acid. The iodate was determined separately by applying the reaction in acetic acid. The working range of the spectrophotometric method was 1–15μM, the limit of the determination was 0.7μM for iodate and for iodate plus bromate and the enrichment factor in the preconcentration step was 50.

36.1.1.4 Molybdate

Vasquez-Gonzalez et al. [10] have described a method for preconcentrating and determining molybdate by electrothermal atomisation atomic absorption spectrometry after preconcentration by means of anion exchange using Amberlite IRA–400 resin in citrate form. The optimal analytical parameters were established by drying, carbonisation, charring, atomisation and cleaning in a graphite furnace. The precision and accuracy of the method were investigated. Less than 0.2μg L $^{-1}$ molybdate could be determined by this procedure.

36.1.1.5 Phosphate

In one procedure [11,12] phosphate is adsorbed onto anion exchange resin. Orthophosphate is quantitatively adsorbed by A,G Dowex 1–X8 anion exchange resin, then eluted and reacted with an acid molybdate reagent for estimation. Arsenic and organic phosphorus compounds did not interfere with the estimation of orthophosphate while polyphosphates did interfere if present in equal or greater amounts than orthophosphate. It is concluded that the use of the anion exchange technique results in a more valid estimate than direct reaction with the acid molybdate reagent.

36.1.1.6 Selenite

Selenite has been preconcentrated on a bismuthiol(II) modified anion exchange resin (Amberlite 1RA–400) [13] followed by fluorometric estimation using diamino–naphthalene. Selenite adsorbed on the column as selenotrisulphate was desorbed with a small volume of 0.1mol L^{-1} penicillamine or 0.1mol L^{-1} cysteine prior to fluorometric determination of selenium [14].

36.1.1.7 Sulphide

Sulphide has been preconcentrated on a column of Amberlite IRA 400 anion exchange resin [15]. The sulphide is removed from the column with 4M sodium hydroxide and determined spectrophotometrically by the N,N'–dimethyl-p-phenylene diamine method. Down to 0.1µg L^{-1} sulphide can be determined by this procedure.

36.1.2 Seawater

36.1.2.1 Molybdate

Shriadal et al. [16] determined molybdenum(VI) in seawater spectrophotometrically after enrichment as the Tiron complex on a thin layer of anion exchange resin. There were no interferences from trace elements or major constituents of seawater except for chromium and vanadium. These were reduced by the addition of ascorbic acid. The concentration of dissolved molybdenum (VI) determined in Japanese seawater was 11.5µg L^{-1} with a relative standard deviation of 1.1%.

Kuroda et al. [17] preconcentrated trace amounts of molybdenum from acidified seawater an a strongly basic anion exchange resin (Bio–Rad AgI, X8 in the chloride form) by treating the water with sodium azide. Molybdenum(VI) complexes with azide were stripped from the resin by elution with ammonium chloride/ammonium hydroxide solution (2mol L^{-1}/mol L^{-1}). Relative standard deviations of better than 8% at levels of 10µg L^{-1} were attained for seawater using graphite furnace atomic absorption spectrometry.

36.1.2.2 Rhenate

This is discussed in Section 24.3.3.

36.2 Cation exchange resins

36.2.1 Non saline waters

36.2.1.1 Selenate, arsenate, molybdate, vanadate, tungstate and chromate

Fung and Dao [19] used Chelex–100 resin to remove common interfering inorganic anions from non saline waters, and retained selenate, arsenate, molybdate, vanadate, tungstate and chromate. These species left on the column could then be determined at sub ppb detection limits.

References

1 Takahasi, Y. *Bunseki Kagaku*, **36**, 693 (1987).
2 Duchateau, N.L., Verbruggen, A., Hendrickx, F. and De Bieure, P. *Analytica Chimica Acta*, **196**, 41 (1987).
3 Jun, Z. Oshima, M. and Motomizu, S. *Analyst (London)*, **113**, 1631 (1988).
4 Parkow, J.F., Lieta, D.P., Lin, J.W. et al. *Science of the Total Environment*, **7**, 17 (1977).
5 Parkow, J.F. and Janauer, G.E. *Analytica Chimica Acta*, **69**, 97 (1974).
6 Mazzucotelli, A., Minoia, C., Pozzoli, L. and Ariati, L. *Applied Spectroscopy*, **4**, 182 (1982).
7 Aoyama, M., Hobo, T. and Suzuki, S. *Analytica Chimica Acta*, **129**, 237 (1981).
8 Butler, E.C.V. and Gershey, R. *Analytica Chimica Acta*, **164**, 153 (1984).
9 Carlsson, A., Lundstrom, U. and Olin, A. *Talanta*, **34**, 615 (1987).
10 Vasquez-Gonzalez, J.F., Bermejo-Barrera, P. and Bermejo-Martinez, F. *Atomic Spectroscopy*, **8**, 159 (1987).
11 Westland, A.D. and Bouchair, I. *Water Research*, **8**, 467 (1974).
12 Blanchar, W. and Riego, D. *Journal of Environmental Quality*, **4**, 45 (1975).
13 Wu, T.L., Lambert, L., Hastings, D. and Banning, D. *Bulletin of Environmental Contamination and Toxicology*, **24**, 411 (1980).
14 Smith, D.H. and Christie, W.H. *International Journal of Environmental Analytical Chemistry*, **8**, 241 (1980).
15 Paez, D.M. and Guagnini, O.A. *Mikrochimica Acta*, **2**, 220 (1971).
16 Shriadal, H.M.A., Katoaka, M. and Ohzeki, K. *Analyst (London)*, **110**, 125 (1985).
17 Kuroda, R., Matsumoto, N. and Ogmura, K. *Fresenius Zeitschrift für Analytische Chemie*, **330**, 111 (1988).
18 Matthews, A.D. and Riley, J.P. *Analytica Chimica Acta*, **51**, 455 (1970).
19 Fung, Y.S. and Dao, K.I. *Analytica Chimica Acta*, **309**, 173 (1995).

Chapter 37

Anions: Adsorption on metal oxides

37.1 Alumina

37.1.1 Non saline waters

37.1.1.1 Chromate

Alumina has been used to preconcentrate hexavalent chromium before determination by inductively coupled plasma atomic emission spectrometry [1,2]. Trivalent chromium is not adsorbed under these conditions. A detection limit of $0.2 \mu g$ L^{-1} was achieved by this procedure. Hydrous metal oxides have also been used to preconcentrate chromate [3].

An innovation that has been used for metals and also for chromate preconcentration consists of a modification of the flow injection analysis technique whereby the samples pass through a microcolumn containing an adsorbent for the ion of interest, thereby achieving a concentration factor. Following an automatic switch to an acidic reagent, the adsorbed anion is desorbed in a sharp pulse of the flowing reagent and then passes on to the detection system. Systems such as this, therefore, combine preconcentration and automation. Syty et al. [2] separated chromium(III) from chromium(VI) on a microcolumn of alumina which preconcentrated chromium(VI). Chromium(VI) was then flushed from the column with a small volume of acid before determination by inductively coupled plasma atomic emission spectrometry in amounts down to $0.2 \mu g$ L^{-1}.

37.2 Mercapto modified silica gel

37.2.1 Non saline waters

37.2.1.1 Arsenite

Howard et al. [4] electively preconcentrated arsenite onto mercapto modified silica gel. Arsenate, monomethylarsonate and dimethyl arsenite, which commonly occur in arsenic contaminated non saline water samples, do not interfere in this procedure.

37.3 Cadmium exchanged zeolites

37.3.1 Non saline water

37.3.1.1 Sulphide

Desalvo and Street [5] preconcentrated sulphide from well water on a column of cadmium exchanged zeolite sorbent prior to determination by the methylene blue visible spectrophotometric method. This method had an appreciably lower detection limit than conventional sulphide methods.

References

1 Cox, A.G., Cook, I.G. and McLeod, C.W. *Analyst (London)*, **110**, 331 (1985).
2 Syty, A., Christenson, R.G. and Rains, T.C. *Atomic Spectroscopy*, **7**, 89 (1986).
3 Music, O., Ristic, M. and Tonkovic, H. Z. *Wasser Abwasser Forsch.*, **19**, 186 (1986).
4 Howard, A.G., Volkan, N. and Ataman, Y. *Analyst (London)*, **112**, 159 (1987).
5 Desalvo, D.P. and Street, K.W. *Analyst (London)*, **111**, 1307 (1986).

Chapter 38

Anions: Coprecipitation procedures

38.1 Ferric hydroxide

38.1.1 Non saline waters

38.1.1.1 Chromate

Chromium and chromium(VI) [1] have been preconcentrated by procedures based on coprecipitation with ferric hydroxide. The mechanism of the sorption of chromate ions onto amorphous ferric hydroxide, ferric oxide and magnetite particles has been examined and the effects of several variables including the pH of the solution and the presence of competing ions (chloride, sulphate, molybdate) determined. The sorption of hexavalent chromium onto hydrous ferric oxide can be explained in terms of ligand exchange, hydroxyl ions being released from the surface of the hydrated oxide particles. For magnetic particles, however, the sorption effect was explained by the reduction of traces of hexavalent chromium at the magnetite/water interface, as a result of which a very fine ferric hydroxide/chromium hydroxide coating was produced which excluded further reduction of hexavalent chromium.

38.1.1.2 Phosphate and arsenate

Hori *et al*. [2] examined the adsorption behaviour of various phosphorus containing anions on ferric hydroxide as a function of solution pH. Adsorbed phosphorus compounds were determined spectrophotometrically and percent adsorption calculated from adsorbed and initial amounts. This was plotted against solution pH. Orthophosphate was adsorbed quantitatively at pH4.0–8.0; triphosphate and pyrophosphate at pH4.0–9.3 and monomethylphosphate, phosphite and α and β glycerophosphates at pH4.0–6.8. Dimethylphosphate and hypophosphite were only slightly adsorbed at all pH values examined. The adsorbed ions were desorbed readily into a small volume of alkaline solution.

A solution of iron containing ferric chloride and acidified with hydrochloric acid is treated with aqueous ammonia to precipitate ferric hydroxide coprecipitated with ferric phosphate produced by any phosphate ions in the sample [3]. The precipitate is filtered off and dissolved in dilute hydrochloric acid and ammonia and uranyl acetate added to produce uranyl phosphate which is estimated polarographically. To determine phosphate plus arsenate the ferric hydroxide precipitate, containing coprecipitated iron phosphate and arsenite, is made acid with hydrochloric acid and potassium iodide added. The arsenic trichloride produced is extracted with carbon tetrachloride and this solution back extracted into hydrochloric acid. Application of the ammonia ferric chloride precipitation technique to this gives a preconcentrate containing arsenic only.

38.1.2 Seawater

38.1.2.1 Chromate

Mullins [4] has described a procedure for determining the concentrations of dissolved chromium species in seawater. Chromium(III) and chromium(VI) separated by coprecipitation with hydrated iron(III) oxide and total dissolved chromium are determined separately by conversion to chromium(VI), extraction with ammonium pyrrolidinedithio-carbamate into methyl isobutyl ketone and determination by atomic absorption spectrometry. The detection limit is 40ng L $^{-1}$ Cr. The dissolved chromium not amenable to separation and direct extraction is calculated by difference. In the waters investigated, total concentrations were relatively high (1–5µg L $^{-1}$) with chromium(VI) the predominant species in all areas sampled with one exception, where organically bound chromium was the major species. A standard contact time of 4h was found to be necessary for the quantitative coprecipitation of chromium on ferric oxide. The rsd values for the determination of chromium(III), chromium(VI) and total dissolved chromium were generally 10.0, 4.0 and 5.0% respectively. From these results the rsd for the calculated concentration of the bound species was 20%.

Boniforti et al. [5] compared several preconcentration methods for the determination of metals in seawater. A comparison was made of ammonium pyrrolidinedithiocarbamate-8-quinolinol complexation followed by extraction with methyl isobutyl ketone or Freon–113, coprecipitation with magnesium hydroxide or iron(II) hydroxide or chelating by batch treatment with Chelex–100 for the determination of chromium. Atomic absorption spectrometry and inductively coupled plasma atomic emission spectrometry were used for analysis. Interferences, recovery, precision, accuracy and detection limits were compared. The Chelex–100 resin method was most suitable for the preconcentration of all determinands except chromium, whereas preconcentration of chromium(III) and chromium(V) was achieved only by coprecipitation with iron(II) hydroxide.

38.2 Alumina

38.2.1 Non saline waters

38.2.1.1 Phosphate

Coprecipitation with aluminium hydroxide has been used to preconcentrate phosphate ion in non saline waters [6]. This method involves coprecipitation with aluminium hydroxide, flotation of the precipitate which is then dissolved in sulphuric acid and determination of phosphate by a conventional molybdenum blue method. Recoveries are better than 95% and the relative standard deviation is 1% with samples in the range 5–150µg L $^{-1}$.

38.3 Barium sulphate

38.3.1 Non saline waters

38.3.1.1 Chromate and dichromate

The preconcentration of chromium(IV) (chromate) in non saline waters by coprecipitation with barium sulphate has been shown to be a selective and accurate procedure [7]. Spectrophotometric analysis of the preconcentrate enables down to 0.02µg L $^{-1}$ of chromium to be determined. The method is based on the similarity between the solubility products of barium sulphate and barium chromate. Interference from ferric iron, aluminium and trivalent chromium was overcome by using salicylic acid as a masking agent. The method was used on river water samples and the results obtained were in good agreement with those obtained by an aluminium hydroxide coprecipitation method. In order to prevent contamination, the membrane filter, the sample bottle, and other vessels were carefully pre-cleaned with concentrated hydrochloric acid and redistilled water. The coprecipitation step for chromium(VI) was carried out as soon as possible after sampling to minimise the sample storage problem.

38.4 Bismuth oxide

38.4.1 Seawater

38.4.1.1 Chromate

Nakayama et al. [8] have described a method for the determination of chromium(III), chromium(VI) and organically bound chromium in seawater. They found that seawater in the sea of Japan contained about 9

$\times 10^{-9}$mol L^{-1} dissolved chromium. This is shown to be divided as about 15% inorganic Cr(III), about 25% inorganic Cr(IV) and about 60% organically bound chromium. These workers studied the coprecipitation behaviours of chromium species with hydrated iron(III) and bismuth oxides.

The collection behaviour of chromium species was examined as follows. Seawater (400ml), spiked with 10^{-8}mol L^{-1} Cr(III), chromium(VI) and chromium(III) organic complexes labelled with ^{51}Cr, was adjusted to the desired pH with hydrochloric acid or sodium hydroxide. An appropriate amount of hydrated iron(III) or bismuth oxide was added; the oxide precipitates were prepared separately and washed thoroughly with distilled water before use [9]. After about 24h, the samples were filtered on 0.4μm Nuclepore filters. The separated precipitates were dissolved with hydrochloric acid and the solutions thus obtained were used for γ-activity measurements. In the examination of solvent extraction, chromium was measured by using ^{51}Cr, while iron and bismuth were measured by electrothermal atomic absorption spectrometry. The decomposition of organic complexes and other procedures were also examined by electrothermal atomic absorption spectrometry.

The percentage collection of chromium(III) with hydrated iron(III) oxide may decrease considerably in the neutral pH range when organic materials capable of combining with Cr(III), such as citric acid and certain amino acids, are added to the seawater [10]. Moreover, synthesised organic chromium(III) complexes are poorly collected with hydrated iron(III) oxide over a wide pH range [10]. As it was not known what kind of organic matter acts as the major ligand for chromium in seawater Nakayama et al. [8] used EDTA and 8-quinolinol-5-sulphonic acid to examine the collection and decomposition of organic chromium species, because these ligands form quite stable water-soluble complexes with chromium(III) although they are not actually present in seawater. The organic chromium species were then decomposed to inorganic chromium(III) and chromium(VI) species by boiling with 1g ammonium persulphate per 400ml sea water and acidified to 0.1mol L^{-1} of acid with hydrochloric acid. Iron and bismuth, which would interfere in atomic absorption spectrometry, were 99.9% removed by extraction from 2mol L^{-1} hydrochloric acid solution with a p-xylene solution of 5% tri-iso-octylamine. Chromium(III) remained almost quantitatively in the aqueous phase in the concentration range 10^{-9}–10^{-6}mol L^{-1} whether or not iron or bismuth was present. However, as about 95% of chromium(VI) was extracted by the same method, samples which may contain chromium(VI) should be treated with ascorbic acid before extraction so as to reduce chromium(VI) to chromium(III).

When the residue obtained by the evaporation of the aqueous phase after the extraction was dissolved in 0.1mol L^{-1} nitric acid and the resulting solution was used for electrothermal atomic absorption spectrometry, a negative interference, which was seemingly due to residual organic matter, was observed. This interference was successfully removed by digesting the residue on a hot plate with 1ml of concentrated hydrochloric acid and 3ml of concentrated nitric acid. This process had the advantage that the interference of chloride in the atomic absorption spectroscopy was eliminated during the heating with nitric acid.

Cranston and Murray [11,12] took the samples in polyethylene bottles that had been precleaned at 20°C for 4d with 1% hydrochloric acid. This acid had been previous distilled to reduce metal impurity levels. Total chromium, chromium(IV) + chromium(III) + Cr_p (particulate chromium) was coprecipitated with iron(II) hydroxide and reduced chromium (Cr(III) + Cr_p) was coprecipitated with iron(III) hydroxide. These coprecipitation steps were completed within minutes of sample collection to minimise storage problems. The iron hydroxide precipitates were filtered through 0.4µm Nuclepore filters and stored in polyethylene vials for later analyses in the laboratory. Particulate chromium was also obtained by filtering unaltered samples through 0.4µm filters. In the laboratory the iron hydroxide coprecipitates were dissolved in 6mol L^{-1} hydrochloric acid and analysed by flameless atomic absorption. The limit of detection of this method is about 0.1–0.2nmol L^{-1}. Precision is about 5%.

38.4.1.2 Molybdate

Hidalgo *et al.* [13] preconcentrated molybdenum(VI) in potable water and seawater by co-flotation on iron(III) hydroxide. Co-flotation was achieved by means of surfactants; octadecylamine for tap water samples, hexadecyltrimethylammonium bromide for non saline waters, and both hexadecyltrimethylammonium bromide and octadecylamine for seawater samples. Differential pulse polarography using the catalytic wave caused by molybdenum(VI) in nitrate medium was applied to the preconcentrate. Good reproducibility was obtained with mean values of 0.7µg L^{-1} and 5.7µg L^{-1} for molybdenum(VI) in tap water and seawater, respectively.

References

1 Rozanski, L. *Chimica Analit.*, **17**, 55 (1972).
2 Hori, T., Moriguchi, M., Sassaki, K., Kitagawa, S. and Munakato, M. *Analytica Chimica Acta*, **173**, 299 (1985).
3 Jeffrey, L.M. and Hood, D.W. *Journal of Marine Research*, **17**, 247 (1958).
4 Mullins, T.L. *Analytica Chimica Acta*, **165**, 97 (1984).
5 Boniforti, R., Ferraroli, R., Frigieri, P., Heltai, D. and Queirezza, G. *Analytica Chimica Acta*, **162**, 33 (1984).
6 Aoyama, H., Hobo, T. and Suzuki, S. *Analytica Chimica Acta*, **153**, 291 (1983).

7 Yamazaki, H. *Analytica Chimica Acta*, **113**, 131 (1980).
8 Nakayama, E., Kuwamoto, T., Takoro, H. and Fujinaka, F. *Analytica Chimica Acta*, **131**, 247 (1981).
9 Nakayama, E., Kuwamoto, T., Takoro, H. and Fujinaka, F. *Analytica Chimica Acta*, **130**, 401 (1981).
10 Nakayama, E., Kuwamoto, T., Takoro, H. and Fujinaka, F. *Analytica Chimica Acta*, **130**, 289 (1981).
11 Cranston, R.E. and Murray, J.W. *Analytica Chimica Acta*, **99**, 275 (1978).
12 Cranston, R.E. and Murray, J.W. *Limnology and Oceanography*, **25**, 1104 (1980).
13 Hidalgo, J.L., Gomez, M.A., Caballero, M., Cela, R. and Perez-Bustamante, A. *Talanta*, **35**, 301 (1988).

Chapter 39

Anions: Miscellaneous preconcentration procedures

39.1 Electrochemical method

39.1.1 Non saline waters

39.1.1.1 Sulphide

Bain [1] preconcentrated sulphide in non saline waters at levels of 1–10000ppb, at 0.20V (vs Ag/AgCl) and determined by cathodic stripping voltammetry. Over the concentration range of 0–40µg L^{-1} the stripping peak height is linearly proportional to sulphide concentration, over the range of 40–10000µg L^{-1} and concentrations are obtained from the calibration plot. The detection limits are 0.20 and 0.10µg L^{-1} for dc and differential pulse stripping analysis, respectively, and the coefficients of variation are 3.7% and 2.6% for 10 and 100µg L^{-1} sulphate.

39.2 Solvent sublation

39.2.1 Non saline waters

39.2.1.1 Iodide

Iodide in non saline waters has been preconcentrated by transport extraction based on solvent sublimation [2].

39.3 Ion flotation

39.3.1 Non saline water

39.3.3.1 Chromate

Aoyama et al. [3] have described a rapid method for the determination of down to 3µg L^{-1} of chromate in which hexavalent chromium is complexed with diphenylcarbazide, and this complex is floated with a sodium lauryl sulphate anionic surfactant. After dilution the subsided foam is measured spectrophotometrically. The continuous flotation procedure is as follows.

Place the water (1L), previously acidified to 0.1M with sulphuric acid, in the separation tube, pass nitrogen into the tube at about 105ml min $^{-1}$, and pump in the 1% sodium lauryl sulphate solution at 0.39ml min $^{-1}$. Continuously supply the sample pH and the 1% diphenylcarbazide solution to the separation tube at the rate of 2L h $^{-1}$ and 0.230ml min $^{-1}$ respectively. A steady state flotation was achieved 1h after flotation had started. Mix the sample and diphenyl carbazide solutions in the mixing chambers before entering the tube. Collect the foam subsided in the collector and intermittently transfer to flasks and dilute to 50ml with water, and measure absorbances as described above.

The effect of diverse ions on the flotation of chromium(VI) were investigated. Only copper(II) and iron(III) in 10-fold amounts, and vanadium(V) in 200-fold amounts caused fading of the colour. Even in those cases, however, chromium(VI) could be measured by a standard addition method.

39.3.1.2 Phosphate

Ion flotation has also been applied to the preconcentration of phosphate [4].

Shida and Matsuo [5] reported a very sensitive spectrophotometric method based on formation of the ion associate of molybdosphosphate with methylene blue, flotation of this ion pair between the aqueous phase and cyclohexane-4-methylpentan-2-one and displacement of methylene blue from the ion pair by tetradecyldimethylbenzylammonium ion.

39.4 Colloid flotation

39.4.1 Non saline waters

39.4.1.1 Molybdate

Murthey and Ryan [6] used colloid flotation as a means of pre-concentration prior to neutron activation analysis for molybdenum. Hydrous iron(III) is floated in the presence of sodium dodecyl sulphate with small nitrogen bubbles from 1L of seawater at pH5.7. Recoveries of molybdenum were better than 95%. This method has been used [7] to precipitate traces of molybdenum from water samples. The trace element was concentrated by coprecipitation with thionalide at pH9.1. Coprecipitation with thionalide allowed the concentration of both ions and colloids.

39.5 Gas diffusion methods

39.5.1 Non saline waters

39.5.1.1 Sulphide

Milosalvjevic *et al.* [8] have described a flow injection–gas diffusion method for the preconcentration and determination of µg L $^{-1}$ sulphide in non saline waters. This method employs an amperometric detector. Only cyanide interferes. Accumulation and preconcentration of the analyte is accomplished using the acceptor stream of a diffusion unit in a closed-loop recirculating mode. The detection limit from a 5ml sample is 0.15µg L $^{-1}$ level is 4%.

39.6 Miscellaneous

39.6.1 Potable waters

39.6.1.1 Chromate

High performance flow flame atomic absorption spectrometry has been used for the automated determination of chromium(III)/chromium(VI) and preconcentration of chromium(VI) [9]. A high performance liquid chromatography integration at the output of the flame atomic absorption spectrometer renders possible the simultaneous signal processing of both oxidation states of chromium. Detection limits are 0.03µg L $^{-1}$ for chromium(III) and 0.02µg L $^{-1}$ for chromium(VI).

References

1 Bain, G. *Fenxi Huaxue*, **14**, 618 (1986).
2 Palagyi, S. and Braun, T. *Fresenius Journal of Analytical Chemistry*, **346**, 905 (1993).
3 Aoyama, M., Hobo, T. and Susuki, S. *Analytica Chimica Acta*, **129**, 237 (1981).
4 Auoyama, M., Hobo, T. and Suzuki, S. *Analytica Chimica Acta*, **153**, 291 (1983).
5 Shida, J. and Matsuo, T. *Bulletin of the Chemical Society of Japan*, **53**, 2868 (1980).
6 Murthey, R.S.S. and Ryan, D.E. *Analytical Chemistry*, **55**, 682 (1983).
7 Zmijewska, W., Polkowska Motrenko, H. and Stakowska, H. *Journal of Radioanalytical and Nuclear Chemistry Articles*, **84**, 319 (1984).
8 Milosalvjevic, E.B., Solujic, L., Hendrix, J.L. and Nelson, J.H. *Analytical Chemistry*, **60**, 2791 (1988).
9 Posta, J., Berndt, H., Luo, S.K. and Scholdach, G. *Analytical Chemistry*, **65**, 2590 (1993).

Chapter 40

Rationale, preconcentration of anions

As an aid to the quick location in the text of suitable preconcentration methods for particular anions, methods used for non saline and seawaters are reviewed respectively in Tables 40.1 and 40.2. Acceptably low detection limits have been achieved for most of the anions listed ranging from as low as 0.001µg L $^{-1}$ (molybdate), 0.005µg L $^{-1}$ (phosphate) to 8µg L $^{-1}$ (iodide).

Table 40.1 Preconcentration of anions in non saline waters

Key to preconcentration method: chelation–solvent extraction (1); immobilised chelation (2); immobilised non chelation (3); active carbon adsorbtion (45); anion exchange resins (5); cation exchange resins (6); adsorption on metals and metal oxides (7); coprecipitation (8); on-line methods (9); gas evolution methods (10); electrolytic methods (11); electrochemical methods (12); cathodic deposition methods (13); osmosis (14); dialysis (15); flotation (16); stripping methods (17); solvent sublation (18); evaporation (19); aeration (20); co-crystallisation (21); freeze drying (22); ion exclusion (23)

Key to analytical finish: atomic absorption spectrometry (1); spectrophotometric (2); inductively coupled plasma atomic emission spectrometry (3); inductively coupled plasma mass spectrometry (4); neutron activation analysis (5); X-ray fluorescence spectroscopy (6); emission spectrometry (7); supercritical fluid methods (8); anodic stripping voltammetry (9); gas chromatography (10); scinatillation counting (11); X-ray spectrometry (12); high performance liquid chromatography (13); differential pulse polarography (14); radiochemical methods (15); flow injection analysis (16); electron spin resonance spectroscopy (17); spectrophotometric methods (18); voltammetry (19); isotope dilution (20); ultraviolet spectroscopy (21); mass spectrometry (22); γ ray spectrometry (23); molecular emission cavity analysis (24); titration (25); luminescence methods (26); spectrofluorometry (27); fluorophotometry (28); laser enhanced ionisation (29)

Anion	Preconcn. method	Analytical finish	Detection limit*	Section No.	[Ref.]
SbO_4	1	1		34.10.1	[1]
	1	5	1	34.1.1	[2,3]
AsO_4	1	5		34.1.1	[2,3]
	1	10		34.1.1	[5]
	1	15	0.2	34.1.1	[4]
	1	18		34.1.1	[6]
	8	18		38.1.1.2	[2]
	23			36.2.1.1.	[19]
AsO_3	1	18		34.1.1	[6]
	1	10		34.1.1	[5]
	2			37.2.1.1	[4]
BO_3	2	18		35.1.1.1	[1]
	5	13		36.1.1.1	[3]
	5	20		36.1.1.1	[2]
	5	3		36.1.1.1	[1]
Br, I	5	18		36.1.1.3	[9]
CrO_4	1	1		34.1.2	[7,8]
	1	1	0.2	34.1.2	[9]
	1	10	0.05	34.1.2	[10]
	3			35.5.1.1	[15,16]
	3			35.1.1.2	[4]
	3			37.1.1.1	[3]
	3			35.2.1.1	[5]
	3	3	0.2	37.1.1.1	[1,2]
	3	10	0.05	35.1.1.2	[3]
	3	13		35.1.1.2	[4]
	5			36.1.1.2	[5,6]
	5	1		36.1.1.2	[4]
	5			36.2.1.1	[20]
	8	18	0.02	38.3.1.1	[7]

*μg L^{-1} unless otherwise stated

Anion	Preconcn. method	Analytical finish	Detection limit*	Section No.	[Ref.]
	8			38.1.1.1	[1]
	16	18	3	39.3.1.1	[3]
			0.02	39.6.1.1	[9]
Cr_2O_7	1	1		34.1.2	[7,8]
	1	18	0.02	38.3.1.1	[7]
F	1	3	0.59	34.1.3	[11]
IO_3	1	3	8.3	34.1.3	[12]
I	1	3	21	34.1.3	[12]
	3	12		35.2.1.2	[6]
	5	18		36.1.1.3	[9]
	5		0.1 nmole L^{-1}	36.1.1.3	[8]
	18			39.2.1.1	[2]
MoO_4	5			36.2.1.1	[19]
	5	1	<0.2	36.1.1.4	[10]
	16			39.4.1.1	[6,7]
PO_4	1	18	4	34.1.4	[30]
	1	18	0–0.6	34.1.4	[23]
	1	18		34.1.4	[17,26]
	1	18		34.1.1	[6]
	1	3	0.005–0.05	34.1.4	[29]
	1	3	0.37	34.1.4	[30]
	1	3		34.1.4	[28,29]
	1	27		34.1.4	[18–22]
	1	14		34.1.4	[27]
	1	16	0.1	34.1.4	[25]
	1		µg L^{-1}	34.1.4	[24]
	3	18	0.02	35.4.1.1	[12]
	3	6	0.25	35.2.1.3	[7]
	3			35.3.1.1	[8]
	5	18		36.1.1.5	[11,12]
	8	18		38.2.1.1	[6]
	8	18		38.1.1.2	[2]
	16			39.3.1.2	[4,5]
WO_4	5			36.2.1.1	[19]
SeO_4	3		0.05	35.4.1.2	[13,14]
	3	6		35.3.1.2	[9–11]
SeO_3	3		0.05	35.4.1.2	[13,14]
	3	6		35.3.1.2	[9–11]
	5	27		36.1.1.6	[13,14]
	6			36.2.1.1	[19]
SiO_3	1	18		34.1.5	[31]
S	3	18		37.3.1.1	[5]
	3	18		36.1.1.7	[15]
	12	19	0.1	39.1.1.1	[1]
	10		0.15	39.5.1.1	[8]
VO_4	5			36.2.1.1	[19]

*µg L^{-1} unless otherwise stated

Source: Own files

Table 40.2 Preconcentration of anions in seawater

Key as in Table 40.1

Anion	Preconcn. method	Analytical finish	Detection limit*	Section No.	[Ref.]
SbO_4	1	18		34.2.2	[6]
AsO_4	1	18		34.2.2	[6]
AsO_3	1	18		34.2.1	[6]
CrO_4	8	1	0.04	38.1.2.1	[4]
	8	1		38.4.1.1	[8.9]
	8	1		38.1.2.1	[5]
	8	3		38.1.2.1	[5]
	8		0.1–0.2nM L^{-1}	38.4.1.1	[11,12]
F	1	3		34.2.1	[11]
MoO_4	5	18	<10	36.1.2.1	[16]
	5	1	<10	36.1.2.1	[17]
	5	5	<5	36.1.2.2	[18]
	8	1	0.04	38.1.2.1	[4]
	8	1		38.4.1.1	[8.9]
	8	1		38.1.2.1	[5]
	8	3		38.1.2.1	[5]
	8		<0.001	38.4.1.2	[13]

*μg L^{-1} unless otherwise stated

Source: Own files

Appendix

Limit of detection

Although the criterion of detection is used for deciding whether or not one can claim to have detected the determinand, it does not follow that detection will always be achieved when the determinand's concentration in a sample equals the criterion of detection. Indeed, in this case, detection would be achieved on only 50% of occasions, that is, $\beta = 0.5$

Consider a method in which single sample and blank determinations are made, and the result R = S – B. If a sample contains a concentration, L, of the determinand, the difference (S – B) will be distributed about L with a standard deviation $\sqrt{2}\sigma_w$ (assuming that, for small concentrations, standard deviation is independent of concentration). Thus, as L increases, it becomes more probable that S – B will exceed the criterion of detection. The Limit of Detection is that concentration for which there is a desirably small probability that the result R, will be less than the criterion of detection. From the normal distribution, it follows that, for $\gamma = 0.95$ ($\beta = 0.05$), the limit of detection is $4.65\sigma_w$ (ie 4.65 within batch standard deviation).

This approach to defining the limit of detection is based on the idea of controlling the risk of an Error of the Second Kind. Thus, if the analysis of a sample and a blank is seen as an experiment whose aim is the detection of the determinand, the power of the experiment is 0.95 or greater provided the sample contains a concentration $4.65\sigma_w$ or greater. This power can only be improved (for a given concentration) or the limit of detection decreased (for a given power) by

(a) decreasing the value of σ_w and/or
(b) analysing more than one portion of the sample and the blank.

As with the criterion of detection, the limit of detection is governed by the standard deviation of results at small concentrations of the determinand. Thus, appropriate allowance can be made in methods of analysis for which the result is not simply the difference between single sample and blank determinations.

The definitions of criterion of detection and limit of detection assume that the within-batch standard deviations of both the blank and samples containing very small concentrations of the determinand are the same. This is likely to be true unless samples are subject to additional sources of random error not present for blanks. The definitions are also not necessarily valid when the analytical response is zero for finite concentrations of the determinand. Finally if the sample and blank are biased with respect to each other (for example, by the presence of interfering substances in the sample and/or the blank), the definitions are not valid. All these limitations can be allowed for statistically but, for simplicity, details are omitted here. The definitions as given will often be valid for water analysis, and will lead to objective numerical values instead of arbitrary and subjective estimates.

Some methods specify calibration curves for which the function of concentration does not allow the point corresponding to the blank to be plotted, for example, when analytical response is plotted against log − (concentration). In such cases, σ_w cannot be estimated in concentration units and the criterion and limit of detection must then be obtained from the calibration curve as the concentrations corresponding to $B^1 + 2.33\sigma_w^1$ and $B^1 + 4.65\sigma_w^1$, respectively, where B^1 is the analytical response for the blank and σ_w^1 is the within-batch standard deviation of the blank (response units). However, it is considered useful, whenever possible, to employ procedures whose calibration curves allow plotting the point corresponding to the blank and which are preferably linear or nearly so.

Index

Non saline waters, preconcentration of organics, acetic acid 258; acetyl chloride 293; alcohols 293; aldehydes 258; aldicarb 109; alicyclic hydrocarbons 257; aliphatic diamines 144; aliphatic hydrocarbons 134, 179, 217–221, 249–257; alkyl benzene sulphonates 152; amines 35, 192, 292; amino acids 99;